Superintegrability in Classical and Quantum Systems

Volume 37

CRM PROCEEDINGS & LECTURE NOTES

Centre de Recherches Mathématiques
Université de Montréal

Superintegrability in Classical and Quantum Systems

P. Tempesta
P. Winternitz
J. Harnad
W. Miller, Jr.
G. Pogosyan
M. Rodriguez
Editors

The Centre de Recherches Mathématiques (CRM) of the Université de Montréal was created in 1968 to promote research in pure and applied mathematics and related disciplines. Among its activities are special theme years, summer schools, workshops, postdoctoral programs, and publishing. The CRM is supported by the Université de Montréal, the Province of Québec (FCAR), and the Natural Sciences and Engineering Research Council of Canada. It is affiliated with the Institut des Sciences Mathématiques (ISM) of Montréal, whose constituent members are Concordia University, McGill University, the Université de Montréal, the Université du Québec à Montréal, and the Ecole Polytechnique. The CRM may be reached on the Web at www.crm.umontreal.ca.

American Mathematical Society
Providence, Rhode Island USA

The production of this volume was supported in part by the Fonds pour la Formation de Chercheurs et l'Aide à la Recherche (Fonds FCAR) and the Natural Sciences and Engineering Research Council of Canada (NSERC).

2000 *Mathematics Subject Classification.* Primary 81–06, 81R12, 37J15, 22E70.

Library of Congress Cataloging-in-Publication Data

Workshop on Superintegrability in Classical and Quantum Systems (2002 : Montréal, Québec)
Superintegrability in classical and quantum systems / P. Tempesta... [et al.], editors.
p. cm. — (CRM proceedings & lecture notes, ISSN 1065-8580 ; v. 37)
Includes bibliographical references.
ISBN 0-8218-3329-4 (alk. paper)
1. Hamiltonian systems—Congresses. 2. Differential equations, Partial—Congresses. 3. Manifolds (Mathematics)—Congresses. 4. Quantum theory—Congresses. 5. Mathematical physics—Congresses. I. Tempesta, P. (Piergiulio), 1972– II. Title. III. Series.

QC20.7.H35+ 2004
pcc 2004055437

∞ The paper used in this book is acid-free and falls within the guidelines
established to ensure permanence and durability.
This volume was submitted to the American Mathematical Society
in camera ready form by the Centre de Recherches Mathématiques.
Visit the AMS home page at `http://www.ams.org/`

10 9 8 7 6 5 4 3 2 1 09 08 07 06 05 04

Contents

Preface

The Workshop on Superintegrability in Classical and Quantum Systems, organized under the auspices of the Centre de recherches mathématiques, Université de Montréal, took place in Montréal (Québec), September 16–21, 2002.

This meeting brought together scientists working in the area of finite-dimensional integrable systems to discuss the new developments registered in this topical field. As a result, these Proceedings constitute an up-to-date monograph which collects contributions based on complementary points of view on the subject, reflecting the rich gamut of approaches recently proposed in the literature.

The properties possessed by these systems are manifold. In classical mechanics, they have stable periodic orbits (all finite orbits are periodic). In quantum mechanics, all known superintegrable systems have been shown to be exactly solvable. Their energy spectrum is degenerate and can be calculated algebraically. The spectra of superintegrable systems may also have other interesting properties, such as the saturation of eigenfunction norm bounds.

The workshop attracted 48 participants from 16 countries (ordered according to the number of participants: Canada (20), Spain (5), USA (4), Czech Republic, France, Israel, Japan, Mexico and Turkey (2), Armenia, Germany, Italy, Poland, Russia, UK, and Ukraine (1)). A total of 35 lectures was presented.

The contributions presented in this volume cover several (overlapping) lines of research:

- *Standard superintegrable systems in classical and quantum mechanics* (that is Hamiltonians with scalar potentials and second-order integrals of motion leading to the separation of variables in the Schrödinger and Hamilton–Jacobi equations): F. J. Herranz, L. Mardoyan, W. Miller, Jr., G. Pogosyan, A. Zhalij;
- *Superintegrable systems with higher-order or nonpolynomial integrals, or velocity dependent forces.* It was shown that new phenomena arise, in particular nontrivial superintegrable systems in quantum mechanics that correspond to free motion in the classical limit: J. Bérubé, S. Gravel, A. Nikitin, M. Rañada, M. Sheftel, R. Smirnov, P. Winternitz;
- *New types of superintegrable systems in classical mechanics*, i.e.. systems possessing completely periodic solutions in a region of phase space and new maximally superintegrable systems: F. Calogero, J.-P. Françoise, C. Gonera;
- *Time discretization of integrable Euler equations and symmetry preserving discretization of superintegrable quantum systems using the finite operator calculus*: A. Penskoi, P. Tempesta.
- *Superintegrability, exact and quasi-exact solvability in standard and PT-symmetric quantum mechanics*: M. A. Rodriguez, M. Znojil;

• *Quantum deformation, Nambu dynamics and algebraic perturbation theory of superintegrable systems*: A. Ballesteros, T. Curtright, A. Turbiner, Y. Uwano;

• *Connection with infinite-dimensional integrable systems, algebraic geometry of Hurwitz spaces, spectral curves, bispectrality and commuting operators, Atiyah–Hitchin brackets*: D. Korotkin, K. Takasaki, M. Yakimov, O. Yermolayeva;

• *Computer assisted classification of integrable equations*: T. Wolf;

• *Supersymmetric superintegrable models and their solutions*: P. Mathieu.

All contributions were refereed.

We believe that this monograph will represent a useful reference book for researchers active in this rapidly developing field.

P. Tempesta and P. Winternitz
Montréal, April 2004

Participants

Angel Ballesteros
Universidad de Burgos
Burgos, Spain

Josée Bérubé
Université de Montréal
Montréal, Canada

Francesco Calogero
University of Roma "La Sapienza"
Roma, Italy

Alexei Cheviakov
Queen's University
Kingston, Canada

Thomas Curtright
University of Miami
Miami, USA

Catherine Cyr-Gagnon
Université de Montréal
Montréal, Canada

Jamil Daboul
Ben-Gurion University of the Negev
Beer-Sheva, Israel

Patrick Desrosiers
Université Laval
Québec, Canada

Evgeny Ferapontov
Loughborough University
Loughborough, UK

Jean-Pierre Françoise
Université Pierre et Marie Curie
Paris, France

Jean-Pierre Gazeau
Université Denis Diderot
Paris, France

David Gomez-Ullate
Université de Montréal
Montréal, Canada

Cezary Gonera
University of Łódź
Łódź, Poland

Galina Goujvina
Moscow State University
Moscow, Russia

Simon Gravel
Université de Montréal
Montréal, Canada

Alfred Michel Grundland
Université du Québec à Trois-Rivières
Trois-Rivières, Canada

John Harnad
Concordia University
Montréal, Canada

Miloslav Havłicek
Czech Technical University in Prague
Prague, Czech Republic

Francisco Jose Herranz
Universidad de Burgos
Burgos, Spain

Dmitry Jakobson
McGill University
Montréal, Canada

Niky Kamran
McGill University
Montréal, Canada

Siaka Kone
Université de Cocody
Abidjan, Côte d'Ivoire

Dmitry Korotkin
Concordia University
Montréal, Canada

Levon Mardoyan
Yerevan State University
Yerevan, Armenia

Pierre Mathieu
Université Laval
Québec, Canada

Raymond G. McLenaghan
University of Waterloo
Waterloo, Canada

John Meloche
University of Waterloo
Waterloo, Canada

Willard Miller, Jr.
University of Minnesota
Minneapolis, USA

Anatoly Nikitin
Ukrainian National Academy of Sciences
Kyuv, Ukraine

Yavuz Nutku
Feza Gürsey Institute
Istanbul, Turkey

Alexei V. Penskoi
Université de Montréal
Montréal, Canada
George Pogosyan
Joint Institute for Nuclear Research
Dubna, Russia
Manuel F. Rañada
Universidad de Zaragoza
Zaragoza, Spain
Thierry Robart
Howard Universtity
Washington, USA
Miguel A. Rodriguez
Universidad Complutense
Madrid, Spain
Mikhail B. Sheftel
Feza Gursey Institute
Istanbul, Turkey
Roman Smirnov
Universität Paderborn
Paderborn, Germany
Kanehisa Takasaki
Kyoto University
Kyoto, Japan
Piergiulio Tempesta
Université de Montréal
Montréal, Canada
Dennis The
University of British Columbia
Vancouver, Canada
John Toth
McGill University
Montréal, Canada
Alexander Turbiner
Universidad Nacional Autónoma de México
Mexico, Mexico
Yoshio Uwano
Kyoto University
Kyoto, Japan
Pavel Winternitz
Université de Montréal
Montréal, Canada
Thomas Wolf
Brock University
St. Catharines, Canada
Milen Yakimov
Cornell University
Ithaca, USA
Alexander Zhalij
Université de Montréal
Montréal, Canada
Miloslav Znojil
Joint Institute for Nuclear Research
Dubna, Russia

Centre de Recherches Mathématiques
CRM Proceedings and Lecture Notes
Volume **37**, 2004

Superintegrable Deformations of the Smorodinsky–Winternitz Hamiltonian

Ángel Ballesteros, Francisco J. Herranz, Fabio Musso, and Orlando Ragnisco

Abstract. A constructive procedure to obtain superintegrable deformations of the classical Smorodinsky–Winternitz Hamiltonian by using quantum deformations of its underlying Poisson sl(2) coalgebra symmetry is introduced. Through this example, the general connection between coalgebra symmetry and quasi-maximal superintegrability is analyzed. The notion of comodule algebra symmetry is also shown to be applicable in order to construct new integrable deformations of certain Smorodinsky–Winternitz systems.

1. Introduction

The aim of this work is to review a (co)algebraic approach to the superintegrability properties of the classical Smorodinsky–Winternitz (SW) Hamiltonian [**12, 18**]. As we shall see, the main consequence of making explicit such sl(2) Poisson coalgebra symmetry is the possibility of constructing superintegrable deformations of the SW Hamiltonian by making use of quantum algebra deformations of sl(2). We would also like to emphasize that such deformation procedure is rather general and can be applied to other superintegrable Hamiltonians [**7**].

In the next section we recall the essentials of coalgebra symmetry [**5**] and also the intrinsic superintegrability properties of the associated Hamiltonian systems [**7**]. Section 3 is devoted to the description of the coalgebra symmetry of the SW system [**1**], and a set of $(2N-2)$ functionally independent constants of the motion (including the Hamiltonian) is deduced by making use of the sl(2) coalgebra. The nonstandard deformation of sl(2) [**10, 16, 19**] is then used (Section 4) in order to construct a family of integrable deformations of the SW Hamiltonian with a common set of $(2N-2)$ functionally independent deformed integrals of the motion. In Section 5, one of these deformations is shown to be of the Stäckel type [**17**], and a

2000 *Mathematics Subject Classification.* Primary 37J35; Secondary 17B37.

This work has been partially supported by the Ministerio de Ciencia y Tecnología, Spain (Project BFM2000-1055).

The fourth author has been partially supported by INFN and by MIUR (COFIN2001 "Geometry and Integrability").

The first two authors are grateful to CRM for hospitality.

This is the final form of the paper.

new set of $(N-1)$ integrals related with this separability property is obtained [**1**]. The notion of comodule algebra symmetry [**4**] is introduced in Section 6, and it is shown that some specific SW systems have such a new type of dynamical symmetry. Once again, this symmetry enables us to construct a new integrable (but perhaps nonsuperintegrable) deformation of the SW Hamiltonian. Finally, some remarks and open problems are briefly commented.

2. Coalgebra symmetry and superintegrability

We recall that a coalgebra (A, Δ) is a (unital, associative) algebra A endowed with a coproduct map [**9, 13**]:

$$\Delta \colon A \to A \otimes A, \tag{2.1}$$

which is coassociative

$$(\Delta \otimes \mathrm{id}) \circ \Delta = (\mathrm{id} \otimes \Delta) \circ \Delta, \tag{2.2}$$

i.e., the following diagram is a commutative one:

$$\begin{array}{ccc} A & \xrightarrow{\ \Delta\ } & A \otimes A \\ {\scriptstyle\Delta}\downarrow & & \downarrow{\scriptstyle\Delta\otimes\mathrm{id}} \\ A \otimes A & \xrightarrow{\ \mathrm{id}\otimes\Delta\ } & A \otimes A \otimes A \end{array}$$

This "two-fold way" for the definition of the objects on $A \otimes A \otimes A$ will be essential as far as superintegrability is concerned. Note that, in addition, Δ has to be an algebra homomorphism from A to $A \otimes A$:

$$\Delta(ab) = \Delta(a)\Delta(b), \quad \forall a, b \in A. \tag{2.3}$$

Moreover, if A is a Poisson algebra and

$$\Delta(\{a, b\}_A) = \{\Delta(a), \Delta(b)\}_{A\otimes A}, \quad \forall a, b \in A, \tag{2.4}$$

we shall say that (A, Δ) is a Poisson coalgebra, which will be the relevant object for the construction [**5**] of classical integrable systems that is summarized in the sequel (see [**1, 2, 6, 14, 15**] for different applications to classical and quantum systems).

Let (A, Δ) be a Poisson coalgebra with generators X_i $(i = 1, \ldots, l)$, Casimir function $\mathcal{C}(X_1, \ldots, X_l)$ and coassociative coproduct $\Delta \equiv \Delta^{(2)}$ which is a Poisson map with respect to the Poisson bracket on $A \otimes A$ given by:

$$\{X_i \otimes X_j, X_r \otimes X_s\}_{A\otimes A} = \{X_i, X_r\}_A \otimes X_j X_s + X_i X_r \otimes \{X_j, X_s\}_A. \tag{2.5}$$

The mth coproduct map $\Delta_{\mathrm{L}}^{(m)} : A \to A \otimes A \otimes \cdots^{m)} \otimes A$ can be defined by applying recursively the coproduct $\Delta^{(2)}$ in the form

$$\Delta_{\mathrm{L}}^{(m)} := \bigl(\mathrm{id} \otimes \mathrm{id} \otimes \cdots^{m-2)} \otimes \mathrm{id} \otimes \Delta^{(2)}\bigr) \circ \Delta_{\mathrm{L}}^{(m-1)}. \tag{2.6}$$

Such an induction ensures that $\Delta_{\mathrm{L}}^{(m)}$ is also a Poisson map. As a consequence of the definition of $\Delta_{\mathrm{L}}^{(m)}$, for any smooth function $\mathcal{H}(X_1, \ldots, X_l)$ we can define a N-sites Hamiltonian as the Nth coproduct of $\mathcal{H}$:

$$H^{(N)} := \Delta_{\mathrm{L}}^{(N)}\bigl(\mathcal{H}(X_1, \ldots, X_l)\bigr) = \mathcal{H}\bigl(\Delta_{\mathrm{L}}^{(N)}(X_1), \ldots, \Delta_{\mathrm{L}}^{(N)}(X_l)\bigr). \tag{2.7}$$

By construction, it can be proved that the $(N-1)$ functions given by $(m = 2, \ldots, N)$

$$C^{(m)} := \Delta_{\mathrm{L}}^{(m)}\bigl(\mathcal{C}(X_1, \ldots, X_l)\bigr) = \mathcal{C}\bigl(\Delta_{\mathrm{L}}^{(m)}(X_1), \ldots, \Delta_{\mathrm{L}}^{(m)}(X_l)\bigr), \tag{2.8}$$

Poisson-commute with the Hamiltonian:

$$\{C^{(m)}, H^{(N)}\}_{A\otimes A\otimes\cdots N)\otimes A} = 0, \quad m = 2, \ldots, N. \tag{2.9}$$

Moreover, all these integrals of the motion are mutually in involution:

$$\{C^{(m)}, C^{(n)}\}_{A\otimes A\otimes\cdots N)\otimes A} = 0, \quad m, n = 2, \ldots, N. \tag{2.10}$$

When the Hamiltonians $\mathcal{H}$ are defined on Poisson–Lie algebras, the coproduct is "primitive": $\Delta(X_i) = X_i \otimes 1 + 1 \otimes X_i$. However, the Poisson analogues of quantum algebras and groups [**9, 13**] are also (deformed) coalgebras (A_z, Δ_z) (where z is the deformation parameter). Consequently, any function of the generators of a given "quantum" Poisson algebra (with Casimir element C_z) will provide, under an appropriate (deformed) symplectic representation, an integrable deformation of the Hamiltonian defined on the coalgebra (A, Δ).

2.1. Coalgebras and quasi-maximal superintegrability. Instead of (2.6), another recursion relation for the mth coproduct map can be defined:

$$\Delta_{\mathrm{R}}^{(m)} := (\Delta^{(2)} \otimes \mathrm{id} \otimes \cdots^{m-2)} \otimes \mathrm{id}) \circ \Delta_{\mathrm{R}}^{(m-1)}. \tag{2.11}$$

Due to the coassociativity property of the coproduct, this new expression will provide exactly the same expressions for the Nth coproduct of any generator [**5**]. However, if we label from 1 to N the sites of the chain of N copies of A, lower-dimensional coproducts $\Delta^{(m)}$ (with $m < N$) will be "different" in the sense that $\Delta_{\mathrm{L}}^{(m)}$ will contain objects living on the tensor product space $1 \otimes 2 \otimes \cdots \otimes m$, whilst $\Delta_{\mathrm{R}}^{(m)}$ will be defined on the sites $(N - m + 1) \otimes (N - m) \otimes \cdots \otimes N$. Therefore, for rank-one coalgebras the coalgebra symmetry of a given Hamiltonian gives rise to two sets of $(N - 1)$ integrals of the motion that Poisson-commute with $H^{(N)}$ [**7**]:

- A set of "left" integrals $\{C^{(m)} = \Delta_{\mathrm{L}}^{(m)}(\mathcal{C}), m = 2, \ldots, N\}$:

$$\begin{aligned}
C^{(2)} &\equiv \Delta_{\mathrm{L}}^{(2)}(\mathcal{C}) \quad \text{which is defined on the space } 1 \otimes 2\\
C^{(3)} &\equiv \Delta_{\mathrm{L}}^{(3)}(\mathcal{C}) \quad \text{which is defined on the space } 1 \otimes 2 \otimes 3\\
&\ \ \vdots \qquad\qquad\qquad\qquad\qquad\qquad\qquad\qquad \vdots\\
C^{(N)} &\equiv \Delta_{\mathrm{L}}^{(N)}(\mathcal{C}) \quad \text{which is defined on the space } 1 \otimes 2 \otimes \cdots \otimes N
\end{aligned}$$

- A set of "right" integrals $\{I^{(m)} = \Delta_{\mathrm{R}}^{(m)}(\mathcal{C}), m = 2, \ldots, N\}$:

$$\begin{aligned}
I^{(2)} &\equiv \Delta_{\mathrm{R}}^{(2)}(\mathcal{C}) \quad \text{on the space} \qquad (N-1) \otimes N\\
I^{(3)} &\equiv \Delta_{\mathrm{R}}^{(3)}(\mathcal{C}) \quad \text{on the space} \qquad (N-2) \otimes (N-1) \otimes N\\
&\ \ \vdots \qquad\qquad\qquad\qquad\qquad\qquad\qquad\qquad \vdots\\
I^{(N)} &\equiv \Delta_{\mathrm{R}}^{(N)}(\mathcal{C}) \quad \text{on the space } 1 \otimes \cdots \otimes (N-2) \otimes (N-1) \otimes N
\end{aligned}$$

Note that $C^{(N)} \equiv I^{(N)}$. Thus, if all these integrals are functionally independent, we obtain an explicit construction of "quasi-maximally superintegrable systems," since the coalgebra generates a set of $(2N - 2)$ functions in involution

$$\{H^{(N)}, C^{(2)}, \ldots, C^{(N-1)}, C^{(N)} \equiv I^{(N)}, I^{(N-1)}, \ldots, I^{(2)}\}. \tag{2.12}$$

We remark that, in some cases, one more independent integral could exist (leading to a maximally superintegrable system), but such remaining constant of the motion cannot be deduced from the coalgebra symmetry.

3. Coalgebra symmetry of the SW Hamiltonian

Let us consider the sl(2) Poisson coalgebra [**1**]:

$$\begin{gathered}\{J_3, J_+\} = 2J_+, \qquad \{J_3, J_-\} = -2J_-, \qquad \{J_-, J_+\} = 4J_3, \\ \Delta(J_i) = 1 \otimes J_i + J_i \otimes 1, \quad i = +, -, 3,\end{gathered} \tag{3.1}$$

with Casimir function $\mathcal{C} = J_3^2 - J_- J_+$. A one-particle symplectic realization of this coalgebra is given by

$$D(J_-) = q_1^2, \qquad D(J_+) = p_1^2 + \frac{b_1}{q_1^2}, \qquad D(J_3) = q_1 p_1, \tag{3.2}$$

where $\{q_1, p_1\} = 1$. Note that, under this realization, $D(\mathcal{C}) = -b_1$.

If we consider the following Hamiltonian function:

$$\mathcal{H} = J_+ + \omega^2 J_-, \tag{3.3}$$

its one-particle realization is just

$$H^{(1)} = D(\mathcal{H}) = p_1^2 + \omega^2 q_1^2 + \frac{b_1}{q_1^2}. \tag{3.4}$$

The 2-particle realization of the coalgebra is obtained through the coproduct:

$$\begin{aligned}(D \otimes D)(\Delta^{(2)}(J_-)) &= f_-^{(2)} = q_1^2 + q_2^2, \\ (D \otimes D)(\Delta^{(2)}(J_+)) &= f_+^{(2)} = p_1^2 + p_2^2 + \frac{b_1}{q_1^2} + \frac{b_2}{q_2^2}, \\ (D \otimes D)(\Delta^{(2)}(J_3)) &= f_3^{(2)} = q_1 p_1 + q_2 p_2.\end{aligned} \tag{3.5}$$

Hence the associated 2-particle Hamiltonian is

$$H^{(2)} = (D \otimes D)(\Delta^{(2)}(\mathcal{H})) = \sum_{i=1}^{2} \left(p_i^2 + \omega^2 q_i^2 + \frac{b_i}{q_i^2} \right), \tag{3.6}$$

which is just the $N = 2$ SW Hamiltonian. A (both left and right) constant of the motion for $H^{(2)}$ is given by the coproduct of the Casimir:

$$C^{(2)} = (D \otimes D)(\Delta^{(2)}(\mathcal{C})) = -(q_1 p_2 - q_2 p_1)^2 - \left(b_1 \frac{q_2^2}{q_1^2} + b_2 \frac{q_1^2}{q_2^2} \right) - \sum_{i=1}^{2} b_i. \tag{3.7}$$

In general, the N-particle realization is obtained by applying the $\Delta^{(N)}$ map:

$$\begin{aligned}(D \otimes D \otimes \cdots^{N)} \otimes D)(\Delta_{\mathrm{L}}^{(N)}(J_-)) &= f_-^{(N)} = \sum_{i=1}^{N} q_i^2, \\ (D \otimes D \otimes \cdots^{N)} \otimes D)(\Delta_{\mathrm{L}}^{(N)}(J_+)) &= f_+^{(N)} = \sum_{i=1}^{N} \left(p_i^2 + \frac{b_i}{q_i^2} \right), \\ (D \otimes D \otimes \cdots^{N)} \otimes D)(\Delta_{\mathrm{L}}^{(N)}(J_3)) &= f_3^{(N)} = \sum_{i=1}^{N} q_i p_i,\end{aligned} \tag{3.8}$$

and the N-particle Hamiltonian given by the coalgebra is just the SW system:

$$H^{(N)} = (D \otimes D \otimes \cdots^{N)} \otimes D)(\Delta_{\mathrm{L}}^{(N)}(\mathcal{H})) = \sum_{i=1}^{N} \left(p_i^2 + \omega^2 q_i^2 + \frac{b_i}{q_i^2} \right). \tag{3.9}$$

The first set of $(N-1)$ (left) constants of the motion in involution turns out to be $(m = 2, \ldots, N)$:

$$C^{(m)} = (D \otimes D \otimes \cdots^{m)} \otimes D)(\Delta_{\mathrm{L}}^{(m)}(\mathcal{C})) = -\sum_{i<j}^{m} I_{ij} - \sum_{i=1}^{m} b_i, \tag{3.10}$$

where

$$I_{ij} = (q_i p_j - q_j p_i)^2 + \left(b_i \frac{q_j^2}{q_i^2} + b_j \frac{q_i^2}{q_j^2} \right). \tag{3.11}$$

In this way, the complete integrability of the SW Hamiltonian is extracted from the coalgebra symmetry of the model [**1**].

3.1. Coalgebraic superintegrability. Further to the integrability, the coalgebra symmetry also underlies the superintegrability of the SW Hamiltonian since besides the "left integrals" $C^{(m)}$ (3.10), there exists a set of "right" ones $I^{(m)}$ given by $(m = 2, \ldots, N)$:

$$I^{(m)} = (D \otimes D \otimes \cdots^{m)} \otimes D)(\Delta_{\mathrm{R}}^{(m)}(\mathcal{C})) = -\sum_{N-m+1 \leq i<j}^{N} I_{ij} - \sum_{i=N-m+1}^{N} b_i. \tag{3.12}$$

The functional independence of all these integrals follows from the properties of their I_{ij} building blocks. Let us firstly consider the $N = 3$ integrals:

$$C^{(2)} = -I_{12} - (b_1 + b_2), \qquad I^{(2)} = -I_{23} - (b_2 + b_3),$$
$$C^{(3)} \equiv I^{(3)} = -I_{12} - I_{13} - I_{23} - (b_1 + b_2 + b_3),$$

which are functionally independent, since $C^{(3)}$ contains the I_{13} term. Similarly for the $N = 4$ case, where the integrals coming from the coalgebra read

$$C^{(2)} = -I_{12} - (b_1 + b_2), \qquad I^{(2)} = -I_{34} - (b_3 + b_4),$$
$$C^{(3)} = -I_{12} - I_{13} - I_{23} - (b_1 + b_2 + b_3), \; I^{(3)} = -I_{23} - I_{24} - I_{34} - (b_2 + b_3 + b_4),$$
$$C^{(4)} \equiv I^{(4)} = -I_{12} - I_{13} - I_{14} - I_{23} - I_{24} - I_{34} - (b_1 + b_2 + b_3 + b_4).$$

Once again, the fact that I_{14} does appear within $C^{(4)}$ implies the functional independence of the full set of integrals. In the N-dimensional case, by following the same construction, the functional independence is proved by considering that $C^{(N)} \equiv I^{(N)}$ is the only integral that contains the I_{1N} term. Since the N-dimensional SW Hamiltonian is, by construction, functionally independent of the $C^{(N)}$ integral, the quasi-maximal superintegrability of the SW Hamiltonian is proved. Finally, in this particular (separable) case we can take any of the one-particle SW Hamiltonians as the remaining independent integral leading to the full maximal superintegrability of the system.

Furthermore, we stress that a much more general family of coalgebra-symmetric quasi-maximally superintegrable Hamiltonians than (3.9) can also be defined [**1**]. For instance, let us consider the Hamiltonian function

$$\mathcal{H} = J_+ + \mathcal{F}(J_-), \tag{3.13}$$

where $\mathcal{F}(J_-)$ is an arbitrary smooth function of J_-. By construction, any N-particle Hamiltonian of the form

$$H^{(N)} = f_+^{(N)} + \mathcal{F}(f_-^{(N)}) = \sum_{i=1}^{N}\left(p_i^2 + \frac{b_i}{q_i^2}\right) + \mathcal{F}\left(\sum_{i=1}^{N} q_i^2\right), \tag{3.14}$$

is completely integrable (moreover, quasi-maximally superintegrable), and its constants of the motion are the previous sets $C^{(m)}$ and $I^{(m)}$. Note that in the case $N = 3$, this system is just one of the superintegrable potentials given by Evans [**11**].

4. A superintegrable deformation of the SW Hamiltonian

Now we consider the Poisson analogue [**1**] of the "nonstandard" deformation of $\mathrm{sl}_z(2)$ [**16**]:

$$\begin{aligned} \{J_3, J_+\} &= 2J_+ \cosh zJ_-, \qquad \{J_3, J_-\} = -2\frac{\sinh zJ_-}{z}, \\ \{J_-, J_+\} &= 4J_3. \end{aligned} \tag{4.1}$$

A deformed Casimir function for $\mathrm{sl}_z(2)$ is found to be:

$$\mathcal{C}_z = J_3^2 - \frac{\sinh zJ_-}{z} J_+. \tag{4.2}$$

The deformed coproduct map $\Delta_z : sl_z(2) \to sl_z(2) \otimes sl_z(2)$ is given by:

$$\begin{aligned} \Delta_z(J_-) &= J_- \otimes 1 + 1 \otimes J_-, \qquad \Delta_z(J_+) = J_+ \otimes \mathrm{e}^{zJ_-} + \mathrm{e}^{-zJ_-} \otimes J_+, \\ \Delta_z(J_3) &= J_3 \otimes \mathrm{e}^{zJ_-} + \mathrm{e}^{-zJ_-} \otimes J_3. \end{aligned} \tag{4.3}$$

Let us mimic the construction performed in Section 3 by taking again the function (3.3) for $\mathcal{H}$. A one-particle deformed symplectic realization of $\mathrm{sl}_z(2)$ is:

$$\begin{aligned} D_z(J_-) &= q_1^2, \qquad D_z(J_+) = \frac{\sinh zq_1^2}{zq_1^2} p_1^2 + \frac{zb_1}{\sinh zq_1^2}, \\ D_z(J_3) &= \frac{\sinh zq_1^2}{zq_1^2} q_1 p_1, \end{aligned} \tag{4.4}$$

which is characterized by the Casimir function $C_z^{(1)} = D_z(\mathcal{C}_z) = -b_1$. The associated one-particle Hamiltonian is just:

$$H_z^{(1)} = D_z(\mathcal{H}) = \frac{\sinh zq_1^2}{zq_1^2} p_1^2 + \frac{zb_1}{\sinh zq_1^2} + \omega^2 q_1^2, \tag{4.5}$$

and the 2-particle symplectic realization is obtained through Δ_z:

$$\begin{aligned} (D_z \otimes D_z)(\Delta_z^{(2)}(J_-)) &= \tilde{f}_-^{(2)} = q_1^2 + q_2^2, \\ (D_z \otimes D_z)(\Delta_z^{(2)}(J_+)) &= \tilde{f}_+^{(2)} = \left(\frac{\sinh zq_1^2}{zq_1^2} p_1^2 + \frac{zb_1}{\sinh zq_1^2}\right)\mathrm{e}^{zq_2^2} \\ &\qquad + \left(\frac{\sinh zq_2^2}{zq_2^2} p_2^2 + \frac{zb_2}{\sinh zq_2^2}\right)\mathrm{e}^{-zq_1^2}, \\ (D_z \otimes D_z)(\Delta_z^{(2)}(J_3)) &= \tilde{f}_3^{(2)} = \frac{\sinh zq_1^2}{zq_1^2} q_1 p_1 \mathrm{e}^{zq_2^2} + \frac{\sinh zq_2^2}{zq_2^2} q_2 p_2 \mathrm{e}^{-zq_1^2}. \end{aligned} \tag{4.6}$$

As a consequence, a deformation of the 2-particle SW Hamiltonian is obtained as $H_z^{(2)} = (D_z \otimes D_z)(\Delta^{(2)}(\mathcal{H}))$. Namely,

$$
\begin{aligned}
(4.7)\quad H_z^{(2)} &= \left(\frac{\sinh zq_1^2}{zq_1^2}p_1^2 + \frac{zb_1}{\sinh zq_1^2}\right)\mathrm{e}^{zq_2^2} \\
&\qquad + \left(\frac{\sinh zq_2^2}{zq_2^2}p_2^2 + \frac{zb_2}{\sinh zq_2^2}\right)\mathrm{e}^{-zq_1^2} + \omega^2(q_1^2+q_2^2).
\end{aligned}
$$

Note that separability is destroyed under deformation. The corresponding constant of the motion is $C_z^{(2)} = (D_z\otimes D_z)\big(\Delta^{(2)}(\mathcal{C}_z)\big)$. Explicitly,

$$
\begin{aligned}
(4.8)\quad C_z^{(2)} = &-\frac{\sinh zq_1^2}{zq_1^2}\frac{\sinh zq_2^2}{zq_2^2}(q_1p_2-q_2p_1)^2\mathrm{e}^{z(q_2^2-q_1^2)} - b_1\mathrm{e}^{2zq_2^2} - b_2\mathrm{e}^{-2zq_1^2} \\
&- \left(b_1\frac{\sinh zq_2^2}{\sinh zq_1^2} + b_2\frac{\sinh zq_1^2}{\sinh zq_2^2}\right)\mathrm{e}^{z(q_2^2-q_1^2)}.
\end{aligned}
$$

The generic m-particle symplectic realization is then obtained through the (either right or left) mth deformed coproduct and reads:

$$
\begin{aligned}
(D_z\otimes\cdots^{m)}\otimes D_z)\big(\Delta_{z,L}^{(m)}(J_3)\big) &= \tilde{f}_-^{(m)} = \sum_{i=1}^m q_i^2, \\
(D_z\otimes\cdots^{m)}\otimes D_z)\big(\Delta_{z,L}^{(m)}(J_+)\big) &= \tilde{f}_+^{(m)} \\
(4.9)\qquad\qquad &= \sum_{i=1}^m\left(\frac{\sinh zq_i^2}{zq_i^2}p_i^2 + \frac{zb_i}{\sinh zq_i^2}\right)\exp\big\{zK_i^{(m)}(q^2)\big\}, \\
(D_z\otimes\cdots^{m)}\otimes D_z)\big(\Delta_{z,L}^{(m)}(J_-)\big) &= \tilde{f}_3^{(m)} = \sum_{i=1}^m \frac{\sinh zq_i^2}{zq_i^2}q_ip_i\exp\big\{zK_i^{(m)}(q^2)\big\},
\end{aligned}
$$

where the "long-range" interaction is encoded within the functions

$$
\begin{aligned}
K_i^{(m)}(q^2) &= -\sum_{k=1}^{i-1}q_k^2 + \sum_{l=i+1}^m q_l^2, \\
(4.10)\qquad K_{ij}^{(m)}(q^2) &= K_i^{(m)}(q^2) + K_j^{(m)}(q^2) \\
&= -2\sum_{k=1}^{i-1}q_k^2 - q_i^2 + q_j^2 + 2\sum_{l=j+1}^m q_l^2 \quad (i<j).
\end{aligned}
$$

In this way, the N-particle deformed SW Hamiltonian is defined as:

$$
\begin{aligned}
(4.11)\qquad H_z^{(N)} &= \tilde{f}_+^{(N)} + \omega^2\tilde{f}_-^{(N)} \\
&= \sum_{i=1}^N\left(\frac{\sinh zq_i^2}{zq_i^2}p_i^2 + \frac{zb_i}{\sinh zq_i^2}\right)\exp\big\{zK_i^{(N)}(q^2)\big\} + \omega^2\sum_{i=1}^N q_i^2.
\end{aligned}
$$

And the (left) constants of the motion in involution with $H_z^{(N)}$ are:

$$
\begin{aligned}
(4.12)\quad C_z^{(m)} = &-\sum_{i<j}^m \frac{\sinh zq_i^2}{zq_i^2}\frac{\sinh zq_j^2}{zq_j^2}(q_ip_j-q_jp_i)^2\exp\big\{zK_{ij}^{(m)}(q^2)\big\} \\
&-\sum_{i<j}^m\left(b_i\frac{\sinh zq_j^2}{\sinh zq_i^2} + b_j\frac{\sinh zq_i^2}{\sinh zq_j^2}\right)\exp\big\{zK_{ij}^{(m)}(q^2)\big\} \\
&\qquad -\sum_{i=1}^m b_i\exp\big\{2zK_i^{(m)}(q^2)\big\}.
\end{aligned}
$$

These deformed integrals can also be written as

$$C_z^{(m)} = -\sum_{i<j}^{m} I_{ij}^z \exp\{zK_{ij}^{(m)}(q^2)\} - \sum_{i=1}^{m} b_i \exp\{2zK_i^{(m)}(q^2)\}, \tag{4.13}$$

where we have defined the following analogues of the I_{ij} symbols (3.11):

$$I_{ij}^z = \frac{\sinh zq_i^2}{zq_i^2}\frac{\sinh zq_j^2}{zq_j^2}(q_ip_j - q_jp_i)^2 + \left(b_i\frac{\sinh zq_j^2}{\sinh zq_i^2} + b_j\frac{\sinh zq_i^2}{\sinh zq_j^2}\right). \tag{4.14}$$

4.1. Coalgebraic superintegrability of the deformation. By following the very same procedure as in the nondeformed case, the set of deformed right integrals $I_z^{(m)}$ can be easily constructed and reads

$$I_z^{(m)} = -\sum_{N-m+1\le i<j}^{N} I_{ij}^z \exp\{zR_{ij}^{(m)}(q^2)\} - \sum_{i=N-m+1}^{N} b_i \exp\{2zR_i^{(m)}(q^2)\}, \tag{4.15}$$

where the "long-range" interaction R-functions, similar to (4.10), are defined by

$$\begin{aligned} R_i^{(m)}(q^2) &= -\sum_{p=N-m+1}^{i-1} q_p^2 + \sum_{l=i+1}^{N} q_l^2, \\ R_{ij}^{(m)}(q^2) &= R_i^{(m)}(q^2) + R_j^{(m)}(q^2) \\ &= -2\sum_{p=N-m+1}^{i-1} q_p^2 - q_i^2 + q_j^2 + 2\sum_{l=j+1}^{N} q_l^2, \quad (i<j). \end{aligned} \tag{4.16}$$

The functional independence of the left and right deformed integrals follows from the fact that they are analytic in the deformation parameter z:

$$C_z^{(m)} = C^{(m)} + o[z], \qquad I_z^{(m)} = I^{(m)} + o[z].$$

As they are functionally independent at $z = 0$, they they will be so in the whole complex z-plane, up to isolated points.

Let us explicitly write such integrals in the $N = 3$ case:

$$\begin{aligned} C_z^{(2)} &= -I_{12}^z \exp\{zK_{12}^{(2)}(q^2)\} - \sum_{i=1}^{2} b_i \exp\{2zK_i^{(2)}(q^2)\}, \\ I_z^{(2)} &= -I_{23}^z \exp\{zR_{23}^{(2)}(q^2)\} - \sum_{i=2}^{3} b_i \exp\{2zR_i^{(2)}(q^2)\}, \\ C_z^{(3)} \equiv I_z^{(3)} &= -I_{12}^z \exp\{zK_{12}^{(3)}(q^2)\} - I_{13}^z \exp\{zK_{13}^{(3)}(q^2)\} \\ &\quad - I_{23}^z \exp\{zK_{23}^{(3)}(q^2)\} - \sum_{i=1}^{3} b_i \exp\{2zK_i^{(3)}(q^2)\}, \end{aligned}$$

where the K and R-functions involved in the previous expressions read

$$\begin{array}{lll} K_1^{(2)}(q^2) = q_2^2, & K_2^{(2)}(q^2) = -q_1^2, & K_{12}^{(2)}(q^2) = -q_1^2 + q_2^2, \\ K_1^{(3)}(q^2) = q_2^2 + q_3^2, & K_{12}^{(3)}(q^2) = -q_1^2 + q_2^2 + 2q_3^2, & K_2^{(3)}(q^2) = -q_1^2 + q_3^2, \\ K_{13}^{(3)}(q^2) = -q_1^2 + q_3^2, & K_3^{(3)}(q^2) = -q_1^2 - q_2^2, & K_{23}^{(3)}(q^2) = -2q_1^2 - q_2^2 + q_3^2, \\ R_2^{(2)}(q^2) = q_3^2, & R_3^{(2)}(q^2) = -q_2^2, & R_{23}^{(2)}(q^2) = -q_2^2 + q_3^2. \end{array}$$

Moreover, the following family of N-dimensional Hamiltonian systems is also quasi-maximally superintegrable:

$$\mathcal{H} = J_+ + \mathcal{F}(J_-), \tag{4.17}$$

where $\mathcal{F}(J_-)$ is an arbitrary smooth function of J_-. Explicitly,

$$\begin{aligned} H_z^{(N)} &= \tilde{f}_+^{(N)} + \mathcal{F}(\tilde{f}_-^{(N)}) \\ &= \sum_{i=1}^{N} \left(\frac{\sinh zq_i^2}{zq_i^2} p_i^2 + \frac{zb_i}{\sinh zq_i^2} \right) \exp\{zK_i^{(N)}(q^2)\} + \mathcal{F}\left(\sum_{i=1}^{N} q_i^2 \right), \end{aligned} \tag{4.18}$$

will Poisson-commute with all the $C_z^{(m)}$ and $I_z^{(m)}$.

5. A deformation of Stäckel type

It is obvious that the SW Hamiltonian (3.9) is a Liouville system, and another possible set of integrals of motion in involution is given by

$$M_i = p_i^2 + \omega^2 q_i^2 + \frac{b_i}{q_i^2} - \frac{H^{(N)}}{N}, \quad i = 1, \ldots, N, \tag{5.1}$$

where $\sum_{i=1}^{N} M_i = 0$. In order to get the maximal superintegrability of the nondeformed SW Hamiltonian we can take any of these integrals in order to complete, in a functionally independent way, the $C^{(m)}$ and $I^{(m)}$ sets of constants of the motion. On the contrary, in the deformed case (4.11), the separability is broken due to the long-range interaction introduced by the deformation. However, since the coalgebra construction allows for an infinite family of deformed Hamiltonians, and all of them Poisson-commute with the same $C^{(m)}$ and $I^{(m)}$ sets, it could happen that another choice of the dynamical Hamiltonian could fulfill the separability conditions.

This is the case if we consider the Hamiltonian function [**1**]:

$$\mathcal{H} = J_+ \mathrm{e}^{zJ_-} + \omega^2 \left(\frac{\mathrm{e}^{2zJ_-} - 1}{2z} \right). \tag{5.2}$$

By introducing the Nth particle symplectic realization (4.9) we obtain

$$\begin{aligned} H_z^{(N)} = \sum_{i=1}^{N} \frac{\sinh zq_i^2}{zq_i^2} \mathrm{e}^{zq_i^2} \exp\left\{ 2z \sum_{k=i+1}^{N} q_k^2 \right\} \left(p_i^2 + b_i \left(\frac{zq_i}{\sinh zq_i^2} \right)^2 \right) \\ + \omega^2 \left(\frac{\exp\{2z \sum_{j=1}^{N} q_j^2\} - 1}{2z} \right), \end{aligned} \tag{5.3}$$

which has the form of a Stäckel system

$$H_z^{(N)} = \sum_{i=1}^{N} a_i(q_1, \ldots, q_N) \left(\tfrac{1}{2} p_i^2 + U_i(q_i) \right), \tag{5.4}$$

provided that

$$a_i(q_1, \ldots, q_N) = 2 \frac{\sinh zq_i^2}{zq_i^2} \mathrm{e}^{zq_i^2} \exp\left\{ 2z \sum_{k=i+1}^{N} q_k^2 \right\}, \quad i = 1, \ldots, N,$$

$$
\begin{aligned}
U_1(q_1) &= \frac{b_1}{2}\left(\frac{zq_1}{\sinh zq_1^2}\right)^2 + \frac{\omega^2}{4z}\mathrm{e}^{zq_1^2}\frac{zq_1^2}{\sinh zq_1^2},\\
U_i(q_i) &= \frac{b_i}{2}\left(\frac{zq_i}{\sinh zq_i^2}\right)^2, \qquad\qquad i = 2,\dots,N-1,\\
U_N(q_N) &= \frac{b_N}{2}\left(\frac{zq_N}{\sinh zq_N^2}\right)^2 - \frac{\omega^2}{4z}\mathrm{e}^{-zq_N^2}\frac{zq_N^2}{\sinh zq_N^2}.
\end{aligned} \tag{5.5}
$$

Stäckel's theorem claims that a Hamiltonian (5.4) admits separation of variables in the Hamilton–Jacobi equation if and only if there exists an $N \times N$ matrix B with entries $b_{ij}(q_j)$ such that

$$
\det B \neq 0, \qquad \sum_{j=1}^{N} b_{ij}(q_j)a_j(q_1,\dots,q_N) = \delta_{i1}. \tag{5.6}
$$

And this is the case for the new deformed Hamiltonian. The nonvanishing entries of B and its determinant are found to be

$$
\begin{aligned}
b_{1N}(q_N) &= \frac{zq_N^2}{2\sinh zq_N^2}\mathrm{e}^{-zq_N^2},\\
b_{ii-1}(q_{i-1}) &= \frac{zq_{i-1}^2}{\sinh zq_{i-1}^2}\mathrm{e}^{-zq_{i-1}^2}, \quad b_{ii}(q_i) = -\frac{zq_i^2}{\sinh zq_i^2}\mathrm{e}^{zq_i^2}, \quad i = 2,\dots,N,\\
\det B &= \frac{1}{2}\prod_{i=1}^{N}\frac{zq_i^2}{\sinh zq_i^2}\mathrm{e}^{-zq_i^2}.
\end{aligned} \tag{5.7}
$$

As a consequence, Stäckel's theorem gives us a new set of N functionally independent integrals of motion in involution

$$
Z_j = \sum_{i=1}^{N} a_{ij}\left(\tfrac{1}{2}p_i^2 + U_i(q_i)\right), \quad j = 1,\dots,N, \tag{5.8}
$$

where a_{ij} are the entries of B^{-1}. Then $a_{i1} = a_i$, so that the first integral I_1 is just the Hamiltonian. In our case, the nonzero functions a_{ij} turn out to be

$$
\begin{aligned}
a_{i1} &= 2\frac{\sinh zq_i^2}{zq_i^2}\mathrm{e}^{zq_i^2}\exp\left\{2z\sum_{k=i+1}^{N} q_k^2\right\}, \qquad i = 1,\dots,N,\\
a_{ij} &= \frac{\sinh zq_i^2}{zq_i^2}\mathrm{e}^{zq_i^2}\exp\left\{2z\sum_{k=i+1}^{j-1} q_k^2\right\}, \qquad i = 1,\dots,N,\ i < j.
\end{aligned} \tag{5.9}
$$

The new set of $N-1$ conserved quantities is given by ($j = 2,\dots,N$):

$$
\begin{aligned}
Z_j^z = \sum_{i=1}^{j-1}\frac{\sinh zq_i^2}{2zq_i^2}\mathrm{e}^{zq_i^2}\exp\left\{2z\sum_{k=i+1}^{j-1} q_k^2\right\}\left(p_i^2 + b_i\left(\frac{zq_i}{\sinh zq_i^2}\right)^2\right)&\\
+ \frac{\omega^2}{4z}\left(\exp\left\{2z\sum_{k=1}^{j-1} q_k^2\right\} - 1\right)&.
\end{aligned} \tag{5.10}
$$

Thus, for instance, the function Z_2^z can be taken as the remaining constant of the motion, which together with the family of "left" and "right" ones prove the maximal superintegrability of the Hamiltonian (5.3).

In the $z \to 0$ limit, the integrals (5.10) reduce to

$$Z_j^0 = \frac{1}{2}\sum_{i=1}^{j-1}\left(p_i^2 + \frac{b_i}{q_i^2}\right) + \frac{1}{2}\omega^2\sum_{k=1}^{j-1} q_k^2. \tag{5.11}$$

6. Comodule algebra symmetry

The notion of coproduct can be generalized by introducing the so called "coactions" [**13**]. A (right) coaction of a Hopf algebra (A, Δ) on a vector space V is a map $\phi \colon V \to V \otimes A$ such that

$$(\phi \otimes id) \circ \phi = (id \otimes \Delta) \circ \phi, \tag{6.1}$$

that is, if the following diagram is commutative:

$$\begin{array}{ccc} V & \xrightarrow{\ \phi\ } & V \otimes A \\ \phi \big\downarrow & & \big\downarrow \phi\otimes \mathrm{id} \\ V \otimes A & \xrightarrow{\ \mathrm{id}\otimes\Delta\ } & V \otimes A \otimes A. \end{array}$$

If V is an algebra, we shall say that V is an A-comodule algebra if the coaction ϕ is a homomorphism on V

$$\phi(ab) = \phi(a)\phi(b), \quad \forall a, b \in V. \tag{6.2}$$

Moreover, if V is endowed with a Poisson structure and A is a Poisson-Hopf algebra, V will also be a Poisson A-comodule algebra if:

$$\phi(\{a, b\}_V) = \{\phi(a), \phi(b)\}_{V\otimes A}, \quad \forall a, b \in V. \tag{6.3}$$

Note that any Hopf algebra A is an A-comodule algebra with respect to A provided that $\phi \equiv \Delta$. The construction of integrable systems by making use of comodule algebras has been recently introduced in [**4**] by defining recursively the Nth coaction as a homomorphism that maps V within $V \otimes A \otimes \cdots^{(N-1)} \otimes A$. Let $\{X_1, \ldots, X_l\}$ be the generators of V and let C be a Casimir function/operator of V. It can be proved [**4**] that the Hamiltonian

$$H^{(N)} := \phi^{(N)}(\mathcal{H}(X_1, \ldots, X_l)) = \mathcal{H}(\phi^{(N)}(X_1), \ldots, \phi^{(N)}(X_l)), \tag{6.4}$$

together with the following iterated (left) coactions of the Casimir are a set of N functions in involution and functionally independent ($m = 2, \ldots, N$):

$$C^{(m)} := \phi^{(m)}(C(X_1, \ldots, X_l)) = C(\phi^{(m)}(X_1), \ldots, \phi^{(m)}(X_l)). \tag{6.5}$$

We stress that the "right" integrals cannot be defined in this approach, thus the superintegrability of comodule symmetric systems cannot be ensured algebraically, and it has to be analyzed in each particular case.

6.1. Comodule algebra symmetry of the SW Hamiltonian.

Let us now describe an integrable deformation of the $N = 2$ SW Hamiltonian (3.6) with comodule algebra symmetry. We take as the Poisson–Hopf algebra A the following

Poisson version of a nonstandard deformation [**3**] of the Schrödinger algebra h_6^σ [**8**]:

$$\begin{aligned} &\{\mathcal{D},\mathcal{P}\}=-\mathcal{P}, && \{\mathcal{D},\mathcal{K}\}=\mathcal{K}, && \{\mathcal{K},\mathcal{P}\}=\mathcal{M}, && \{\mathcal{M},\cdot\}=0,\\ &\{\mathcal{D},\mathcal{H}\}=-2\mathcal{H}, && \{\mathcal{D},\mathcal{C}\}=2\mathcal{C}, && \{\mathcal{H},\mathcal{C}\}=\mathcal{D}, && \{\mathcal{H},\mathcal{P}\}=0,\\ &\{\mathcal{P},\mathcal{C}\}=-\mathcal{K}, && \{\mathcal{K},\mathcal{H}\}=\mathcal{P}, && \{\mathcal{K},\mathcal{C}\}=0, \end{aligned} \tag{6.6}$$

$$\begin{aligned} \Delta(\mathcal{M}) &= 1\otimes\mathcal{M}+\mathcal{M}\otimes 1,\\ \Delta(\mathcal{H}) &= 1\otimes\mathcal{H}+\mathcal{H}\otimes(1+\sigma\mathcal{P})^2,\\ \Delta(\mathcal{D}) &= 1\otimes\mathcal{D}+\mathcal{D}\otimes\frac{1}{1+\sigma\mathcal{P}}-\frac{1}{2}\mathcal{M}\otimes\frac{\sigma\mathcal{P}}{1+\sigma\mathcal{P}},\\ \Delta(\mathcal{C}) &= 1\otimes\mathcal{C}+\mathcal{C}\otimes\frac{1}{(1+\sigma\mathcal{P})^2}+\sigma\mathcal{D}'\otimes\frac{1}{1+\sigma\mathcal{P}}\mathcal{K}\\ &\qquad+\frac{\sigma^2}{2}(\mathcal{D}')^2\otimes\frac{\mathcal{M}}{(1+\sigma\mathcal{P})^2},\\ \Delta(\mathcal{P}) &= 1\otimes\mathcal{P}+\mathcal{P}\otimes 1+\sigma\mathcal{P}\otimes\mathcal{P},\\ \Delta(\mathcal{K}) &= 1\otimes\mathcal{K}+\mathcal{K}\otimes\frac{1}{1+\sigma\mathcal{P}}+\sigma\mathcal{D}'\otimes\frac{\mathcal{M}}{1+\sigma\mathcal{P}}, \end{aligned} \tag{6.7}$$

where $\mathcal{D}'=\mathcal{D}+\frac{1}{2}\mathcal{M}$.

The Poisson-gl(2) subalgebra generated by $\{\mathcal{M},\mathcal{H},\mathcal{D},\mathcal{C}\}$ is a Schrödinger comodule algebra V and the coaction $\phi^{(2)}\colon \mathrm{gl}(2)\to\mathrm{gl}(2)\otimes\mathrm{h}_6^\sigma$ is given by the restriction to the gl(2) subalgebra of the full coproduct map in h_6^σ [**4**]:

$$\phi^{(2)}(X):=\Delta(X),\quad X\in\{\mathcal{M},\mathcal{H},\mathcal{D},\mathcal{C}\}. \tag{6.8}$$

We take the following symplectic realization of h_6^σ:

$$\begin{aligned} &S(\mathcal{C})=\frac{q_1^2}{2}, && S(\mathcal{H})=\frac{p_1^2}{2}, && S(\mathcal{D})=-p_1\,q_1,\\ &S(\mathcal{M})=\lambda_1^2, && S(\mathcal{K})=\lambda_1 q_1, && S(\mathcal{P})=\lambda_1 p_1, \end{aligned} \tag{6.9}$$

and we consider a different symplectic realization for the gl(2) subalgebra

$$T(\mathcal{C})=\frac{q_1^2}{2},\qquad T(\mathcal{H})=\frac{p_1^2}{2}+\frac{b_1}{q_1^2},\qquad T(\mathcal{D})=-p_1q_1. \tag{6.10}$$

If we take as Hamiltonian function $H=\mathcal{H}+\mathcal{C}$ we find that

$$H_\sigma^{(1)}=T(H)=T(\mathcal{H})+T(\mathcal{C})=\frac{p_1^2}{2}+\frac{q_1^2}{2}+\frac{b_1}{q_1^2}, \tag{6.11}$$

is equivalent to the undeformed SW Hamiltonian (3.4) with $\omega^2=1$. But the 2-particle case provides in a straightforward way a new integrable deformation of the SW Hamiltonian with comodule algebra symmetry:

$$\begin{aligned} H_\sigma^{(2)} &= (T\otimes S)\big(\phi^{(2)}(H)\big)=(T\otimes S)\big(\phi^{(2)}(\mathcal{H})+\phi^{(2)}(\mathcal{C})\big)\\ &= \frac{1}{2}(p_1^2+p_2^2)+\frac{b_1}{q_1^2}+\frac{q_1^2}{2(1+\sigma\lambda_2p_2)^2}+\frac{q_2^2}{2}\\ &\quad+\sigma\lambda_2\left(2\left(\frac{p_1^2}{2}+\frac{b_1}{q_1^2}\right)p_2+\frac{q_2(\lambda_1^2-2q_1p_1)}{2(1+\sigma\lambda_2p_2)}\right)\\ &\qquad+\sigma^2\lambda_2^2\left(\left(\frac{p_1^2}{2}+\frac{b_1}{q_1^2}\right)p_2^2+\frac{(\lambda_1^2-2q_1p_1)^2}{8(1+\sigma\lambda_2p_2)^2}\right). \end{aligned} \tag{6.12}$$

By considering the gl(2) Casimir function $C_V = \frac{1}{4}\mathcal{D}^2 - \mathcal{H}\mathcal{C}$, the constant of the motion in involution with $H_\sigma^{(2)}$ is obtained:

$$C_\sigma^{(2)} = (T \otimes S)\big(\phi^{(2)}(C_V)\big) = (T \otimes S)\big(\tfrac{1}{4}\Delta(\mathcal{D})^2 - \Delta(\mathcal{H})\big). \tag{6.13}$$

As expected, the limit $\sigma \to 0$ of $C_\sigma^{(2)}$ is just

$$C_0^{(2)} = -\frac{1}{4}(p_2 q_1 - p_1 q_2)^2 - \frac{b_1}{2}\left(1 + \frac{q_2^2}{q_1^2}\right).$$

Further iterations of the coaction map would provide the corresponding integrable deformation in N dimensions, but in any case the only nonvanishing centrifugal term would be the one that corresponds to b_1.

We end with some remarks and open problems. Firstly, note that for higher rank coalgebras, we have a set of right and left integrals coming from each of the Casimir functions of the Poisson algebra. In general, these sets could not be functionally independent under an arbitrary symplectic realization and the number of independent integrals coming from the coalgebra has to be fixed for each individual realization. We also mention that subcoalgebras can also be used in order to extract superintegrability properties, as it was pointed out in [**2**]. Finally, we think that the search for explicit solutions of the deformed SW Hamiltonians and the corresponding deformed Lax formalism are worthy to be considered in the future, as well as the construction and analysis of the quantum mechanical analogues of the deformed SW Hamiltonians here introduced.

References

1. Á. Ballesteros and F. J. Herranz, *Integrable deformations of oscillator chains from quantum algebras*, J. Phys. A **32** (1999), no. 50,8851–8862.
2. ———, *Two-photon algebra and integrable Hamiltonian systems*, J. Nonlinear Math. Phys. **8** (2001), suppl., 18–22.
3. Á. Ballesteros, F. J. Herranz, J. Negro, and L. M. Nieto, *Twist maps for nonstandard quantum algebras and discrete Schrödinger symmetries*, J. Phys. A **33** (2000), no. 27, 4859–4870.
4. Á. Ballesteros, F. Musso, and O. Ragnisco, *Comodule algebras and integrable systems*, J. Phys. A **35** (2002), no. 37, 8197–8211.
5. Á. Ballesteros and O. Ragnisco, *A systematic construction of integrable Hamiltonians from coalgebras*, J. Phys. A **31** (1998), no. 16, 3791–3813.
6. ———, *Classical Hamiltonian systems with* sl(2) *coalgebra symmetry and their integrable deformations*, J. Math. Phys. **43** (2002), no. 2, 954–969.
7. ———, *Coalgebra symmetry and superintegrability* (in preparation).
8. G. Burdet, J. Patera, M. Perrin, and P. Winternitz, *The optical group and its subgroups*, J. Math. Phys. **19** (1978), no. 8, 1758–1780.
9. V. Chari and A. Pressley, *A guide to quantum groups*, Cambridge Univ. Press, Cambridge, 1994.
10. E. E. Demidov, Yu. I. Manin, E. E. Mukhin, and D. V. Zhdanovich, *Nonstandard quantum deformations of* GL(N) *and constants solutions of the Yang–Baxter equation*, Common Trends in Mathematics and Quantum Field Theories (Kyoto, 1990), Progr. Theoret. Phys. Suppl., vol. 102, Progress of Theoretical Physics, Kyoto, 1990, pp. 203–218.
11. N. W. Evans, *Superintegrability in classical mechanics*, Phys. Rev. A (3) **41** (1990), no. 10, 5666–5676.
12. ———, *Superintegrability of the Winternitz system*, Phys. Lett. A **147** (1990), no. 8-9, 483–486.
13. S. Majid, *Foundations of quantum group theory*, Cambridge Univ. Press, Cambridge, 1995.
14. F. Musso and O. Ragnisco, *Exact solution of the quantum Calogero–Gaudin system and of its q deformation*, J. Math. Phys. **41** (2000), no. 11, 7386–7401.

15. ———, *The spin-$\frac{1}{2}$ Calogero–Gaudin system and its q-deformation*, J. Phys. A **34** (2001), no. 12, 2625–2635.
16. C. Ohn, *A $*$-product on SL(2) and the corresponding nonstandard quantum $U(\mathfrak{sl}(2))$*, Lett. Math. Phys. **25** (1992), no. 2, 85–88.
17. A. M. Perelomov, *Integrable systems of classical mechanics and Lie algebras*. I, Birkhäuser, Basel, 1990.
18. P. Winternitz, Ya. A. Smorodinsky, M. Uhliř, and I. Friš, *Symmetry groups in classical and quantum mechanics*, Jadernaja Fiz. **4** (1966), 625–635 (Russian); English transl., Soviet J. Nuclear Phys. **4** (1967), 444–450.
19. S. Zakrzewski, *A Hopf star-algebra of polynomials on the quantum $\mathrm{SL}(2,\mathbb{R})$ for a "unitary" R-matrix*, Lett. Math. Phys. **22** (1991), no. 4, 287–289.

Departamento de Física, Facultad de Ciencias, Universidad de Burgos, Pza. Misael Bañuelos s.n., 09001 Burgos, Spain
E-mail address: angelb@ubu.es

Departamento de Física, Escuela Politécnica Superior, Universidad de Burgos, Avda. Cantabria, 09006 Burgos, Spain
E-mail address: fjherranz@ubu.es

Scuola Internazionale Superiore di Studi Avanzati, Via Beirut 4, 34014 Trieste, Italy
E-mail address: musso@sissa.it

Dipartimento di Fisica, Terza Università di Roma, Via della Vasca Navale 84, 00146 Roma, Italy
E-mail address: ragnisco@fis.uniroma3.it

Centre de Recherches Mathématiques
CRM Proceedings and Lecture Notes
Volume **37**, 2004

Isochronous Motions Galore: Nonlinearly Coupled Oscillators with Lots of Isochronous Solutions

F. Calogero and J.-P. Françoise

ABSTRACT. We consider $N + M$ complex nonlinear oscillators the time-evolution of which is characterized by the system of $N + M$ complex first-order equations of motion

$$\dot{x}_n - ip_n x_n = f_n(\boldsymbol{x}; \boldsymbol{y}), \quad n = 1, \dots, N;$$
$$\dot{y}_m + iq_m y_m = g_m(\boldsymbol{x}; \boldsymbol{y}), \quad m = 1, \dots, M,$$

where N and M are two arbitrary *nonnegative* integers, the $N + M$ complex functions of the *real* independent variable t ("time") $x_n \equiv x_n(t)$, $y_m \equiv y_m(t)$, are the dependent variables, i is of course the imaginary unit, $i^2 = -1$, the $N + M$ numbers p_n, q_m are arbitrary *nonnegative integers* (not all vanishing), and the $N + M$ functions f_n, g_m depend arbitrarily on their $N + M$ arguments except for the following restrictions: (i) they are *holomorphic* in all their $N + M$ arguments x_n, y_m at $\boldsymbol{x} = \boldsymbol{0}$, $\boldsymbol{y} = \boldsymbol{0}$; (ii) they vanish there faster than linearly; (iii) they are polynomial (at least) in all the variables x_n *or* in all the variables y_m (this condition can of course be ignored if M or N vanishes), and (iv) they satisfy certain additional restrictions constraining additionally their behavior when their arguments vanish.

We then show that there exists, in the neighborhood of the equilibrium configuration $\boldsymbol{x} = \boldsymbol{0}$, $\boldsymbol{y} = \boldsymbol{0}$, an open ball of ("small") initial data $x_n(0)$, $y_m(0)$, characterized by the restrictions $|x_n(0)| < r$, $|y_m(0)| < r$ with a positive r, $r > 0$, such that *all* solutions emerging from it are *completely periodic* ("isochronous") with period (at least) 2π, $\boldsymbol{x}(t + 2\pi) = \boldsymbol{x}(t)$, $\boldsymbol{y}(t + 2\pi) = \boldsymbol{y}(t)$.

1. Introduction and main result

Recently the potentialities of a simple trick, consisting essentially of a change of variables, have been noted, especially to identify dynamical systems and evolution equations possessing lots of *completely periodic* (indeed "isochronous") solutions [**1–16**]. This trick—as we shall explain in Section 3—relates the time-evolution of a dynamical system (with the time t being of course considered the *real* independent variable) to the properties of a related system, the independent time-like variable τ of which is instead complex: in particular an (open) domain in which the solutions of the second system are *holomorphic* in τ corresponds to

2000 *Mathematics Subject Classification.* 34C15, 34C20, 37J35.
This is the final form of the paper.

an (open) domain of the first system in which the evolution is *completely periodic* ("isochronous") in t. Heretofore this methodology was mainly (although not exclusively [**12**]) applied to relate *autonomous* dynamical systems evolving in the *real* time t, to systems "evolving" in the independent complex variable τ which were also *autonomous*, and this constraint entailed that consideration was restricted to dynamical systems characterized by certain properties of rescaling [**8, 13**]. In this paper we drop this restriction, by extending our consideration to *autonomous* dynamical systems evolving in the real time variable t which get related via the trick to systems which may feature an explicit dependence on the complex variable τ, namely to systems which generally are *no more autonomous*. This allows of course to capture within our approach a larger class of dynamical systems, while allowing nevertheless to obtain our main result.

In this paper we focus on the system of $N + M$ complex nonlinear oscillators the time-evolution of which is characterized by the system of $N + M$ first-order equations of motion

$$\dot{x}_n - ip_n x_n = f_n(\boldsymbol{x}; \boldsymbol{y}), \quad n = 1, \dots, N, \tag{1.1a}$$

$$\dot{y}_m + iq_m y_m = g_m(\boldsymbol{x}; \boldsymbol{y}), \quad m = 1, \dots, M, \tag{1.1b}$$

where N and M are two arbitrary *nonnegative* integers, the $N + M$ quantities $x_n \equiv x_n(t)$, $y_m \equiv y_m(t)$ are the dependent variables, i is of course the imaginary unit, $i^2 = -1$, the $N + M$ numbers p_n, q_m are arbitrary *nonnegative integers* (not all vanishing), and the $N + M$ functions f_n, g_m depend arbitrarily on their $N + M$ arguments (namely on the N components of the N-vector $\boldsymbol{x} \equiv (x_1, \dots, x_n)$ and on the M components of the M-vector $\boldsymbol{y} \equiv (y_1, \dots, y_m)$) except for the following restrictions: (i) they are *holomorphic* in all their $N + M$ arguments x_n, y_m at $\boldsymbol{x} = \boldsymbol{0}$, $\boldsymbol{y} = \boldsymbol{0}$; (ii) they vanish there faster than linearly,

$$\lim_{\varepsilon \to 0} [\varepsilon^{-1} f_n(\varepsilon \boldsymbol{x}; \varepsilon \boldsymbol{y})] = 0, \quad n = 1, \dots, N, \tag{1.2a}$$

$$\lim_{\varepsilon \to 0} [\varepsilon^{-1} g_m(\varepsilon \boldsymbol{x}; \varepsilon \boldsymbol{y})] = 0, \quad m = 1, \dots, M; \tag{1.2b}$$

(iiiA) they are polynomial in all the variables y_m,

$$f_n(\boldsymbol{x}; \boldsymbol{y}) = \sum_{\boldsymbol{k}=\boldsymbol{0}}^{\boldsymbol{K}} \prod_{\mu=1}^{M} (y_\mu)^{k_\mu} f_{n\mu}(\boldsymbol{x}), \quad n = 1, \dots, N, \tag{1.3a}$$

$$g_m(\boldsymbol{x}; \boldsymbol{y}) = \sum_{\boldsymbol{l}=\boldsymbol{0}}^{\boldsymbol{L}} \prod_{\mu=1}^{M} (y_\mu)^{l_\mu} g_{m\mu}(\boldsymbol{x}), \quad m = 1, \dots, M, \tag{1.3b}$$

and (ivA) they satisfy for all (fixed) values of $\boldsymbol{x}$ and $\boldsymbol{y}$ the following restrictions:

$$\lim_{\epsilon \to 0} [\epsilon^{-1-p_n} f_n(\dots, \epsilon^{p_\nu} x_\nu, \dots; \dots, \epsilon^{-q_\mu} y_\mu, \dots)] = \text{finite}, \quad n = 1, \dots, N, \tag{1.4a}$$

$$\lim_{\epsilon \to 0} [\epsilon^{-1+q_m} g_m(\dots, \epsilon^{p_\nu} x_\nu, \dots; \dots, \epsilon^{-q_\mu} y_\mu, \dots)] = \text{finite}, \quad m = 1, \dots, M, \tag{1.4b}$$

or (iiiB) they are polynomial in all the variables x_n,

$$f_n(\boldsymbol{x}; \boldsymbol{y}) = \sum_{\boldsymbol{k}=\boldsymbol{0}}^{\boldsymbol{K}} \prod_{\nu=1}^{N} (x_\nu)^{k_\nu} f_{n\nu}(\boldsymbol{y}), \quad n = 1, \dots, N, \tag{1.5a}$$

$$g_m(\boldsymbol{x}; \boldsymbol{y}) = \sum_{\boldsymbol{l}=\boldsymbol{0}}^{\boldsymbol{L}} \prod_{\nu=1}^{N} (x_\nu)^{l_\nu} g_{m\nu}(\boldsymbol{y}), \quad m = 1, \dots, M, \tag{1.5b}$$

and (ivB) they satisfy for all (fixed) values of $\boldsymbol{x}$ and $\boldsymbol{y}$ the following restrictions:

$$\lim_{\epsilon \to 0}[\epsilon^{-1+p_n} f_n(\ldots, \epsilon^{-p_\nu} x_\nu, \ldots; \ldots, \epsilon^{q_\mu} y_\mu, \ldots)] = \text{finite}, \quad n = 1, \ldots, N, \tag{1.6a}$$

$$\lim_{\epsilon \to 0}[\epsilon^{-1-q_m} g_m(\ldots, \epsilon^{-p_\nu} x_\nu, \ldots; \ldots, \epsilon^{q_\mu} y_\mu, \ldots)] = \text{finite}, \quad m = 1, \ldots, M. \tag{1.6b}$$

We trust the notation used above to be clear; note in particular that, in (1.3), $\boldsymbol{k}$, $\boldsymbol{K}$, $\boldsymbol{l}$, $\boldsymbol{L}$ are of course (nonnegative integer)-valued M-vectors and $\mathbf{0}$ is the null M-vector, while in (1.5) $\boldsymbol{k}$, $\boldsymbol{K}$, $\boldsymbol{l}$, $\boldsymbol{L}$ are instead (nonnegative integer)-valued N-vectors and $\mathbf{0}$ is the null N-vector. Note that the condition (iiiA), see (1.3) (but not (ivA), see (1.4)) disappears if M vanishes, $M = 0$, and likewise the condition (iiiB), see (1.5) (but not (ivB), see (1.6)) disappears if N vanishes, $N = 0$.

We then show that there exists, in the neighborhood of the equilibrium configuration $\boldsymbol{x} = \mathbf{0}$, $\boldsymbol{y} = \mathbf{0}$, an open ball of ("small") initial data $x_n(0)$, $y_m(0)$, characterized by the restrictions

$$|x_n(0)| < r, \; n = 1, \ldots, N; \qquad |y_m(0)| < r, \; m = 1, \ldots, M, \tag{1.7}$$

with a positive r, $r > 0$, such that *all* solutions emerging from it are *completely periodic* ("isochronous") with period (at least) 2π,

$$x_n(t + 2\pi) = x_n(t), \; n = 1, \ldots, N; \quad y_m(t + 2\pi) = y_m(t), \; m = 1, \ldots, M. \tag{1.8}$$

This result is proven in Section 3 (and in the appendix, where we confine some mathematical developments), after the presentation in Section 2 of some reformulations and generalizations of it, as well as some examples. Section 4 contains a discussion of this result in the context of more standard approaches to this type of problems.

2. Reformulations, generalizations, examples

In this section we collect some observations on the system of equations of motions (1.1), and we exhibit certain more explicit versions of it ("examples").

The first remark is that, while the equations of motion (1.1) are obviously complex (see the second term on their left-hand sides), real versions of these equations of motion can easily be obtained by introducing the real and imaginary parts of the dependent variables, say

$$x_n(t) = x_n^{(R)}(t) + i x_n^{(I)}(t), \qquad y_m(t) = y_m^{(R)}(t) + i y_m^{(I)}(t), \tag{2.1}$$

and by then taking the real and imaginary parts of the equations that obtain from (1.1) via this assignment.

The second remark is to note that the more general system of (complex) equations of motion

$$\dot{z}_j - i r_j z_j = h_j(\boldsymbol{z}), \quad j = 1, \ldots, J, \tag{2.2}$$

where J is an arbitrary positive integer, the J (complex) functions $z_j \equiv z_j(t)$ are the dependent variables, and the J numbers r_j are *real* and *rational* but otherwise arbitrary (not all vanishing, but without any restriction on their signs), can be reduced to (1.1) via trivial transformations (including a rescaling of the time variable and some merely notational changes). And the previous remark applies of course as well to this system.

Thirdly we conveniently reformulate the conditions (1.3) and (1.4), and likewise (see below) (1.5) and (1.6), as follows. Instead of (1.3) we write

$$f_n(\boldsymbol{x};\boldsymbol{y}) = \sum_{\boldsymbol{k}=\boldsymbol{0}}^{\boldsymbol{K}} \prod_{\mu=1}^{M} \left[(y_\mu)^{k_\mu} \sum_{\boldsymbol{u}_\mu=\boldsymbol{0}}^{\boldsymbol{U}_\mu} \prod_{\nu=1}^{N} (x_\nu)^{u_{\mu\nu}} \varphi_{nk_\mu u_{\mu\nu}}(\boldsymbol{x}) \right], \quad n=1,\ldots,N, \tag{2.3a}$$

$$g_m(\boldsymbol{x};\boldsymbol{y}) = \sum_{\boldsymbol{l}=\boldsymbol{0}}^{\boldsymbol{L}} \prod_{\mu=1}^{M} \left[(y_\mu)^{l_\mu} \sum_{\boldsymbol{w}_\mu=\boldsymbol{0}}^{\boldsymbol{W}_\mu} \prod_{\nu=1}^{N} (x_\nu)^{w_{\mu\nu}} \gamma_{ml_\mu w_{\mu\nu}}(\boldsymbol{x}) \right], \quad m=1,\ldots,M. \tag{2.3b}$$

Here we trust the notation to be self-evident: in particular, $\boldsymbol{k}, \boldsymbol{K}, \boldsymbol{l}, \boldsymbol{L}$ are clearly M-vectors, while the $4M$ vectors $\boldsymbol{u}_\mu$, $\boldsymbol{U}_\mu$, $\boldsymbol{w}_\mu$, $\boldsymbol{W}_\mu$, $\mu=1,\ldots,M$, are N-vectors, and the components of all these vectors are *nonnegative* integers; the functions $\varphi_{nk_\mu u_{\mu\nu}}(\boldsymbol{x})$, respectively the functions $\gamma_{ml_\mu w_{\mu\nu}}(\boldsymbol{x})$, are *holomorphic* in all their N arguments at $\boldsymbol{x}=\boldsymbol{0}$. It is then plain that the conditions (1.2) are automatically satisfied if these functions are required to vanish identically unless their indices n, k_μ, $u_{\mu\nu}$ respectively m, l_μ, $w_{\mu\nu}$ satisfy the following inequalities:

$$\sum_{\mu=1}^{M} \left[k_\mu + \sum_{\nu=1}^{N} u_{\mu\nu} \right] > 1, \qquad \sum_{\mu=1}^{M} \left[l_\mu + \sum_{\nu=1}^{N} w_{\mu\nu} \right] > 1, \tag{2.4}$$

and likewise the conditions (1.4) are automatically satisfied if these functions are required to vanish identically unless their indices n, k_μ, $u_{\mu\nu}$ respectively m, l_μ, $w_{\mu\nu}$ satisfy the following inequalities:

$$\sum_{\mu=1}^{M} \left[-q_\mu k_\mu + \sum_{\nu=1}^{N} p_\nu u_{\mu\nu} \right] > p_n, \qquad n=1,\ldots,N, \tag{2.5a}$$

$$\sum_{\mu=1}^{M} \left[-q_\mu l_\mu + \sum_{\nu=1}^{N} p_\nu w_{\mu\nu} \right] > -q_m, \quad m=1,\ldots,M. \tag{2.5b}$$

EXAMPLE 1. Let us display the above case with $N=M=1$. Then the equations of motion read (see (1.1))

$$\dot{x} - ipx = f(x;y), \qquad \dot{y} + iqy = g(x;y); \tag{2.6}$$

here we assume p to be a *nonnegative integer*, $p \geq 0$, and q to be an arbitrary *positive integer*, $q > 0$. It is then easily seen that the restrictions on the functions $f(x;y)$ and $g(x;y)$ sufficient to guarantee the existence of an open set of *isochronous* solutions with period (at least) 2π, $x(t+2\pi)=x(t)$, $y(t+2\pi)=y(t)$, in an appropriately small neighborhood of the origin $x=y=0$, read then

$$f(x;y) = \sum_{u=2}^{U} \left[x^u \sum_{k=0}^{\{\{(u-1)p/q\}\}} y^k \varphi_{ku}(x) \right], \tag{2.7a}$$

$$g(x;y) = \sum_{w=1}^{W} \left[x^w \sum_{l=\max(0,2-w)}^{\{\{wp/q\}\}} y^l \gamma_{lw}(x) \right], \tag{2.7b}$$

with the only requirement that the functions $\varphi_{ku}(x)$, $\gamma_{lw}(x)$ be *holomorphic* at $x=0$. Here the double braces identify the integer part of a number, and the two integers U and W are arbitrary (except for the obvious restrictions $U \geq 2$, $W \geq 1$).

The analogous formulas, and the corresponding example, that provide a more explicit version of the conditions (1.5) and (1.6), rather than (1.3) and (1.4), read as follows:

(2.8a) $$f_n(\boldsymbol{x};\boldsymbol{y}) = \sum_{\boldsymbol{k}=\boldsymbol{0}}^{\boldsymbol{K}} \prod_{\nu=1}^{N} \left[(x_\nu)^{k_\nu} \sum_{\boldsymbol{u}_\nu=\boldsymbol{0}}^{\boldsymbol{U}_\nu} \prod_{\mu=1}^{M} (y_\mu)^{u_{\nu\mu}} \varphi_{nk_\nu u_{\nu\mu}}(\boldsymbol{y}) \right], \quad n=1,\ldots,N,$$

(2.8b) $$g_m(\boldsymbol{x};\boldsymbol{y}) = \sum_{\boldsymbol{l}=\boldsymbol{0}}^{\boldsymbol{L}} \prod_{\nu=1}^{N} \left[(x_\nu)^{l_\nu} \sum_{\boldsymbol{w}_\nu=\boldsymbol{0}}^{\boldsymbol{W}_\nu} \prod_{\mu=1}^{M} M(y_\mu)^{w_{\nu\mu}} \gamma_{ml_\nu w_{\nu\mu}}(\boldsymbol{y}) \right], m=1,\ldots,M,$$

with the functions $\varphi_{nk_\nu u_{\nu\mu}}(\boldsymbol{y})$, $\gamma_{ml_\nu w_{\nu\mu}}(\boldsymbol{y})$ required again to be *holomorphic* at $\boldsymbol{y}=\boldsymbol{0}$, and to vanish identically unless their indices satisfy the inequalities

(2.9) $$\sum_{\nu=1}^{N}\left[k_\nu + \sum_{\mu=1}^{M} u_{\nu\mu}\right] > 1, \qquad \sum_{\nu=1}^{N}\left[l_\nu + \sum_{\mu=1}^{M} w_{\nu\mu}\right] > 1,$$

(2.10a) $$\sum_{\nu=1}^{N}\left[-p_\nu k_\nu + \sum_{\mu=1}^{M} q_\mu u_{\nu\mu}\right] > -p_n, \quad n=1,\ldots,N,$$

(2.10b) $$\sum_{\nu=1}^{N}\left[-p_\nu l_\nu + \sum_{\mu=1}^{M} q_\mu w_{\nu\mu}\right] > q_m, \quad m=1,\ldots,M.$$

EXAMPLE 2. Let us again display this case with $N = M = 1$. Then the equations of motion take again the simple form (2.6), where we now assume however p to be a *positive integer*, $p > 0$, and q to be an arbitrary *nonnegative integer*, $q \geq 0$. The restrictions on the functions $f(x;y)$ and $g(x,y)$ sufficient to guarantee the existence of an open set of isochronous solutions with period (at least) 2π, $x(t+2\pi) = x(t)$, $y(t+2\pi) = y(t)$, in an appropriately small neighborhood of the origin $x = y = 0$, read now

(2.11a) $$f(x;y) = \sum_{u=1}^{U}\left[y^u \sum_{k=\max(0,2-u)}^{\{\{uq/p\}\}} x^k \varphi_{ku}(y)\right],$$

(2.11b) $$g(x;y) = \sum_{w=2}^{W}\left[y^w \sum_{l=0}^{\{\{(w-1)q/p\}\}} x^l \gamma_{lw}(y)\right],$$

with the only requirement that the functions $\varphi_{ku}(y)$, $\gamma_{lw}(y)$ be *holomorphic* at $y = 0$. Here the double braces identify again the integer part of a number, and the two integers U and W are arbitrary (except for the obvious restrictions $U \geq 1$, $W \geq 2$).

Hamiltonian case. Let us consider the special case with

(2.12) $$N = M, \qquad p_n = q_n, \quad n = 1,\ldots,N,$$

assuming moreover that

(2.13a) $$f_n(\boldsymbol{x};\boldsymbol{y}) = \partial h(\boldsymbol{x};\boldsymbol{y})/\partial y_n, \quad n=1,\ldots,N,$$

(2.13b) $$g_n(\boldsymbol{x};\boldsymbol{y}) = -\partial h(\boldsymbol{x};\boldsymbol{y})/\partial x_n, n=1,\ldots,N.$$

It is then clear that the (complex) equations of motion (1.1) obtain in the standard manner from the Hamiltonian

$$H(\boldsymbol{x};\boldsymbol{y}) = i\sum_{n=1}^{N} p_n x_n y_n + h(\boldsymbol{x};\boldsymbol{y}), \tag{2.14}$$

where we now interpret x_n respectively y_n as canonically conjugate variables respectively momenta.

It is then of interest to see which conditions on the (single) function $h(\boldsymbol{x};\boldsymbol{y})$ are *sufficient* to guarantee our main result, namely the existence for these Hamiltonian equations of motion, (1.1) with (2.12) and (2.13), of an open set of solutions characterized by the initial conditions (1.7), that are then guaranteed to be *isochronous*, see (1.8).

There are two possibilities, corresponding to (2.3) respectively (2.8). In the first case (and using the same notation as above)

$$h(\boldsymbol{x};\boldsymbol{y}) = \sum_{\boldsymbol{k}=\boldsymbol{0}}^{\boldsymbol{K}} \prod_{\mu=1}^{N} \left[(y_\mu)^{k_\mu} \sum_{\boldsymbol{u}_\mu=\boldsymbol{0}}^{\boldsymbol{U}_\mu} \prod_{\nu=1}^{N} (x_\nu)^{u_{\mu\nu}} \varphi_{k_\mu u_{\mu\nu}}(\boldsymbol{x}) \right], \tag{2.15}$$

with the condition that $\varphi_{k_\mu u_{\mu\nu}}(\boldsymbol{x})$ be *holomorphic* at $\boldsymbol{x} = \boldsymbol{0}$ and vanish identically unless the following two conditions hold:

$$\sum_{\mu=1}^{N} \left[k_\mu + \sum_{\nu=1}^{M} u_{\mu\nu} \right] > 2, \tag{2.16}$$

$$\sum_{\mu=1}^{N} \left[-p_\mu k_\mu + \sum_{\nu=1}^{N} p_\nu u_{\mu\nu} \right] > 1 + p_n, \qquad n = 1, \ldots, N. \tag{2.17}$$

In the second case

$$h(\boldsymbol{x};\boldsymbol{y}) = \sum_{\boldsymbol{k}=\boldsymbol{0}}^{\boldsymbol{K}} \prod_{\nu=1}^{N} \left[(x_\nu)^{k_\nu} \sum_{\boldsymbol{u}_\nu=\boldsymbol{0}}^{\boldsymbol{U}_\nu} \prod_{\mu=1}^{N} (y_\mu)^{u_{\nu\mu}} \varphi_{k_\nu u_{\nu\mu}}(\boldsymbol{y}) \right], \tag{2.18}$$

with the condition that $\varphi_{k_\nu u_{\nu\mu}}(\boldsymbol{y})$ vanish identically unless there hold, again, the two inequalities (2.16) and (2.17).

Example 3. We display the simplest Hamiltonian case, namely that with $N = 1$. Then

$$H(x;y) = ipxy + h(x;y), \tag{2.19}$$

and either

$$h(x;y) = \sum_{u=3}^{U} x^u \sum_{k=0}^{u-1} y^k \varphi_{ku}(x) \tag{2.20a}$$

or

$$h(x;y) = \sum_{u=3}^{U} y^u \sum_{k=0}^{u-1} x^k \varphi_{ku}(y), \tag{2.20b}$$

with, in both cases, U an arbitrary integer no less than 3 and $\varphi_{ku}(z)$ *holomorphic* at $z = 0$ but otherwise arbitrary. Note that this is a complex Hamiltonian problem, but it can be turned into a real Hamiltonian problem via the assignments **[2]**

$$x = q_1 + iq_2, \qquad y = p_1 - ip_2, \tag{2.21}$$

where $q_1 \equiv q_1(t), q_2 \equiv q_2(t)$ are two *real* canonical coordinates and $p_1 \equiv p_1(t), p_2 \equiv p_2(t)$ are the two corresponding, of course as well *real*, canonical momenta, while the real Hamiltonian $H^{(R)}(q_1, q_2; p_1, p_2)$ reads

$$H^{(R)}(q_1, q_2; p_1, p_2) = p(q_1 p_2 - q_2 p_1) + \mathrm{Re}[h(q_1 + iq_2; p_1 - ip_2)], \tag{2.22}$$

where, let us recall, p is a *positive integer* and $h(x, y)$ is given either by (2.20a) or (2.20b).

3. Proof

In this section we prove our main result, as formulated in Section 1. To this end we introduce the following version of the trick:

$$x_n(t) = \exp(ip_n t)\xi_n(\tau), \qquad n = 1, \ldots, N, \tag{3.1a}$$

$$y_m(t) = \exp(-iq_m t)\eta_m(\tau), \quad m = 1, \ldots, M, \tag{3.1b}$$

$$\tau = [\exp(ist) - 1]/(is), \tag{3.2}$$

where s is a sign, $s = \pm$, which we reserve to choose later, see below. Note that this change of dependent and independent variables entails the following simple relation,

$$x_n(0) = \xi_n(0), \qquad n = 1, \ldots, N, \tag{3.3a}$$

$$y_m(0) = \eta_m(0), \quad m = 1, \ldots, M, \tag{3.3b}$$

among the "initial data." The new dependent variables $\xi_n(\tau)$, $\eta_m(\tau)$ satisfy then, see (1.1), the following system of (*nonautonomous*) ODEs:

$$\xi_n' = (1 + is\tau)^{-1-sp_n} f_n[\ldots, (1 + is\tau)^{sp_\nu}\xi_\nu, \ldots; \ldots, (1 + is\tau)^{-sq_\mu}\eta_\mu, \ldots], \quad n = 1, \ldots, N, \tag{3.4a}$$

$$\eta_m' = (1 + is\tau)^{-1+sq_m} g_m[\ldots, (1 + is\tau)^{sp_\nu}\xi_\nu, \ldots; \ldots, (1 + is\tau)^{-sq_\mu}\eta_\mu, \ldots], \quad m = 1, \ldots, M, \tag{3.4b}$$

where the appended prime denotes of course differentiation with respect to the new (complex) independent variable τ. It is then easy to prove (see Appendix), using the properties specified above of the functions f_n, g_m (see (1.2), and either (1.3) with (1.4), in which case we set $s = +$, or (1.5) with (1.6), in which case we set $s = -$), that there exists a *positive* number $r > 0$ such that the $N + M$ conditions

$$|\xi_n(0)| < r, \quad n = 1, \ldots, N; \qquad |\eta_m(0)| < r, \quad m = 1, \ldots, M \tag{3.5}$$

entail that the $N+M$ functions $\xi_n(\tau)$, $\eta_m(\tau)$ are *holomorphic* for $|\tau| \leq 2$, hence that these dependent variables $\xi_n(\tau)$, $\eta_m(\tau)$, considered as functions of the real variable t, are periodic with period 2π (see (3.2)), hence that all the solutions $x_n(t), y_m(t)$ of (1.1) with initial data that satisfy the restrictions (1.7) are as well (see (3.5) and (3.3)) *completely periodic* ("isochronous") with period 2π.

Note that this entails that, for small enough initial data, the system (1.1) with the restrictions on the functions f_n, g_m specified in Section 1 retains the isochronicity property of its linear part, namely of the system that obtains from (1.1) by equating its left-hand side to zero, the solutions of which read obviously

$$x_n(t) = x_n(0)\exp(ip_n t), \qquad y_n(t) = y_n(0)\exp(-iq_n t), \tag{3.6}$$

and are therefore *completely periodic* with period 2π.

4. Outlook

In this section we discuss tersely the relation of the result reported in this paper with more standard approaches.

A classical method to treat systems of evolution ODEs of the type we consider, see (1.1), is via reduction to normal forms. Let us mention in this respect that, for the kind of systems we consider, see (1.1), no difficulty due to "small denominators" arises, due to the integer character of the relevant frequencies p_n, q_m, which entail that the moduli of the relevant denominators D_n, d_m,

$$D_n = p_n - \sum_{\nu=1}^{N} p_\nu k_\nu + \sum_{\mu=1}^{M} q_\mu l_\mu, \qquad n = 1, \ldots, N, \tag{4.1a}$$

$$d_m = q_m + \sum_{\nu=1}^{N} p_\nu k_\nu - \sum_{\mu=1}^{M} q_\mu l_\mu, \quad m = 1, \ldots, M \tag{4.1b}$$

(where the $N + M$ "exponents" k_ν, l_μ are of course arbitrary nonnegative integers) have at least unity as their minimum value—provided they do not vanish.

But the problem is that they might vanish. It is easily seen that this cannot happen if only nonnegative or nonpositive frequencies are present (hence if either $M = 0$ or $N = 0$, see (1.1)) and they are moreover *all equal*, say $p_n = p$ or $q_m = q$; then the result reported in this paper can just as easily be obtained via the standard technique of reduction to normal forms *à la Poincaré* [**18, 21**] and it might be therefore justified to rate it as trivial (although our approach might still be preferable if one wishes to get some estimate of the value of r, see (1.7); indeed in the Appendix we obtain such an explicit estimate, although no effort is made there to obtain an accurate one).

If instead the relevant denominators do vanish, as it might happen even in the special cases $M = 0$ or $N = 0$ if the frequencies p_n or q_m featured by the linear part of the equations of motion (1.1) are different, then the process of reduction yields normal forms containing as many nonlinear terms as there are "resonances," namely sets of values of the relevant "exponents" $k_1, \ldots, k_n$ (if $M = 0$) or $l_1, \ldots, l_m$ (if $N = 0$) for which some denominators D_n or d_m, see (4.1), do vanish. But even in such cases it has been shown by Walcher [**22**] and by Gaeta [**19**] that a further process of reduction is possible, yielding again finally only *linear* ODEs. Thus, as long as the number of resonances is *finite*, as is indeed the case if all the frequencies have the same sign (namely in the special cases $M = 0$ or $N = 0$), the result proven in this paper can presumably be as well obtained via the more standard route of reduction to normal forms *à la Poincaré–Dulac* [**17, 18, 21**].

However, in the general case treated in this paper, namely when neither M nor N vanish (see (1.1)), then clearly the approach via reduction to normal forms runs into the difficulty of encountering an *infinite* number of resonances, because in this case, for any assignment of the *nonnegative integers* p_n, q_m there is generally an *infinite* number of sets of values of the *nonnegative integers* k_ν, l_μ that cause some of the denominators D_n or d_m, see (4.1), to vanish. It is therefore far from clear that in this general case the result proven in this paper via the trick could be as well obtained via the standard approach based on reduction to normal forms. Indeed doubts about the easy feasibility of such a program are reinforced by the observation that, in order to reach our result, certain *global* properties of the right-hand sides of the equations of motion (1.1) must be posited, see (1.3) or (1.5), while the reduction to normal forms is an essentially *local* approach.

An analogous comparison with the standard approach can be made in the Hamiltonian case, see the last part of Section 2. In this Hamiltonian case the procedure, which is generally associated with the name of Birkhoff (see for instance [**18**]) , consists in identifying a *symplectic* change of variables (defined by a formal series which generally does not converge) that brings the system into Birkhoff normal form. In the special one-degree-of-freedom case, as considered in Example 3 of Section 2, the transition to Birkhoff normal form is always *holomorphic*, being defined, in this case, by a *convergent* series, and it amounts to introducing a new canonical coordinate and its conjugate momentum, say $\tilde{x}$ and $\tilde{y}$, such that the Hamiltonian depend only on their product $\tilde{x}\tilde{y}$, $H(x,y) \equiv \widetilde{H}(\tilde{x}\tilde{y})$. It is easily seen how this can be done in the very special subcase of Example 3, see (2.19) with (2.20a), characterized by the assignments $U = 3$ and $\varphi_{k3}(x) = \delta_{k0}\varphi(x)$ entailing (see (2.19))

$$H(x;y) = ipxy + x^3\varphi(x), \tag{4.2}$$

when the symplectic transformation $\tilde{x} = x$, $\tilde{y} = y - ix^2\varphi(x)$ indeed yields

$$H(x,y) \equiv \widetilde{H}(\tilde{x}\tilde{y}) = ip\tilde{x}\tilde{y}. \tag{4.3}$$

Note that in this case the transformed Hamiltonian is actually linear in the product $\tilde{x}\tilde{y}$, consistently with the isochronous character of the motion. Indeed, as a consequence of our result guaranteeing isochronicity, it is clear that reduction to the linear form (4.3) shall be a general outcome of the Birkhoff procedure in the general N-degree-of-freedom Hamiltonian case (2.14), provided $h(\boldsymbol{x};\boldsymbol{y})$ is characterized by the restrictions following (2.14) hence belongs to the class to which our approach is applicable. But it is far from obvious how this result (including the convergence of the relevant expansion) can be proven via the standard approach in this more general case.

In conclusion, let us mention that it is for these reasons that in this paper, even at the cost of some notational repetitiveness, we presented our results in a manner intended to emphasize, see (1.1), that our treatment encompasses the general case with both *positive* and *negative* frequencies, which also includes, as explained in the last part of Section 2, the Hamiltonian case.

Acknowledgements. The new idea underlying the results reported in this paper emerged in June 2002 during a visit of FC to the Université Pierre et Marie Curie (Paris 6). A very special case of these results, namely the case with $M = 0$ and $p_n = 1$, see (1.1), was then reported by FC at the international meeting held in Pontignano near Sienna in June 2002 to celebrate the 60th birthdate of Francesco Guerra; it is a pleasure to acknowledge fruitful discussions held there with Giovanni Gallavotti (and later, via email, with Luigi Galgani), who pointed out that such results could not be considered new (much as they did look surprising at first sight), inasmuch as they could be as well obtained via the standard approach (as discussed in Section 4). The draft of this paper was then discussed during a visit of JPF to the Università di Roma "La Sapienza" on the occasion of the international meeting held there at the beginning of September 2002 to celebrate the 60th birthdate of Giovanni Gallavotti, and completed when both authors participated in the meeting on "Superintegrability in classical and quantum systems" held in mid-September 2002 at the Université de Montréal, Canada. It is moreover a pleasure to acknowledge enlightening conversations with Giuseppe Gaeta.

Appendix A

In this appendix we prove the result mentioned in Section 3, namely that (every component $\xi_n(\tau)$, $\eta_m(\tau)$ of) the solution of the (*nonautonomous*) system of evolution equations,

$$\text{(A.1a)} \quad \xi_n' = (1+i\tau)^{-1-p_n} f_n[\ldots,(1+i\tau)^{p_\nu}\xi_\nu,\ldots;\ldots,(1+i\tau)^{-q_\mu}\eta_\mu,\ldots], \quad n=1,\ldots,N,$$

$$\text{(A.1b)} \quad \eta_m' = (1+i\tau)^{-1+q_m} g_m[\ldots,(1+i\tau)^{p_\nu}\xi_\nu,\ldots;\ldots,(1+i\tau)^{-q_\mu}\eta_\mu,\ldots], \quad m=1,\ldots,M,$$

is a *holomorphic* function of the complex independent variable τ in a circular disk centered at $\tau = 0$ in the complex τ-plane, the radius ρ of which can be made arbitrarily large by requiring the initial data, see (3.3), to satisfy the restrictions (3.5), with a sufficiently small value of r – provided of course there hold the properties of the $N + M$ functions $f_n(\boldsymbol{x},\boldsymbol{y}), g_m(\boldsymbol{x},\boldsymbol{y})$ detailed in Section 1, namely (i) these functions are *holomorphic* in all their $N+M$ arguments x_n, y_m at $\boldsymbol{x} = \boldsymbol{0}$, $\boldsymbol{y} = \boldsymbol{0}$,(ii) they vanish there faster than linearly, see (1.2), (iii) they are polynomial in all the variables y_m, see (1.3), and (iv) they satisfy the restrictions (1.4). Note that the evolution equations we just wrote, (A.1), correspond to (3.4) with $s = +$, and correspondingly we are assuming validity of the restrictions (1.3) and (1.4). If instead the valid properties were (1.5) and (1.6), we should have set $s = -$, to then proceed in a completely analogous manner in order to complete the proof (see below)—a task that is left for the diligent reader. Of course if (1.3) with (1.4), as well as (1.5) with (1.6), are both valid, the isochronicity property holds a fortiori.

So, what we now prove is that (every component of) the solution of (A.1), under these conditions, is a *holomorphic* function of the (complex) independent variable τ (at least) in the circular disk

$$\text{(A.2)} \qquad |\tau| < \rho,$$

the radius ρ of which can be made arbitrarily large by choosing r sufficiently small, namely we prove that, as r tends to zero, $r \to 0$, the radius ρ diverges, $\rho \to \infty$.

The proof is based on the following standard theorem (see for instance [**20**, Section 12.21]): let a system of analytic evolution equations read

$$\text{(A.3)} \qquad w_j' = F_j(\boldsymbol{w},\tau), \quad j=1,\ldots,J$$

with the solutions $w_j \equiv w_j(\tau)$ characterized by the simple initial data

$$\text{(A.4)} \qquad w_j(0) = 0, \quad j=1,\ldots,J.$$

Here of course $\boldsymbol{w}$ is a J-vector, $\boldsymbol{w} \equiv (w_1,\ldots,w_J)$. The J functions $F_j(\boldsymbol{w},\tau)$ are analytic in all their $J+1$ arguments, and *holomorphic* at $\boldsymbol{w} = \boldsymbol{0}$ and $\tau = 0$ (namely where the initial data are assigned, see (A.4)). Then the solutions $w_j(\tau)$ of (A.3) are *holomorphic* functions of the complex variable τ (at least) in the disk $|\tau| < \rho$, see (A.2), with the positive quantity ρ bounded below by the formula

$$\text{(A.5)} \qquad \rho \geqslant a[1 - \exp\{-b/[(J+1)aF]\}].$$

Here the *positive* quantities a and b are characterized by the requirement that all the functions $F_j(\boldsymbol{w}, \tau)$ be *holomorphic* in all their $J+1$ arguments provided

$$|\tau| \leq a \tag{A.6}$$

and

$$|w_j| \leq b, \quad j = 1, \ldots, J, \tag{A.7}$$

while the *positive* quantity F is defined by the formula

$$F = \max_{\substack{j=1,\ldots,J \\ |\tau|\leq a \\ |w_k|\leq b,\ k=1,\ldots,J}} |F_j(\boldsymbol{w}, \tau)|. \tag{A.8}$$

Note that the right-hand side of (A.5) is an increasing function of a and of b, and a decreasing function of F, hence the lower bound (A.5) is valid a fortiori if we underestimate a and b, and we overestimate F, as we shall do below.

To apply this theorem to our case we set $J = N + M$ and

$$w_n(\tau) = \xi_n(\tau) - \xi_n(0), \qquad n = 1, \ldots, N; \tag{A.9a}$$

$$w_{N+m}(\tau) = \eta_m(\tau) - \eta_m(0), \quad m = 1, \ldots, M \tag{A.9b}$$

so that the conditions (A.4) are now automatically satisfied, while our evolution equations, (A.1), can now be identified with the standard evolution equations (A.3) by setting (in self-evident notation)

$$\begin{aligned} F_n(\boldsymbol{w}, \tau) = (1+i\tau)^{-1-p_n} f_n(\ldots, (1+i\tau)^{p_\nu}[\xi_\nu(0) + w_\nu(\tau)], \ldots; \\ \ldots, (1+i\tau)^{-q_\mu}[\eta_\mu(0) + w_{N+\mu}(\tau)], \ldots), \\ n = 1, \ldots, N, \end{aligned} \tag{A.10a}$$

$$\begin{aligned} F_{N+m}(\boldsymbol{w}, \tau) = (1+i\tau)^{-1+q_m} g_m(\ldots, (1+i\tau)^{p_\nu}[\xi_\nu(0) + w_\nu(\tau)], \ldots; \\ \ldots, (1+i\tau)^{-q_\mu}[\eta_\mu(0) + w_{N+\mu}(\tau)], \ldots), \\ m = 1, \ldots, M. \end{aligned} \tag{A.10b}$$

We now note that, thanks to the assumption (1.4), these functions $F_j(\boldsymbol{w}, \tau)$ are *nonsingular* at $\tau = i$; moreover, by recalling the polynomial (hence *nonsingular*) character of their dependence on the second set of their arguments, see (1.3), as well as the fact that the numbers p_n, q_m are all *nonnegative*, we see that it is now justified to set

$$a = \alpha |\log(\delta)|, \tag{A.11}$$

$$b = \beta\delta, \tag{A.12}$$

and

$$r = \gamma\delta, \tag{A.13}$$

with α, β and γ all *positive*, $\alpha > 0$, $\beta > 0$, $\gamma > 0$, and all independent of δ, and to treat δ as a small *positive* parameter, namely to consider the limit $\delta \to 0$ (which entails $a \to \infty$, $b \to 0$, $r \to 0$), and to then conclude that in this limit the right-hand side of (A.5) diverges. Note that indeed, in this limit, these assignments, (A.11), (A.12) and (A.13), are justified, see (A.6), (3.5), (A.12). On the other hand the conclusion we just mentioned clearly entails validity of the result we must prove, namely the divergence of ρ as $r \to 0$.

To show that the right-hand side of (A.5) diverges in the $\delta \to 0$ limit we must still find an upper bound to F, see (A.5) and (A.8). To obtain it, it is convenient to insert in the definition (A.8) of F, the definitions (A.10) of $F_n(\boldsymbol{w}, \tau)$ and $F_{N+m}(\boldsymbol{w}, \tau)$ with the explicit representations (2.3) of $f_n(\boldsymbol{x}; \boldsymbol{y})$ and $g_m(\boldsymbol{x}; \boldsymbol{y})$. It is then clear that, thanks to the restrictions (2.4) and (2.5), as $\delta \to 0$ (see (A.7) with (A.12), and restrict attention to values of τ which satisfy (A.6) with (A.11) so that as $\delta \to 0$ the arguments of the functions f_n and g_m in the right-hand sides of (A.10) tends to zero,

$$F < \delta^2[A|1 + i\tau|^p + B|1 + i\tau|^P], \tag{A.14}$$

where A and B are two positive (finite!) numbers (independent of τ and of δ), p is a *nonnegative integer*, $p \geq 0$, and P is a *positive integer*, $P > 0$. The two terms in the right-hand side of this inequality, (A.14), characterized by the exponents p respectively P, take care of the possibilities that the quantity $|1 + i\tau|$ be small respectively large (in the latter case, within the limitation entailed by (A.6) with (A.11)). Note that the nonnegative character of p is essential: it is a consequence of the inequality (2.5), for the validity of which the assumed polynomial character of $f_n(\boldsymbol{x}; \boldsymbol{y})$ and $g_m(\boldsymbol{x}; \boldsymbol{y})$ is of course essential (indeed unless all the components of the two M-vectors $\boldsymbol{K}$ and $\boldsymbol{L}$ in (2.3) were finite, the inequalities (2.5) could not hold). As for P, it might be large but it is certainly finite, as entailed by the finiteness of all the components of the $2M$ N-vectors $\boldsymbol{U}_\mu$, $\boldsymbol{W}_\mu$, see (2.3).

Via (A.6) with (A.11) it is easily seen that this inequality, (A.14), entails for small δ the following upper bound to F,

$$F < \delta^2[C + D|\log(\delta)|^P], \tag{A.15}$$

where again C and D are two *finite* (of course *positive*) constants independent of δ, and P is some (*finite*!) *positive* integer.

Insertion of this majorization, (A.15), together with (A.11) and (A.12), in (A.5) clearly entails that, in the $\delta \to 0$ limit, the right-hand side of this inequality diverges (logarithmically—but this is good enough, and we could have done better had we cared to).

References

1. F. Calogero, *A class of integrable Hamiltonian systems whose solutions are (perhaps) all completely periodic*, J. Math. Phys. **38** (1997), no. 11, 5711–5719.
2. ______, *Classical many-body problems amenable to exact treatments*, Lecture Notes in Phys. New Ser. m Monogr., vol. 66, Springer, Berlin, 2001.
3. ______, *Periodic solutions of a system of complex ODEs*, Phys. Lett. A **293** (2002), no. 3-4, 146–150.
4. ______, *On a modified version of a solvable ODE due to Painlevé*, J. Phys. A **35** (2002), no. 4, 985–992.
5. ______, *A complex deformation of the classical gravitational many-body problem that features a lot of completely periodic motions*, J. Phys. A **35** (2002), no. 16, 3619–3627.
6. ______, *On modified versions of some solvable ordinary differential equations due to Chazy*, J. Phys. A **35** (2002), no. 19, 4249–4256.
7. ______, *Solvable three-body problem and Painlevé conjectures*, Theoret. and Math. Phys. **133** (2002), no. 2, 1445–1454.
8. ______, *Differential equations featuring many periodic solutions*, Geometry and Integrability (L. Mason and Y. Nutku, eds.), London Math. Soc. Lecture Notes, vol. 295, Cambridge Univ. Press, Cambridge, 2003, pp. 9–21.

9. ———, *Partially superintegrable (indeed isochronous) systems are not rare*, New Trends in Integrability and Partial Solvability (Cadiz, 2002), NATO Sci. Ser. II Math. Phys. Chem, vol. 132, Kluwer, Dordrecht, 2004, pp. 49–77.
10. ———, *General olution of a three-body problem in the plane*, J. Phys. A **36** (2003), no. 26, 7291–7299.
11. F. Calogero and J.-P. Françoise, *Periodic solutions of a many-rotator problem in the plane*, Inverse Problems **17** (2001), no. 4, 871–878.
12. ———, *Periodic motions galore: how to modify nonlinear evolution equations so that they feature a lot of periodic solutions*, J. Nonlinear Math. Phys. **9** (2002), no. 1, 99–125.
13. ———, *Nonlinear evolution ODEs featuring many periodic solutions*, Theoret. and Math. Phys. **137** (2003), no. 3, 1663–1675.
14. F. Calogero, J.-P. Françoise, and M. Sommacal, *Periodic solutions of a many-rotator problem in the plane. II. Analysis of various motions*, J. Nonlinear Math. Phys. **10** (2003), no. 2, 157–214.
15. F. Calogero and V. I. Inozemtsev, *Nonlinear harmonic oscillators*, J. Phys. A **35** (2002), no. 48, 10365–10375.
16. F. Calogero and M. Sommacal,*Periodic solutions of a system of complex ODEs. II. Higher periods*, J. Nonlinear Math. Phys. **9** (2002), no. 4, 483–516.
17. H. Dulac, *Solutions d' un système d' équations différentielles dans le voisinage des valeurs singuliéres*, Bull. Soc. Mat. France **40** (1912), 324–383.
18. J.-P. Françoise, *Géométrie analytique et systèmes dynamiques*, Mathématiques, Presses Universitaires de France, Paris, 1995.
19. G. Gaeta, *Resonant Poincaré–Dulac normal forms as constrained linear systems*, Modern Phys. Lett. A **17** (2002), no. 10, 583–597.
20. E. L. Ince, *Ordinary differential equations*, Dover, New York, 1956.
21. É. Picard, *Traité d'analyse.* Tome III, Grands Class. Gauthier-Villars, Éd. Jacques Gabay, Paris, 1991, Chapter I.
22. S. Walcher, *On differential equations in normal forms*, Math. Ann. **291** (1991), no. 2, 293–314.

Dipartimento di Fisica, Università di Roma "La Sapienza," 00185 Roma, Italy and Istituto Nazionale di Fisica Nucleare, Sezione di Roma

E-mail address: francesco.calogero@roma1.infn.it, francesco.calogero@uniroma1.it

Géométrie différentielle, Systèmes différentiels, Imagerie géométrique et Biomédicale (GSIB), Université Pierre et Marie Curie, 175 rue du Chevaleret, 75013 Paris, France

E-mail address: jpf@ccr.jussieu.fr

Centre de Recherches Mathématiques
CRM Proceedings and Lecture Notes
Volume **37**, 2004

Nambu Dynamics, Deformation Quantization, and Superintegrability

THOMAS L. CURTRIGHT and Cosmas K. Zachos

This paper is dedicated to Milton Hendrix (1930–2000).

ABSTRACT. Phase space is a framework ideally suited for quantizing superintegrable systems through the use of deformation methods, as illustrated here by applications to de Sitter and chiral particles. Within this framework, Nambu brackets elegantly incorporate the additional quantum invariants of such models. New results are presented for the non-Abelian quantization of these brackets.

1. Introduction

For systems with velocity-dependent potentials, when quantization of the classical system presents operator ordering ambiguities involving x and p, the general consensus has long been [**6, 22, 27, 29, 31**–**33, 40, 42, 43, 45**] to select those orderings in the quantum Hamiltonian which maximally preserve the symmetries present in the corresponding classical Hamiltonian. However, even for simple systems such constructions may become involved and quite technical.

Recently it was emphasized [**11, 47**] that in contrast to conventional operator quantization, the problem of selecting the quantum Hamiltonian which maximally preserves integrability is addressed very directly in Moyal's phase-space quantization formulation [**8,35,46**]. Basically, the reason for this is that the kernels involved ("kernel functions" or "Weyl transforms of operators") are c-number functions and have an interpretation analogous to that of the classical phase-space theory, although in general they involve $\hbar$-corrections ("deformations"). This stand-alone formulation of quantum mechanics is based on the Wigner Function (WF), a quasi-probability distribution function in phase-space which comprises the kernel function of the density matrix. Observables and transition amplitudes are phase-space integrals of kernel functions weighted by the WF, in analogy to statistical mechanics.

2000 *Mathematics Subject Classification.* Primary 53D55, 81R60; Secondary 81R50, 37J35.

The first author was supported in part by NSF Award 0073390.

The second author was supported in part by the US Department of Energy, Division of High Energy Physics, Contract W-31-109-ENG-38.

This is the final form of the paper.

There is no free lunch, however. Kernel functions compose through the $\star$-product, a noncommutative, associative, pseudodifferential operation, which encodes the entire quantum mechanical action and whose antisymmetrization (commutator) is the Moyal Bracket (MB) [**8, 35, 46**]. Any arbitrary operator ordering can be brought to Weyl-ordering format, by use of Heisenberg commutations, and through Weyl's transform corresponds invertibly to a specific $\hbar$-deformation of the classical kernel [**19, 30, 44**]. Thus, two operators of different orderings correspond to kernel functions differing in their deformation terms of $O(\hbar)$, and the problem of selecting the correct ordering reduces to a purely $\star$-product algebraic one.

Previously, Hietarinta [**20**] has investigated in this phase-space quantization language the simplest integrable systems of velocity-dependent potentials. In each system, he has promoted the vanishing of the Poisson Bracket (PB) of the (one) classical invariant I (conserved integral) with the Hamiltonian, $\{H, I\} = 0$, to the vanishing of its (quantum) Moyal Bracket (MB) with the Hamiltonian, $\{H_{qm}, I_{qm}\}_{\text{MB}} = 0$. This dictates quantum corrections, addressed perturbatively in $\hbar$: he has found $O(\hbar^2)$ corrections to the Is and H (V), needed for quantum symmetry. The expressions found are quite simple, as the systems chosen are such that the polynomial character of the ps, or suitable balanced combinations of ps and qs, ensure collapse or subleading termination of the MBs. The specification of the symmetric Hamiltonian then is complete, since the quantum Hamiltonian in terms of classical phase-space variables corresponds uniquely to the Weyl-ordered expression for these variables in operator language. Berry [**4**] has also studied the WFs of integrable systems.

In this contribution we shall review the work in [**11, 47**] and make some extensions of it. We shall discuss nonlinear σ-models to argue for the general principles of power and convenience in isometry-preserving quantization in phase space, particularly for superintegrable situations [**6, 27, 40, 42, 45**]. Briefly, we find that the symmetry generator invariants are undeformed by quantization, but the Casimir invariants of their MB algebras are deformed. Hence the Hamiltonians are also deformed by terms $O(\hbar^2)$, as they consist of quadratic Casimir invariants, but nonetheless their energy spectra can be read off through the usual group theoretic techniques once these are properly adapted to phase space. The basic principles are illustrated for the two-sphere, in Section 2, and then applied to larger classes of symmetric manifolds such as N-spheres, in Section 3, and chiral models, in Section 4.

Finally, in Section 5, the evolution of such systems is formulated in phase-space through the use of Nambu Brackets (NBs) [**7, 17, 18, 21, 36–38, 41**]. The quantization of Nambu's approach is discussed in detail and compared to the standard Moyal deformation quantization. This last section is the most original part of the paper, and in it we present several new results involving Quantum Nambu Brackets (QNBs). An appendix indicates how the QNB results can also be expressed in the more traditional language of operators acting on Hilbert space.

2. Basic principles and the 2-sphere

A Hamiltonian system with N degrees of freedom is *integrable* if it has N invariants in involution (globally defined and independent), and *superintegrable* [**6, 27, 40, 42, 45**] if it has additional conservation laws up to a maximum $2N - 1$ invariants. For example, consider a particle constrained to the surface of a unit radius two-sphere S^2, but otherwise moving freely. Thus $N = 2$ and $2N - 1 = 3$. Three independent invariants of this maximally superintegrable system are the angular momenta about the center of the sphere:

$$L_x, \qquad L_y, \qquad L_z.$$

Actually, no two of these are in involution, but this is quickly remedied, and moreover it isn't much of a hindrance as we shall see below in the Nambu approach to mechanics, an approach that places all invariants on a more equal footing.

To be more explicit, we may coordinate the upper and lower ($\pm$) hemispheres by projecting the particle's location onto the equatorial disk, $\{(x, y) \mid x^2 + y^2 \leq 1\}$. The invariants are then

$$L_z = xp_y - yp_x, \qquad L_y = \pm\sqrt{1 - x^2 - y^2}\, p_x, \qquad L_x = \mp\sqrt{1 - x^2 - y^2}\, p_y.$$

The last two are the de Sitter momenta, or nonlinearly realized axial charges, corresponding to the x, y "pions" of this truncated σ-model.

The Poisson Brackets (PBs) of these expressions close into the expected so(3).

$$\{L_x, L_y\}_{\text{PB}} = L_z, \qquad \{L_y, L_z\}_{\text{PB}} = L_x, \qquad \{L_z, L_x\}_{\text{PB}} = L_y.$$

The usual Hamiltonian of the free particle system is the Casimir invariant.

$$H = \tfrac{1}{2}(L_x L_x + L_y L_y + L_z L_z).$$

Thus, it immediately follows algebraically that PBs of H with the $\mathbf{L}$ vanish, and time-invariance holds.

$$\frac{d}{dt}\mathbf{L} = \{\mathbf{L}, H\}_{\text{PB}} = 0.$$

So any one of the Ls and this Casimir constitute a pair of invariants in involution.

In quantizing this system using operators some simple ordering issues arise because the system has an effective momentum-dependent potential.

$$H = \tfrac{1}{2}(1 - x^2)p_x^2 + \tfrac{1}{2}(1 - y^2)p_y^2 - xyp_x p_y.$$

But in the deformation quantization of phase-space one may just insert Groenewold's [**19, 30**] noncommutative but associative $\star$-products,[1] defined as

$$\star \equiv \exp \frac{i\hbar}{2}(\overleftarrow{\partial}_x \overrightarrow{\partial}_{p_x} - \overleftarrow{\partial}_{p_x} \overrightarrow{\partial}_x + \overleftarrow{\partial}_y \overrightarrow{\partial}_{p_y} - \overleftarrow{\partial}_{p_y} \overrightarrow{\partial}_y),$$

at strategic points in the above classical expressions and achieve full quantum integrability. That is, the classical Poisson bracket statements,

$$\{I, H\}_{\text{PB}} = 0 \quad \text{(for invariants)},$$

may be promoted to quantum Moyal bracket statements,

$$\{I_{qm}, H_{qm}\}_{\text{MB}} \equiv \frac{1}{i\hbar}(I_{qm} \star H_{qm} - H_{qm} \star I_{qm}) = 0 \quad \text{(for invariants)},$$

with simple choices for I_{qm} and H_{qm}. As $\hbar \to 0$, the MB reduces to the PB1. More importantly, the MB provides the unique route around the Groenewold-van Hove theorem.[2] The $\star$ product is the unique (up to equivalence) one-parameter associative deformation of ordinary products, and the MB is the corresponding deformation of the PB.

---

[1]A general discussion of star products on curved spaces may be found in [**16**] and [**39**].

[2]Groenewold introduced the $\star$-product to evade his "no-go" theorem in the same 1946 paper wherein he proved it, thereby immediately reducing the theorem to a mere appurtenance.

Maximally symmetric phase-space quantization is achieved here by a Hamiltonian of the form

$$H_{qm} = \tfrac{1}{2}(L_x \star L_x + L_y \star L_y + L_z \star L_z).$$

The reason is that, in this realization, there are no $O(\hbar)$ corrections to the individual $\mathbf{L}$s. These particular invariants are undeformed by quantization.

$$\mathbf{L} = \mathbf{L}_{qm}.$$

The algebra of the $\mathbf{L}$s is then promoted to the corresponding MB expression *without any modification*, since all "$\mathbf{L}$ with $\mathbf{L}$" MBs collapse to PBs by the linearity in momenta of the arguments. Consequently, given associativity for $\star$, the corresponding quantum quadratic Casimir invariant $\mathbf{L} \cdot \star\mathbf{L}$ has vanishing MBs with $\mathbf{L}$, and automatically generates a symmetry-preserving time-evolution. The $\star$-product in this Hamiltonian trivially evaluates to expose a quantum correction to the classical phase-space energy:

$$H_{qm} = H + \frac{\hbar^2}{8}\left(\frac{1}{1 - x^2 - y^2} - 3\right).$$

Thus we have

$$\{\mathbf{L}, H_{qm}\}_{\text{MB}} = 0 \neq \{\mathbf{L}, H\}_{\text{MB}}.$$

Also note there will be quantum corrections to the classical equations of motion for (p_x, p_y), but not (x, y).

In phase-space quantization [**8,35,46**], the Wigner function (WF) $f(x, p_x, y, p_y)$, the Weyl kernel function of the density operator, evolves according to Moyal's equation [**35**]

$$\frac{\partial f}{\partial t} = \{H_{qm}, f\}_{\text{MB}}.$$

In addition to it, the WFs for pure stationary states also satisfy Fairlie's "$\star$-genvalue" equations [**8, 15**] specifying the spectrum

$$\begin{aligned} H_{qm}(x,p) \star f(x,p) &= f(x,p) \star H_{qm}(x,p) \\ &= H_{qm}\left(x + \frac{i\hbar}{2}\overrightarrow{\partial}_p, p - \frac{i\hbar}{2}\overrightarrow{\partial}_x\right) f(x,p) = \mathrm{E}\, f(x,p). \end{aligned}$$

Eigenvalue problems of this type also occur in spectral theory for (special) Jordan algebras, a point to be re-emphasized below.

The spectrum of this Hamiltonian, then, is proportional to the $\hbar^2 l(l+1)$ spectrum of the so(3) Casimir invariant $\mathbf{L} \cdot \star\mathbf{L} = L_+ \star L_- + L_z \star L_z - \hbar L_z$, for integer l [**2, 3**]. This can be proved algebraically by the identical standard recursive ladder operations in quantum phase-space which are used in the operator formalism Fock space,

$$L_z \star L_+ - L_+ \star L_z = \hbar L_+,$$

where $L_\pm \equiv L_x \pm i L_y$. From the *real $\star$-square theorem* [**10**], it follows that

$$\langle \mathbf{L} \cdot \star\mathbf{L} - L_z \star L_z \rangle = \langle L_x \star L_x + L_y \star L_y \rangle \geq 0.$$

The $\star$-genvalues of L_z, m, are thus bounded, $|m| \leq l < \sqrt{\langle \mathbf{L} \cdot \star\mathbf{L}\rangle}/\hbar$, necessitating $L_- \star f_{m=-l} = 0$. Hence

$$L_+ \star L_- \star f_{-l} = 0 = (\mathbf{L} \cdot \star\mathbf{L} - L_z \star L_z + \hbar L_z) \star f_{-l},$$

and consequently $\langle \mathbf{L} \cdot \star\mathbf{L}\rangle = \hbar^2 l(l+1)$. Similar $\star$-ladder arguments and inequalities apply directly in phase space to all Lie algebras.

Classical Hamiltonians are scalars under canonical transformations, but it should not be assumed that the quantum mechanical expression above is a canonical scalar [**11**]. For canonical transformations in phase-space quantization see [**8**]. The $\star$-product and WFs are also not invariant under canonical transformations, in general, but transform in a suitable quantum covariant way [**8**], so as to yield an identical MB algebra and $\star$-genvalue equations, and thus spectrum, following from the identical group theoretical construction.

3. Quantum N-sphere

For the generic sphere models, S^N, the maximally symmetric Hamiltonians are the quadratic Casimir invariants of so$(N+1)$,

$$H = \tfrac{1}{2} P_a P_a + \tfrac{1}{4} L_{ab} L_{ab},$$

where again in terms of equatorial plane coordinates

$$L_{ab} = q^a p_b - q^b p_a, \qquad\qquad P_a = \sqrt{1-q^2}\, p_a,$$

for $a = 1, \ldots, N$, $q^2 = \sum_{a=1}^N q^a q^a$. These are the usual angular and de Sitter momenta of so$(N+1)/$ so(N). All of these $N(N+1)/2$ sphere transformations are symmetries of the classical Hamiltonian. (There are more of them than the $2N-1$ allowed *independent* invariants on the phase-space, if $N > 2$. We will explain how to select $2N-1$ independent invariants later, when we discuss Nambu brackets.)

Quantization proceeds as in S^2, maintaining conservation of all P_a and L_{ab},

$$H_{qm} = \tfrac{1}{2} P_a \star P_a + \tfrac{1}{4} L_{ab} \star L_{ab},$$

and hence the quantum correction is

$$H_{qm} - H = \frac{\hbar^2}{8} \left(\frac{1}{1-q^2} - 1 - N(N-1) \right).$$

The spectra are proportional to the Casimir eigenvalues $l(l+N-1)$ for integer l [**2, 3**]. (For $N=3$, this form is reconciled with the Casimir expression in the next section as $l = 2j$, and agrees with [**22, 29, 31, 32**]).

Can the above quantum Hamiltonian be expressed geometrically through the use of tangent-space methods? The classical description is indeed simple upon utilization of Vielbeine

$$g_{ab} = \delta_{ij} V_a^i V_b^j = \delta_{ab} + \frac{1}{1-q^2} q^a q^b, \qquad g^{ab} = \delta_{ab} - q^a q^b, \qquad g^{ab} V_a^i V_b^j = \delta^{ij}.$$

For the generic sphere models, S^N, standard choices for the Vielbeine are

$$V_a^i = \delta_{ai} - \frac{q^a q^i}{q^2} \left(1 \pm \frac{1}{\sqrt{1-q^2}} \right), \qquad V^{ai} = \delta_{ai} - \frac{q^a q^i}{q^2} (1 \pm \sqrt{1-q^2}).$$

The classical Hamiltonian equals

$$H = \tfrac{1}{2} (p_a V^{ai})(V^{bi} p_b),$$

but the quantum Hamiltonian is *not* equal to the obvious guess

$$H_{\text{other}} \equiv \tfrac{1}{2} (p_a V^{ai}) \star (V^{bi} p_b) \neq H_{qm}.$$

Although we will see below how this simple form *does* apply in chiral models, particularly for the S^3 case, through a different choice of Dreibeine.

The MBs of cotangent bundle currents, for a general manifold, do not close among the Vielbein-currents, but instead,

$$\{V^{aj}p_a, V^{bk}p_b\}_{\rm MB} = \omega^{a[jk]}p_a = (V^{bk}\partial_b V^{aj} - V^{bj}\partial_b V^{ak})p_a,$$

where for the N-sphere

$$\omega^{a[ij]} = (\delta^{ai}q^j - \delta^{aj}q^i)w, \quad w \equiv \left(1-\sqrt{1-q^2}\right)/q^2,$$

choosing the $-$ sign in the definition of the Vielbeine so $V^{aj} = \delta_{aj} - q^a q^j w$. It follows that

$$H_{qm} - H_{\rm other} = \frac{1}{8}\hbar^2(N-1)(1-2w-N).$$

$H_{\rm other}$ corresponds to a different operator ordering in the conventional Hilbert space formulation, and has less symmetry than H_{qm}. $H_{\rm other}$ conserves the rotations L_{ab} (i.e., it is symmetric under the SO(N) stability subgroup for S^N). However, it does not conserve the Vielbein-currents on the N-sphere, nor does it conserve the de Sitter momenta. This last statement follows from the above difference $H_{qm} - H_{\rm other}$ dictating,

$$\{H_{\rm other}, P_c\}_{\rm MB} = \{H_{\rm other} - H_{qm}, P_c\}_{\rm MB} = \hbar^2 q^c(N-1)\left(\frac{2w-1}{4q^2}\right) \neq 0.$$

Nevertheless, it can be shown that a $\star$-similarity transformation compensates for the difference in these Hamiltonians. Consider

$$\begin{aligned} w^{-(N-1)/2} \star p_a V^{aj} \star w^{(N-1)/2} &= w^{1-N} \star (p_a V^{aj} w^{N-1}) \\ &= V^{aj}p_a - \tfrac{1}{2}i\hbar(N-1)V^{aj}\partial_a \ln w, \end{aligned}$$

and the complex conjugate transformation

$$\begin{aligned} w^{(N-1)/2} \star p_a V^{aj} \star w^{-(N-1)/2} &= (w^{N-1}V^{aj}p_a) \star w^{1-N} \\ &= V^{aj}p_a + \tfrac{1}{2}i\hbar(N-1)V^{aj}\partial_a \ln w. \end{aligned}$$

Associativity of the $\star$-product then allows the maximally symmetric real Hamiltonian to be written as

$$H_{qm} = \tfrac{1}{2}(w^{N-1}V^{aj}p_a) \star w^{-2(N-1)} \star (w^{N-1}V^{bj}p_b).$$

Using homogeneous coordinates on the sphere, $w = 1/(1+\cos\theta)$, where θ is the polar angle.

4. Chiral models

The treatment of the 3-sphere S^3 also accords to the standard chiral model technology using left- and right-invariant Vielbeine. Specifically, the two choices for such Dreibeine for the 3-sphere are [**9**]: $q^2 = x^2 + y^2 + z^2$

$$^{(\pm)}V_a^i = \epsilon^{iab}q^b \pm \sqrt{1-q^2}g_{ai}, \qquad ^{(\pm)}V^{ai} = \epsilon^{iab}q^b \pm \sqrt{1-q^2}\delta^{ai}.$$

The corresponding right and left conserved charges (left- and right-invariant, respectively) then are

$$\mathcal{R}^i = {}^{(+)}V_a^i \frac{d}{dt}q^a = {}^{(+)}V^{ai}p_a, \qquad \mathcal{L}^i = {}^{(-)}V_a^i \frac{d}{dt}q^a = {}^{(-)}V^{ai}p_a.$$

More intuitive than those for S^2 are the linear combinations into Axial and Isospin charges (again linear in the momenta),

$$\tfrac{1}{2}(\mathcal{R}-\mathcal{L})=\sqrt{1-q^2}\,\mathbf{p}\equiv\mathbf{A}, \qquad \tfrac{1}{2}(\mathcal{R}+\mathcal{L})=\mathbf{q}\times\mathbf{p}\equiv\mathbf{I}.$$

It can easily be seen that the $\mathcal{L}$s and the $\mathcal{R}$s have PBs closing into standard su(2) $\otimes$ su(2), i.e., su(2) relations within each set, and vanishing between the two sets. Thus they are seen to be constant, since the Hamiltonian (and also the Lagrangian) can, in fact, be written in terms of either quadratic Casimir invariant.

$$H=\tfrac{1}{2}\mathcal{L}\cdot\mathcal{L}=\tfrac{1}{2}\mathcal{R}\cdot\mathcal{R}.$$

Quantization consistent with integrability thus proceeds as above for the 2-sphere, since the MB algebra collapses to PBs again, and so the quantum invariants $\mathcal{L}$ and $\mathcal{R}$ again coincide with the classical ones, without deformation (quantum corrections). The $\star$-product is now the obvious generalization to 6-dimensional phase-space. The eigenvalues of the relevant Casimir invariant are now $j(j+1)$, for half-integer j. However, for this chiral model the symmetric quantum Hamiltonian is simpler, geometrically, than that of the previous N-sphere, since it can also be written as

$$H_{qm}=\frac{1}{2}(p_aV^{ai})\star(V^{bi}p_b)=\frac{1}{2}\left(g^{ab}p_ap_b+\frac{\hbar^2}{4}\partial_aV^{bi}\partial_bV^{ai}\right)=\frac{1}{2}\mathcal{L}\cdot\star\mathcal{L}=\frac{1}{2}\mathcal{R}\cdot\star\mathcal{R}.$$

No $\star$-similarity transformations are needed for chiral models, unlike the general N-sphere. The Dreibeine throughout this formula can be either ${}^{+}V_a^i$ or ${}^{-}V_a^i$, corresponding to either the right- or the left-acting quadratic Casimir invariant. The quantum correction then amounts to

$$H_{qm}-H=\frac{\hbar^2}{8}\left(\frac{1}{1-q^2}-7\right).$$

This expression again is not canonically invariant. E.g., in gnomonic PR$_N$ coordinates,[3] it is $\frac{3}{4}\hbar^2(Q^2-1)$, i.e., it has not transformed as a canonical scalar [**8, 11**]. Note, however, that either of these quantum corrections are $\geq -\frac{3}{4}\hbar^2$ on the manifold.

In general, the above discussion also applies to all chiral models, with the algebra for a chiral group G replacing su(2) above. I.e., the Vielbein-momenta combinations $V^{aj}p_a$ represent algebra generator invariants, whose quadratic Casimir group invariants yield the respective Hamiltonians, and whence the properly $\star$-ordered quantum Hamiltonians as above.

That is to say, for [**5**] group matrices U generated by exponentiated constant group algebra matrices T weighted by functions of the particle coordinates q, we have

$$iU^{-1}\frac{d}{dt}U={}^{(+)}V_a^jT_j\frac{d}{dt}q^a={}^{(+)}V^{aj}p_aT_j, \qquad iU\frac{d}{dt}U^{-1}={}^{(-)}V^{aj}p_aT_j,$$

It follows that PBs of left- and right-invariant charges,

$${}^{(\pm)}V^{aj}p_a=\frac{i}{2}\,\mathrm{Tr}\left(T_jU^{\mp 1}\frac{d}{dt}U^{\pm 1}\right),$$

close to the identical Lie algebras,

$$\{{}^{(\pm)}V^{aj}p_a,{}^{(\pm)}V^{bk}p_b\}_{\mathrm{PB}}=-2f^{jkn}\,{}^{(\pm)}V^{an}p_a,$$

[3]The inverse gnomonic Vielbein is polynomial, $\mathsf{V}^{aj}=\delta^{aj}+Q^jQ^a+\epsilon^{jab}Q^b$. See [**22**].

and PB commute with each other,

$$\{^{(+)}V^{aj}p_a, {}^{(-)}V^{bk}p_b\}_{\text{PB}} = 0.$$

These two statements are implicit in [**5**] as well as throughout the literature, and are fully explicated in [**11**].

MBs collapse to PBs by linearity in momenta as before, and the Hamiltonian is again the simple form

$$H_{qm} = \tfrac{1}{2}(p_a V^{ai}) \star (V^{bi} p_b).$$

The spectra are given by the Casimir eigenvalues for the relevant algebras and representations. The quantum correction is now found to be

$$H_{qm} - H = \frac{\hbar^2}{8}(\Gamma^b_{ac} g^{cd} \Gamma^a_{bd} - f_{ijk} f_{ijk}),$$

(reducing to the previous result for S^3). In operator language, this Hamiltonian amounts to an obvious Weyl-ordering of all products [**11**].

5. Nambu dynamics

All the models considered above have extra invariants beyond the number of conserved quantities in involution (mutually commuting) required for integrability in the Liouville sense. The most systematic way of accounting for such additional invariants, and placing them all on a more equal footing, even when they do not all simultaneously commute, is the Nambu bracket formalism.

5.1. Classical Nambu mechanics. For example, the classical mechanics of a particle on an N-sphere as discussed above may be summarized elegantly through Nambu mechanics in phase space [**17, 37, 41**]. Specifically, [**7, 18, 21, 36, 38**], in an N-dimensional space, and thus $2N$-dimensional phase space, motion is confined on the constant surfaces specified by the algebraically independent integrals of the motion (eg, L_x, L_y, L_z for S^2 above.) Therefore, the phase-space velocity $\mathbf{v} = (\dot{\mathbf{q}}, \dot{\mathbf{p}})$ is always perpendicular to the $2N$-dimensional phase-space gradients $\nabla = (\partial_{\mathbf{q}}, \partial_{\mathbf{p}})$ of all these integrals of the motion.

As a consequence, if there are $2N-1$ algebraically independent such integrals (ie, the system is maximally superintegrable [**6, 27, 40, 42, 45**]), the phase-space velocity must be proportional [**7, 21, 36**] to the cross-product of all those gradients, and hence the motion is fully specified for any phase-space function $k(\mathbf{q}, \mathbf{p})$ by a phase-space Jacobian which amounts to the Nambu Bracket.

$$\frac{dk}{dt} \equiv \{k, I_1, \ldots, I_{2N-1}\}_{\text{NB}} = \mathbf{v} \cdot \nabla k$$
$$\propto \partial_{i_1} k \epsilon^{i_1 i_2 \cdots i_{2N}} \partial_{i_2} I_1 \cdots \partial_{i_{2N}} I_{2N-1} = \frac{\partial(k, I_1, \ldots, I_{2N-1})}{\partial(q_1, p_1, q_2, p_2, \ldots, q_N, p_N)}.$$

For instance, consider the above S^2 case to find the concise result

$$\frac{dk}{dt} = \frac{\partial(k, L_x, L_y, L_z)}{\partial(x, p_x, y, p_y)}.$$

For the more general S^N, one now has a choice of $2N-1$ of the $N(N+1)/2$ invariants of $so(N+1)$. One of several possible expressions is

$$\frac{dk}{dt} = \frac{(-1)^{N-1}}{P_2 P_3 \cdots P_{N-1}} \frac{\partial(k, P_1, L_{12}, P_2, L_{23}, P_3, \ldots, P_{N-1}, L_{N-1\,N}, P_N)}{\partial(x_1, p_1, x_2, p_2, \ldots, x_N, p_N)},$$

where $P_a = \sqrt{1-q^2}p_a$, for $a = 1, \dots, N$, and $L_{a,a+1} = q^a p_{a+1} - q^{a+1} p_a$, for $a = 1, \dots, N-1$. In general [**17, 41**], classical NBs are Jacobian determinants and possess all antisymmetries of such. As they are linear in all derivatives, they also obey the Leibniz rule of partial differentiation.

$$\{k(L,M), f_1, f_2, \dots\}_{\text{NB}} = \frac{\partial k}{\partial L}\{L, f_1, f_2, \dots\}_{\text{NB}} + \frac{\partial k}{\partial M}\{M, f_1, f_2, \dots\}_{\text{NB}}.$$

Thus, an entry in the NB algebraically dependent on the rest leads to a vanishing bracket. For example, it is seen directly from above that the Hamiltonian is constant,

$$\frac{dH}{dt} = \left\{\frac{\mathbf{L}\cdot\mathbf{L}}{2}, \dots\right\}_{\text{NB}} = 0,$$

since each term of this NB vanishes. This also applies to all explicit examples discussed here, as they are all maximally superintegrable.

The impossibility to antisymmetrize more than $2N$ indices in $2N$-dimensional phase space,

$$\epsilon^{ab\cdots c[i}\epsilon^{j_1 j_2 \cdots j_{2N}]} = 0,$$

leads to the so-called "fundamental identity" [**17, 41**], slightly generalized here. For any f_i, g_j, and V,

$$\begin{aligned}
&\{V\{g_1, \dots, g_{2N-1}, f_1\}_{\text{NB}}, f_2, \dots, f_{2N}\}_{\text{NB}} \\
&\qquad + \{f_1, V\{g_1, \dots, g_{2N-1}, f_2\}_{\text{NB}}, f_3, \dots, f_{2N}\}_{\text{NB}} \\
&\qquad + \cdots + \{f_1, \dots, f_{2N-1}, V\{g_1, \dots, g_{2N-1}, f_{2N}\}_{\text{NB}}\}_{\text{NB}} \\
&\qquad\qquad = \{g_1, \dots, g_{2N-1}, V\{f_1, f_2, \dots, f_{2N}\}_{\text{NB}}\}_{\text{NB}}.
\end{aligned}$$

The proportionality factor V in

$$\frac{dk}{dt} = V\{k, I_1, \dots, I_{2N-1}\}_{\text{NB}},$$

has to be a time-invariant [**18, 38**] if it has no *explicit* time dependence. This is seen from consistency of

$$\begin{aligned}
\frac{d}{dt}(V\{f_1, \dots, f_{2N}\}_{\text{NB}}) &= \dot{V}\{f_1, \dots, f_{2N}\}_{\text{NB}} \\
&\quad + V\{\dot{f}_1, \dots, f_{2N}\}_{\text{NB}} + \cdots + V\{f_1, \dots, \dot{f}_{2N}\}_{\text{NB}}.
\end{aligned}$$

where $\dot{V} = dV/dt$, etc. This yields

$$\begin{aligned}
&V\{V\{f_1, \dots, f_{2N}\}_{\text{NB}}, I_1, \dots, I_{2N-1}\}_{\text{NB}} \\
&\quad = \dot{V}\{f_1, \dots, f_{2N}\}_{\text{NB}} + V\{V\{f_1, I_1, \dots, I_{2N-1}\}_{\text{NB}}, \dots, f_{2N}\}_{\text{NB}} \\
&\qquad + \cdots + V\{f_1, \dots, V\{f_{2N}, I_1, \dots, I_{2N-1}\}_{\text{NB}}\}_{\text{NB}},
\end{aligned}$$

and, by virtue of the FI, $dV/dt = 0$ follows [**11**].

Actually, PBs result from a maximal reduction of NBs, by inserting $2N-2$ phase-space coordinates and summing over them, thereby taking *symplectic traces*,

$$\{L, M\}_{\text{PB}} = \frac{1}{(N-1)!}\{L, M, x_{i_1}, p_{i_1}, \dots, x_{i_{N-1}}, p_{i_{N-1}}\}_{\text{NB}},$$

where summation over all $N-1$ pairs of repeated indices is understood. Fewer traces lead to relations between NBs of maximal rank, $2N$, and those of lesser rank, $2k$,

$$\{L_1,\ldots,L_{2k}\}_{\mathrm{NB}} = \frac{1}{(N-k)!}\{L_1,\ldots,L_{2k},x_{i_1},p_{i_1},\ldots,x_{i_{N-k}},p_{i_{N-k}}\}_{\mathrm{NB}},$$

(which is one way to define the lower rank NBs for $k \neq 1$), or between two lesser rank NBs. A complete theory of these relations has not been developed; but, naively, $\{L_1,\ldots,L_{2k}\}_{\mathrm{NB}}$ acts like a Dirac Bracket (DB) up to a normalization, $\{L_1,L_2\}_{\mathrm{DB}}$, where the fixed additional entries $L_3,\ldots,L_{2k}$ in the NB play the role of the constraints in the DB. (In effect, this has been previously observed, e.g., [**17, 18, 38, 41**], for the extreme case $N=k$, without symplectic traces.)

By applying this symplectic trace to a general system—not only a superintegrable one—Hamilton's equations admit an NB expression different than Nambu's original one, namely

$$\frac{dk}{dt} = \{k,H\}_{\mathrm{PB}} = \frac{1}{(N-1)!}\{k,H,x_{i_1},p_{i_1},\ldots,x_{i_{N-1}},p_{i_{N-1}}\}_{\mathrm{NB}}.$$

5.2. Quantum Nambu mechanics. Despite considerable interest in the problem for nearly 30 years, the quantization of the Nambu formalism was not completely settled until recently. We believe a transparent, user-friendly technique is now at hand (see [**11, 47**]). The problem of quantizing Nambu brackets *might* be solved by the Abelian deformation method [**1, 13, 14, 34**], but this is not yet clear.

Remarkably, however, the quantization is completely solved by Nambu's original method, when consistently applied to the phase-space formalism. Define quantum Nambu brackets (QNBs)

$$[A_1,A_2,\ldots,A_k]_\star \equiv \sum_{\substack{\text{all } k! \text{ perms } \{p_1,p_2,\ldots,p_k\}\\ \text{of the indices } \{1,2,\ldots,k\}}} (-1)^{\pi(p)} A_{p_1}\star A_{p_2}\star\cdots\star A_{p_k}.$$

Use these antisymmetrized $\star$-products in the quantum theory instead of the previous Jacobians. This approach grants in an obvious way only one of three mathematical desiderata: *full antisymmetry*. With these definitions, the Leibniz property and the Fundamental Identity are not manifestly satisfied. But to some extent, the apparent loss of the latter two properties is a subjective shortcoming, and dependent on the specific application context.

Even order QNBs may always be resolved into sums of products of commutators: $[A,B]_\star = A\star B - B\star A$. For instance,

$$\begin{aligned}[A,B,C,D]_\star = [A,B]_\star\star[C,D]_\star - [A,C]_\star\star[B,D]_\star - [A,D]_\star\star[C,B]_\star \\ + [C,D]_\star\star[A,B]_\star - [B,D]_\star\star[A,C]_\star - [C,B]_\star\star[A,D]_\star.\end{aligned}$$

Let us use this fact and re-consider the S^2 example. The result is a combinatoric identity relating 4 brackets to commutators.

$$[A,L_x,L_y,L_z]_\star = i\hbar[A, L_x\star L_x + L_y\star L_y + L_z\star L_z]_\star,$$

as follows from the su(2) MB algebra and the commutator resolution of the 4-bracket. Therefore in this case the Leibniz properties (and the particular FIs corresponding to them) *are* satisfied, leading to an effective fundamental identity (EFI).

Explicitly,

$$[A \star B, L_x, L_y, L_z]_\star = A \star [B, L_x, L_y, L_z]_\star + [A, L_x, L_y, L_z]_\star \star B.$$

As a consequence, time evolution of any WF for S^2 is given by the QNB expression:

$$\frac{\partial f}{\partial t} = \frac{1}{2(i\hbar)^2}[L_x, L_y, L_z, f]_\star.$$

In particular, this approach is in agreement with the $\star$-product quantization of the equations of motion.

$$\frac{dx}{dt} = \frac{-1}{2\hbar^2}[x, L_x, L_y, L_z]_\star, \qquad \frac{dp_x}{dt} = \frac{-1}{2\hbar^2}[p_x, L_x, L_y, L_z]_\star,$$

where the second of these does indeed incorporate the previously given quantum correction.

How does the use of 4-brackets extend to other examples? Any Lie algebra will allow a commutator with a quadratic Casimir to be rewritten as a *sum* of 4-brackets. Suppose

$$[Q_a, Q_b]_\star = i\hbar f_{abc} Q_c,$$

in a basis where f_{abc} is totally antisymmetric (this particular choice of basis is helpful, but not crucial). Then for a *sum* (over all repeated indices) of quantum Nambu 4-brackets we have

$$f_{abc}[A, Q_a, Q_b, Q_c]_\star = \big[A, f_{abc}[Q_a, Q_b, Q_c]_\star\big]_\star.$$

Only a commutator with the trilinear invariant survives. Moreover, this trilinear invariant reduces to the quadratic Casimir.

$$f_{abc}[Q_a, Q_b, Q_c]_\star = 3 f_{abc} Q_a \star [Q_b, Q_c]_\star = 3i\hbar f_{abc} f_{bcd} Q_a \star Q_d.$$

For simple Lie algebras (with appropriately normalized charges) we have

$$f_{abc} f_{bcd} = c_{\text{adjoint}}\, \delta_{ad},$$

where c_{adjoint} is a *number*. For example, $c_{\text{adjoint}} = N$ for su(N).

So we obtain the Casimir $Q_a \star Q_a$

$$f_{abc}[Q_a, Q_b, Q_c]_\star = 3i\hbar c_{\text{adjoint}}\, Q_a \star Q_a,$$

and we conclude that

$$f_{abc}[A, Q_a, Q_b, Q_c]_\star = 3i\hbar c_{\text{adjoint}}[A, Q_a \star Q_a]_\star.$$

As a corollary, we again have the effective 4-bracket fundamental identity (EFI)

$$\begin{aligned} f_{abc}&\big[Q_a, Q_b, Q_c, [A, B, \ldots, D]_\star\big]_\star \\ &= f_{abc}\big[[Q_a, Q_b, Q_c, A]_\star, B, \ldots, D\big]_\star + f_{abc}\big[A, [Q_a, Q_b, Q_c, B]_\star, \ldots, D\big]_\star \\ &\qquad + \cdots + f_{abc}\big[A, B, \ldots, [Q_a, Q_b, Q_c, D]_\star\big]_\star. \end{aligned}$$

By using this result, *all* the models above can be quantized through the use of QNBs to describe the time evolution of their WFs.

More than three invariants may also be incorporated into a QNB. Only a careful and complete physical interpretation of the result is needed in the general case. For example, for S^N with $N > 2$ a full set of $2N-1$ invariants leads from the previously given classical Jacobian to a QNB:

$$\frac{\partial(f, P_1, L_{12}, P_2, \ldots, P_{N-1}, L_{N-1\,N}, P_N)}{\partial(x_1, p_1, x_2, p_2, \ldots, x_N, p_N)} = (-1)^{N-1} P_2 \ldots P_{N-1} \frac{df}{dt}$$
$$= (-1)^{N-1} \frac{d}{dt}(P_2 \cdots P_{N-1} f),$$

becomes, for $\hbar \neq 0$,

$$\frac{1}{(i\hbar)^N N}[f, P_1, L_{12}, P_2, \ldots, P_{N-1}, L_{N-1\,N}, P_N]_\star$$
$$= (-1)^{N-1} \frac{d}{dt}\{P_2, \ldots, P_{N-1}, f\}_\star + \text{quantum-connection terms.}$$

The invariants appearing in the totally symmetric generalization of the Jordan product [**24, 28**],

$$\{A_1, A_2, \ldots, A_k\}_\star \equiv \sum_{\substack{\text{all } k! \text{ perms } \{p_1, p_2, \ldots, p_k\} \\ \text{of the indices } \{1, 2, \ldots, k\}}} A_{p_1} \star A_{p_2} \star \cdots \star A_{p_k},$$

on the above RHS,

$$\left\{P_2, P_3, \ldots, P_{N-1}, \frac{df}{dt}\right\}_\star = \frac{d}{dt}\{P_2, P_3, \ldots, P_{N-1}, f\}_\star,$$

will have the effect to set, through their net (but *not* simultaneous, individual) eigenvalues, a variable, dynamical time scale for evolution of the various eigen-WFs.

In addition, the "quantum-connection terms" represent higher order corrections, in powers of $\hbar$, corresponding to group rotations of f that will be described fully elsewhere. For example, if f is the bilinear

$$f_{ab} \equiv (L_a + P_a) \star (L_b - P_b).$$

then $df_{ab}/dt = 0$ for a particle moving freely on the surface of the 3-sphere, but the corresponding group rotation is *not* zero. Hence the six bracket reduces entirely to these quantum connection terms. Explicitly we find

$$[f_{ab}, P_x, L_z, P_y, L_x, P_z]_\star = 4i\hbar^5 \sum_c (\varepsilon_{b2c} f_{ac} - \varepsilon_{a2c} f_{cb}).$$

Note that the rotation on the RHS here is a quantum effect, and vanishes in the classical limit as given by

$$\lim_{\hbar \to 0} [f_{ab}, P_x, L_z, P_y, L_x, P_z]_\star / \hbar^3 = 0.$$

We believe such a complete physical interpretation explains the perceived failure of the Leibniz rules and FI in the general situation, in a transparent way, and is the only additional ingredient required for a successful *non-Abelian* quantum implementation of the most general Nambu brackets. *A priori*, this approach could conceivably be equivalent to the Abelian deformation approach [**1, 13, 14, 34**], but no one has shown this.

These points are worth re-emphasizing for the 3-sphere expressed in the chiral language. Again, let us use $\mathcal{L}_i$ and $\mathcal{R}_i$ ($i = 1, 2, 3 \sim x, y, z$) for the mutually commuting su(2) charges.

$$[\mathcal{L}_i, \mathcal{L}_j]_\star = i\hbar\varepsilon_{ijk}\mathcal{L}_k, \qquad [\mathcal{R}_i, \mathcal{R}_j]_\star = i\hbar\varepsilon_{ijk}\mathcal{R}_k, \qquad [\mathcal{L}_i, \mathcal{R}_j]_\star = 0.$$

Define the usual quadratic Casimirs for the left and right su(2)'s:

$$I_\mathcal{L} = \sum_i \mathcal{L}_i \star \mathcal{L}_i, \qquad I_\mathcal{R} = \sum_i \mathcal{R}_i \star \mathcal{R}_i.$$

We find the simple result:

$$\begin{aligned}[f, F(I_\mathcal{L}, I_\mathcal{R}), \mathcal{R}_x, \mathcal{R}_y, \mathcal{L}_x, \mathcal{L}_y]_\star &= (i\hbar)^2\{[f, F(I_\mathcal{L}, I_\mathcal{R})]_\star, \mathcal{L}_z, \mathcal{R}_z\}_\star \\ &= (i\hbar)^2[\{f, \mathcal{L}_z, \mathcal{R}_z\}_\star, F(I_\mathcal{L}, I_\mathcal{R})]_\star,\end{aligned}$$

where $F(I_\mathcal{L}, I_\mathcal{R})$ is *any* $\star$*-function* of the left and right Casimirs. Physically, this Nambu bracket is simply time evolution generated by the Hamiltonian $H \equiv F(I_\mathcal{L}, I_\mathcal{R})$ with the Jordan-like eigenvalues σ, of

$$\{f, \mathcal{L}_z, \mathcal{R}_z\}_\star = \sigma f,$$

setting the time scales for the various sectors of the theory: i.e., *dynamical time scales.* In particular this is true for the choices $F(I_\mathcal{L}, I_\mathcal{R}) = I_\mathcal{L}$ or $F(I_\mathcal{L}, I_\mathcal{R}) = I_\mathcal{R}$, choices relevant for the particle on the 3-sphere.

A complete set of solutions consists of all WFs of the form $f_{\lambda_1\rho_1,\lambda_2\rho_2}$ where

$$\begin{aligned}\mathcal{L}_z \star f_{\lambda_1\rho_1,\lambda_2\rho_2} &= \lambda_1 f_{\lambda_1\rho_1,\lambda_2\rho_2}, \qquad & f_{\lambda_1\rho_1,\lambda_2\rho_2} \star \mathcal{L}_z &= \lambda_2 f_{\lambda_1\rho_1,\lambda_2\rho_2}, \\ \mathcal{R}_z \star f_{\lambda_1\rho_1,\lambda_2\rho_2} &= \rho_1 f_{\lambda_1\rho_1,\lambda_2\rho_2}, \qquad & f_{\lambda_1\rho_1,\lambda_2\rho_2} \star \mathcal{R}_z &= \rho_2 f_{\lambda_1\rho_1,\lambda_2\rho_2}.\end{aligned}$$

Hence $\{f, \mathcal{L}_z, \mathcal{R}_z\}_\star = \sigma_{12} f$ with

$$\sigma_{12} = 2\lambda_1\rho_1 + \lambda_1\rho_2 + \rho_1\lambda_2 + 2\lambda_2\rho_2,$$

since

$$\{f, \mathcal{L}_z, \mathcal{R}_z\}_\star = \{\mathcal{L}_z, \mathcal{R}_z\}_\star \star f + \mathcal{L}_z \star f \star \mathcal{R}_z + \mathcal{R}_z \star f \star \mathcal{L}_z + f \star \{\mathcal{L}_z, \mathcal{R}_z\}_\star.$$

So the time scale is indeed dynamical, and given by σ_{12}.

The simple Leibniz rule for generic f and g, that would equate

$$[f \star g, F(I_\mathcal{L}, I_\mathcal{R}), \mathcal{R}_x, \mathcal{R}_y, \mathcal{L}_x, \mathcal{L}_y]_\star$$

and

$$f \star [g, F(I_\mathcal{L}, I_\mathcal{R}), \mathcal{R}_x, \mathcal{R}_y, \mathcal{L}_x, \mathcal{L}_y]_\star + [f, F(I_\mathcal{L}, I_\mathcal{R}), \mathcal{R}_x, \mathcal{R}_y, \mathcal{L}_x, \mathcal{L}_y]_\star \star g,$$

will fail for products $f_{\lambda_1\rho_1,\lambda_2\rho_2} \star g_{\lambda_2\rho_2,\lambda_3\rho_3}$ unless

$$\sigma_{12} = \sigma_{23} = \sigma_{13}.$$

There is no quantum-connection term in this particular case due to our choice for the invariants in the bracket $[f, F(I_\mathcal{L}, I_\mathcal{R}), \mathcal{R}_x, \mathcal{R}_y, \mathcal{L}_x, \mathcal{L}_y]_\star$. The more general situation is revealed by a different choice as follows:

$$[f, \mathcal{L}_x, \mathcal{L}_y, \mathcal{L}_z, \mathcal{R}_x, \mathcal{R}_y]_\star = \frac{3}{2}(i\hbar)^2[\{f, \mathcal{R}_z\}_\star, I_\mathcal{L}]_\star + \frac{1}{2}(i\hbar)^2 \sum_i \Big[\big[[f, \mathcal{L}_i]_\star, \mathcal{L}_i\big]_\star, \mathcal{R}_z\Big]_\star,$$

or equivalently

$$[f, \mathcal{R}_x, \mathcal{R}_y, \mathcal{R}_z, \mathcal{L}_x, \mathcal{L}_y]_\star = \frac{3}{2}(i\hbar)^2[\{f, \mathcal{L}_z\}_\star, I_\mathcal{R}]_\star + \frac{1}{2}(i\hbar)^2 \sum_i \Big[\big[[f, \mathcal{R}_i]_\star, \mathcal{R}_i\big]_\star, \mathcal{L}_z\Big]_\star.$$

These are the exact results for these QNBs. The first terms (single commutators) on the RHS's are inherently $O(\hbar^3)$, and encode the expected time evolution for Nambu quantum dynamics, while the second terms (triple commutators) are $O(\hbar^5)$. One is tempted to interpret these extra terms as some type of covariant completion for this particular example of Nambu time evolution, with "quantum connections" as the given higher order effects in $\hbar$.

As a test case, again take f to be a bilinear $f_{ab} \equiv \mathcal{L}_a \star \mathcal{R}_b$ of specific left and right charges. Since $\star$ commutators are indeed $\star$ derivations, all $\star$ functions of the six possible $\mathcal{L}_a$ and $\mathcal{R}_b$ charges commute with the Casimirs, so the first terms on the RHS's of the last two equations vanish for $f = f_{ab}$ (i.e., f_{ab} for a particle moving freely on the surface of a 3-sphere has no time derivatives). However, the second terms on the RHS's do *not* vanish for $f = f_{ab}$ but are just rotations of the $\mathcal{L}_a$ and $\mathcal{R}_b$ charges about the z axis.

$$\sum_i \Big[[[f_{ab}, \mathcal{L}_i]_\star, \mathcal{L}_i]_\star, \mathcal{R}_z \Big]_\star = 2i\hbar^3 \sum_c \varepsilon_{b3c} f_{ac}.$$

$$\sum_i \Big[[[f_{ab}, \mathcal{R}_i]_\star, \mathcal{R}_i \big]_\star, \mathcal{L}_z \Big]_\star = 2i\hbar^3 \sum_{c=1,2,3} \varepsilon_{a3c} f_{cb}.$$

For this particular example, then, the total effects of the chosen invariants in the 6-brackets are

$$[f_{ab}, \mathcal{L}_x, \mathcal{L}_y, \mathcal{L}_z, \mathcal{R}_x, \mathcal{R}_y]_\star = -i\hbar^5 \sum_c \varepsilon_{b3c} f_{ac},$$

$$[f_{ab}, \mathcal{R}_x, \mathcal{R}_y, \mathcal{R}_z, \mathcal{L}_x, \mathcal{L}_y]_\star = -i\hbar^5 \sum_c \varepsilon_{a3c} f_{cb}.$$

It remains to link up the QNB approach with the $\star$ eigenvalue equation for static WFs, as needed to develop the spectral theory in such a formalism. This must be described in detail elsewhere. However an essential point is contained in the previous formulas. A complete spectral theory in the NB framework requires solving totally symmetrized bracket eigenvalue equations, of the form

$$\lambda f = \{I_1, I_2, \ldots, I_k, f\}_\star,$$

for the allowed *net* eigenvalues λ, where in general $[I_i, I_j]_\star \neq 0$. These net eigenvalue equations set the different time scales under NB evolution as expressed above. They are natural generalizations of the anticommutator eigenvalue equations encountered in the standard spectral theory of (special) Jordan algebras [**23, 25, 26**]. Nevertheless, we have not found a complete discussion of such equations in the literature.

Acknowledgments. We thank Y. Nambu and Y. Nutku for helpful discussions. We also thank the organizers of the superintegrability workshop, especially P. Winternitz, for the opportunity to present this research. The warm hospitality of the Centre de recherches mathématiques, Université de Montréal, is most gratefully acknowledged.

Appendix A. Nambu quantum mechanics of the isotropic oscillator

The analysis above can also be carried out using conventional Hilbert space operator techniques. Rather than repeat our previous discussion, we take another example to illustrate operator methods. (For a thorough discussion of the ideas in this appendix, see [**12**]. This later paper contains a more complete list of references.)

Consider the n dimensional oscillator using the standard lowering/raising operator basis that gives the commutator algebra

$$[a_i, b_j] = \hbar\delta_{ij}, \qquad [a_i, a_j] = 0 = [b_i, b_j].$$

Construct the usual bilinear charges that realize the u(n) algebra

$$N_{ij} \equiv b_i a_j, \qquad [N_{ij}, N_{kl}] = \hbar(N_{il}\delta_{jk} - N_{kj}\delta_{il}).$$

Then the isotropic Hamiltonian is

$$H = \omega \sum_{i=1}^{n} (N_i + \tfrac{1}{2}\hbar), \quad N_i \equiv N_{ii},$$

which gives the n^2 conservation laws $[H, N_{ij}] = 0$. Now when we consider the isotropic oscillator dynamics using quantum Nambu brackets (QNBs) we obtain the main result for oscillator $2n$-brackets in the form of a theorem.

Isotropic Oscillator Theorem (reductio ad dimidium). Let $\mathsf{N} = N_1 + N_2 + \cdots + N_n$ and intercalate the $n-1$ nondiagonal operators $N_{i\,i+1}$ for $i = 1, \ldots, n-1$ into a $2n$-bracket with the n mutually commuting N_j for $j = 1, \ldots, n$ to find

$$\begin{aligned}[f, N_1, N_{12}, N_2, N_{23}, \ldots, N_{n-1}, N_{n-1\,n}, N_n] &= \hbar^{n-1}\{[f, \mathsf{N}], N_{12}, N_{23}, \ldots, N_{n-1\,n}\} \\ &= \hbar^{n-1}[\{f, N_{12}, N_{23}, \ldots, N_{n-1\,n}\}, \mathsf{N}].\end{aligned}$$

Here we have used a fully symmetrized Jordan operator product [**24**] as generalized by Kurosh [**28**]

$$\{A_1, A_2, \ldots, A_k\} \equiv \sum_{\substack{\text{all } k! \text{ perms } \{p_1, p_2, \ldots, p_k\} \\ \text{of the indices } \{1, 2, \ldots, k\}}} A_{p_1} A_{p_2} \cdots A_{p_k}.$$

This was introduced by Pascual Jordan, in the bilinear form, to render non-Abelian algebras into Abelian algebras at the expense of nonassociativity. We have also used the fully antisymmetrized Nambu operator product, or QNB [**37**],

$$[A_1, A_2, \ldots, A_k] \equiv \sum_{\substack{\text{all } k! \text{ perms } \{p_1, p_2, \ldots, p_k\} \\ \text{of the indices } \{1, 2, \ldots, k\}}} (-1)^{\pi(p)} A_{p_1} A_{p_2} \cdots A_{p_k}.$$

The nondiagonal operators $N_{i\,i+1}$ do not all commute among themselves, nor with all the N_j, but their non-Abelian properties are encountered in the above Jordan and Nambu products in a rather minimal way. Only adjacent entries in the list N_{12}, N_{23}, $N_{34}, \ldots, N_{n-1\,n}$ fail to commute. Also in the list of $2n-1$ generators within the original QNB, each N_j fails to commute only with the adjacent $N_{j-1\,j}$ and $N_{j\,j+1}$. Such a list of invariants constitutes a "Hamiltonian path" through the algebra.[4]

Proof of the Oscillator Theorem. Linearity in each argument and total antisymmetry of the Nambu bracket allow us to replace any one of the N_i by the

[4]There are other Hamiltonian paths through the algebra. A different set of $2n-1$ invariants which leads to an equivalent theorem can be obtained just by taking an arbitrarily ordered list of the mutually commuting N_i, and then intercalating nondiagonal generators to match adjacent indices on the N_i. That is, for any permutation of the indices $\{p_1, \ldots, p_n\}$, we have: $[f, N_{p_1}, N_{p_1 p_2}, N_{p_2}, N_{p_2 p_3}, N_{p_3}, \ldots, N_{p_{n-1}}, N_{p_{n-1} p_n}, N_{p_n}] = \hbar^{n-1}\{[f, \mathsf{N}], N_{p_1 p_2}, N_{p_2 p_3}, \ldots, N_{p_{n-1} p_n}\} = \hbar^{n-1}[\{f, N_{p_1 p_2}, N_{p_2 p_3}, \ldots, N_{p_{n-1} p_n}\}, \mathsf{N}]$.

sum N. We choose to replace $N_n \to \mathsf{N}$, hence to obtain

$$[f, N_1, N_{12}, N_2, \ldots, N_{n-1}, N_{n-1\,n}, N_n] = [f, N_1, N_{12}, N_2, \ldots, N_{n-1}, N_{n-1\,n}, \mathsf{N}].$$

Now since $[\mathsf{N}, N_{ij}] = 0$, the commutator resolution of the $2n$-bracket implies that N must appear "locked" in a commutator with f, and therefore f cannot appear in any other commutator. But then N_1 commutes with all the remaining free N_{ij} except N_{12}. So N_1 must be locked in $[N_1, N_{12}]$. Continuing in this way, N_2 must be locked in $[N_2, N_{23}]$, etc., until finally N_{n-1} is locked in $[N_{n-1}, N_{n-1\,n}]$. Thus all $2n$ entries have been paired and locked in the indicated n commutators, i.e., they are all "zipped-up." Moreover, these n commutators can and will appear as products ordered in all $n!$ possible ways with coefficients $+1$ since interchanging a pair of commutators requires interchanging two pairs of the original entries in the bracket. Thus we conclude

$$[f, N_1, N_{12}, N_2, \ldots, N_{n-1}, N_{n-1\,n}, N_n] = \{[f, \mathsf{N}], [N_1, N_{12},], \ldots, [N_{n-1}, N_{n-1\,n}]\}.$$

Now all the paired N_{ij} commutators evaluate as $[N_{i-1}, N_{i-1\,i}] = \hbar N_{i-1\,i}$, so we have

$$[f, N_1, N_{12}, N_2, \ldots, N_{n-1}, N_{n-1\,n}, N_n] = \hbar^{n-1}\{[f, \mathsf{N}], N_{12}, \ldots, N_{n-1\,n}\}.$$

Finally the commutator with N may be performed either before or after the Jordan product of f with all the $N_{i-1\,i}$, since again $[\mathsf{N}, N_{ij}] = 0$. Hence

$$\{[f, \mathsf{N}], N_{12}, \ldots, N_{n-1\,n}\} = [\{f, N_{12}, \ldots, N_{n-1\,n}\}, \mathsf{N}]. \qquad \square$$

The oscillator theorem is a remarkable relation. The invariants which are in involution (i.e., the Cartan subalgebra of $\mathrm{u}(n)$) are separated out of the QNB into a single commutator involving their sum, hence H/ω, to yield a time derivative, while the invariants which do not commute ($n-1$ of them, corresponding in number to the rank of $\mathrm{su}(n)$) are swept into a Jordan product. As in the N-sphere examples, this helps to clarify why the Leibniz rules fail when time evolution is expressed using QNBs. As is fairly well-known, evolution under the QNB does not satisfy the trivial Leibniz property: $[fg, N_1, N_{12}, N_2, \ldots, N_{n-1\,n}, N_n] \neq f[g, N_1, N_{12}, N_2, \ldots, N_{n-1\,n}, N_n] + [f, N_1, N_{12}, N_2, \ldots, N_{n-1\,n}, N_n]g$. But here the failure of this Leibniz rule for the Nambu bracket has been linked to the intervention of a generalized Jordan product involving noncommuting invariants.

References

1. H. Awata, M. Li, D. Minic, and T. Yoneya, *On the quantization of Nambu brackets*, J. High Energy Phys. (2001), no. 2, Paper 13.
2. F. Bayen, M. Flato, C. Fronsdal, A. Lichnerowicz, and D. Sternheimer, *Quantum mechanics as a deformation of classical mechanics*, Lett Math Phys **1** (1977), no. 6, 521–530.
3. ______, *Deformation theory and quantization.* I. *Deformations of symplectic structures*, Ann Physics **111** (1978), no. 1, 61–110; II. *Physical applications*, 111–151.
4. M. V. Berry, *Semi-classical mechanics in phase space: a study of Wigner's function*, Philos. Trans. Roy. Soc. London Ser. A **287** (1977), no. 1343, 237–271.
5. E. Braaten, T. L. Curtright, and C. K. Zachos, *Torsion and geometrostasis in nonlinear sigma models*, Nuclear Phys. B **260** (1985), no. 3-4, 630–688.
6. R. Caseiro, J.-P. Françoise, and R. Sasaki, *Quadratic algebra associated with rational Calogero–Moser models*, J. Math. Phys. **42** (2001), no. 11, 5329–5340.
7. R. Chatterjee, *Dynamical symmetries and Nambu mechanics*, Lett. Math. Phys. **36** (1996), no. 2, 117–126.
8. T. Curtright, D. Fairlie, and C. K. Zachos, *Features of time-independent Wigner functions*, Phys. Rev. D (3) **58** (1998), no. 2, 025002.

9. T. Curtright, T. Uematsu, and C. Zachos, *Geometry and duality in supersymmetric σ-models*, Nuclear Phys. B **469** (1996), no. 3, 488–512.
10. T. Curtright and C. Zachos, *Negative probability and uncertainty relations*, Modern Phys. Lett. A **16** (2001), no. 37, 2381–2385.
11. ———, *Deformation quantization of superintegrable systems and Nambu mechanics*, New J. Phys. **4** (2002) 83.1-83.16 (electronic).
12. ———, *Classical and quantum Nambu mechanics*, Phys. Rev. D **68** (2003), no. 8, 085001.
13. G. Dito and M. Flato, *Generalized abelian deformations: application to Nambu mechanics*, Lett Math Phys **39** (1997), no. 2, 107–125.
14. G. Dito, M. Flato, D. Sternheimer, and L. Takhtajan, *Deformation quantization and Nambu mechanics*, Comm. Math. Phys. **183** (1997), no. 1, 1–22.
15. D. B. Fairlie, *The formulation of quantum mechanics in terms of phase space functions*, Proc. Cambridge Philos. Soc. **60** (1964), 581–586.
16. B. V. Fedosov, *A simple geometrical construction of deformation quantization*, J. Differential Geom. **40** (1994), no. 2, 213–238.
17. V. T Filippov, n-Lie algebras, Sibirsk. Mat. Zh. **26** (1985), no. 6, 126–140 (Russian); English transl., Siberian Math. J. **26** (1985), no. 6, 879–891.
18. C. Gonera and Y. Nutku, *Super-integrable Calogero-type systems admit maximal number of Poisson structures*, Phys. Lett. A **285** (2001), no. 5-6, 301–306.
19. H. J. Groenewold, *On the principles of elementary quantum mechanics*, Physica **12** (1946), 405–460.
20. J. Hietarinta, *Classical versus quantum integrability*, J. Math. Phys. **25** (1984), no. 6, 1833–1840.
21. ———, *Nambu tensors and commuting vector fields*, J. Phys. A **30** (1997), no. 3, L27–L33.
22. P. W. Higgs, *Dynamical symmetries in a spherical geometry.* I, J. Phys. A **12** (1979), no. 3, 309–323.
23. P. Jordan, *Über eine Klasse nichtassoziativer hyperkomplexer Algebren*, Göttinger Nachr. (1932), 569–575.
24. ———, *Über die Multiplikation quantenmechanischer Größen*, Z. Phys. **80** (1933), 285–291.
25. ———, *Über Verallgemeinerungsmglichkeiten des Formalismus der Quantenmechanik*, Göttinger Nachr. (1933), 209–217.
26. P. Jordan, J. von Neumann, and E. Wigner, *On an algebraic generalization of the quantum mechanical formalism*, Ann. Math **35** (1934), no. 1, 29–64.
27. E. G. Kalnins, J. M. Kress, and P. Winternitz, *Superintegrability in a two-dimensional space of nonconstant curvature*, J. Math. Phys. **43** (2002), no. 2, 970–983.
28. A. G. Kuroš, *Multiple operator rings and algebras*, Uspehi Mat. Nauk **24** (1969), no. 1, 3–15 (Russian); English transl., Russian Math. Surveys **24** (1969), no. 1, 1–13.
29. M. Lakshmanan and K. Eswaran, *Quantum dynamics of a solvable nonlinear chiral model*, J. Phys. A **8** (1975), no. 10, 1658–1669.
30. B. Leaf, *Weyl transformation and the classical limit of quantum mechanics*, J. Mathematical Phys. **9** (1968), no. 1, 65–72.
31. H. I. Leemon, *Dynamical symmetries in a spherical geometry.* II, J. Phys. A **12** (1979), no. 4, 489–501.
32. A. N. Leznov, *Some integrable models in quantized spaces*, Nuclear Phys. B **40** (2002), no. 3, 469–480.
33. H. E. Lin, W. C. Lin, and R. Sugano, *On velocity-dependent potentials in quantum mechanics*, Nuclear Phys. B **16** (1970), no. 2, 431–449.
34. D. Minic and C.-H. Tze, *Nambu quantum mechanics: a nonlinear generalization of geometric quantum mechanics*, Phys. Lett. B **536** (2002), no. 3-4, 305–314.
35. J. E. Moyal, *Quantum mechanics as a statistical theory*, Proc. Cambridge Philos. Soc. **45** (1949), 99–124.
36. N. Mukunda and E. C. G. Sudarshan, *Relation between Nambu and Hamiltonian mechanics*, Phys Rev D (3) **13** (1976), no. 10, 2846–2850.
37. Y. Nambu, *Generalized Hamiltonian mechanics*, Phys Rev D (3) **7** (1973), 2405–2412
38. Y. Nutku, *Quantization of superintegrable systems with Nambu-Poisson brackets*, in this volume.

39. N. Reshetikhin and M. Yakimov, *Deformation quantization of Lagrangian fiber bundles*, Conférence Moshé Flato 1999, Vol. II (Dijon), Math. Phys. Stud., vol. 22, Kluwer Acad. Publ., Dordrecht, 2000, pp. 263–287.
40. M. A. Rodríguez and P. Winternitz, *Quantum superintegrability and exact solvability in n dimensions*, J. Math. Phys. **43** (2002), no. 3, 1309–1322.
41. L. Takhtajan, *On foundation of the generalized Nambu mechanics*, Comm. Math. Phys. **160** (1994), no. 2, 295–316.
42. P. Tempesta, A. V. Turbiner, and P. Winternitz, *Exact solvability of superintegrable systems*, J. Math. Phys. **42** (2001), no. 9, 4248–4257.
43. G. Velo and J. Wess, *A solvable quantum-mechanical model with nonlinear transformation laws*, Nuovo Cimento A (11) **1** (1971) 177–187.
44. H. Weyl, *Quantenmechanik und Gruppentheorie*, Z. Phys. **46** (1927) 1–33; review, The Theory of Groups and Quantum Mechanics, Dover, New York, 1931, pp. 272–280.
45. S. Wojciechowski, *Superintegrability of the Calogero–Moser system*, Phys Lett A **95** (1983), no. 6, 279–281.
46. C. Zachos, *Deformation quantization: quantum mechanics lives and works in phase-space*, Internat. J. Modern Phys. A **17** (2002), no. 3, 297–316.
47. C. K. Zachos and T. L. Curtright, *Deformation quantization, superintegrability, and Nambu mechanics*, Acta Phys. Hungar. **19** (2004), 199–203.

Department of Physics, University of Miami, Coral Gables, FL 33124-8046, USA
E-mail address: curtright@physics.miami.edu

High Energy Physics Division, Argonne National Laboratory, Argonne, IL 60439-4815, USA
E-mail address: zachos@hep.anl.gov

Centre de Recherches Mathématiques
CRM Proceedings and Lecture Notes
Volume **37**, 2004

Maximally Superintegrable Systems of Winternitz Type

Cezary Gonera

ABSTRACT. A simple general scheme of constructing maximally superintegrable systems of Winternitz type is presented and illustrated by an example of a model admitting an explicit construction of angle variables and additional integrals of motion.

Superintegrable systems, that is integrable systems which admit more integrals of motion than degrees of freedom (in the particular case of maximal superintegrability in addition to n Liouville integrals, there are $n-1$ functionally independent and global ones) form an interesting class of Hamiltonian systems. However, the list of known superintegrable models is rather limited. The aim of the present note, based on [**4, 5**], is to present a general scheme of constructing superintegrable models with completely separated Hamiltonians generalizing the Winternitz system [**1–3, 7, 9**].

In order to explain our motivation and the main points we would like to address here, let us start with the elementary example of a system of two oscillators given by

$$H=\sum_{i=1}^{2}\left(\frac{p_i^2}{2\mu_i}+\frac{1}{2}w_i\mu_i x_i^2\right)\equiv\sum_{i=1}^{2}H_i(x_i,p_i). \tag{0.1}$$

This model is in an obvious way integrable and solvable with Liouville integrals given for example by $H_i(x_i,p_i)$, $i=1,2$. However, as is well known, the properties of the model depend in a crucial way on the ratio ω_1/ω_2. If this ratio is a rational number all trajectories are closed, the motion is strictly periodic and in addition to Liouville integrals there exists an extra "super-integral" S (in the simplest case when $\omega_1=\omega_2$, $S=x_2p_1-x_1p_2$). In other words we are dealing with a superintegrable model. On the other hand, if the ratio is not a rational number, the trajectories are not closed, the motion is quasi-periodic and there is no additional integral of motion. The system is not a superintegrable one.

The main lesson we can learn from this elementary example is that complete separability of a Hamiltonian system, implying in an obvious way its integrability, does not imply automatically its superintegrability. This leads to a natural general

---

2000 *Mathematics Subject Classification.* Primary 70H12; Secondary 70H06.
The author is supported by KBN grant, no. 5 P03B 060 21.
The extended version of this paper will appear in J. Phys. A **37** (2004).

question: when is a dynamical system with a completely separated Hamiltonian

$$H = \sum_{i=1}^{N} \left(\frac{p_i^2}{2\mu_i} + U_i(x_i) \right) \tag{0.2}$$

(maximally) superintegrable. This is really the main point of the note. It should be recalled here that an elegant particular answer to our question is provided by the Winternitz model with

$$U_i(x_i) = \frac{\mu_i \omega_i^2}{2} x_i^2 + \frac{\gamma_i}{x_i^2}, \quad \gamma_i = \text{const.}$$

Before addressing our main point let us collect some basic facts concerning (maximal) superintegrability.

First of all due to the Liouville–Arnold theorem the integrability of a Hamiltonian system with N degrees of freedom implies that:

(a) The level surfaces of the integrals of motion $F_i = \alpha_i$, $i = 1, \ldots, N$, are N-dimensional tori T^N (in the case of confining forces, which is the most interesting one when superintegrability is considered).

(b) One can introduce action-angle coordinates $(J_k, \phi_k)\phi_k \in (0, 2\pi)$, $k = 1, \ldots, N$ such that:

(i) They define a canonical symplectic structure

$$\{J_k, J_p\} = 0 = \{\phi_k, \phi_p\}, \qquad \{\phi_k, J_p\} = \delta_{kp} \tag{0.3}$$

(ii)

$$J_k = J_k(F) = \frac{1}{2\Pi} \int_{\gamma_k} \sum_{i}^{N} p_i(x, F)\, dx^i \tag{0.4}$$

γ_k is the kth basic cycle on torus T^k

$$\phi_k = \int_{x_0}^{x} \sum_{i=1}^{N} \frac{\partial p^i(x, J)}{\partial J_k}\, dx^i \equiv \frac{\partial S(x, J)}{\partial J_k} \qquad S(x, J) = \int_{x_0}^{x} \sum_{i=1}^{N} p^i(x, J) dx^i. \tag{0.5}$$

(c) The equations of motion take a simple form in the action-angle variables

$$\dot{J}_k = 0 \qquad \dot{\phi}_k = \frac{\partial H(J)}{\partial J_k} \equiv \omega_k(J_1, \ldots, J_N). \tag{0.6}$$

The following fact is important for our construction: an integrable system with N degree of freedom is maximally superintegrable iff its Hamiltonian in action-angle variables is a function of a linear combination with integer coefficients of action variables, i.e. H in these variables takes the form

$$H = H(J) = H\left(\sum_{k=1}^{N} m_k J_k \right) \tag{0.7}$$

where m_k are integers.

Such a form of H implies that the quantities $S_k = \sin((m/m_k)\phi_k - (m/m_1)\phi_1)$ where m is the least common multiple of m_k and $k = 2, \ldots, N$, provide $N-1$ global, independent integrals of motion.

On the other hand, maximal superintegrability implies that all trajectories of the system are compact common intersections of $2N - 1$ hypersurfaces and hence they are closed. This is possible if and only if all ratios $\omega_k(J)/\omega_j(J)$ of the frequencies $\omega_k(J) \equiv \partial H/\partial J_k$ are rational numbers which in turn means that we have

$$\omega_k(J) = m_k \omega(J), \tag{0.8}$$

where $\omega(J)$ is a fixed function of the action variables. The general form of the Hamiltonian leading to (0.8) is given by (0.7).

Now, we can come back to our main question of constructing maximally superintegrable models of Winternitz type. Our starting point is a system with completely separated Hamiltonian given by (0.2) and our aim is the construction of the potentials $U_k(x)$ leading to maximally superintegrable dynamics.

Due to the form of the Hamiltonian (0.2) our system is trivially integrable with N integrals of motion being just the Hamiltonians H_k and Liouville tori given by $H_k = \alpha_k$, $k = 1, \ldots, N$. Action variables take the form $J_k = \oint p_k \, dx^k$ which means that each H_k is a function of J_k only. So, in terms of action variables our Hamiltonian (0.2) takes the form $H = \sum_{k=1}^N H_k(J_k)$. On the other hand according to the previous discussion H will correspond to maximally superintegrable dynamics provided it is of the form given by (0.7). In the case under consideration this is possible iff

$$H_k(J_k) = \alpha m_k J_k \tag{0.9}$$

with some real constant α.

Taking into account the definition of the J_k action variable, we conclude that a system described by Hamiltonian (0.2) is maximally superintegrable iff the following equation takes place

$$T_k \equiv 2\pi \frac{dJ_k}{d\alpha_k} \equiv 2\Pi\sqrt{2\mu_k} \int_{x_{1k}}^{x_{2k}} \frac{dx^k}{\sqrt{\alpha_k - U_k(x^k)}} = \frac{2\pi}{\alpha m_k}, \tag{0.10}$$

where

$$U(x_{1k}) = \alpha_k = U(x_{2k}).$$

The meaning of (0.10) is clear. Its LHS determines the period T_k of one dimensional motion in the confining potential U_k. In the case of maximally superintegrable dynamics these periods should be energy independent and their ratios should be rational numbers.

So, our problem reduces to solving (0.10) with respect to U_k. This solution is, however, well known [**6**]. Assume that $U(x)$ is such that:

(i) 0 is the absolute minimum of $U(x)$

(ii) $U(x) = E$ has exactly two solutions $x_{1,2}(E)$ for any $E > 0$.

Then given the period $T(E)$ as a function of energy, one can find all $U(x)$ which produces $T(E)$, from the equation

$$x_2(E) - x_1(E) = \frac{1}{\pi\sqrt{2\mu}} \int_0^E \frac{T(\varepsilon)\, d\varepsilon}{\sqrt{E - \varepsilon}}. \tag{0.11}$$

In our case $T_k = 2\pi/\alpha m_k$, so $U_k(x)$ is given by

$$x_2(E) - x_1(E) = \frac{4}{\alpha m_k \sqrt{2\mu_k}} \sqrt{E}, \tag{0.12}$$

i.e.,

$$E = \beta_k^2 \big(x_2(E) - x_1(E)\big)^2, \quad \beta_k^2 = \tfrac{1}{8}\alpha^2 m_k^2 \mu_k^2. \tag{0.13}$$

Now, there are exactly two solutions to $U_k(x) = E$ so one can put

$$x_2(E) = \varphi_k\bigl(x_1(E)\bigr) \tag{0.14}$$

with some, yet unspecified, function $\varphi_k(x)$.

Then $U_k(x)$ reads

$$U_k(x) = \beta_k^2\bigl(\varphi_k(x) - x\bigr)^2. \tag{0.15}$$

Due to the condition $U_k(x) = U_k\bigl(\varphi_k(x)\bigr)$ we get an equation for the $\varphi_k(x)$ function

$$\varphi_k\bigl(\varphi_k(x)\bigr) = x. \tag{0.16}$$

The choice $\varphi_k(x) = -x$, resp. $\varphi_k(x) = \gamma_k/x$, correponds to harmonic oscillators, or to the Winternitz system, respectively. However, (0.16) admits much more solutions. In fact, all solutions except $\varphi_k(x) = x$ are given by

$$\varphi_k(x) = F\bigl(c - F^{-1}(x)\bigr) \tag{0.17}$$

where $F\colon \mathbb{R} \to \mathbb{R}$ is a continuously differentiable one-to-one mapping of $\mathbb{R}$ onto $\mathbb{R}$ and c is some constant.

This general solutions (0.17) provides a rather wide class of maximally superintegrable systems with Hamiltonians

$$H = \sum_{k=1}^{N}\left(\frac{p_k^2}{2\mu_k} + \beta_k^2\bigl(\varphi_k(x_k) - x_k\bigr)^2\right). \tag{0.18}$$

It is worth noting here that as mentioned at the beginning, in order to find additional integrals of motion in explicit form one needs angle variables. However, an explicit integration necessary to obtain these variables in general cannot be performed. Therefore, our method gives a very wide class of maximally superintegrable systems with rather complicated additional integrals of motion which, in general are not quadratic in momenta. It means, in particular that the freedom in the choice of variables which separate the Hamilton–Jacobi equation appears only at the level of general canonical (rather than point) transformations.

On the other hand there is a large freedom in the choice of $\varphi_k(x)$ functions which can be used to design maximally superintegrable systems enjoying angle variables and additional integrals of motion given explicitely in terms of elementary functions of the original canonical variables.

If for example we have

$$F(x) = a\sinh\left(\frac{x}{a}\right), \tag{0.19}$$

then the relevant Hamiltonian is $(\sigma_k \equiv c_k/a_k)$

$$\begin{aligned} H = \sum_{k=1}^{N}\Bigg(\frac{p_k^2}{2\mu_k} + \beta_k^2 a_k^2\bigg(&2(1+\cosh\sigma_k)\cosh\sigma_k\frac{x_k^2}{a_k^2} \\ &- 2\sinh\sigma_k(1+\cosh\sigma_k)\frac{x_k}{a_k}\sqrt{1+\frac{x_k^2}{a_k^2}+\sinh^2\sigma_k}\bigg)\Bigg). \end{aligned} \tag{0.20}$$

The advantage of this model is that one can find the angle variables by elementary integration [**4**]. In particular, in the simplest case when $N = 2$, $m_1 = m_2 = 1$,

$\mu_1 = \mu_2 = \mu$, $a_1 = a_2 = a$, $\sigma_1 = \sigma_2 = \sigma = \ln 3$ there exists an integral of motion polynomially expressible [**4**] in terms of:

$$\begin{gathered}\frac{4\beta a}{\sqrt{3H_k}}\left(\frac{2x_k}{\sqrt{3}a} - \frac{1}{\sqrt{3}}\sqrt{1+\frac{x_k^2}{a^2}}\right),\\ \frac{4\beta a}{\sqrt{3H_k + 16\beta^2 a^2}}\left(\frac{2}{\sqrt{3}}\sqrt{1+\frac{x_k^2}{a^2}} - \frac{1}{\sqrt{3}}\frac{x_k}{a}\right),\\ \frac{p_k}{\sqrt{2\mu H_k}}, \qquad \sqrt{\frac{p_k^2 + (32/3)\beta^2 a^2 \mu}{2\mu H_k}}.\end{gathered} \tag{0.21}$$

Let us note that even in this simplest case the additional integral can not be chosen as a polynomial function of the momenta.

We would like to finish this note with two remarks. First, our discussion can be generalized to semiaxes [**4**, **5**]. The counterpart of (0.17) then reads

$$\phi(x) = F\left(\frac{c}{F^{-1}(x)}\right) \tag{0.22}$$

where $F\colon \mathbb{R}_+ \to \mathbb{R}_+$ is a continuously differentiable isomorphism. In particular, $F(x) \sim x^\lambda$ leads to Winternitz model, but again a variety of systems can be constructed by selecting various F.

Second, our method is not restricted to the Hamiltonians of the form given by (0.2). For example, it can be used to show that the Liouville model [**8**] given by

$$\begin{gathered}H = \frac{\sum_{k=1}^N \left((p_k^2/2\mu_k) + 2\beta_k^2 x_k^2\right)}{1 + \sum_{k=1}^N (x_k/a_k)\sqrt{1+(x_k^2/a_k^2)}},\\ \beta_k^2 = \frac{\alpha^2 \mu_k m_k^2}{8}, \quad \beta_1^2 a_1^2 = \cdots = \beta_N^2 a_N^2\end{gathered} \tag{0.23}$$

where m_k are arbitrary integers, is superintegrable [**4**].

As a special example we can consider the case $m_k = 1$ (then $\beta_1 = \cdots = \beta_N = \beta$, $a_1 = \cdots = a_N = a$) to obtain a deformation of the isotropic oscillator which is still superintegrable.

Acknowledgements. I would like to thank the organizers and especially Professor P. Winternitz for the invitation and support. Numerous very useful discussions with P. Kosinski and P. Maslanka are gratefully acknowledged.

References

1. N. W. Evans, *Superintegrability of the Winternitz system*, Phys. Lett. A **147** (1990), no. 8-9, 483–486.
2. ———, *Group theory of the Smorodinsky–Winternitz system*, J. Math. Phys. **32** (1991), no. 12, 3369–3375.
3. J. Friš, V. Mandrosov, Ya. A. Smorodinsky, M. Uhlíř, and P. Winternitz, *On higher symmetries in quantum mechanics*, Phys. Lett. **16** (1965), 354–356.
4. C. Gonera, *More about generalized maximally superintegrable systems of Winternitz type*, hep-th/0207182.
5. C. Gonera, P. Kosiński, and P. Maślanka, *Superintegrable models of Winternitz type*, Phys. Lett. A **289** (2001), no. 1-2, 66–68.
6. L. D. Landau and E. M. Lifshitz, *Course of theoretical physics. Vol. 1. Mechanics*, Pergamon Press, Oxford–New York–Toronto, 1976.
7. A. Makarov, Ya. A. Smorodinsky, Kh. Valiev, and P. Winternitz, *A systematic search for non-relativistic systems with dynamical symmetries*, Nuovo Cimento A **52** (1967), 1061–1084.

8. A. M. Perelomov, *Integrable systems of classical mechanics and Lie algebras*. I, Birkhäuser, Basel, 1990.
9. P. Winternitz, Ya. A. Smorodinsky, M. Uhlíř, and J. Friš, *Symmetry groups in classical and quantum mechanics*, Jadernaja Fiz. **4** (1966), 625–635 (Russian); English transl., Soviet J. Nuclear Phys. **4** (1967), 444–450.

DEPT. OF THEORETICAL PHYSICS II, UNIVERISTY OF ŁÓDŹ, UL. POMORSKA 149/153, 90-236 ŁÓDŹ, POLAND
E-mail address: cgonera@krysia.uni.lodz.pl

Centre de Recherches Mathématiques
CRM Proceedings and Lecture Notes
Volume **37**, 2004

Cubic Integrals of Motion and Quantum Superintegrability

Simon Gravel

ABSTRACT. We present here new superintegrable Hamiltonians in $E(2)$ that are separable in Cartesian coordinates and that admit a third-order integral, both in quantum and in classical mechanics. Considerable differences between the quantum and classical cases are observed, for many quantum systems can share the same nontrivial classical limit. It is found that some superintegrable systems can be expressed in terms of the Painlevé transcendents. We also discuss problems related to the definition of quantum integrability.

1. Introduction

In classical mechanics, an n-dimensional Hamiltonian system is called Liouville integrable if it allows n functionally independent integrals of motion in involution (including the Hamiltonian), that is

$$\begin{aligned} &\{H, X_i\}_{\mathrm{PB}} = 0, \\ &\{X_i, X_j\}_{\mathrm{PB}} = 0, \quad \forall i, j, \end{aligned} \tag{1.1}$$

where $\{\ \}_{\mathrm{PB}}$ is the Poisson bracket.

The Hamiltonian $H = H(x_1, \dots, x_n, p_1, \dots, p_n)$ and the integrals of motion $X_i = X_i(x_1, \dots, x_n, p_1, \dots, p_n)$ must be well defined functions on phase space [**1,8**]. The system is superintegrable if it allows more than n functionally independent integrals, n of them in involution. It is called maximally superintegrable if it allows $2n - 1$ integrals of motion.

In quantum mechanics, a Hamiltonian system is said to be integrable if there exists a set $\{X_i\}$ of n well defined, algebraically independent operators (including the Hamiltonian) that commute pairwise. It is superintegrable if it possesses further independent operators, $\{Y_j\}$ that commute with the Hamiltonian. The Y_j do not necessarily commute with each other, nor with the X_i.

This definition of quantum integrability assumes that the independence of quantum operators is a well-defined concept, although in this context no perfectly satisfying definition of independence is known. We will discuss this problem, and other questions related to the definition of integrability that will arise in this paper, in

2000 *Mathematics Subject Classification.* 81R12, 17B80, 70E40.

The author is supported by a NSERC ES-A fellowship.
Grateful thanks to Pavel Winternitz for his many valuable suggestions and comments.
This is the final form of the paper.

Section 5. Since we are dealing here with polynomial differential operators only, we will assume that integrals are independent if their classical counterpart is.

Most of the papers about integrable systems were concerned with classical systems with second-order integrals in 2- or n-dimensional flat space [**3–5, 20, 21, 23, 28**], or in spaces with constant [**17, 22**] or nonconstant [**18**] curvature. Quantum integrability was found to be equivalent to classical integrability in this context, as long as the Hamiltonian did not include magnetic terms. Moreover, it was found that there was a strong link between this "quadratic superintegrability," separability of the Schrödinger and Hamilton–Jacobi equations, exact solvability [**25**] and generalized symmetries [**24**].

When people started looking for third- or higher-order integrals in classical mechanics [**6, 7, 9, 16, 19, 26**], following a well-known paper by Drach [**2**], they encountered nonlinear equations, which did not happen for lower-order integrals. When higher-order integrals were considered in the quantum case, it was found that quantum and classical integrability were no longer equivalent [**12, 13, 15**]. In spite of the considerable interest that was dedicated to this topic in the last years, few cases of superintegrable potentials with higher-order integrals are known.

In order to get a better picture of the distribution of these superintegrable systems among integrable ones, and to understand the consequences of the existence of higher-order integrals, a systematic search for superintegrable systems with third-order integrals was started in [**12**]. This search is also expected to yield new superintegrable systems, and such examples may be useful to determine an appropriate definition for quantum integrability.

In [**12**], all potentials of the form

$$H = \tfrac{1}{2}(p_x^2 + p_y^2) + V(x, y)$$

that admit both a first- and a third-order integral were found, and they are discussed in Section 3.

As the conditions for the existence of first-order integrals are fairly restrictive, few systems were found to have this property. A logical sequel to this work is to look for systems that have a third-order integral together with a second-order one. The existence of a second-order integral of motion forces the Hamiltonian to be separable in Cartesian, polar, parabolic or elliptic coordinates, which leads to considerable simplifications in the determining equations for the existence of third-order integrals.

In Section 4, preliminary results are presented for systems that are separable in Cartesian coordinates,

$$H = \tfrac{1}{2}(p_x^2 + p_y^2) + V_1(x) + V_2(y),$$

Complete results will be presented elsewhere (see [**11**]).

2. Existence of a third-order integral

In quantum and classical mechanics, the search for systems that admit integrals that are third-order polynomials in the momenta can be considerably simplified by a few preliminary remarks. In both cases a third-order operator may be written as

$$X = \sum_{i+j=0}^{3} P_{ij}(x, y) p_1^i p_2^j.$$

When we ask that $\{H, X\}_{\mathrm{PB}} = 0$, or $[H, X] = 0$ in quantum mechanics, we find a set of determining equations for the functions P_{ij} and the potential V. Some of the equations are independent of V and determine completely the leading-order terms in X (that is terms of order two and three). These equations can be readily solved and yield that the leading-order terms must lie in $e(2)$, the enveloping algebra of the Lie algebra of the Euclidean group. This can be understood as a consequence of the fact that these equations must hold also for the free motion potential whose third-order integrals are exactly the third-order elements of the enveloping algebra of $e(2)$.

Moreover, in classical mechanics, none of the equations involve simultaneously terms with odd and even powers in the momenta. This means odd and even terms must form separate integrals, and we can therefore set the even terms to 0 without loss of generality,

$$X = \sum_{\substack{i,j,k \\ i+j+k=3}} A_{ijk} L_3^i p_1^j p_2^k + g_1(x,y)p_1 + g_2(x,y)p_2. \tag{2.1}$$

A similar procedure is used in quantum mechanics. Let us first write the operator X as

$$\begin{aligned} X = \sum_{\substack{i,j,k \\ i+j+k=3}} A_{ijk}\{L_3^i, p_1^j p_2^k\} + \sum_{\substack{i,j,k \\ i+j+k=2}} B_{ijk}\{L_3^i, p_1^j p_2^k\} &+ \{g_1(x,y), p_1\} \\ &+ \{g_2(x,y), p_2\} + h(x,y) \\ L_3 = xp_2 - yp_1, \end{aligned}$$

where the bracket stands for the anticommutator.

We can then use the self-adjointness of the Hamiltonian to show that the odd and even terms in the operator must commute separately with the Hamiltonian. We are then left with

$$X = \sum_{\substack{i,j,k \\ i+j+k=3}} A_{ijk}\{L_3^i, p_1^j p_2^k\} + \{g_1(x,y), p_1\} + \{g_2(x,y), p_2\}, \tag{2.2}$$

where the A_{ijk} are real constants, and the g_i real functions.

Once these simplifications are made, the determining equation for $\{H, X\}_{\mathrm{PB}} = 0$ read

$$0 = g_1 V_x + g_2 V_y \tag{2.3}$$

$$(g_1)_x = 3f_1(y)V_x + f_2(x,y)V_y \equiv h_1, \tag{2.4}$$

$$(g_2)_y = f_3(x,y)V_x + 3f_4(x)V_y \equiv h_2, \tag{2.5}$$

$$(g_1)_y + (g_2)_x = 2\left(f_2(x,y)V_x + f_3(x,y)V_y\right) \equiv h_3. \tag{2.6}$$

where

$$\begin{aligned} f_1(y) &= -A_{300}y^3 + A_{210}y^2 - A_{120}y + A_{030}, \\ f_2(x,y) &= 3A_{300}xy^2 - 2A_{210}xy + A_{201}y^2 + A_{120}x - A_{111}y + A_{021}, \\ f_3(x,y) &= -3A_{300}x^2y + A_{210}x^2 - 2A_{201}xy + A_{111}x - A_{102}y + A_{012}, \\ f_4(x) &= A_{300}x^3 + A_{201}x^2 + A_{102}x + A_{003}. \end{aligned}$$

The determining equations for $[H,X]=0$ are very similar:

$$0 = g_1V_x + g_2V_y - \frac{\hbar^2}{4}\big(f_1V_{xxx} + f_2V_{xxy} + f_3V_{xyy} + f_4V_{yyy} + 8A_{300}(xV_y - yV_x) + 2(A_{210}V_x + A_{201}V_y)\big), \tag{2.7}$$

$$(g_1)_x = 3f_1(y)V_x + f_2(x,y)V_y \equiv h_1, \tag{2.8}$$

$$(g_2)_y = f_3(x,y)V_x + 3f_4(x)V_y \equiv h_2, \tag{2.9}$$

$$(g_1)_y + (g_2)_x = 2(f_2(x,y)V_x + f_3(x,y)V_y) \equiv h_3. \tag{2.10}$$

The only difference between these equations and the classical ones is a term in (2.7) that is proportional to $\hbar^2$, and therefore vanishes in the classical limit. In both cases the first equation is nonlinear, which was not the case for lower-order integrals.

We can find a third-order linear compatibility condition for V from the three last equations, that are identical in quantum and classical mechanics,

$$\begin{aligned} 0 &= h_{3xy} - h_{1yy} - h_{2xx} \\ &= -f_3V_{xxx} + (2f_2 - 3f_4)V_{xxy} + (-3f_1 + 2f_3)V_{xyy} - f_2V_{yyy} \\ &\quad + 2(f_{2y} - f_{3x})V_{xx} + 2(-3f_{1y} + f_{2x} + f_{3y} - 3f_{4x})V_{xy} \\ &\quad + 2(-f_{2y} + f_{3x})V_{yy} + (-3f_{1yy} + 2f_{2xy} - f_{3xx})V_x \\ &\quad + (-f_{2yy} + 2f_{3xy} - 3f_{4xx})V_y. \end{aligned} \tag{2.11}$$

Further compatibility conditions for the potentials can be obtained from (2.7) to (2.10), and for convenience we introduce

$$\begin{aligned} \phi_1 &= \frac{V_y}{V_x}, \\ \phi_2 &= -\frac{\hbar^2}{4V_x}\big(f_1V_{xxx} + f_2V_{xxy} + f_3V_{xyy} + f_4V_{yyy} \\ &\qquad + 8A_{300}(xV_y - yV_x) + 2(A_{210}V_x + A_{201}V_y)\big). \end{aligned}$$

We have also introduced h_1, h_2 and h_3 in (2.8) to (2.10) above. The three algebraically independent nonlinear compatibility conditions now read

$$-\phi_{2x} + \left(\frac{\phi_1(h_3\phi_1 + h_2\phi_1^2 + \phi_1\phi_{2y} + \phi_{2x} + h_1)}{\phi_{1x} + \phi_1\phi_{1y}}\right)_x = h_1, \tag{2.12}$$

$$\left(\frac{\phi_1^2h_2 + \phi_1\phi_{2y} + \phi_1h_3 + \phi_{2x} + h_1}{\phi_{1x} + \phi_1\phi_{1y}}\right)_y = -h_2, \tag{2.13}$$

$$\begin{aligned} &h_1(\phi_{1xy} + \phi_{1y}^2) + h_2(\phi_1^2\phi_{1xy} - \phi_{1x}^2 - 2\phi_1\phi_{1x}\phi_{1y}) \\ &\quad + h_3(\phi_1\phi_{1xy} - \phi_{1x}\phi_{1y}) - (h_{1y} + \phi_1h_{2x})(\phi_{1x} + \phi_1\phi_{1y}) \\ &= -\phi_{2x}(\phi_{1xy} + \phi_{1y}^2) + \phi_{2y}(\phi_{1x}\phi_{1y} - \phi_{1xy}\phi_1) + \phi_{2xy}(\phi_{1x} + \phi_1\phi_{1y}). \end{aligned} \tag{2.14}$$

In deriving these, we have assumed

$$\phi_{1x} + \phi_1\phi_{1y} \neq 0, \qquad \phi_{1xy}\phi_1 - \phi_{1x}\phi_{1y} \neq 0. \tag{2.15}$$

which will not be the case for potentials considered in this paper. Equations (2.12), (2.13) and (2.14) are quite complicated fifth-order nonlinear equations (in quantum mechanics). When dealing with special cases, we will generally not need to use such complicated equations, but the nonlinearity will remain.

3. Potentials that admit first- and third-order integrals

The conditions for the existence of a first-order integral impose that the potential be a function of only one variable which, up to rotations and translations, can be chosen to be the radius $r = \sqrt{x^2 + y^2}$ or one of the Cartesian coordinates. The first-order integral of motion is then the angular momentum, or the linear momentum (in the complementary direction), respectively.

One can look for the potentials that admit further second-order integrals that are independent of the first-order one and of the Hamiltonian. Four potentials fulfill these requirements, two of which are radial,

$$V = \frac{\alpha}{r}, \qquad\qquad V = \omega^2 r^2.$$

This result is related to Bertrand's theorem, which states that the harmonic oscillator and the Coulomb potential are the only two radial potentials for which all finite trajectories are closed. Indeed these are also the only radial potentials that admit third-order integrals, both in quantum and in classical mechanics.

The two translationally invariant potentials with second-order integrals are

$$V = ax \tag{3.1}$$

$$V = \frac{a}{x^2}. \tag{3.2}$$

In classical mechanics, no other translationally invariant potential admits third-order integrals. In quantum mechanics, though, new superintegrable potentials arise, and known superintegrable potentials gain additional integrals.

For example, $V = a/x^2$ admits three third-order integrals,

$$X_1 = \{L_3^2, p_2\} + 2a\left\{\frac{y^2}{x^2}, p_2\right\},$$

$$X_2 = \{L_3, p_1 p_2\} - 2a\left\{\frac{y}{x^2}, p_2\right\},$$

$$X_3 = 2p_1^2 p_2 + 2a\left\{\frac{1}{x^2}, p_2\right\},$$

that are consequences of the lower-order integrals and have classical analogues. When $\alpha = \hbar^2$, that is $V = \hbar^2/x^2$, there are four additional integrals;

$$X_4 = 2L_3^3 + \hbar^2\left\{\frac{3y^2}{x} + 2x, p_2\right\} + \hbar^2\left\{\frac{-3y^3}{x^2} - 2y, p_1\right\},$$

$$X_5 = \{L_3^2, p_1\} - 2\hbar^2\left\{\frac{y}{x}, p_2\right\} + \frac{\hbar^2}{2}\left\{\frac{6y^2}{x^2} + 1, p_1\right\},$$

$$X_6 = \{L_3, p_1^2\} + \hbar^2\left\{\frac{1}{x}, p_2\right\} - 3\hbar^2\left\{\frac{y}{x^2}, p_1\right\},$$

$$X_7 = 2p_1^3 + 3\hbar^2\left\{\frac{1}{x^2}, p_1\right\}.$$

These integrals are not independent of each other, but as they have nontrivial algebraic relations with the three first ones and with the lower-order integrals, they are nevertheless instructive.

There are also new superintegrable potential that satisfy

$$\hbar^2 V_{xxx} = 4(3V + \alpha)V_x, \tag{3.3}$$

which is equivalent, if we set $\alpha = 0$ by translating V and put $\hbar = 1$, to a defining equation of the Weierstrass elliptic function,

$$V_x^2 = 4V^3 - g_2 V - g_3, \tag{3.4}$$

and therefore we can write the integrable potentials as

$$V = \hbar^2 \mathcal{P}(x).$$

Special cases can be expressed in terms of elementary functions. For example we find the "soliton" potential

$$V_{1a} = -\frac{(\hbar\omega)^2}{\cosh^2(\omega x)},$$

and also singular periodic, or nonperiodic potential, respectively, such as

$$V_{2a} = \frac{(\hbar\omega)^2}{\sin^2(\omega x)}, \qquad V_{2b} = \frac{(\hbar\omega)^2}{\sinh^2(\omega x)}.$$

The integrals of motion are

$$\begin{aligned} X_1 &= \{L_3, p_1^2\} - \{(3V(x) + \alpha)y, p_1\} + \left\{\alpha + 2xV(x) + \int V(x)\,dx, p_2\right\} \\ X_2 &= p_1^3 + \tfrac{1}{2}\{3V(x) + \alpha, p_1\}, \end{aligned} \tag{3.5}$$

where α is determined by equation (3.3).

This completes the list of potentials with first- and third-order integrals.

4. Potentials separable with respect to Cartesian coordinates

If we assume $V = V_1(x) + V_2(y)$, equations (2.7) to (2.10) become

$$\begin{aligned} 0 = g_1 V_{1x} + g_2 V_{2y} - \frac{\hbar^2}{4}(f_1 V_{1xxx} + f_4 V_{2yyy} &+ 8A_{300}(xV_{2y} - yV_{1x}) \\ &+ 2(A_{210}V_{1x} + A_{201}V_{2y})), \end{aligned} \tag{4.1}$$

$$(g_1)_x = 3f_1(y)V_{1x} + f_2(x,y)V_{2y}, \tag{4.2}$$

$$(g_2)_y = f_3(x,y)V_{1x} + 3f_4(x)V_{2y}, \tag{4.3}$$

$$(g_1)_y + (g_2)_x = 2(f_2(x,y)V_{1x} + f_3(x,y)V_{2y}). \tag{4.4}$$

A consequence of the linear compatibility condition (2.11) is that

$$\begin{aligned} (A_{210}x^2 + A_{111}x + A_{012})V_1^{(3)}(x) + 4(2A_{210}x + A_{111})V_1''(x) & \\ + 12A_{210}V_1'(x) &= ax + b \\ (A_{201}y^2 - A_{111}y + A_{021})V_2^{(3)}(y) + 4(2A_{201}y - A_{111})V_2''(y) & \\ + 12A_{201}V_2'(y) &= cy + d. \end{aligned} \tag{4.5}$$

We can use this to find many new integrable systems, both in quantum and classical mechanics. Among these, some "deformation potentials" (that is quantum potentials that contain terms proportional to $\hbar^2$) admit more integrals than others, as was the case for $V = \hbar^2/x^2$. Here many similar potentials may have the same nontrivial classical limit. For example we find the following superintegrable potentials

(i) $V = a(x^2 + y^2) + \dfrac{b}{x^2} + \dfrac{c}{y^2}$

$$X_1 = \{L, p_x p_y\} + \left\{xy\left(-\frac{2b}{x^3} + 2ax\right), p_y\right\} - \left\{xy\left(-\frac{2c}{y^3} + 2ay\right), p_x\right\}$$

(ii) $$V = a(x^2 + y^2) + \frac{\hbar^2}{x^2} + \frac{\hbar^2}{y^2}$$

$$X_1 = 2L^3 - \hbar^2\left\{\frac{3x^2}{y} + 2y + \frac{3y^3}{x^2}, p_x\right\} + \hbar^2\left\{\frac{3y^2}{x} + 2x + \frac{3x^3}{y^2}, p_y\right\}$$

$$X_2 = \{L, p_x p_y\} + \left\{xy\left(-\frac{2\hbar^2}{x^3} + 2ax\right), p_y\right\} - \left\{xy\left(-\frac{2\hbar^2}{y^3} + 2ay\right), p_x\right\}$$

(iii) $$V = a(x^2 + y^2) + \frac{\hbar^2}{y^2}$$

$$X_1 = 2L^3 - \hbar^2\left\{2y + \frac{3x^2}{y}, p_x\right\} + \hbar^2\left\{\frac{3x^3}{y^2} + 2x, p_y\right\}$$

$$X_2 = \{L, p_x p_y\} - 2\left\{axy^2 - \frac{\hbar^2 x}{y^2}, p_x\right\} + 2a\{x^2 y, p_y\}$$

$$X_3 = \{L, p_y^2\} - \left\{2ay^3 + \hbar^2\frac{1}{y}, p_x\right\} + \left\{\hbar^2\frac{3x}{y^2} + 2axy^2, p_y\right\}$$

(iv) $$V = a(x^2 + y^2)$$

$$X_1 = L^3$$

$$X_2 = \{L, p_x p_y\} + a\{2x^2 y, p_y\} - a\{2xy^2, p_x\}$$

$$X_3 = \{L, p_y^2\} + 2a(\{xy^2, p_y\} - \{y^3, p_x\})$$

$$X_4 = \{L, p_x^2\} - 2a(\{x^2 y, p_x\} - \{x^3, p_y\})$$

These potentials all admit three independent first- or second-order integrals, so they are all maximally superintegrable according to the usual definition of integrability. We notice that potentials (ii) and (iii) can be seen either as special cases of the Smorodinsky–Winternitz potential (i), or as deformations of the harmonic oscillator (iv). We have four different quantum mechanical potentials that have the same classical limit (the fourth is given by the permutation $x \leftrightarrow y$ in (iii)).

We find also genuine third-order superintegrable systems, such as the anisotropic harmonic oscillator,

$$\begin{aligned} V &= a(9x^2 + y^2) \\ X_1 &= \{L, p_y^2\} + \tfrac{2}{3}a\{y^3, p_x\} - 6a\{xy^2, p_y\}. \end{aligned} \tag{4.6}$$

New superintegrable classical potentials were also found. Even if they are quite complicated, they can be expressed in a compact way if we write them implicitly, e.g.,

$$\begin{aligned} V &= ay + V_1(x) \\ d &= V_1(x)(V_1(x) - \alpha x)^2, \end{aligned} \tag{4.7}$$

where d is an arbitrary constant, and the integral is

$$X = ap_x^3 - \alpha p_x^2 p_y + a(3V_1(x) - \alpha x)p_x - 2\alpha V_1(x)p_y. \tag{4.8}$$

These potentials are the classical limit of quantum superintegrable potentials that can be written in terms of the second transcendent of Painlevé, such as

$$V(x, y) = V_1(x) + V_2(y) = ay + \alpha x + (2\alpha\hbar)^{2/3} P_2^2\big((2\alpha/\hbar^2)^{1/3}x, 0\big), \tag{4.9}$$

where

$$P_2''(x,\lambda) = 2P_2^3(x,\lambda) + xP(x,\lambda) + \lambda. \tag{4.10}$$

The integral of motion is

$$X = 2ap_x^3 - 2\alpha p_x^2 p_y + a\{3V_1(x) - \alpha x, p_x\} - 2\alpha\{V_1(x), p_y\}. \tag{4.11}$$

5. Comments on the definition of integrability

Independence of integrals in classical mechanics is a well-defined property. The m integrals of motion $\{X_1, \ldots, X_m\}$ are functionally independent if the Jacobian matrix

$$\frac{\partial(X_1, \ldots, X_m)}{\partial(p_1, \ldots, p_n, x_1, \ldots, x_n)} \tag{5.1}$$

is of maximal rank. The phase space is $2n$ dimensional, so if there is to be any motion at all, there cannot exist more than $2n-1$ independent integrals of motion.

When we pass to quantum mechanics by substituting

$$\begin{aligned} x &\to x \\ p &\to -i\hbar\frac{\partial}{\partial x} \\ \{\ ,\ \}_{\mathrm{PB}} &\to \frac{i}{\hbar}[\ ,\ ], \end{aligned} \tag{5.2}$$

we find a close relationship between classical integrals of motion, and quantum integrals, defined as operators that commute with the Hamiltonian. The conditions for the existence of integrals in both cases are very similar. The only question remaining is how to translate functional independence of integrals in quantum mechanics? There is presently no perfectly satisfactory answer to this question. One cannot ask the operators to be functionally independent, for the commutation condition $[H, X] = 0$ defines a functional dependence.

In order to find a quantum mechanical definition for the independence of operators, we can either try to adapt the classical definition to the quantum context, or else try to find an appropriate definition by an analysis of quantum mechanics alone. Both ways should ideally lead to a similar definition, but up to now this hasn't been achieved. In both cases a number of criteria direct our search for a definition of independence. Among these are

(i) The fact that two operators commute should not imply that they are dependent.
(ii) $\{X_1, X_2, \ldots, X_i\}$ and $X_{i+1} = f(X_1, X_2, \ldots, X_i)$ should be considered dependent for a class of functions f to be specified.
(iii) L_3, p_1 and p_2 should be independent integrals for free motion.
(iv) Quantum integrals that have independent classical limits should be independent.
(v) There should be a maximum of $2n-1$ independent integrals, and a maximum of n that commute pairwise.
(vi) The definition should apply to general quantum systems, not only those that have classical analogues.

Not all these criteria are absolutely necessary, but a definition that does not fulfill the two first ones is not likely to give a coherent definition of superintegrability. Furthermore, (iii), (iv) and (v) are useful if we want quantum and classical integrability to be closely related, while (vi) would be a desirable property, although it is not required.

To proceed by analogy with classical mechanics is probably the most appealing possibility. The definition of quantum integrability, after all, follows historically from classical integrability. One of the reasons for introducing such a concept is that classical nonintegrable systems exhibit interesting behavior (they may be chaotic). Thus the concept of quantum integrability may be useful to understand the quantum counterpart of chaos. For these purposes, quantum integrability should correspond to classical integrability in the classical limit, when such a limit exists. The independence of quantum operators should therefore be closely related to that of their classical analogues (when such analogues exist). A simple way to achieve this is to say that quantum operators are independent if their classical counterpart is. This definition works quite well for quantum systems that have classical analogues, that is, as long as we are dealing with integral or differential operators. Some of its drawbacks were analyzed in [**27**]. Among more technical problems were the fact that this definition can hardly be extended to general quantum systems, and that a quantum mechanical concept should ideally be defined without reference to classical mechanics.

We are therefore looking for a definition for the independence of quantum operators that does not rely on classical mechanics, but that has a good classical limit. Different attempts to define a "functional independence" of operators that exclude commutation relations were suggested, but even when the operators are polynomials in the momenta counterexamples could be found that violated one of the three first criterion. If requested for example that a relation between the operators should be a polynomial in the Jordan algebra of the operators, all commuting operators would be dependent, for $-\{H,\{X,H\}\}+2\{H^2,X\}=H[H,X]-[H,X]H=0$.

If we drop the requirement that quantum integrability should give classical integrability in the classical limit, rigorous definitions can be found. It was suggested that operators should be considered independent unless one of them can be expressed as a polynomial in the others, defining the concept of algebraic integrability. This clearly does not fulfill requirements (iii), (iv) and (v), and it will not be appropriate for nonpolynomial operators. The definition is nevertheless quite useful, as was demonstrated in [**14**].

5.1. Different types of integrability. It appears that different concepts of integrability will be needed to take into account the different properties that are related to the number of integrals. The most obvious example is the one-dimensional case, where the usual definition of integrability implies that all potentials are maximally superintegrable. In this case the usual definition of integrability is not useful, while algebraic integrability still provides an interesting classification of the potentials. The Hamiltonian being of second-order in the momenta and one-dimensional, all systems that have a third-order integral are integrable with respect to that definition. It turns out that very few such systems exist, and they were shown [**14**] to satisfy equation (3.3).

These potentials are therefore the same that were found to be superintegrable in the two-dimensional case, in Section 3. There is therefore a relation between algebraic (super)integrability in one dimension and the usual superintegrability in two dimensions.

Some implications of superintegrability of separable systems on the one-dimensional behavior of the separated systems are discussed in [**10**].

6. Conclusion

The class of superintegrable systems that admit second- and third-order integrals is rich in instructive examples of classical and quantum integrable systems.

Note added in proof. The classification of systems separable in Cartesian coordinates and admitting third-order integrals, which was completed in [**11**], indeed providing many new interesting examples.

A systematic search for systems with third- or higher-order integrals is therefore a useful task, even though the complexity of the nonlinear partial differential equations rapidly increases. A straightforward systematic search is not likely to go further than fourth-order integrals, and new methods will have to be introduced to classify and understand such integrals. In order to do so, it is necessary to find one or more appropriate and rigorous definitions of quantum integrability, for wrong or ambiguous definitions are likely to lead to erroneous or incomplete statements.

References

1. V. I. Arnol′d, *Mathematical methods in classical mechanics*, Springer, New York, 1980.
2. J. Drach, *Sur l'intégration logique des équations de la dynamique à deux variables: forces conservatrices. Intégrales cubiques. Mouvements dans le plan*, C. R. Acad. Sci. **200** (1935), 22–26.
3. N. W. Evans, *Superintegrability in classical mechanics*, Phys. Rev. A **41** (1990), no. 10, 5666–5676.
4. ———, *Superintegrability of the Winternitz system*, Phys. Lett. A **147** (1990), no. 8-9, 483–486.
5. ———, *Group theory of the Smorodinsky–Winternitz system*, J. Math. Phys. **32** (1991), no. 12, 3369–3375.
6. A. S. Fokas and P. A. Lagerström, *Quadratic and cubic invariants in classical mechanics*, J. Math. Anal. Appl. **74** (1980), no. 2, 325–341.
7. ———, *On the use of Lie–Bäcklund operators and quantum mechanics*, J. Math. Anal. Appl. **74** (1980), no. 2, 342–358.
8. H. Goldstein, *Classical mechanics*, Addison-Wesley, Reading, MA, 1990.
9. C. Gonera, P. Kosiński, and P. Maślanka, *Superintegrable models of Winternitz type*, Phys. Lett. A **289** (2001), no. 1-2, 66–68.
10. S. Gravel, *Superintegrability, isochronicity, and quantum harmonic behavior*, math-ph/0310004.
11. ———, *Hamiltonians separable in Cartesian coordinates and third-order integrals of motion*, J. Math. Phys. 45 (2004), no. 3, 1003–1019.
12. S. Gravel and P. Winternitz, *Superintegrability with third-order integrals in quantum and classical mechanics*, J. Math. Phys, **43** (2002), no. 12, 5902–5912.
13. J. Hietarinta, *Classical versus quantum integrability*, J. Math. Phys. **25** (1984), no. 6, 1833–1840.
14. ———, *Solvability in quantum mechanics and classically superfluous invariant*, J. Phys. A **22** (1989), no. 5, L143–L147.
15. ———, *Pure quantum integrability*, Phys. Lett. A **246** (1998), no. 1-2, 97–104.
16. C. R. Holt, *Construction of new integrable Hamiltonians in two degrees of freedom*, J. Math. Phys. **23** (1982), no. 6, 1037–1046.
17. E. G. Kalnins, J. M. Kress, G. S. Pogosyan, and W. Miller, Jr., *Completeness of superintegrability in two-dimensional constant-curvature space*, J. Phys. A **34** (2001), no. 22, 4705–4720.
18. E. G. Kalnins, J. M. Kress, and P. Winternitz, *Superintegrability in a two-dimensional space of non-constant curvature*, J. Math. Phys. **43** (2002), no. 2, 970–983.

19. M. Karlovini and K. Rosquist, *A unified treatment of cubic invariants at fixed and arbitrary energies*, J. Math. Phys. **41** (2000), no. 1, 370–384.
20. A. Makarov, Ya. A. Smorodinsky, Kh. Valiev, and P. Winternitz, *A systematic search for non-relativistic systems with dynamical symmetries*, Nuovo Cimento A **52** (1967), 1061–1084.
21. M. F. Rañada, *Superintegrable $n = 2$ systems, quadratic constants of motion and the potentials of Drach*, J. Math. Phys. **38** (1997), no. 8, 4165–4178.
22. M. F. Rañada and M. Santander, *Superintegrable systems on the two-dimensional sphere $\mathbb{S}^2$ and the hyperbolic plane $\mathbb{H}^2$*, J. Math. Phys. **40** (1999), no. 10, 5026–5057.
23. M. A. Rodríguez and P. Winternitz, *Quantum superintegrability and exact solvability in n dimensions*, J.Math. Phys. **43** (2002), no. 3, 1309–1322.
24. M. B. Sheftel, P. Tempesta, and P. Winternitz, *Superintegrable systems in quantum mechanics and classical Lie theory*, J. Math. Phys. **42** (2001), no. 2, 659–673.
25. P. Tempesta, A. V. Turbiner, and P. Winternitz, *Exact solvability of superintegrable systems*, J.Math.Phys. **42** (2001), no. 9, 4248–4257.
26. A. V. Tsiganov, *The Drach superintegrable systems*, J. Phys. A **33** (2000), no. 41, 7407–7422.
27. S. Weigert, *The problem of quantum integrability*, Phys. D **56** (1992), no. 1, 107–119.
28. P. Winternitz, Ya. A. Smorodinsky, M. Uhli ř, and I. Friš, *Symmetry groups in classical and quantum mechanics*, Jadernaja Fiz. **4** (1966), 625–635 (Russian); English transl., Soviet J. Nuclear Phys. **4** (1967), 444–450.

Centre de recherches mathématiques, Université de Montréal, C. P. 6128, succ. Centre-Ville, Montréal, QC H3C 3J7, Canada

Current address: Department of Physics, Cornell University, 109 Clark Hall, Ithaca, NY 14853-2501, USA

E-mail address: sg324@cornell.edu

Centre de Recherches Mathématiques
CRM Proceedings and Lecture Notes
Volume **37**, 2004

Superintegrability, Lax Matrices and Separation of Variables

J. Harnad and O. Yermolayeva

Abstract. We show how the superintegrability of certain systems can be deduced from the presence of multiple parameters in the rational Lax matrix representation. This is also related to the fact that such systems admit a separation of variables in parametric families of coordinate systems.

1. Rational Lax matrix representations of integrable systems

1.1. Classical R-matrix theory of commuting isospectral flows. In the classical R-matrix approach to finite dimensional integrable systems [**2, 4, 9**], there is a Poisson map from the phase space into a space of $r \times r$ Lax matrices $\mathcal{N}(\lambda)$ depending rationally, trigonometrically or elliptically on a spectral parameter λ. The Poisson bracket is defined by the relation

$$\{\mathcal{N}(\lambda) \overset{\otimes}{,} \mathcal{N}(\mu)\} = [r(\lambda - \mu), \mathcal{N}(\lambda) \otimes \mathbf{I} + \mathbf{I} \otimes \mathcal{N}(\mu)], \tag{1.1}$$

where both sides are interpreted as elements of $\mathrm{End}(\mathbf{C}^r \otimes \mathbf{C}^r)$. The symbol $\{\ \overset{\otimes}{,}\ \}$ signifies a simultaneous tensor product in $\mathrm{End}(\mathbf{C}^r) \sim \mathfrak{gl}(r)$ and Poisson brackets in the components, and $r(\lambda - \mu)$ denotes the classical R-matrix. The simplest case is the rational R-matrix,

$$r(\lambda) := \frac{P_{12}}{\lambda}, \qquad P_{12}(\mathbf{u} \otimes \mathbf{v}) := \mathbf{v} \otimes \mathbf{u}, \tag{1.2}$$

with $\mathcal{N}(\lambda)$ a rational function of λ.

$$\mathcal{N}(\lambda) = \mathcal{B}(\lambda) + \sum_{i=1}^{n} \sum_{a=1}^{n_i} \frac{N_{ia}}{(\lambda - \alpha_i)^i} \tag{1.3}$$

$$\mathcal{B}(\lambda) := \sum_{i=1}^{n_0} B_i \lambda^i, \qquad\qquad N_{ia}, B_i \in \mathfrak{gl}(r). \tag{1.4}$$

Equations (1.1), (1.2) define the standard linear, rational R-matrix structure.

2000 *Mathematics Subject Classification.* Primary 35Q40; Secondary 81R12.

Work supported in part by the Natural Sciences and Engineering Research Council of Canada (NSERC) and the Fonds FCAR du Québec.

This is the final form of the paper.

It follows from the properties of classical R-matrices [**2, 9**] that elements of the algebra of spectral invariants $\phi(\mathcal{N}) \in \mathcal{I}(\widetilde{\mathfrak{gl}}(r))$ Poisson commute amongst themselves and generate commuting isospectral flows determined by the Lax equations:

$$\dot{\mathcal{N}} = \pm[(d\phi_{\mathcal{N}})_\pm, \mathcal{N}], \tag{1.5}$$

where $\mathcal{N}$ is here thought of as an element of the loop algebra $\widetilde{\mathfrak{gl}}(r)$, identified in a standard way with its dual $\widetilde{\mathfrak{gl}}^*(r)$ through the trace-residue pairing, and $(\ \cdot\ )_\pm$ denotes projection to the $\pm$ components relative to the usual splitting of the loop algebra into positive and negative components

$$\widetilde{\mathfrak{gl}}(r) = \widetilde{\mathfrak{gl}}(r)_+ + \widetilde{\mathfrak{gl}}(r)_- \tag{1.6}$$

(i.e., those admitting holomorphic continuations to the interior $(+)$ and exterior $(-)$ of the unit circle respectively with the latter normalized to vanish at ∞). The spectral invariants generate a maximal Poisson commuting algebra on generic symplectic leaves, defining completely integrable systems [**3, 4**]; i.e., there are as many functionally independent generators as half the dimension of the leaf.

1.2. 2×2 rational Lax matrices. In the following, we shall limit our discussion to the case of 2×2 Lax matrices, although most of the considerations that follow are easily extended to higher rank. We may without loss of generality take $\mathcal{N}(\lambda)$ to be traceless (since the trace coefficients are Casimirs)

$$\mathcal{N}(\lambda) = \begin{pmatrix} h(\lambda) & e(\lambda) \\ f(\lambda) & -h(\lambda) \end{pmatrix}, \tag{1.7}$$

where the rational functions $e(\lambda)$, $f(\lambda)$, $h(\lambda)$ satisfy the Poisson bracket relations

$$\begin{gathered} \{h(\lambda), e(\mu)\} = \frac{e(\lambda) - e(\mu)}{\lambda - \mu}, \qquad \{h(\lambda), f(\mu)\} = -\frac{f(\lambda) - f(\mu)}{\lambda - \mu}, \\ \{e(\lambda), f(\mu)\} = -2\frac{h(\lambda) - h(\mu)}{\lambda - \mu}. \end{gathered} \tag{1.8}$$

For this case, the ring $\mathcal{I}(\widetilde{\mathfrak{gl}}(2))$ of spectral invariants, when restricted to the symplectic leaves of the R-matrix Poisson structure, is generated by the quadratic trace invariants; i.e., the coefficients determining the numerator of the rational function

$$-\Delta(\lambda) := \tfrac{1}{2}\operatorname{tr}(\mathcal{N}^2(\lambda)) = h^2(\lambda) - \tfrac{1}{2}\big(e(\lambda)f(\lambda) + f(\lambda)e(\lambda)\big). \tag{1.9}$$

(The order in the last two terms is irrelevant of course, but it is written here in a form that will also be valid in the quantum version below.) If, for example, the polynomial part $\mathcal{B}(\lambda)$ of $\mathcal{N}(\lambda)$ is taken to vanish, and only first order poles appear in $\mathcal{N}(\lambda)$, we have

$$e(\lambda) := \sum_{i=1}^n \frac{e_i}{\lambda - \alpha_i}, \qquad f(\lambda) := \sum_{i=1}^n \frac{f_i}{\lambda - \alpha_i}, \qquad h(\lambda) := \sum_{i=1}^n \frac{h_i}{\lambda - \alpha_i}, \tag{1.10}$$

where the quantities $\{e_i, f_i, h_i\}_{i=1,\dots,n}$ are a set of n $\mathbf{sl}(2)$ generators, which may be canonically coordinatized as:

$$e_i := \frac{1}{2}\left(y_i^2 + \frac{\mu_i^2}{x_i^2}\right) \qquad f_i := \frac{1}{2}x_i^2 \qquad h_i := \frac{1}{2}x_i y_i, \qquad i = 1, \dots, n, \tag{1.11}$$

where $\{\mu_i^2\}_{i=1,\dots,n}$ are the values of the $\mathbf{sl}(2)$ Casimir invariants and $\{x_i, y_i\}_{i=1,\dots,n}$ form a set of canonical coordinates on the symplectic leaves.

1.2.1. *Parametric dependence of invariants and superintegrability.* Again, taking the case when the polynomial part $\mathcal{B}(\lambda)$ of $\mathcal{N}(\lambda)$ vanishes (but not necessarily just first order poles), a complete set of generators is given by

$$\phi_{ia} := \operatorname*{res}_{\lambda=\alpha_i} (\lambda - \alpha_i)^a \operatorname{tr}\big(\mathcal{N}^2(\lambda)\big), \quad i = 1, \dots, n, \ a = 0, \dots, n_i - 1. \tag{1.12}$$

These commute amongst themselves, but they each depend upon the pole locations $\{\alpha_i\}_{i=1,\dots,n}$ in $\mathcal{N}(\lambda)$. However, the linear combination:

$$\phi_{\mathrm{SI}} := \sum_{i=1}^{n} \alpha_i \phi_{i0} = \operatorname*{res}_{\lambda=\infty} \operatorname{tr}\big(\mathcal{N}^2(\lambda)\big) \tag{1.13}$$

does not depend on the α_i's. In general, there is no reason for the invariants $\phi_{ia}(\alpha_i)$ to commute with each other for different choices of the α_i's. But, regardless of the values chosen, they will commute with ϕ_{SI}. Since the $\phi_{ia}(\alpha_i)$'s for different choices of α_i's in general do not generate the same algebra of functions, we may conclude that, taken together, for different evaluations of the parameters $\{\alpha_i\}$, there are more functionally independent integrals that Poisson commute with ϕ_{SI} than half the dimension of the symplectic leaf, and hence the Hamiltonian system it generates is superintegrable. (In fact, in many cases, it may be shown to be maximally superintegrable; see the examples below.)

In particular, if we take the case of purely simple poles as above in (1.10), the resulting Hamiltonian is:

$$\phi_{\mathrm{SI}} = \frac{1}{2}\sum_{i=1}^{n} x_i^2 \sum_{j=1}^{n} y_j^2 - \frac{1}{2}\Big(\sum_{i=1}^{n} x_i y_i\Big)^2 + \frac{1}{2}\sum_{i=1}^{n} x_i^2 \sum_{i=1}^{n} \frac{\mu_i^2}{x_i^2}, \tag{1.14}$$

which, when constrained to the (co)tangent bundle of the $n-1$ sphere S^{n-1}

$$\sum_{i=1}^{n} x_i^2 = 1, \qquad \sum_{i=1}^{n} x_i y_i = 0, \tag{1.15}$$

yields the superintegrable system

$$h_{\mathrm{Ros}} = \frac{1}{2}\sum_{j=1}^{n} y_j^2 + \frac{1}{2}\sum_{i=1}^{n} \frac{\mu_i^2}{x_i^2}, \tag{1.16}$$

which is the trivial case of the Rosochatius system (without a harmonic oscillator potential).

1.2.2. *Separation of variables.* Another viewpoint that helps to explain the superintegrability of systems arising in this way is to note that they may be completely separated in a canonical coordinate system determined by the values of the pole parameters $\{\alpha_i\}$ which, for the $\mathbf{sl}(2)$ case with simple poles with the phase space constrained to S^{n-1} as above reduces to the sphero-conical system $\{\lambda_i, \zeta_i\}_{i=1,\dots,n-1}$ defined by:

$$\sum_{i=1}^{n} \frac{x_i^2}{\lambda - \alpha_i} = \frac{\prod_{j=1}^{n-1}(\lambda - \lambda_j)}{\prod_{i=1}^{n}(\lambda - \alpha_i)}, \qquad \zeta_i := \frac{1}{2}\sum_{i=1}^{n} \frac{x_i y_i}{(\lambda - \alpha_i)}. \tag{1.17}$$

These are just the points (λ_i, ζ_i) on the invariant spectral curve

$$\zeta^2 + \tfrac{1}{2}\Delta(\lambda) = 0 \tag{1.18}$$

where the matrix element $f(\lambda)$ vanishes and $\zeta_i = h(\lambda_i)$ are the eigenvalues at these points. These are particular cases of the spectral *Darboux coordinates* of [**1, 3, 4**].

(Note that these become hyperellipsoidal coordinates if there is a constant term added in the definition (1.10) of $f(\lambda)$.)

The point to note is that the separation of variables occurs in these coordinates simultaneously for *all* the invariants ϕ_{ia}, viewed as generators of Hamiltonian flows. But again, since the leading term spectral invariant ϕ_{SI} does not depend on the values of the parameters α_i, it admits a separation of variables in *any* of the family of sphero-conical (or hyperellipsoidal) coordinates obtained by varying these parameters. This simultaneous separability in multiple coordinates may be viewed as an alternative explanation of the origin of the superintegrability of such systems. (In fact, both these viewpoints are a result of the classical r-matrix setting, and in a sense may be considered as equivalent.)

In the examples given below in the following section, the same principle is used to deduce superintegrable systems from $\mathbf{sl}(2)$ Lax matrices satisfying the Poisson bracket relations (1.1)).

1.2.3. *Quantum integrable systems.* The above discussion is easily extended to the canonically quantized version of such systems. All that must be done is to replace the matrix elements defining $\mathcal{N}(\lambda)$ by their quantized forms $\hat{e}(\lambda)$, $\hat{f}(\lambda), \hat{h}(\lambda)$, which must satisfy the commutator analogs of the Poisson bracket relations (1.8)

$$
\begin{aligned}
&[\hat{h}(\lambda), \hat{e}(\mu)] = \frac{\hat{e}(\lambda) - \hat{e}(\mu)}{\lambda - \mu}, \qquad [\hat{h}(\lambda), \hat{f}(\mu)] = -\frac{\hat{f}(\lambda) - \hat{f}(\mu)}{\lambda - \mu}, \\
&[\hat{e}(\lambda), \hat{f}(\mu)] = -2\frac{\hat{h}(\lambda) - \hat{h}(\mu)}{\lambda - \mu},.
\end{aligned} \tag{1.19}
$$

These can be realized by canonical quantization of the underlying classical phase space variables. For example, in the case of simple poles only, with vanishing polynomial term $\mathcal{B}(\lambda)$, we have:

$$
\hat{e}(\lambda) := \sum_{i=1}^{n} \frac{\hat{e}_i}{\lambda - \alpha_i} \qquad \hat{f}(\lambda) := \sum_{i=1}^{n} \frac{\hat{f}_i}{\lambda - \alpha_i} \qquad \hat{h}(\lambda) := \sum_{i=1}^{n} \frac{\hat{h}_i}{\lambda - \alpha_i}, \tag{1.20}
$$

where the $\mathbf{sl}(2)$ generators $\{\hat{e}_i, \hat{f}_i, \hat{h}_i\}$ may be represented by the operators

$$
\hat{e}_i := \frac{1}{2}\left(\frac{\partial^2}{\partial x_i^2} - \frac{\mu_i^2}{x_i^2}\right) \quad \hat{f}_i := \frac{1}{2}x_i^2 \quad \hat{h}_i := \frac{1}{2}\left(x_i\frac{\partial}{\partial x_i} + \frac{1}{2}\right), \quad i = 1, \dots n, \tag{1.21}
$$

and the commuting invariants are similarly given by the coefficients of the numerator polynomial of the quantum spectral invariant:

$$
\hat{\Delta}(\lambda) := \hat{h}^2(\lambda) - \tfrac{1}{2}\big(\hat{e}(\lambda)\hat{f}(\lambda) + \hat{f}(\lambda)\hat{e}(\lambda)\big). \tag{1.22}
$$

The resulting systems are similarly quantum integrable, and separable in the same coordinates as the classical ones **[6]** and, for the same reasons as above, the quantum version of the Hamiltonian ϕ_{SI} is superintegrable.

In the following section, a number of examples of such classical and quantum superintegrable systems will be given.

2. Examples of superintegrable classical and quantum systems

The examples given below arise in the framework of the so-called Krall–Scheffer problem **[8]** of describing all two-dimensional analogs of classical orthogonal polynomials which result in nine classes of second-order partial differential equations on the plane or on constant curvature surfaces. It was shown in **[5, 7]** that all nine cases are connected with superintegrable systems. The following are some illustrative examples.

2.1. The sphere.

2.1.1. *Classical Lax matrix.* The first case corresponds to three simple poles and vanishing $B(\lambda)$. The Lax matrix has the form:

$$N(\lambda) = \frac{N_1}{(\lambda-\alpha)} + \frac{N_2}{(\lambda-\beta)} + \frac{N_3}{(\lambda-\gamma)} = \begin{pmatrix} h(\lambda) & f(\lambda) \\ e(\lambda) & -h(\lambda) \end{pmatrix}, \tag{2.1}$$

where the matrix elements of the N_i generate a Poisson bracket realization of $\left(\mathbf{sl}(2)\right)^3$:

$$\begin{gathered} N_1 = \frac{1}{2}\begin{pmatrix} s_1p_1 & p_1^2+\mu_1^2/s_1^2 \\ -s_1^2 & -s_1p_1 \end{pmatrix}, \qquad N_2 = \frac{1}{2}\begin{pmatrix} s_2p_2 & p_2^2+\mu_2^2/s_2^2 \\ -s_2^2 & -s_2p_2 \end{pmatrix}, \\ N_3 = \frac{1}{2}\begin{pmatrix} s_3p_3 & p_3^2+\mu_3^2/s_3^2 \\ -s_3^2 & -s_3p_3 \end{pmatrix}. \end{gathered} \tag{2.2}$$

Here (p_1, p_2, p_3) are canonically conjugate to (s_1, s_2, s_3) (and these coincide with the coordinates $\{x_i, y_i\}_{i=1,\ldots,n}$ above).

2.1.2. *Commuting invariants.* The invariants are the coefficients of:

$$-\frac{1}{2}\operatorname{tr} N(\lambda)^2 = \frac{H_1}{(\lambda-\alpha)} + \frac{H_2}{(\lambda-\beta)} + \frac{H_3}{(\lambda-\gamma)} + \frac{\mu_1^2}{(\lambda-\alpha)^2} + \frac{\mu_2^2}{(\lambda-\beta)^2} + \frac{\mu_3^2}{(\lambda-\gamma)^2}. \tag{2.3}$$

Note that only two of the integrals H_1, H_2 and H_3 are independent, since their sum is zero. The Hamiltonian of the problem is given by their linear combination:

$$H = \alpha H_1 + \beta H_2 + \gamma H_3 = \frac{1}{2}(p_1^2 + p_2^2 + p_3^2) + \frac{\mu_1^2}{s_1^2} + \frac{\mu_2^2}{s_2^2} + \frac{\mu_3^2}{s_3^2}. \tag{2.4}$$

This describes the Rosochatius system with harmonic oscillator terms absent on the cotangent bundle of a two-sphere in $\mathbb{R}^3$:

$$s_1^2 + s_2^2 + s_3^2 = 1, \qquad\qquad s_1p_1 + s_2p_2 + s_3p_3 = 0. \tag{2.5}$$

The integrals H_1, H_2 and H_3 are as follows:

$$\begin{aligned} H_1 &= -\frac{1}{2}\frac{L_{13}^2+\mu_3^2s_1^2/s_3^2+\mu_1^2s_3^2/s_1^2}{\alpha-\gamma} - \frac{1}{2}\frac{L_{12}^2+\mu_1^2s_2^2/s_1^2+\mu_2^2s_1^2/s_2^2}{\alpha-\beta} \\ H_2 &= -\frac{1}{2}\frac{L_{23}^2+\mu_3^2s_2^2/s_3^2+\mu_2^2s_3^2/s_2^2}{\beta-\gamma} + \frac{1}{2}\frac{L_{12}^2+\mu_1^2s_2^2/s_1^2+\mu_2^2s_1^2/s_2^2}{\alpha-\beta} \\ H_3 &= \frac{1}{2}\frac{L_{23}^2+\mu_3^2s_2^2/s_3^2+\mu_2^2s_3^2/s_2^2}{\beta-\gamma} + \frac{1}{2}\frac{L_{13}^2+\mu_3^2s_1^2/s_3^2+\mu_1^2s_3^2/s_1^2}{\alpha-\gamma}, \end{aligned} \tag{2.6}$$

where $L_{ij} = s_1p_2 - s_2p_1$.

Note that the Hamiltonian H is independent of the parameters (α, β, γ), whereas the invariants H_1, H_2 do depend on them. Therefore, different choices for the parameters give distinct integrals that commute with H, but do not commute with each other.

2.1.3. *Separating coordinates.* The separating coordinates (λ_1, λ_2) in this case are sphero-conical coordinates. The corresponding momenta are denoted (ξ_1, ξ_2). They are related to (s_1, s_2, s_3) and (p_1, p_2, p_3) by:

$$(2.7)\quad s_1^2 = \frac{(\alpha-\lambda_1)(\alpha-\lambda_2)}{(\alpha-\beta)(\alpha-\gamma)}, \qquad \xi_1 = -\frac{1}{2}\left(\frac{s_1p_1}{\lambda_1-\alpha} + \frac{s_2p_2}{\lambda_1-\beta} + \frac{-s_1p_1-s_2p_2}{\lambda_1-\gamma}\right)$$

$$(2.8)\quad s_2^2 = \frac{(\beta-\lambda_1)(\beta-\lambda_2)}{(\beta-\alpha)(\beta-\gamma)}, \qquad \xi_2 = -\frac{1}{2}\left(\frac{s_1p_1}{\lambda_2-\alpha} + \frac{s_2p_2}{\lambda_2-\beta} + \frac{-s_1p_1-s_2p_2}{\lambda_2-\gamma}\right)$$

$$(2.9)\quad s_3^2 = \frac{(\gamma-\lambda_1)(\gamma-\lambda_2)}{(\gamma-\alpha)(\gamma-\beta)}$$

2.1.4. *Quantum system.* The quantum versions of the integrals above, denoted $\widehat{H}_1$, $\widehat{H}_2$, $\widehat{H}_3$, are obtained by replacing the matrix elements of $N(\lambda)$ by the corresponding differential operators, $\hat{e}(\lambda)$, $\hat{f}(\lambda)$, $\hat{h}(\lambda)$, which in the case of simple poles are as in (1.20)–(1.22).

The quantization procedure leads to replacing the L_{ij}'s by their quantum version:

$$(2.10)\quad \hat{L}_{ij} = \sqrt{-1}(s_i\partial/\partial s_j - s_j\partial/\partial s_i).$$

Introducing the functions

$$(2.11)\quad \omega_{jk}^2 := \mu_j^2 s_k^2/s_j^2 + \mu_k^2 s_j^2/s_k^2, \quad j,k = 1,\dots,3,$$

and denoting $\alpha = \alpha_1$, $\beta = \alpha_2$, $\gamma = \alpha_3$, we can present the quantum integrals as

$$(2.12)\quad \widehat{H}_i = -\frac{1}{2}\sum_{k\neq i} \frac{\hat{L}_{ik} + \omega_{ik}^2}{\alpha_i - \alpha_k}, \quad i,k = 1,\dots,3.$$

The quantum Hamiltonian is

$$(2.13)\quad \widehat{H} = -\frac{1}{2}(\partial_1^2 + \partial_2^2 + \partial_3^2) + \frac{\mu_1^2}{s_1^2} + \frac{\mu_2^2}{s_2^2} + \frac{\mu_3^2}{s_3^2}.$$

The separating coordinates are the configuration space part of the ones for the classical case (λ_1, λ_2).

2.2. The hyperboloid.

2.2.1. *Classical Lax matrix.* Consider now a Lax matrix with one first order and one second order pole:

$$(2.14)\quad N(\lambda) = \frac{N_1}{(\lambda-\alpha)} + \frac{N_2}{(\lambda-\alpha)^2} + \frac{N_3}{(\lambda-\beta)},$$

with

$$(2.15)\quad \begin{gathered} N_1 = \frac{1}{2}\begin{pmatrix} s_1p_1 + s_2p_2 & 2p_1p_2 + 2\gamma_1\gamma_2 \\ -2s_1s_2 & -s_1p_1 - s_2p_2 \end{pmatrix}, \\ N_2 = \frac{1}{2}\begin{pmatrix} -s_2p_1 & -p_1^2 - \gamma_2^2 \\ s_2^2 & s_2p_1 \end{pmatrix}, \qquad N_3 = \frac{1}{2}\begin{pmatrix} s_3p_3 & p_3^2 + \gamma_3^2 \\ -s_3^2 & -s_3p_3 \end{pmatrix}. \end{gathered}$$

Here we have introduced the following notations

$$(2.16)\quad 2\gamma_1\gamma_2 := \frac{2\mu_2^2 s_1}{s_2^3} - \frac{2\mu_1\mu_2}{s_2^2}, \qquad \gamma_2^2 := -\frac{\mu_2^2}{s_2^2}, \qquad \gamma_3^2 := \frac{\mu_3^2}{s_3^2}.$$

The matrix elements of (N_1, N_2) generate a Poisson bracket realization of the jet extension $\mathbf{sl}(2)^{(1)*}$ while those of N_3 generate a second $\mathbf{sl}(2)$.

2.2.2. *Commuting invariants.* The trace formula again gives us only two independent commuting invariants H_1 and H_2

$$-\frac{1}{2}\operatorname{tr} N(\lambda)^2 = \frac{H_1}{(\lambda-\alpha)} + \frac{H_2}{(\lambda-\alpha)^2} - \frac{\mu_1\mu_2}{(\lambda-\alpha)^3} + \frac{\mu_2^2}{2(\lambda-\alpha)^4} + \frac{H_3}{(\lambda-\beta)} - \frac{\mu_3^2}{2(\lambda-\beta)^2} \tag{2.17}$$

since, by taking the residue we obtain

$$H_1 + H_3 = 0. \tag{2.18}$$

The superintegrable Hamiltonian in this case is:

$$H = (\alpha-\beta)H_1 + H_2 - \tfrac{1}{2}\mu_3^2 = 2p_1p_2 - p_1^2 + p_3^2 + 2\gamma_1\gamma_2 - \gamma_2^2 + \gamma_3^2. \tag{2.19}$$

The quadratic constraint now defines a hyperboloid

$$2s_1s_2 + s_3^2 = 1. \tag{2.20}$$

In the ambient coordinates the integrals H_1 and H_2 are

$$\begin{aligned} H_1 &= \frac{(s_1p_3 - s_3p_2)(s_3p_1 - s_2p_3) - \gamma_3^2 s_1s_2 - 2\gamma_1\gamma_2 s_3^2}{\alpha-\beta} - \frac{(s_3p_1 - s_2p_3)^2 + \gamma_3^2 s_2^2 + \gamma_2^2 s_3^2}{2(\alpha-\beta)^2} \\ H_2 &= \frac{1}{2}(s_1p_1 - s_2p_2)^2 - 2\gamma_1\gamma_2 s_1 + \frac{(s_3p_1 - s_2p_3)^2 + \gamma_3^2 s_2^2 + \gamma_2^2 s_3^2}{2(\alpha-\beta)}. \end{aligned} \tag{2.21}$$

Again, whereas the Hamiltonian H does not depend on the parameters (α, β) the integrals H_1, H_2 do, which provides an explanation for the superintegrability in this case.

2.2.3. *Separating coordinates.* These are determined by the relations:

$$s_3^2 = \frac{(\beta-\lambda_1)(\beta-\lambda_2)}{(\alpha-\beta)^2}, \quad \xi_1 = -\frac{1}{2}\left(\frac{s_1p_1+s_2p_2}{\lambda_1-\alpha} - \frac{s_2p_1}{(\lambda_1-\alpha)^2} + \frac{s_3p_3}{\lambda_1-\beta}\right), \tag{2.22}$$

$$s_2^2 = -\frac{(\alpha-\lambda_1)(\alpha-\lambda_2)}{(\alpha-\beta)}, \quad \xi_2 = -\frac{1}{2}\left(\frac{s_1p_1+s_2p_2}{\lambda_2-\alpha} - \frac{s_2p_1}{(\lambda_2-\alpha)^2} + \frac{s_3p_3}{\lambda_2-\beta}\right), \tag{2.23}$$

$$s_1s_2 = -\frac{1}{2}\left(\frac{(\beta-\lambda_1)(\beta-\lambda_2)}{(\alpha-\beta)^2} - 1\right). \tag{2.24}$$

2.2.4. *Quantum system.* The quantized integrals $\hat{H}_1$, $\hat{H}_2$,b $\hat{H}_3$ are obtained as before by replacing all conjugate variables by the corresponding differential operators. The quantum integrals may then be expressed as

$$\begin{aligned} \widehat{H}_1 &= -\frac{(s_1\partial_3 - s_3\partial_2)(s_3\partial_1 - s_2\partial_3) + \gamma_3^2 s_1s_2 + 2\gamma_1\gamma_2 s_3^2}{\alpha-\beta} + \frac{(s_3\partial_1 - s_2\partial_3)^2 - \gamma_3^2 s_2^2 - \gamma_2^2 s_3^2}{2(\alpha-\beta)^2}, \\ \widehat{H}_2 &= \frac{1}{2}\hat{L}_{12}^2 - 2\gamma_1\gamma_2 s_1 - \frac{(s_3\partial_{s_1} - s_2\partial_{s_3})^2 - \gamma_3^2 s_2^2 - \gamma_2^2 s_3^2}{2(\alpha-\beta)}, \end{aligned} \tag{2.25}$$

where $\partial_k := \partial/\partial s_k$.

The quantum Hamiltonian is

$$\widehat{H} = 2\partial_1\partial_2 - \partial_1^2 + \partial_3^2 + 2\gamma_1\gamma_2 - \gamma_2^2 + \gamma_3^2, \tag{2.26}$$

and this again separates in the configuration space coordinates (λ_1, λ_2).

2.3. The plane.

2.3.1. *Classical Lax matrix.* For the cases with zero curvature like the example to follow, the polynomial part $\mathcal{B}(\lambda)$ of the Lax matrix does not vanish. The simplest case involves two distinct finite poles in $N(\lambda)$ and constant $B(\lambda)$

$$N(\lambda) = \begin{pmatrix} 0 & -a \\ 1 & 0 \end{pmatrix} + \frac{1}{2(\lambda-\alpha)} \begin{pmatrix} s_1 p_1 & p_1^2 + \mu_1^2/s_1^2 \\ -s_1^2 & -s_1 p_1 \end{pmatrix} + \frac{1}{2(\lambda-\beta)} \begin{pmatrix} s_2 p_2 & p_2^2 + \mu_2^2/s_2^2 \\ -s_2^2 & -s_2 p_2 \end{pmatrix}. \tag{2.27}$$

The matrix elements of the residues N_1, N_2 generate two copies of $\mathbf{sl}(2)$.

2.3.2. *Commuting invariants.* The invariants of motion are defined by:

$$-\frac{1}{2} \operatorname{tr} N(\lambda)^2 = \frac{H_1}{(\lambda-\alpha)} + \frac{H_2}{(\lambda-\beta)} + \frac{\mu_1^2}{(\lambda-\alpha)^2} + \frac{\mu_2^2}{(\lambda-\beta)^2} - a. \tag{2.28}$$

The superintegrable Hamiltonian in this case is given by

$$H = \frac{1}{4} \operatorname*{res}_{\infty} \operatorname{tr} N(\lambda)^2 = \frac{1}{4}\left(p_1^2 + p_2^2 + a(s_1^2 + s_2^2) + \frac{\mu_1^2}{s_1^2} + \frac{\mu_2^2}{s_2^2} \right), \tag{2.29}$$

which gives an isotropic oscillator together with Rosochatius terms. As before (p_1, p_2) are canonically conjugate to (s_1, s_2).

In terms of the ambient space coordinates the integrals H_1 and H_2 are:

$$\begin{aligned} H_1 &= p_1^2 + a s_1^2 + \frac{\mu_1^2}{s_1^2} - \frac{1}{2(\alpha-\beta)}\left(L_{12}^2 + \frac{\mu_1^2 s_2^2}{s_1^2} + \frac{\mu_2^2 s_1^2}{s_2^2} \right) \\ H_2 &= p_2^2 + a s_2^2 - \frac{\mu_2^2}{s_2^2} + \frac{1}{2(\alpha-\beta)}\left(L_{12}^2 + \frac{\mu_1^2 s_2^2}{s_1^2} + \frac{\mu_2^2 s_1^2}{s_2^2} \right). \end{aligned} \tag{2.30}$$

Where $L_{12} := s_1 p_2 - s_2 p_1$ and $H = \frac{1}{4}(H_1 + H_2)$. Here the additional integral results from the parametric dependence on $(\alpha - \beta)$.

2.3.3. *Separating coordinates.* The separating coordinates $(\lambda_1, \lambda_2, \xi_1, \xi_2)$ in this case are defined by

$$s_1^2 = 2\frac{(\beta-\lambda_1)(\beta-\lambda_2)}{(\alpha-\beta)} \qquad \xi_1 = -\frac{1}{2}\left(\frac{s_1 p_1}{\lambda_1 - \alpha} + \frac{s_2 p_2}{\lambda_1 - \beta} \right) \tag{2.31}$$

$$s_2^2 = -2\frac{(\alpha-\lambda_1)(\alpha-\lambda_2)}{(\alpha-\beta)} \qquad \xi_2 = -\frac{1}{2}\left(\frac{s_1 p_1}{\lambda_2 - \alpha} + \frac{s_2 p_2}{\lambda_2 - \beta} \right) \tag{2.32}$$

2.3.4. *Quantum system.* The Hamiltonian of the corresponding quantum problem is

$$\widehat{H} = \frac{1}{4} \operatorname*{res}_{\infty} \operatorname{tr} \hat{N}(\lambda)^2 = \frac{1}{4}\left(\partial_1^2 + \partial_2^2 + a(s_1^2 + s_2^2) + \frac{\mu_1^2}{s_1^2} + \frac{\mu_2^2}{s_2^2} \right) = \frac{1}{4}(\widehat{H}_1 + \widehat{H}_2). \tag{2.33}$$

The quantum integrals $\widehat{H}_1$ and $\widehat{H}_2$ are:

$$\begin{aligned} \widehat{H}_1 &= \partial_1^2 + a s_1^2 + \frac{\mu_1^2}{s_1^2} - \frac{1}{2(\alpha-\beta)}\left(\hat{L}_{12}^2 + \frac{\mu_1^2 s_2^2}{s_1^2} + \frac{\mu_2^2 s_1^2}{s_2^2} \right) \\ \widehat{H}_2 &= \partial_2^2 + a s_2^2 - \frac{\mu_2^2}{s_2^2} + \frac{1}{2(\alpha-\beta)}\left(\hat{L}_{12}^2 + \frac{\mu_1^2 s_2^2}{s_1^2} + \frac{\mu_2^2 s_1^2}{s_2^2} \right) \end{aligned} \tag{2.34}$$

and the separating coordinates are again (λ_1, λ_2), which depend on the additional parameter $(\alpha - \beta)$.

References

1. M. R. Adams, J. Harnad, and J. Hurtubise, *Darboux coordinates and Liouville–Arnol'd integration in loop algebras*, Comm. Math. Phys. **155** (1993), no. 2, 385–413.
2. L. D. Faddeev and L. A. Takhtajan,*Hamiltonian methods in the theory of solitons*, Springer Ser. Soviet Math., Springer-Verlag, Berlin, 1987.
3. J. Harnad, *Isospectral flow and Liouville–Arnol'd integration in loop algebras*, Geometric and Quantum Methods in Integrable Systems (Scheveningen, 1992), Lecture Notes in Phys., vol. 424, Springer-Verlag, Berlin, 1993, pp. 1–42.
4. ———, *Loop groups, R-matrices and separation of variables*, Integrable Systems: From Classical to Quantum (Montréal, 1999), CRM Proc. Lecture Notes, vol. 26, Amer. Math. Soc., Providence, RI, 2000, pp. 21–54.
5. J. Harnad, L. Vinet, O. Yermolaeva, and A. Zhedanov, *Two-dimensional Krall–Sheffer polynomials and integrable systems*, J. Phys. A **34** (2001), no. 48, 10619–10625.
6. J. Harnad and P. Winternitz, *Classical and quantum integrable systems in* $\widetilde{\mathfrak{gl}}(2)^{+*}$ *and separation of variables*, Comm. Math. Phys. **172** (1995), 263–285.
7. J. Harnad, A. Zhedanov, and O. Yermolaeva, *R-matrix approach to the Krall–Sheffer problem*, Symmetry in Nonlinear Mathematical Physics, Part 1, 2 (Kiev, 2001), Pr. Inst. Mat. Nats. Akad. Nauk Ukr. Mat. Zastos., vol. 43, Part 1, 2, Natsīonal. Akad. Nauk Ukraïni, Īnst. Mat., Kiev, 2002, pp. 314–320.
8. H. Krall and I. M. Sheffer, *Orthogonal polynomials in two variables*, Ann. Mat. Pura Appl. (4) **76** (1967), 325–376.
9. M. A. Semenov-Tyan-Shanskiĭ, *What a classical r-matrix is*, Funktsional. Anal. i Prilozhen. **17** (1983), no. 4, 17–33 (Russian); English transl., Functional Anal. Appl. **17** (1983), no. 4, 259–272.

Centre de recherches mathématiques, Université de Montréal, C.P. 6128, succ. Centre-ville, Montréal, QC H3C 3J7, Canada and Department of Mathematics and Statistics, Concordia University, 7141 Sherbrooke West, Montréal, QC H4B 1R6, Canada

E-mail address: harnad@crm.umontreal.ca

Centre de recherches mathématiques, Université de Montréal, C.P. 6128, succ. Centre-ville, Montréal, QC H3C 3J7, Canada and Department of Mathematics and Statistics, Concordia University, 7141 Sherbrooke West, Montréal, QC H4B 1R6, Canada

Current address: Scuola Internazionale Superiore di Studi Avanzati, via Beirut 4, 34014 Trieste, Italy

E-mail address: yermolae@crm.umontreal.ca

Centre de Recherches Mathématiques
CRM Proceedings and Lecture Notes
Volume **37**, 2004

Maximally Superintegrable Smorodinsky–Winternitz Systems on the N-Dimensional Sphere and Hyperbolic Spaces

Francisco J. Herranz, Ángel Ballesteros, Mariano Santander, and Teresa Sanz-Gil

Abstract. The classical Smorodinsky–Winternitz systems on the ND sphere, Euclidean and hyperbolic spaces $\mathbb{S}^N$, $\mathbb{E}^N$ and $\mathbb{H}^N$ are simultaneously approached starting from the Lie algebras $\mathfrak{so}_\kappa(N+1)$, which include a parametric dependence on the curvature. General expressions for the Hamiltonian and its integrals of motion are given in terms of intrinsic geodesic coordinate systems. Each Lie algebra generator gives rise to an integral of motion, so that a set of $N(N+1)/2$ integrals is obtained. Furthermore, $2N-1$ functionally independent ones are identified which, in turn, shows that the well-known maximal superintegrability of the Smorodinsky–Winternitz system on $\mathbb{E}^N$ is preserved when curvature arises. On both $\mathbb{S}^N$ and $\mathbb{H}^N$, the resulting system can be interpreted as a superposition of an "actual" oscillator and N "ideal" oscillators (for the sphere, these are alike the actual ones), which can also be understood as N "centrifugal terms"; this is the form seen in the Euclidean limiting case.

1. Introduction

Superintegrable systems on the two- and three-dimensional (3D) Euclidean spaces have been classified in [**6, 25**], and also extended to the 2D and 3D spheres [**11**] as well as to the 2D hyperbolic plane [**19, 20**] and 3D hyperbolic space [**12**]. Recent classifications of superintegrable systems for these 2D Riemannian spaces have been presented in [**18, 21, 24, 27**]. In particular, in the 2D sphere $\mathbb{S}^2$ there are two (maximally) superintegrable families: the harmonic oscillator and the Kepler–Coulomb potential, both of them with some "additional" terms. Let us consider $\mathbb{S}^2$ as embedded through $s_0^2+s_1^2+s_2^2=1$ in an ambient space $\mathbb{R}^3=(s_0,s_1,s_2)$; we set the geodesic polar coordinates (r,θ) such that $s_0=\cos r$, $s_1=\sin r\cos\theta$, $s_2=\sin r\sin\theta$. Following the notation and results given in [**27**], the first classical

2000 *Mathematics Subject Classification.* Primary 37J15; Secondary 37J35.

This work was partially supported by the Ministerio de Ciencia y Tecnología, Spain (Projects BFM2000-1055 and BFM2002-3773).

The first two authors are also grateful to G. S. Pogosyan for helpful discussions and to the CRM for hospitality.

This is the final form of the paper.

superintegrable family on $\mathbb{S}^2$ is given by

$$\mathcal{U}_{\text{ho}} = \beta_0 \left(\frac{s_1^2 + s_2^2}{s_0^2} \right) + \frac{\beta_1}{s_1^2} + \frac{\beta_2}{s_2^2} = \beta_0 \tan^2 r + \frac{\beta_1}{\sin^2 r \cos^2 \theta} + \frac{\beta_2}{\sin^2 r \sin^2 \theta}, \tag{1.1}$$

where β_i are real constants. The first term, $\tan^2 r$, is the *spherical oscillator* or Higgs potential [**16, 26**]. Under contraction to the Euclidean plane, this reduces to the usual harmonic oscillator, r^2, while the two remaining terms give rise to two "centrifugal barriers." However, very recently, all the three terms in $\mathcal{U}_{\text{ho}}$ have been interpreted as spherical oscillators with different centers [**29**].

On the other hand, the second superintegrable family on $\mathbb{S}^2$ turns out to be

$$\mathcal{U}_{\text{KC}} = \beta_0 \frac{s_0}{\sqrt{s_1^2 + s_2^2}} + \beta_1 \frac{s_1}{s_2^2 \sqrt{s_1^2 + s_2^2}} + \frac{\beta_2}{s_2^2} = \beta_0 \frac{1}{\tan r} + \beta_1 \frac{\cos \theta}{\sin^2 r \sin^2 \theta} + \frac{\beta_2}{\sin^2 r \sin^2 \theta}, \tag{1.2}$$

where $1/\tan r$ is the "spherical" Kepler–Coulomb potential, first studied by Schrödinger [**30**]; the two potentials $\tan^2 r$ and $1/\tan r$ are mutually related [**22, 23**].

The aim of this contribution is to study the maximal superintegrability of the generalization of the potential (1.1) on the ND spaces $\mathbb{S}^N$, $\mathbb{E}^N$ and $\mathbb{H}^N$ from a group theoretical standpoint. This family, depending on the curvature κ as a parameter, includes for $\kappa = 0$ the well known maximally superintegrable Euclidean Smorodinsky–Winternitz (SW) system [**7–10**], the Hamiltonian of which reads

$$\mathcal{H} = \frac{1}{2} \sum_{i=1}^{N} \left(p_i^2 + 2\beta_0 q_i^2 + \frac{2\beta_i}{q_i^2} \right), \tag{1.3}$$

where $\sum_i q_i^2 \equiv r^2$ is the harmonic oscillator potential and each $1/q_i^2$ is a "centrifugal term." Two sets of integrals of motion for $\mathcal{H}$ are given by ($i < j$; $i, j = 1, \ldots, N$):

$$\begin{aligned} I_{0i} &= \tilde{P}_i^2 + 2\beta_0 q_i^2 + 2\beta_i \frac{1}{q_i^2}, &\quad \text{with } \tilde{P}_i = p_i, \\ I_{ij} &= \tilde{J}_{ij}^2 + 2\beta_i \frac{q_j^2}{q_i^2} + 2\beta_j \frac{q_i^2}{q_j^2}, &\quad \text{with } \tilde{J}_{ij} = q_i p_j - q_j p_i. \end{aligned} \tag{1.4}$$

The first set comes from the separability of the Hamiltonian $2\mathcal{H} = \sum_i I_{0i}$, while the second one has a term quadratic in momenta (the square of the components of the angular momentum tensor) plus some additional terms depending on coordinates. The functions $\tilde{P}_i$, $\tilde{J}_{ij}$ close in phase space the commutation relations of the Euclidean algebra $\mathfrak{iso}(N)$ with respect to the canonical Lie–Poisson bracket:

$$\{f, g\} = \sum_{i=1}^{N} \left(\frac{\partial f}{\partial q_i} \frac{\partial g}{\partial p_i} - \frac{\partial g}{\partial q_i} \frac{\partial f}{\partial p_i} \right). \tag{1.5}$$

The structure of this contribution is as follows. The next section contains the details required (related to their maximal groups of isometries) on the three ND classical Riemannian spaces with constant curvature κ. Geodesic motion is obtained in Section 3 starting from the metric on these spaces; this is the kinetic energy term to which possible potentials can be added. The generalization of the SW family to these spaces (any κ) is performed in Section 4. General expressions for the Hamiltonian and its integrals of motion are explicitly given both in terms of the (Weierstrass) coordinates in an ambient linear space $\mathbb{R}^{N+1}$ as well as by means of two sets of intrinsic coordinates, the non-zero curvature versions of the Euclidean Cartesian and polar ones. Moreover, $2N-1$ integrals, including the Hamiltonian, are shown to be functionally independent, thus proving that the SW system on curved spaces is also maximally superintegrable [**4**]. As an example, we apply in Section 5 the general expressions to the $N=4$ case.

2. The sphere $\mathbb{S}^N$, Euclidean $\mathbb{E}^N$, and hyperbolic $\mathbb{H}^N$ spaces

Let $\mathfrak{so}_\kappa(N+1)$ be the Lie algebra of the motion group $\mathrm{SO}_\kappa(N+1)$ on a generic ND real Riemannian space with constant curvature κ, denoted $S^N_{[\kappa]}$. In the basis $\{J_{0i}\equiv P_i, J_{ij}\}$ $(i,j=1,\dots,N;\ i<j)$, the nonvanishing commutation relations of $\mathfrak{so}_\kappa(N+1)$ are given by

$$(2.1)\qquad \begin{aligned} &[J_{ij},J_{ik}]=J_{jk}, && [J_{ij},J_{jk}]=-J_{ik}, && [J_{ik},J_{jk}]=J_{ij},\\ &[J_{ij},P_i]=P_j, && [J_{ij},P_j]=-P_i, && [P_i,P_j]=\kappa J_{ij}, \end{aligned}$$

with $i<j<k$. The Lie algebra $\mathfrak{so}_\kappa(N+1)$ is isomorphic to either $\mathfrak{so}(N+1)$ for $\kappa>0$, $\mathfrak{iso}(N)$ for $\kappa=0$, or $\mathfrak{so}(N,1)$ for $\kappa<0$. Notice that by scaling the generators P_i, any value of κ can always be reduced to either $+1$, 0 or -1. The involutive automorphism defined by

$$(2.2)\qquad \Theta\colon J_{ij}\to J_{ij},\ \ P_i\mapsto -P_i,\quad i,j=1,\dots,N,$$

provides the following Cartan decomposition of $\mathfrak{so}_\kappa(N+1)$:

$$(2.3)\qquad \mathfrak{so}_\kappa(N+1)=\mathfrak{h}\oplus\mathfrak{p},\quad \mathfrak{h}=\langle J_{ij}\rangle=\mathfrak{so}(N),\ \mathfrak{p}=\langle P_i\rangle.$$

The generators invariant under Θ span the subalgebra $\mathfrak{h}$ of the Lie subgroup $H\simeq \mathrm{SO}(N)$, so that $S^N_{[\kappa]}=\mathrm{SO}_\kappa(N+1)/\mathrm{SO}(N)$ is a family, parametrized by κ, of ND symmetric homogeneous rank-one spaces [**13**]. The generators J_{ij} leave an *origin* point $\mathcal{O}$ invariant, thus acting as rotations around $\mathcal{O}$, while the remaining P_i generate translations that move $\mathcal{O}$ along N basic geodesics l_i orthogonal at $\mathcal{O}$. The space $S^N_{[\kappa]}$ comprises the three classical Riemannian spaces:

$\kappa>0$,	Sphere,	$S^N_{[+]}\equiv \mathrm{SO}(N+1)/\mathrm{SO}(N)\equiv\mathbb{S}^N$;
$\kappa=0$,	Euclidean,	$S^N_{[0]}\equiv \mathrm{ISO}(N)/\mathrm{SO}(N)\equiv\mathbb{E}^N$;
$\kappa<0$,	Hyperbolic,	$S^N_{[-]}\equiv \mathrm{SO}(N,1)/\mathrm{SO}(N)\equiv\mathbb{H}^N$.

The curvature κ can also be interpreted as a graded contraction parameter coming from the $\mathbb{Z}_2$-grading of $\mathfrak{so}_\kappa(N+1)$ determined by Θ; the value $\kappa=0$ corresponds to the contraction $\mathbb{S}^N\to\mathbb{E}^N\leftarrow\mathbb{H}^N$ around a point (the origin $\mathcal{O}$) of the spaces.

The Killing–Cartan form of $\mathfrak{so}_\kappa(N+1)$, $g^{\mathrm{KC}}(J_{ij},J_{kl})=\mathrm{Trace}(\mathrm{ad}\,J_{ij}\cdot\mathrm{ad}\,J_{kl})$, is diagonal with the following nonzero elements

$$(2.4)\qquad g^{\mathrm{KC}}(P_i,P_i)=-2(N-1)\kappa,\qquad g^{\mathrm{KC}}(J_{ij},J_{ij})=-2(N-1);$$

hence the restriction of g^{KC} to the subspace $\mathfrak{p}$, can be written as:

$$(2.5)\qquad g^{\mathrm{KC}}\big|_{\mathfrak{p}}=-2(N-1)\kappa\, g\big|_{\mathcal{O}},\qquad g\big|_{\mathcal{O}}(P_i,P_j)=\delta_{ij},$$

where $g\big|_{\mathcal{O}}$ is to be considered as the metric in the tangent space $\mathfrak{p}$ at $\mathcal{O}$, represented by the $N\times N$ matrix $\mathrm{diag}(+,+,\dots,+)$. This metric can be translated to all points of $S^N_{[\kappa]}$ by the group action. Even if $g^{\rm KC}$ vanishes in the Euclidean case, the choice $g \propto g^{\rm KC}/\kappa$ ensures a nondegenerate metric in $S^N_{[\kappa]}$ $\forall\kappa$.

2.1. Vector model and Weierstrass coordinates. The *vector representation* of $\mathfrak{so}_\kappa(N+1)$ is given by $(N+1)\times(N+1)$ real matrices fulfilling (2.1):

$$P_i = -\kappa e_{0i} + e_{i0}, \qquad J_{ij} = -e_{ij} + e_{ji}, \tag{2.6}$$

where e_{ij} is the matrix with a single nonzero entry 1, at row i and column j. Their exponential gives rise to the following one-parametric subgroups of $\mathrm{SO}_\kappa(N+1)$:

$$\begin{aligned} \exp\{xP_i\} &= \sum_{\substack{s=1\\ s\neq i}}^{N} e_{ss} + \mathrm{C}_\kappa(x)e_{00} + \mathrm{C}_\kappa(x)e_{ii} - \kappa\,\mathrm{S}_\kappa(x)e_{0i} + \mathrm{S}_\kappa(x)e_{i0}\\ &= \mathbb{I} + P_i\,\mathrm{S}_\kappa(x) + P_i^2\,\mathrm{V}_\kappa(x),\\ \exp\{xJ_{ij}\} &= \sum_{\substack{s=0\\ s\neq i,j}}^{N} e_{ss} + \cos x e_{ii} + \cos x e_{jj} - \sin x e_{ij} + \sin x e_{ji}\\ &= \mathbb{I} + J_{ij}\sin x + J_{ij}^2(1-\cos x), \end{aligned} \tag{2.7}$$

where $\mathbb{I}$ is the $(N+1)\times(N+1)$ identity matrix. The curvature-dependent cosine $\mathrm{C}_\kappa(x)$ and sine $\mathrm{S}_\kappa(x)$ functions are defined by [**3**]:

$$\begin{aligned} \mathrm{C}_\kappa(x) &= \sum_{l=0}^{\infty}(-\kappa)^l\frac{x^{2l}}{(2l)!} = \begin{cases} \cos\sqrt{\kappa}x, & \kappa>0;\\ 1, & \kappa=0;\\ \cosh\sqrt{-\kappa}x, & \kappa<0. \end{cases}\\ \mathrm{S}_\kappa(x) &= \sum_{l=0}^{\infty}(-\kappa)^l\frac{x^{2l+1}}{(2l+1)!} = \begin{cases} (\sin\sqrt{\kappa}x)/\sqrt{\kappa}, & \kappa>0;\\ x, & \kappa=0;\\ (\sinh\sqrt{-\kappa}x)/\sqrt{-\kappa}, & \kappa<0. \end{cases} \end{aligned} \tag{2.8}$$

From them, we define the "versed sine" or versine $\mathrm{V}_\kappa(x)$ and the tangent $\mathrm{T}_\kappa(x)$:

$$\mathrm{V}_\kappa(x) = \frac{1}{\kappa}(1-\mathrm{C}_\kappa(x)), \qquad \mathrm{T}_\kappa(x) = \frac{\mathrm{S}_\kappa(x)}{\mathrm{C}_\kappa(x)}. \tag{2.9}$$

These κ-trigonometric functions coincide with the usual elliptic (circular) and hyperbolic ones for $\kappa=+1$ and $\kappa=-1$, respectively. The "flat" case with $\kappa=0$ gives the parabolic functions: $\mathrm{C}_0(x)=1$, $\mathrm{S}_0(x)=x$ and $\mathrm{V}_0(x)=x^2/2$. In this sense, they can be interpreted as the *curvature κ-deformed versions* of 1, x, and x^2. All the known trigonometric relations (necessary in the further computations) also extend for these κ-functions [**14, 15**] such as, for instance,

$$\mathrm{C}_\kappa^2(x) + \kappa\,\mathrm{S}_\kappa^2(x) = 1, \qquad \frac{\mathrm{d}}{\mathrm{d}x}\mathrm{C}_\kappa(x) = -\kappa\,\mathrm{S}_\kappa(x), \qquad \frac{\mathrm{d}}{\mathrm{d}x}\mathrm{S}_\kappa(x) = \mathrm{C}_\kappa(x).$$

In the vector representation (2.6), any generator X satisfies the relation

$$X^T\mathbb{I}_\kappa + \mathbb{I}_\kappa X = 0, \quad \mathbb{I}_\kappa = e_{00} + \kappa\sum_{i=1}^{N} e_{ii} = \mathrm{diag}(1,\kappa,\ldots,\kappa), \tag{2.10}$$

(X^T is the transpose matrix of X), so that any element $R\in\mathrm{SO}_\kappa(N+1)$ verifies $R^T\mathbb{I}_\kappa R = \mathbb{I}_\kappa$. In this way, $\mathrm{SO}_\kappa(N+1)$ can be seen as a group of linear isometries of the bilinear form $\mathbb{I}_\kappa$ acting on $\mathbb{R}^{N+1} = (s_0, s_1, \ldots, s_N)$ via matrix multiplication.

The action of $\mathrm{SO}_\kappa(N+1)$ on $\mathbb{R}^{N+1}$ is linear but not transitive, since it conserves the quadratic form $s_0^2 + \kappa\sum_{i=1}^N s_i^2$ provided by $\mathbb{I}_\kappa$, and $H \simeq \mathrm{SO}(N) = \langle J_{ij}\rangle$ (2.3) is the isotopy subgroup of the point $\mathcal{O} = (1, 0, \ldots, 0) \in \mathbb{R}^{N+1}$ which will be taken as the *origin* in the space $S^N_{[\kappa]}$. The action becomes transitive if we restrict to the orbit in $\mathbb{R}^{N+1}$ of the point $\mathcal{O}$, which is contained in the "sphere" Σ:

$$\Sigma \equiv s_0^2 + \kappa\sum_{i=1}^N s_i^2 = 1, \tag{2.11}$$

which reproduces the whole hypersphere, two hyperplanes and a two-sheeted hyperboloid for $\kappa >, =, < 0$, respectively. This orbit is identified with the space $S^N_{[\kappa]}$, and $(s_0, s_1, \ldots, s_N)$, fulfilling the "sphere" constraint (2.11), are called *Weierstrass coordinates*. In terms of these, the metric on $S^N_{[\kappa]}$ comes from the flat ambient metric in $\mathbb{R}^{N+1}$ divided by κ and restricted to Σ:

$$\mathrm{d}\sigma^2 = \frac{1}{\kappa}\left(\mathrm{d}s_0^2 + \kappa\sum_{i=1}^N \mathrm{d}s_i^2\right)\bigg|_\Sigma = \sum_{i=1}^N \mathrm{d}s_i^2 + \kappa\frac{(\sum_{i=1}^N s_i\,\mathrm{d}s_i)^2}{1-\kappa\sum_{i=1}^N \mathrm{d}s_i^2}, \tag{2.12}$$

which reduces to the Euclidean one for $\kappa = 0$. A differential realization of the generators as first-order vector fields in $\mathbb{R}^{N+1}$ with $\partial_i = \partial/\partial s_i$ is (see (2.6)):

$$P_i = \kappa s_i\partial_0 - s_0\partial_i, \qquad J_{ij} = s_j\partial_i - s_i\partial_j. \tag{2.13}$$

2.2. Geodesic coordinate systems. Let us consider in the vector model the origin $\mathcal{O} = (1, 0, \ldots, 0)$ and a generic point $Q \in S^N_{[\kappa]}$ with Weierstrass coordinates $\mathbf{s} = (s_0, s_1, \ldots, s_N) \in \mathbb{R}^{N+1}$. Starting from $\mathcal{O}$, the point Q can be reached in different ways through the action of N one-parametric subgroups (2.7), that is, by means of motions in the space $S^N_{[\kappa]}$. In this way, we introduce coordinates which are intrinsic quantities to the space $S^N_{[\kappa]}$ itself, the associated Weierstrass coordinates automatically fulfilling the condition (2.11). We shall consider two possibilities.

2.2.1. *Geodesic parallel coordinates $a = (a_1, \ldots, a_N)$.* We move the origin $\mathcal{O}$ by using the N translations subgroups as

$$\mathbf{s}(a) = \exp(a_1P_1)\exp(a_2P_2)\cdots\exp(a_NP_N)\mathcal{O}$$

$$\begin{pmatrix} s_0\\ s_1\\ s_2\\ \vdots\\ s_{N-1}\\ s_N\end{pmatrix} = \begin{pmatrix} \mathrm{C}_\kappa(a_1)\,\mathrm{C}_\kappa(a_2)\,\mathrm{C}_\kappa(a_3)\cdots\mathrm{C}_\kappa(a_N)\\ \mathrm{S}_\kappa(a_1)\,\mathrm{C}_\kappa(a_2)\,\mathrm{C}_\kappa(a_3)\cdots\mathrm{C}_\kappa(a_N)\\ \mathrm{S}_\kappa(a_2)\,\mathrm{C}_\kappa(a_3)\cdots\mathrm{C}_\kappa(a_N)\\ \vdots\\ \mathrm{S}_\kappa(a_{N-1})\,\mathrm{C}_\kappa(a_N)\\ \mathrm{S}_\kappa(a_N)\end{pmatrix}. \tag{2.14}$$

Each coordinate a_i is associated to the generator P_i and has dimensions of *length*. If $l_1, l_2, \ldots, l_N$ are the N (oriented) basic geodesics in $S^N_{[\kappa]}$ orthogonal at $\mathcal{O}$, then the first coordinate of a point Q is the geodesic distance a_1 between $\mathcal{O}$ and Q_1 (the orthogonal projection of Q on l_1), measured along l_1. The second coordinate is the distance a_2 between Q_1 and another point Q_2, measured along a geodesic l_2' orthogonal to l_1 through Q_1 and parallel to l_2 (in the sense of parallel transport) and so on, up to reaching Q [**15**]. This is depicted in figure 1 for $\mathbb{S}^2$ and $\mathbb{H}^2$. By introducing (2.14) in (2.12), we obtain the metric:

$$\mathrm{d}\sigma^2 = \sum_{i=1}^{N-1} \Big(\prod_{l=i+1}^{N} \mathrm{C}_\kappa^2(a_l) \Big) \mathrm{d}a_i^2 + \mathrm{d}a_N^2. \tag{2.15}$$

From it we may compute the Levi-Civita connections, whose nonzero Christoffel symbols are ($i = 1, \ldots, N-1$; $i+1 \le j \le N$):

$$\Gamma^i{}_{ij} = -\kappa\, \mathrm{T}_\kappa(a_j), \qquad \Gamma^j{}_{ii} = \kappa\, \mathrm{T}_\kappa(a_j) \prod_{l=i+1}^{j} \mathrm{C}_\kappa^2(a_l). \tag{2.16}$$

2.2.2. *Geodesic polar coordinates* $\theta = (r, \theta_2, \ldots, \theta_N)$. In this case, we move $\mathcal{O}$ by using the $N-1$ rotations $J_{i i+1}$ as well as the first translation $J_{01} \equiv P_1$:

$$\mathbf{s}(\theta) = \exp(\theta_N J_{N-1\,N}) \cdots \exp(\theta_2 J_{12}) \exp(r P_1) \mathcal{O}$$

$$\begin{pmatrix} s_0 \\ s_1 \\ s_2 \\ \vdots \\ s_{N-1} \\ s_N \end{pmatrix} = \begin{pmatrix} \mathrm{C}_\kappa(r) \\ \mathrm{S}_\kappa(r)\cos\theta_2 \\ \mathrm{S}_\kappa(r)\sin\theta_2\cos\theta_3 \\ \vdots \\ \mathrm{S}_\kappa(r)\sin\theta_2\cdots\sin\theta_{N-1}\cos\theta_N \\ \mathrm{S}_\kappa(r)\sin\theta_2\cdots\sin\theta_{N-1}\sin\theta_N \end{pmatrix}. \tag{2.17}$$

The "radial" coordinate r associated to P_1 has dimensions of *length* and is the distance between $\mathcal{O}$ and Q measured along the geodesic l joining both points. The remaining $\theta_2, \ldots, \theta_N$ are ordinary *angles* parameterizing the orientation of l with respect to the reference flag at $\mathcal{O}$ spanned by $\{l_1, l_2, \ldots, l_{i-1}\}$ (see Figure 1 for $N = 2$). On the sphere $\mathbb{S}^N$ of radius R and curvature $\kappa = 1/R^2$, all the usual spherical coordinates are angles, and these differ from the geodesic polar ones only in the first coordinate [**28**], which is commonly taken as the dimensionless quantity $\theta_1 \equiv r/R$. While these conventional spherical coordinates require an explicit contraction to the Euclidean ones, as done for instance in [**17**], our choice

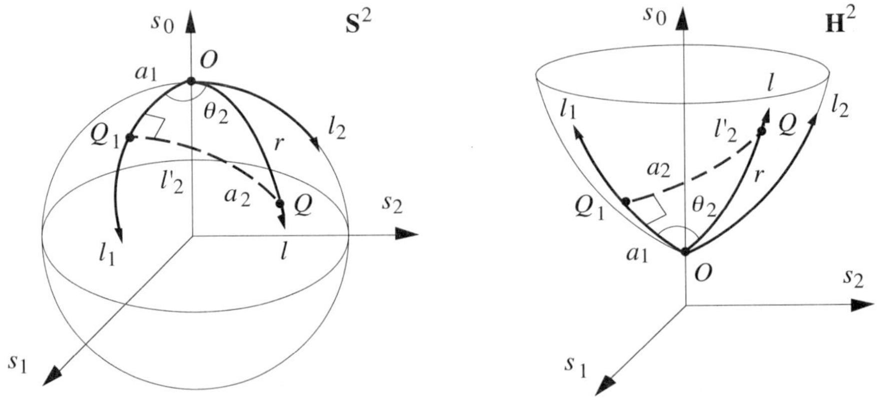

FIGURE 1. Geodesic parallel (a_1, a_2) and polar (r, θ_2) coordinates in $\mathbb{S}^2$ and $\mathbb{H}^2$.

works for *any* value of κ and when $\kappa = 0$ we recover directly, without any limiting procedure, the polar (and Cartesian) coordinates in $\mathbb{E}^N$.

In polar coordinates the metric (2.12) turns out to be

$$\mathrm{d}\sigma^2 = \mathrm{d}r^2 + \mathrm{S}_\kappa^2(r)\left(\mathrm{d}\theta_2^2 + \sum_{i=3}^{N}\left(\prod_{l=2}^{i-1}\sin^2\theta_l\right)\mathrm{d}\theta_i^2\right). \tag{2.18}$$

The components of the Riemann and Ricci tensors can be computed in either geodesic coordinate system; the scalar and the sectional curvatures of $S^N_{[\kappa]}$ are $R = N(N-1)\kappa$ and $K = \kappa$, both constant, respectively. When $\kappa = 0$, the expressions (2.15) and (2.18) give the Euclidean metric

$$\mathrm{d}\sigma^2 = \sum_{i=1}^{N}\mathrm{d}a_i^2 = \mathrm{d}r^2 + r^2\left(\mathrm{d}\theta_2^2 + \sum_{i=3}^{N}\left(\prod_{l=2}^{i-1}\sin^2\theta_l\right)\mathrm{d}\theta_i^2\right), \tag{2.19}$$

and all the Christoffel symbols $\Gamma^i_{\ jk}$ (2.16) vanish in parallel coordinates.

3. Geodesic motion on $S^N_{[\kappa]}$ and phase space realization of $\mathfrak{so}_\kappa(N+1)$

If we now adopt a dynamical viewpoint, the expressions of the metric in $S^N_{[\kappa]}$ (2.15) and (2.18) provide the kinetic energy $\mathcal{T}$ of a particle in terms of the velocities, generically denoted as $\dot{q}$ (either $\dot{a}$ or $\dot{\theta}$) in the coordinate systems (2.14) and (2.17):

$$2\mathcal{T} = \sum_{i=1}^{N-1}\left(\prod_{l=i+1}^{N}\mathrm{C}_\kappa^2(a_l)\right)\dot{a}_i^2 + \dot{a}_N^2 = \dot{r}^2 + \mathrm{S}_\kappa^2(r)\left(\dot{\theta}_2^2 + \sum_{i=3}^{N}\left(\prod_{l=2}^{i-1}\sin^2\theta_l\right)\dot{\theta}_i^2\right), \tag{3.1}$$

which is the Lagrangian $\mathcal{L} \equiv \mathcal{T}$ of the geodesic motion on $S^N_{[\kappa]}$. The canonical momenta, $p = \partial\mathcal{L}/\partial\dot{q}$, (denoted p, π for $q = a, \theta$) are given by:

$$\begin{gathered} p_i = \left(\prod_{l=i+1}^{N}\mathrm{C}_\kappa^2(a_l)\right)\dot{a}_i, \quad i = 1, \ldots, N-1; \qquad p_N = \dot{a}_N, \\ \pi_1 = \dot{r}, \qquad \pi_2 = \mathrm{S}_\kappa^2(r)\dot{\theta}_2, \qquad \pi_j = \mathrm{S}_\kappa^2(r)\left(\prod_{l=2}^{j-1}\sin^2\theta_l\right)\dot{\theta}_j, \ j = 3, \ldots, N. \end{gathered} \tag{3.2}$$

By introducing these momenta in (3.1), we obtain the free Hamiltonian $\mathcal{H} \equiv \mathcal{T}$ in the phase space of motions in $S^N_{[\kappa]}$ expressed in either "parallel" (a, p) or "polar" (θ, π) canonical coordinates and momenta:

$$2\mathcal{T} = \sum_{i=1}^{N-1}\frac{p_i^2}{\prod_{l=i+1}^{N}\mathrm{C}_\kappa^2(a_l)} + p_N^2 = \pi_1^2 + \frac{1}{\mathrm{S}_\kappa^2(r)}\left(\pi_2^2 + \sum_{i=3}^{N}\frac{\pi_i^2}{\prod_{l=2}^{i-1}\sin^2\theta_l}\right). \tag{3.3}$$

On the other hand, the Lie generators are expressed in terms of Weierstrass coordinates as:

$$\tilde{P}_i(s(q), \dot{s}(q,p)) = s_0\dot{s}_i - s_i\dot{s}_0, \qquad \tilde{J}_{ij}(s(q), \dot{s}(q,p)) = s_i\dot{s}_j - s_j\dot{s}_i, \tag{3.4}$$

and thus we get an N-particle realization of $\mathfrak{so}_\kappa(N+1)$ in the phase space simply by rewriting everything either in terms of (a, p) or (θ, π). The time derivatives of $\mathbf{s}$

are obtained from either (2.14) in terms of parallel coordinates and velocities $(a, \dot{a})$ $(i = 1, \ldots, N-1)$:

$$\begin{aligned} &\dot{s}_0 = -\kappa \prod_{m=1}^{N} \mathrm{C}_\kappa(a_m) \sum_{l=1}^{N} \mathrm{T}_\kappa(a_l)\dot{a}_l, \\ &\dot{s}_i = \prod_{m=i}^{N} \mathrm{C}_\kappa(a_m)\Big(\dot{a}_i - \kappa\, \mathrm{T}_\kappa(a_i) \sum_{l=i+1}^{N} \mathrm{T}_\kappa(a_l)\dot{a}_l\Big), \qquad \dot{s}_N = \mathrm{C}_\kappa(a_N)\dot{a}_N, \end{aligned} \tag{3.5}$$

or from (2.17) in terms of polar coordinates and velocities $(\theta, \dot{\theta})$ $(j = 2, \ldots, N-1)$:

$$\begin{aligned} &\dot{s}_0 = -\kappa\, \mathrm{S}_\kappa(r)\dot{r}, \qquad \dot{s}_1 = \mathrm{S}_\kappa(r)\sin\theta_2\Big(\frac{\dot{r}}{\mathrm{T}_\kappa(r)\tan\theta_2} - \dot{\theta}_2\Big), \\ &\dot{s}_j = \mathrm{S}_\kappa(r)\prod_{m=2}^{j+1}\sin\theta_m\Big(\frac{\dot{r}}{\mathrm{T}_\kappa(r)\tan\theta_{j+1}} + \sum_{l=2}^{j}\frac{\dot{\theta}_l}{\tan\theta_l\tan\theta_{j+1}} - \dot{\theta}_{j+1}\Big), \\ &\dot{s}_N = \mathrm{S}_\kappa(r)\prod_{m=2}^{N}\sin\theta_m\Big(\frac{\dot{r}}{\mathrm{T}_\kappa(r)} + \sum_{l=2}^{N}\frac{\dot{\theta}_l}{\tan\theta_l}\Big). \end{aligned} \tag{3.6}$$

We now introduce the parameterizations (2.14), (2.17), velocities (3.5), (3.6) as well as the momenta (3.2) in (3.4), obtaining a phase space realization of the generators of $\mathfrak{so}_\kappa(N+1)$ given in "parallel canonical" coordinates by $(i, j = 1, \ldots, N)$:

$$\begin{aligned} &\tilde{P}_i = \prod_{k=1}^{i} \mathrm{C}_\kappa(a_k)\, \mathrm{C}_\kappa(a_i)p_i + \kappa\, \mathrm{S}_\kappa(a_i)\sum_{s=1}^{i}\mathrm{S}_\kappa(a_s)\frac{\prod_{m=1}^{s}\mathrm{C}_\kappa(a_m)}{\prod_{l=s}^{i}\mathrm{C}_\kappa(a_l)}p_s, \\ &\tilde{J}_{ij} = \mathrm{S}_\kappa(a_i)\,\mathrm{C}_\kappa(a_j)\prod_{s=i+1}^{j}\mathrm{C}_\kappa(a_s)p_j - \frac{\mathrm{C}_\kappa(a_i)\,\mathrm{S}_\kappa(a_j)}{\prod_{k=i+1}^{j}\mathrm{C}_\kappa(a_k)}p_i \\ &\qquad + \kappa\, \mathrm{S}_\kappa(a_i)\,\mathrm{S}_\kappa(a_j)\sum_{s=i+1}^{j}\mathrm{S}_\kappa(a_s)\frac{\prod_{m=i+1}^{s}\mathrm{C}_\kappa(a_m)}{\prod_{l=s}^{j}\mathrm{C}_\kappa(a_l)}p_s, \end{aligned} \tag{3.7}$$

and in geodesic polar coordinates and momenta (θ, π) by $(i, j = 1, \ldots, N-1)$:

$$\begin{aligned} &\tilde{P}_i = \frac{\prod_{k=2}^{i+1}\sin\theta_k}{\tan\theta_{i+1}}\pi_1 + \sum_{s=2}^{i+1}\frac{\prod_{m=s}^{i+1}\sin\theta_m\cos\theta_s\pi_s}{\mathrm{T}_\kappa(r)\tan\theta_{i+1}\prod_{l=2}^{s}\sin\theta_l} - \frac{\pi_{i+1}}{\mathrm{T}_\kappa(r)\prod_{l=2}^{i+1}\sin\theta_l}, \\ &\tilde{P}_N = \prod_{k=2}^{N}\sin\theta_k\pi_1 + \sum_{s=2}^{N}\frac{\prod_{m=s}^{N}\sin\theta_m\cos\theta_s}{\mathrm{T}_\kappa(r)\prod_{l=2}^{s}\sin\theta_l}\pi_s, \\ &\tilde{J}_{ij} = \sin\theta_{i+1}\cos\theta_{j+1}\prod_{k=i+1}^{j}\sin\theta_k\pi_{i+1} - \frac{\cos\theta_{i+1}\sin\theta_{j+1}}{\prod_{l=i+1}^{j}\sin\theta_l}\pi_{j+1} \\ &\qquad + \cos\theta_{i+1}\cos\theta_{j+1}\sum_{s=i+1}^{j}\frac{\prod_{m=s}^{j}\sin\theta_m\cos\theta_s}{\prod_{l=i+1}^{s}\sin\theta_l}\pi_s, \\ &\tilde{J}_{iN} = \sin\theta_{i+1}\prod_{k=i+1}^{N}\sin\theta_k\pi_{i+1} + \cos\theta_{i+1}\sum_{s=i+1}^{N}\frac{\prod_{m=s}^{N}\sin\theta_m\cos\theta_s}{\prod_{l=i+1}^{s}\sin\theta_l}\pi_s. \end{aligned} \tag{3.8}$$

Then the following results follow [**4**].

PROPOSITION 3.1. *Both sets of generators* (3.7) *and* (3.8) *fulfill the commutation rules* (2.1) *of* $\mathfrak{so}_\kappa(N+1)$ *with respect to the canonical Poisson bracket* (1.5).

PROPOSITION 3.2. *Any generator* (3.7) *and* (3.8) *Poisson-commutes with the kinetic energy* $\mathcal{T}$ (3.3).

The second statement is straightforward as $\mathcal{T}$ is related with a phase space realization of the second-order Casimir $\mathcal{C}$ of $so_\kappa(N+1)$ through

$$2\mathcal{T} = \tilde{\mathcal{C}} = \sum_{i=1}^{N} \tilde{P}_i^2 + \kappa \sum_{i,j=1}^{N} \tilde{J}_{ij}^2. \tag{3.9}$$

In fact, the geodesic motion is maximally superintegrable and its integrals of motion come from any function of a phase space realization of the Lie generators.

4. Smorodinsky–Winternitz system on curved spaces

The problem now is to find which potentials $\mathcal{U}(q)$ can be added to $\mathcal{T}$ in such a manner that the new Hamiltonian $\mathcal{H} = \mathcal{T} + \mathcal{U}$ preserves the maximal superintegrability. This requires to add "some" terms to "some" functions of the generators in order to ensure their involutivity with respect to $\mathcal{H}$.

4.1. Potential. As far as the 2D potential (1.1) is concerned, the generalization to the space $S^N_{[\kappa]}$ is straightforward and reads

$$\mathcal{U}(s(q)) = \beta_0 \frac{\sum_{l=1}^{N} s_l^2}{s_0^2} + \sum_{i=1}^{N} \frac{\beta_i}{s_i^2}, \tag{4.1}$$

which in geodesic parallel and polar coordinates turns out to be

$$\begin{aligned} \mathcal{U} &= \beta_0 \sum_{i=1}^{N} \frac{\mathrm{S}_\kappa^2(a_i)}{\prod_{l=1}^{i} \mathrm{C}_\kappa^2(a_l)} + \sum_{i=1}^{N-1} \frac{\beta_i}{\mathrm{S}_\kappa^2(a_i) \prod_{l=i+1}^{N} \mathrm{C}_\kappa^2(a_l)} + \frac{\beta_N}{\mathrm{S}_\kappa^2(a_N)} \\ &= \beta_0 \mathrm{T}_\kappa^2(r) + \frac{1}{\mathrm{S}_\kappa^2(r)} \left(\frac{\beta_1}{\cos^2\theta_2} + \sum_{i=2}^{N-1} \frac{\beta_i}{\cos^2\theta_{i+1} \prod_{l=2}^{i} \sin^2\theta_l} + \frac{\beta_N}{\prod_{l=2}^{N} \sin^2\theta_l} \right). \end{aligned} \tag{4.2}$$

This potential, which coincides with (1.1) for $\kappa = 1$, has been interpreted on $\mathbb{S}^2$ as three Higgs spherical oscillators, whose "centers" are placed at the three vertices of a sphere's octant [**29**] (recall that the Higgs oscillator has two antipodal "centers"); this interpretation can straightforwardly be extended to the sphere $\mathbb{S}^N$ with $\kappa = 1/R^2 > 0$. Let O_i be the points placed along the basic geodesics l_i and a quadrant apart from $\mathcal{O}$ (each pair taken from $\mathcal{O}$, O_i are mutually separated a quadrant distance $\pi/(2\sqrt{\kappa}) = R\pi/2$ on $\mathbb{S}^N$). If we set $\kappa = 1$ and denote r, r_i the distances between any generic point Q and $\mathcal{O}$, O_i along the joining geodesics, we find $s_0 = \cos r$, $s_i = \cos r_i$ $(i = 1, \ldots, N)$, and (4.2) can be rewritten as

$$\mathcal{U} = \beta_0 \tan^2 r + \sum_{i=1}^{N} \frac{\beta_i}{\cos^2 r_i} = \beta_0 \tan^2 r + \sum_{i=1}^{N} \beta_i \tan^2 r_i + \sum_{i=1}^{N} \beta_i, \tag{4.3}$$

which can clearly be recognized under this form as the joint potential due to a set of $N+1$ spherical oscillators whose centers are at the $N+1$ points $\mathcal{O}$, O_i. Alternatively, if $x_i = \pi/2 - r_i$, then each "β_i" term in (4.3) can be described, as usual, as the spherical "centrifugal" barriers $\beta_i/\sin^2 x_i$. Under the contraction $\kappa = 0$, $\mathbb{S}^N \to \mathbb{E}^N$,

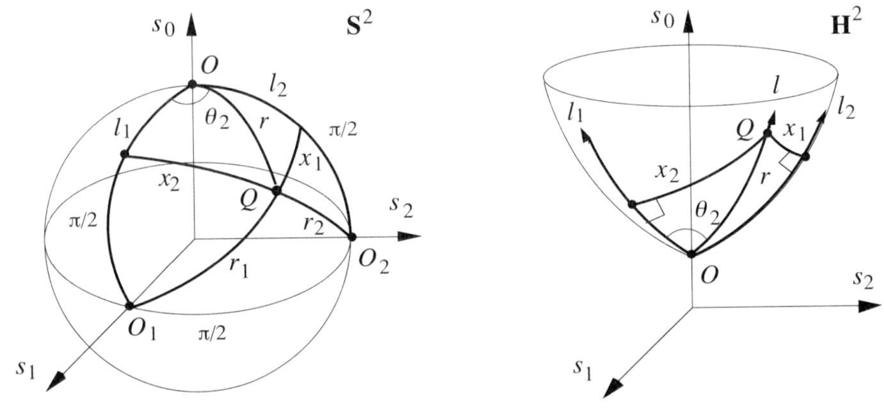

FIGURE 2. Distances involved in the SW potential on $\mathbb{S}^2$ and $\mathbb{H}^2$.

the Higgs-term gives rise to the "flat" harmonic oscillator $\beta_0 r^2 = \beta_0 \sum_i a_i^2$, while the N remaining oscillators (whose centers would be now "at infinity") leave the Euclidean "centrifugal" barriers $\beta_i/x_i^2 \equiv \beta_i/a_i^2$ as their imprints.

On the hyperbolic space $\mathbb{H}^N$ the potential (4.2) can similarly be interpreted. Let x_i be the distance between the generic point Q and the totally geodesic codimension one submanifold through $\mathcal{O}$ orthogonal to the geodesic l_i. When $\kappa = -1$ we find that $s_0 = \cosh r$, $s_i = \sinh x_i$ $(i = 1, \ldots, N)$, so the potential can be written as:

$$\mathcal{U} = \beta_0 \tanh^2 r + \sum_{i=1}^{N} \frac{\beta_i}{\sinh^2 x_i} = \beta_0 \tanh^2 r + \sum_{i=1}^{N} \frac{\beta_i}{\tanh^2 x_i} - \sum_{i=1}^{N} \beta_i. \tag{4.4}$$

The first term is an "actual" hyperbolic oscillator with center at $\mathcal{O}$ (whose potential is bounded in the whole hyperbolic space), while each of the N remaining terms can be interpreted, as done conventionally, as some kind of hyperbolic "centrifugal" potentials, but also as "ideal" hyperbolic oscillators, whose centers would be beyond infinity, that is, in the exterior region of the hyperbolic space.

4.2. Integrals of motion. By taking into account the results given in **[27]** for the integrals of motion of (1.1), let us consider the following functions, with quadratic dependence on momenta, given in the ambient space $\mathbb{R}^{N+1}$ by $(i < j$, $i, j = 0, 1, \ldots, N)$:

$$I_{ij} = (s_i \dot{s}_j - s_j \dot{s}_i)^2 + 2\beta_i \frac{s_j^2}{s_i^2} + 2\beta_j \frac{s_i^2}{s_j^2}. \tag{4.5}$$

Therefore we have $N(N+1)/2$ phase space functions coming from the Lie generators, which in the geodesic parallel canonical coordinates (3.7) turn out to be

$$\begin{aligned} I_{0i} &= \tilde{P}_i^2 + 2\beta_0 \frac{\mathrm{S}_\kappa^2(a_i)}{\prod_{l=1}^{i} \mathrm{C}_\kappa^2(a_l)} + 2\beta_i \frac{\prod_{l=1}^{i} \mathrm{C}_\kappa^2(a_l)}{\mathrm{S}_\kappa^2(a_i)}, \\ I_{ij} &= \tilde{J}_{ij}^2 + 2\beta_i \frac{\mathrm{S}_\kappa^2(a_j)}{\mathrm{S}_\kappa^2(a_i) \prod_{l=i+1}^{j} \mathrm{C}_\kappa^2(a_l)} + 2\beta_j \frac{\mathrm{S}_\kappa^2(a_i) \prod_{l=i+1}^{j} \mathrm{C}_\kappa^2(a_l)}{\mathrm{S}_\kappa^2(a_j)}. \end{aligned} \tag{4.6}$$

The same quantities read in the geodesic polar canonical coordinates (3.8):

$$
\begin{aligned}
I_{0i} &= \tilde{P}_i^2 + 2\beta_0 \frac{\mathrm{T}_\kappa^2(r)\prod_{l=2}^{i+1}\sin^2\theta_l}{\tan^2\theta_{i+1}} + 2\beta_i \frac{\tan^2\theta_{i+1}}{\mathrm{T}_\kappa^2(r)\prod_{l=2}^{i+1}\sin^2\theta_l},\\
I_{0N} &= \tilde{P}_N^2 + 2\beta_0 \mathrm{T}_\kappa^2(r)\prod_{l=2}^{N}\sin^2\theta_l + 2\beta_N \frac{1}{\mathrm{T}_\kappa^2(r)\prod_{l=2}^{N}\sin^2\theta_l},\\
I_{ij} &= \tilde{J}_{ij}^2 + 2\beta_i \frac{\cos^2\theta_{j+1}\prod_{l=i+1}^{j}\sin^2\theta_l}{\cos^2\theta_{i+1}} + \frac{2\beta_j\cos^2\theta_{i+1}}{\cos^2\theta_{j+1}\prod_{l=i+1}^{j}\sin^2\theta_l},\\
I_{iN} &= \tilde{J}_{iN}^2 + 2\beta_i \frac{\prod_{l=i+1}^{N}\sin^2\theta_l}{\cos^2\theta_{i+1}} + 2\beta_N \frac{\cos^2\theta_{i+1}}{\prod_{l=i+1}^{N}\sin^2\theta_l}.
\end{aligned}
\tag{4.7}
$$

A first property for these quantities is given by:

PROPOSITION 4.1. *The $N(N+1)/2$ functions given by either* (4.6) *or* (4.7) *fulfill $\{I_{ij}, I_{lm}\} = 0$ whenever all four indices $i < j$; $l < m$, are different.*

From now on we consider the Hamiltonian $\mathcal{H} = \mathcal{T} + \mathcal{U}$ with $\mathcal{T}$ and $\mathcal{U}$ given in (3.3) and (4.2). Then we find that:

PROPOSITION 4.2. *The $N(N+1)/2$ functions either* (4.6) *or* (4.7) *are integrals of the motion for the Hamiltonian, that is, $\{I_{ij}, \mathcal{H}\} = 0\ \forall i, j$.*

We remark that the property analogous to (3.9) is now given by

$$
2\mathcal{H} = \sum_{i=1}^{N} I_{0i} + \kappa \sum_{i,j=1}^{N} I_{ij} + 2\kappa \sum_{i=1}^{N} \beta_i. \tag{4.8}
$$

Notice also that when $\kappa = 0$, the Hamiltonian expressed in parallel coordinates and the integrals (4.6) with generators (3.7) directly reduce to the flat SW system characterized by (1.3) and (1.4).

4.3. Maximal superintegrability. The last step is firstly to identify, within the initial set of $N(N+1)/2$ integrals of the motion together with the Hamiltonian, N which are functionally independent and in involution in order to prove the complete integrability of $\mathcal{H}$, and secondly to find out how many are functionally independent thus analyzing its superintegrability.

Let us choose two subsets of $N-1$ integrals $Q^{(l)}$ and $Q_{(l)}$ ($l = 2, \ldots, N$) obtained starting from the integrals I_{ij} associated to the rotation generators as:

$$
Q^{(l)} = \sum_{i,j=1}^{l} I_{ij}, \qquad Q_{(l)} = \sum_{i,j=N-l+1}^{N} I_{ij}, \tag{4.9}
$$

which share the element $Q^{(N)} \equiv Q_{(N)}$. The generators that determine the quadratic terms in the momenta in the first set $Q^{(l)}$ are associated to a chain of orthogonal subalgebras within $\mathfrak{h} = \mathfrak{so}(N) = \langle J_{ij} \rangle$ (2.3) starting "upwards" from $\mathfrak{so}(2) = \langle J_{12} \rangle$; likewise for $Q_{(l)}$ but starting "backwards" from $\mathfrak{so}(2) = \langle J_{N-1\,N} \rangle$:

$$
\begin{array}{ll}
Q^{(2)} \subset \cdots \subset Q^{(l)} \subset \cdots \subset Q^{(N)} & \qquad Q_{(N)} \supset \cdots \supset Q_{(l)} \supset \cdots \supset Q_{(2)}\\
\mathfrak{so}(2) \subset \cdots \subset \mathfrak{so}(l) \subset \cdots \subset \mathfrak{so}(N) & \qquad \mathfrak{so}(N) \supset \cdots \supset \mathfrak{so}(l) \supset \cdots \supset \mathfrak{so}(2)
\end{array}
$$

We remark that the SW Hamiltonian on $\mathbb{E}^N$, $\mathcal{H}\big|_{\kappa=0}$, can also be constructed through a coalgebra approach [**5**] by means of N copies of $\mathfrak{sp}(2,\mathbb{R}) \simeq \mathfrak{sl}(2,\mathbb{R})$ [**1**]. In this flat case, each integral of motion $Q^{(l)}\big|_{\kappa=0}$ in Cartesian coordinates *coincides* with an N-particle phase space realization of the kth order left-coproduct of the Casimir of $\mathfrak{sl}(2,\mathbb{R})$; similarly for $Q_{(l)}\big|_{\kappa=0}$ using the right-coproduct (these are called left- and right-integrals, respectively). Such coalgebra method shows that the N functions $\{Q^{(l)}\big|_{\kappa=0}, \mathcal{H}\big|_{\kappa=0}\}$ are mutually in involution (the right-set as well), and the $2N-2$ functions $\{Q^{(l)}\big|_{\kappa=0}, Q_{(l)}\big|_{\kappa=0}, \mathcal{H}\big|_{\kappa=0}\}$ are functionally independent [**2**]. For the general case with arbitrary curvature κ, we have the following result.

THEOREM 4.3. (i) *The N functions $\{Q^{(2)}, Q^{(3)}, \dots, Q^{(N)}, \mathcal{H}\}$ are mutually in involution. The same property holds for the second set $\{Q_{(N)}, \dots, Q_{(3)}, Q_{(2)}, \mathcal{H}\}$.*

(ii) *The $2N-1$ functions $\{Q^{(2)}, Q^{(3)}, \dots, Q^{(N)} \equiv Q_{(N)}, \dots, Q_{(3)}, Q_{(2)}, I_{0i}, \mathcal{H}\}$, where i is fixed, are functionally independent, thus $\mathcal{H}$ is a maximally superintegrable Hamiltonian.*

Notice that if $\kappa \neq 0$, each $Q^{(l)}$ (or $Q_{(l)}$) can be seen as a smooth function on the curvature: $Q^{(l)} = Q^{(l)}\big|_{\kappa=0} + \kappa Q_1^{(l)} + o(\kappa^2)$. The same holds if one introduces a quantum deformation of the Euclidean SW system in terms of a deformation parameter z [**1, 2**]: superintegrability is preserved and the deformed integrals $Q_z^{(l)}$ (or $Q_{(l)}^z$) can be written as power series of z: $Q_z^{(l)} = Q_z^{(l)}\big|_{z=0} + zQ_{z,1}^{(l)} + o(z^2)$. Then, $Q^{(l)}\big|_{\kappa=0} \equiv Q_z^{(l)}\big|_{z=0}$ are the integrals of the SW system on $\mathbb{E}^N$. This suggests some kind of duality between quantum deformations (z) and curvature (κ) [**3**].

5. The $N = 4$ case

In what follows we illustrate the general expressions obtained in the previous sections by applying them to the case $N = 4$ in terms of geodesic parallel coordinates and momenta. The SW Hamiltonian on $\mathbb{S}^4$, $\mathbb{E}^4$ and $\mathbb{H}^4$ reads

$$
\begin{aligned}
(5.1)\quad \mathcal{H} = \frac{1}{2}\biggl(&\frac{p_1^2}{\mathrm{C}_\kappa^2(a_2)\,\mathrm{C}_\kappa^2(a_3)\,\mathrm{C}_\kappa^2(a_4)} + \frac{p_2^2}{\mathrm{C}_\kappa^2(a_3)\,\mathrm{C}_\kappa^2(a_4)} + \frac{p_3^2}{\mathrm{C}_\kappa^2(a_4)} + p_4^2\biggr)\\
&+ \beta_0\biggl(\mathrm{T}_\kappa^2(a_1) + \frac{\mathrm{T}_\kappa^2(a_2)}{\mathrm{C}_\kappa^2(a_1)} + \frac{\mathrm{T}_\kappa^2(a_3)}{\mathrm{C}_\kappa^2(a_1)\,\mathrm{C}_\kappa^2(a_2)} + \frac{\mathrm{T}_\kappa^2(a_4)}{\mathrm{C}_\kappa^2(a_1)\,\mathrm{C}_\kappa^2(a_2)\,\mathrm{C}_\kappa^2(a_3)}\biggr)\\
&+ \frac{\beta_1}{\mathrm{S}_\kappa^2(a_1)\,\mathrm{C}_\kappa^2(a_2)\,\mathrm{C}_\kappa^2(a_3)\,\mathrm{C}_\kappa^2(a_4)} + \frac{\beta_2}{\mathrm{S}_\kappa^2(a_2)\,\mathrm{C}_\kappa^2(a_3)\,\mathrm{C}_\kappa^2(a_4)}\\
&\qquad + \frac{\beta_3}{\mathrm{S}_\kappa^2(a_3)\,\mathrm{C}_\kappa^2(a_4)} + \frac{\beta_4}{\mathrm{S}_\kappa^2(a_4)}.
\end{aligned}
$$

The phase space realization (3.7) for $\mathfrak{so}_\kappa(5)$ is given by four translations

$$
\begin{aligned}
&\tilde{P}_1 = p_1, \qquad \tilde{P}_2 = \mathrm{C}_\kappa(a_1)p_2 + \kappa\,\mathrm{S}_\kappa(a_1)\,\mathrm{T}_\kappa(a_2)p_1,\\
&\tilde{P}_3 = \mathrm{C}_\kappa(a_1)\,\mathrm{C}_\kappa(a_2)p_3 + \kappa\,\mathrm{T}_\kappa(a_3)\biggl(\frac{\mathrm{S}_\kappa(a_1)}{\mathrm{C}_\kappa(a_2)}p_1 + \mathrm{C}_\kappa(a_1)\,\mathrm{S}_\kappa(a_2)p_2\biggr),\\
(5.2)\quad &\tilde{P}_4 = \mathrm{C}_\kappa(a_1)\,\mathrm{C}_\kappa(a_2)\,\mathrm{C}_\kappa(a_3)p_4\\
&\qquad + \kappa\,\mathrm{T}_\kappa(a_4)\biggl(\frac{\mathrm{S}_\kappa(a_1)}{\mathrm{C}_\kappa(a_2)\,\mathrm{C}_\kappa(a_3)}p_1 + \frac{\mathrm{C}_\kappa(a_1)\,\mathrm{S}_\kappa(a_2)}{\mathrm{C}_\kappa(a_3)}p_2\\
&\qquad\qquad + \mathrm{C}_\kappa(a_1)\,\mathrm{C}_\kappa(a_2)\,\mathrm{S}_\kappa(a_3)p_3\biggr),
\end{aligned}
$$

together with six rotation generators:

$$
\begin{aligned}
\tilde J_{12} &= \mathrm{S}_\kappa(a_1)p_2 - \mathrm{C}_\kappa(a_1)\,\mathrm{T}_\kappa(a_2)p_1,\\
\tilde J_{13} &= \mathrm{S}_\kappa(a_1)\,\mathrm{C}_\kappa(a_2)p_3 - \frac{\mathrm{C}_\kappa(a_1)}{\mathrm{C}_\kappa(a_2)}\,\mathrm{T}_\kappa(a_3)p_1 + \kappa\,\mathrm{S}_\kappa(a_1)\,\mathrm{S}_\kappa(a_2)\,\mathrm{T}_\kappa(a_3)p_2,\\
\tilde J_{14} &= \mathrm{S}_\kappa(a_1)\,\mathrm{C}_\kappa(a_2)\,\mathrm{C}_\kappa(a_3)p_4 - \frac{\mathrm{C}_\kappa(a_1)}{\mathrm{C}_\kappa(a_2)\,\mathrm{C}_\kappa(a_3)}\,\mathrm{T}_\kappa(a_4)p_1\\
&\qquad + \kappa\,\mathrm{S}_\kappa(a_1)\,\mathrm{T}_\kappa(a_4)\left(\frac{\mathrm{S}_\kappa(a_2)}{\mathrm{C}_\kappa(a_3)}p_2 + \mathrm{C}_\kappa(a_2)\,\mathrm{S}_\kappa(a_3)p_3\right),\\
\tilde J_{23} &= \mathrm{S}_\kappa(a_2)p_3 - \mathrm{C}_\kappa(a_2)\,\mathrm{T}_\kappa(a_3)p_2,\\
\tilde J_{24} &= \mathrm{S}_\kappa(a_2)\,\mathrm{C}_\kappa(a_3)p_4 - \frac{\mathrm{C}_\kappa(a_2)}{\mathrm{C}_\kappa(a_3)}\,\mathrm{T}_\kappa(a_4)p_2 + \kappa\,\mathrm{S}_\kappa(a_2)\,\mathrm{S}_\kappa(a_3)\,\mathrm{T}_\kappa(a_4)p_3,\\
\tilde J_{34} &= \mathrm{S}_\kappa(a_3)p_4 - \mathrm{C}_\kappa(a_3)\,\mathrm{T}_\kappa(a_4)p_3.
\end{aligned}
\tag{5.3}
$$

From these generators we construct ten integrals of motion (4.6):

$$
\begin{array}{cccc}
\tilde P_1 & \tilde P_2 & \tilde P_3 & \tilde P_4\\
\hline
 & \tilde J_{12} & \tilde J_{13} & \tilde J_{14}\\
 & & \tilde J_{23} & \tilde J_{24}\\
 & & & \tilde J_{34}
\end{array}
\quad\Longrightarrow\quad
\begin{array}{cccc}
I_{01} & I_{02} & I_{03} & I_{04}\\
\hline
 & I_{12} & I_{13} & I_{14}\\
 & & I_{23} & I_{24}\\
 & & & I_{34}
\end{array}
$$

Hence from (5.2) we obtain four "translation-like" integrals given by

$$
\begin{aligned}
I_{01} &= \tilde P_1^2 + 2\beta_0\,\mathrm{T}_\kappa^2(a_1) + 2\beta_1\frac{1}{\mathrm{T}_\kappa^2(a_1)},\\
I_{02} &= \tilde P_2^2 + 2\beta_0\frac{\mathrm{T}_\kappa^2(a_2)}{\mathrm{C}_\kappa^2(a_1)} + 2\beta_2\frac{\mathrm{C}_\kappa^2(a_1)}{\mathrm{T}_\kappa^2(a_2)},\\
I_{03} &= \tilde P_3^2 + 2\beta_0\frac{\mathrm{T}_\kappa^2(a_3)}{\mathrm{C}_\kappa^2(a_1)\,\mathrm{C}_\kappa^2(a_2)} + 2\beta_3\frac{\mathrm{C}_\kappa^2(a_1)\,\mathrm{C}_\kappa^2(a_2)}{\mathrm{T}_\kappa^2(a_3)},\\
I_{04} &= \tilde P_4^2 + 2\beta_0\frac{\mathrm{T}_\kappa^2(a_4)}{\mathrm{C}_\kappa^2(a_1)\,\mathrm{C}_\kappa^2(a_2)\,\mathrm{C}_\kappa^2(a_3)} + 2\beta_4\frac{\mathrm{C}_\kappa^2(a_1)\,\mathrm{C}_\kappa^2(a_2)\,\mathrm{C}_\kappa^2(a_3)}{\mathrm{T}_\kappa^2(a_4)},
\end{aligned}
\tag{5.4}
$$

together with six "rotation-like" ones coming from (5.3):

$$
\begin{aligned}
I_{12} &= \tilde J_{12}^2 + 2\beta_1\frac{\mathrm{T}_\kappa^2(a_2)}{\mathrm{S}_\kappa^2(a_1)} + 2\beta_2\frac{\mathrm{S}_\kappa^2(a_1)}{\mathrm{T}_\kappa^2(a_2)},\\
I_{13} &= \tilde J_{13}^2 + 2\beta_1\frac{\mathrm{T}_\kappa^2(a_3)}{\mathrm{S}_\kappa^2(a_1)\,\mathrm{C}_\kappa^2(a_2)} + 2\beta_3\frac{\mathrm{S}_\kappa^2(a_1)\,\mathrm{C}_\kappa^2(a_2)}{\mathrm{T}_\kappa^2(a_3)},\\
I_{14} &= \tilde J_{14}^2 + 2\beta_1\frac{\mathrm{T}_\kappa^2(a_4)}{\mathrm{S}_\kappa^2(a_1)\,\mathrm{C}_\kappa^2(a_2)\,\mathrm{C}_\kappa^2(a_3)} + 2\beta_4\frac{\mathrm{S}_\kappa^2(a_1)\,\mathrm{C}_\kappa^2(a_2)\,\mathrm{C}_\kappa^2(a_3)}{\mathrm{T}_\kappa^2(a_4)},\\
I_{23} &= \tilde J_{23}^2 + 2\beta_2\frac{\mathrm{T}_\kappa^2(a_3)}{\mathrm{S}_\kappa^2(a_2)} + 2\beta_3\frac{\mathrm{S}_\kappa^2(a_2)}{\mathrm{T}_\kappa^2(a_3)},\\
I_{24} &= \tilde J_{24}^2 + 2\beta_2\frac{\mathrm{T}_\kappa^2(a_4)}{\mathrm{S}_\kappa^2(a_2)\,\mathrm{C}_\kappa^2(a_3)} + 2\beta_4\frac{\mathrm{S}_\kappa^2(a_2)\,\mathrm{C}_\kappa^2(a_3)}{\mathrm{T}_\kappa^2(a_4)},\\
I_{34} &= \tilde J_{34}^2 + 2\beta_3\frac{\mathrm{T}_\kappa^2(a_4)}{\mathrm{S}_\kappa^2(a_3)} + 2\beta_4\frac{\mathrm{S}_\kappa^2(a_3)}{\mathrm{T}_\kappa^2(a_4)}.
\end{aligned}
\tag{5.5}
$$

Thus, in this case, within the subset of "rotation-like" integrals, we find three "upwards-integrals" $Q^{(2)}$, $Q^{(3)}$, $Q^{(4)}$, associated to $\mathfrak{so}(2) \subset \mathfrak{so}(3) \subset \mathfrak{so}(4)$, that determine a completely integrable Hamiltonian:

$$Q^{(2)} = I_{12} \qquad Q^{(3)} = I_{12} + I_{13} + I_{23} \qquad Q^{(4)} = Q_{(4)} = I_{12} + I_{13} + I_{14} + I_{23} + I_{24} + I_{34}$$

These three functions together with the two "backwards-integrals" $Q_{(2)}$, $Q_{(3)}$, associated to $\mathfrak{so}(2) \subset \mathfrak{so}(3)$, and one additional "translation-like" one, say I_{01}, characterize a maximally superintegrable Hamiltonian (5.1):

$$Q_{(2)} = I_{34} \qquad Q_{(3)} = I_{23} + I_{24} + I_{34} \qquad I_{01}$$

To end with, we would like to point out that a similar algebraic construction may likely be applied to the curved version of the "Kepler–Coulomb" superintegrable family (1.2). Furthermore, the consideration of a second contraction parameter, say κ_2, that enables to take into account indefinite metrics of Lorentzian signature [**14, 15**], would allow one to obtain superintegrable systems on different spacetimes. The application of these maximally superintegrable systems in quantum mechanics also deserves a further study.

References

1. Á. Ballesteros and F. J. Herranz, *Integrable deformations of oscillator chains from quantum algebras*, J. Phys. A **32** (1999), no. 50, 8851–8862.
2. Á. Ballesteros, F. J. Herranz, F. Musso, and O. Ragnisco, *Superintegrable deformations of the Smorodinsky–Winternitz Hamiltonian* (in this volume).
3. Á. Ballesteros, F. J. Herranz, M. A. del Olmo, and M. Santander, *Classical deformations, Poisson–Lie contractions, and quantization of dual Lie bialgebras*, J. Math. Phys. **36** (1995), no. 2, 631–640.
4. Á. Ballesteros, F. J. Herranz, M. Santander, and T. Sanz-Gil, *Maximal superintegrability on N-dimensional curved spaces*, J. Phys. A **36** (2003), no. 7, L93–L99.
5. Á. Ballesteros and O. Ragnisco, *A systematic construction of integrable Hamiltonians from coalgebras*, J. Phys. A **31** (1998), no. 16, 3791–3813.
6. N. W. Evans, *Superintegrability in classical mechanics*, Phys. Rev. A **41** (1990), no. 10, 5666–5676.
7. ———, *Superintegrability of the Winternitz system*, Phys. Lett. A **147** (1990), no. 8-9, 483–486.
8. ———, *Group theory of the Smorodinsky–Winternitz system*, J. Math. Phys. **32** (1991), no. 12, 3369–3375.
9. J. Friš, V. Mandrosov, Ya. A. Smorodinsky, M. Uhlíř, and P. Winternitz, *On higher symmetries in quantum mechanics*, Phys. Lett. **16** (1965), 354–356.
10. C. Grosche, G. S. Pogosyan, and A. N. Sissakian, *Path integral discussion for Smorodinsky-Winternitz potentials. I. Two- and three-dimensional Euclidean space*, Fortschr. Phys. **43** (1995), no. 6, 453–521.
11. ———, *Path integral discussion for Smorodinsky-Winternitz potentials.* II. *The two- and three-dimensional sphere*, Fortschr. Phys. **43** (1995), no. 6 523–563.
12. ———, *Path integral approach for superintegrable potentials on the three-dimensional hyperboloid*, Phys. Particles Nuclei **28** (1997), 486–519.
13. S. Helgason, *Differential geometry and symmetric spaces*, Pure Appl. Math., vol. 12, Academic Press, New York–London, 1962.
14. F. J. Herranz, R. Ortega, and M. Santander, *Trigonometry of spacetimes: a new self-dual approach to a curvature/signature (in)dependent trigonometry*, J. Phys. A **33** (2000), no. 24, 4525–4551.

15. F. J. Herranz and M. Santander, *Conformal symmetries of spacetimes*, J. Phys. A **35** (2002), no. 31, 6601–6618.
16. P. W. Higgs, *Dynamical symmetries in a spherical geometry.* I, J. Phys. A **12** (1979), no. 3, 309–323.
17. A. A. Izmest′ev, G. S. Pogosyan, A. N. Sissakian, and P. Winternitz, *Contractions of Lie algebras and separation of variables. The n-dimensional sphere*, J. Math. Phys. **40** (1999), no. 3, 1549–1573.
18. E. G. Kalnins, J. M. Kress, G. S. Pogosyan, and W. Miller, Jr., *Completeness of superintegrability in two-dimensional constant-curvature spaces*, J. Phys. A **34** (2001), no. 22, 4705–4720.
19. E. G. Kalnins, W. Miller, Jr., Ye. M. Hakobyan, and G. S. Pogosyan, *Superintegrability on the two-dimensional hyperboloid.* II, J. Math. Phys. **40** (1999), no. 5, 2291–2306.
20. E. G. Kalnins, W. Miller, Jr., and G. S. Pogosyan, *Superintegrability of the two-dimensional hyperboloid*, J. Math. Phys. **38** (1997), no. 10, 5416–5433.
21. ———, *Completeness of multiseparable superintegrability on the complex 2-sphere*, J. Phys. A **33** (2000), no. 38, 6791–6806.
22. ———, *Coulomb-oscillator duality in spaces of constant curvature*, J. Math. Phys. **41** (2000), no. 5, 2629–2657.
23. ———, *The Coulomb-oscillator relation on n-dimensional spheres and hyperboloids*, Phys. Atomic Nuclei **65** (2002), no. 6, 1086–1094.
24. E. G. Kalnins, G. S. Pogosyan, and W. Miller, Jr., *Completeness of multiseparable superintegrability in two dimensions*, Phys. Atomic Nuclei **65** (2002), no. 6, 1033–1035.
25. E. G. Kalnins, G. C. Williams, W. Miller, Jr., and G. S. Pogosyan, *Superintegrability in three-dimensional Euclidean space*, J. Math. Phys. **40** (1999), no. 2, 708–725.
26. H. I. Leemon, *Dynamical symmetries in a spherical geometry.* II, J. Phys. A **12** (1979), no. 4, 489–501.
27. M. F. Rañada and M. Santander, *Superintegrable systems on the two-dimensional sphere S^2 and the hyperbolic plane H^2*, J. Math. Phys. **40** (1999), no. 10, 5026–5057.
28. ———, *On harmonic oscillators on the two-dimensional sphere S^2 and the hyperbolic plane H^2*, J. Math. Phys. **43** (2002), no. 1, 431–451.
29. M. F. Rañada, M. Santander, and T. Sanz-Gil, *Superintegrable potentials and the superposition of Higgs oscillators on the sphere S^2*, Banach Center Publ., Polish Acad. Sci., Warsaw (to appear).
30. E. Schrödinger, *A method of determining quantum mechanical eigenvalues and eigenfunctions*, Proc. Roy. Irish Acad. Sect. A **46** (1940), 9–16.

Departamento de Física, Escuela Politécnica Superior, Universidad de Burgos, 09006 Burgos, Spain

E-mail address: `fjherranz@ubu.es`

Departamento de Física, Facultad de Ciencias, Universidad de Burgos, 09001 Burgos, Spain

E-mail address: `angelb@ubu.es`

Departamento de Física Teórica, Facultad de Ciencias, Universidad de Valladolid, 47011 Valladolid, Spain

E-mail address, M. Santander: `santander@fta.uva.es`

E-mail address, T. Sanz-Gil: `sanzgil@fta.uva.es`

Centre de Recherches Mathématiques
CRM Proceedings and Lecture Notes
Volume **37**, 2004

Invariant Wirtinger Projective Connection and Tau-Functions on Spaces of Branched Coverings

A. KOKOTOV and D. Korotkin

ABSTRACT. Using the invariant Wirtinger projective connection on a branched covering of the Riemann sphere, we construct a flat line bundle over the Hurwitz space $H_{g,N}(1,\dots,1)$. The covariant constant holomorphic section of this bundle is called the Wirtinger tau-function of the covering. In genera 0 and 1 we obtain the explicit formulas for the Wirtinger tau-function, in higher genus we compute its modulus square in terms of the appropriately regularized Dirichlet integral.

The spaces of meromorphic functions on Riemann surfaces (Hurwitz spaces) play an important role in modern mathematical physics. It is enough to mention the theory of Frobenius manifolds and WDVV equations, where Hurwitz spaces (or, more precisely, their coverings) arise among the most interesting examples. At the same time Hurwitz spaces are connected with deep problems in combinatorics.

The geometry of Hurwitz spaces considered as complex manifolds is also of great interest. To study this geometry it may be useful to look at global objects on Hurwitz spaces. To this end we introduce the so called Wirtinger tau-function—a global holomorphic object on the Hurwitz space, which is actually a section of a holomorphic line bundle over it. The main goal of these notes is to calculate the Wirtinger tau-function (or, at least, its modulus squared) explicitly.

Let $H_{g,N}$ be the Hurwitz space of branched coverings $\mathcal{L}$ of the Riemann sphere with given genus and number of sheets, i.e., the space of meromorphic functions of given degree N on Riemann surfaces of fixed genus g. In what follows we shall work within the bulk $H_{g,N}(1,\dots,1)$ of the Hurwitz space $H_{g,N}$ consisting of coverings with only simple branch points, none of which projects to infinity. It is known that the family of meromorphic functions $H_{g,N}(1,\dots,1)$ is a connected complex manifold.

2000 *Mathematics Subject Classification.* 32G99.

This work is partially supported by the grant of Fonds pour la Formation de Chercheurs et l'Aide à la Recherche du Québec, the grant of Natural Sciences and Engineering Research Council of Canada and Faculty Research Development Program of Concordia University.

This is the final form of the paper.

We use the projections $\lambda_1, \dots, \lambda_M$ of branch points $P_1, \dots, P_M$ of the covering $\mathcal{L}$ as local coordinates on the space $H_{g,N}(1, \dots, 1)$; according to the Riemann–Hurwitz formula $M = 2g + 2N - 2$.

Let λ be the coordinate of the projection of a point $P \in \mathcal{L}$ to $\mathbb{P}$. In a neighborhood of a branch point P_m we may introduce the local coordinate $x_m = \sqrt{\lambda - \lambda_m}$, where λ_m is the projection of P_m.

Besides the Hurwitz space $H_{g,N}(1, \dots, 1)$, we shall use the "punctured" Hurwitz space $H'_{g,N}(1, \dots, 1)$, which is obtained from $H_{g,N}(1, \dots, 1)$ by excluding all branch coverings which have at least one vanishing theta-constant. Unless $g = 0, 1, 2$ or $N = 2$ and $g > 2$, these exceptional coverings form a subspace of codimension 1 in $H_{g,N}(1, \dots, 1)$ (for genus greater than 2 and $N = 2$ at least one of the theta-constants always vanishes; in genera 0, 1, 2 there are no vanishing theta-constants).

In the trivial bundle $H'_{g,N}(1, \dots, 1) \times \mathbb{C}$ we introduce the connection

$$d_{\mathrm{W}} = d - \sum_{m=1}^{M} \mathcal{A}_m \, d\lambda_m, \tag{1}$$

where d is the external differentiation operator including both holomorphic and antiholomorphic parts; connection coefficients are expressed in terms of the invariant Wirtinger projective connection S_{W} on the covering $\mathcal{L}$ as follows:

$$\mathcal{A}_m = -\frac{1}{12} S_{\mathrm{W}}(x_m)|_{x_m=0}, \qquad m = 1, \dots M. \tag{2}$$

The connection coefficients $\mathcal{A}_m$ are holomorphic with respect to λ_m and correctly defined for all coverings $\mathcal{L}$ of the "punctured" Hurwitz space $H'_{g,N}(1, \dots, 1)$.

Connection (1) turns out to be flat; therefore, it determines a character of the fundamental group of $H'_{g,N}(1, \dots, 1)$; this character defines a flat holomorphic line bundle $\mathcal{T}_{\mathrm{W}}$ over $H'_{g,N}(1, \dots, 1)$. We call this bundle the "Wirtinger line bundle" over Hurwitz space; its horizontal holomorphic section we call the "Wirtinger tau-function."

We obtain explicit formulas for the modulus squared of the Wirtinger tau-function in genus greater than 1; in genera 0 and 1 we perform the "holomorphic factorization" and derive explicit formulas for the Wirtinger tau-function itself.

Consider the Wirtinger bidifferential on the branched covering $\mathcal{L}$:

$$W(P, Q) = B(P, Q) + \frac{2}{4^g + 2^g} \sum_{i,j=1}^{g} v_i(P) v_j(Q) \frac{\partial^2}{\partial z_i \partial z_j} \ln \prod_{\beta \text{ even}} \Theta[\beta](z \mid \mathbb{B})|_{z=0}, \tag{3}$$

where $B(P, Q) = d_P \, d_Q \ln E(P, Q)$ is the Bergmann kernel on $\mathcal{L}$; $E(P, Q)$ is the prime-form; $v_i(P)$, $i = 1, \dots, g$ is the basis of normalized holomorphic differentials, $\mathbb{B} = \|\mathbb{B}_{ij}\|_{i,j=1}^{g}$ is the matrix of b-periods of $\mathcal{L}$, β runs through the set of all even characteristics (see [**2, 6**]).

An important advantage of the Wirtinger bidifferential in comparison to the Bergmann kernel is that the former does not depend on the choice of canonical basic cycles (a_k, b_k) on the covering $\mathcal{L}$. The disadvantage is that the Wirtinger bidifferential is undefined on Riemann surfaces on which there exists at least one non-vanishing theta-constant.

The Wirtinger bidifferential has the following asymptotics near the diagonal:

$$W(P, Q) = \left\{ \frac{1}{\big(x(P) - x(Q)\big)^2} + \frac{1}{6} S_{\mathrm{W}}\big(x(P)\big) + o(1) \right\} dx(P) \, dx(Q) \tag{4}$$

as $P \to Q$, where $x(P)$ is a local coordinate on $\mathcal{L}$. The quantity $S_{\rm W}$ is a projective connection on $\mathcal{L}$ called the invariant Wirtinger projective connection.

Like the Wirtinger bidifferential itself, the Wirtinger projective connection does not depend on the choice of basic cycles on $\mathcal{L}$.

We recall that $H'_{g,N}(1,\dots,1)$ denotes the set of branched coverings of the Hurwitz space $H_{g,N}(1,\dots,1)$ for which none of the theta-constants vanishes. Introduce the connection

$$d_{\rm W} = d - \sum_{m=1}^{M} \mathcal{A}_m \, d\lambda_m, \tag{5}$$

acting in the trivial bundle $H'_{g,N}(1,\dots,1) \times \mathbb{C}$, where

$$d = \sum_{m=1}^{M} \frac{\partial}{\partial \lambda_m} \, d\lambda_m + \sum_{m=1}^{M} \frac{\partial}{\partial \bar{\lambda}_m} \, d\bar{\lambda}_m$$

is the external differentiation; the connection coefficients are defined as follows:

$$\mathcal{A}_m = -\frac{1}{12} S_{\rm W}(x_m)|_{x_m=0}, \qquad m = 1,\dots M. \tag{6}$$

Here $x_m = \sqrt{\lambda - \lambda_m}$ is the local parameter near the branch point P_m of the covering $\mathcal{L}$.

The connection $d_{\rm W}$ is correctly defined for all coverings $\mathcal{L}$ of $H_{g,N}(1,\dots,1)$, except coverings which have at least one vanishing theta-constant.

If we choose another global holomorphic coordinate μ on $\mathbb{P}$,

$$\lambda = \frac{a\mu + b}{c\mu + d}, \qquad \text{where } ad - bc = 1,$$

then the connection $d_{\rm W}$ turns into a gauge equivalent connection. Namely, let μ_m be the new coordinates of the branch points,

$$\lambda_m = \frac{a\mu_m + b}{c\mu_m + d}; \tag{7}$$

then the gauge transformation of connection $d_{\rm W}$ in local coordinates looks as follows

$$d_{\rm W} \mapsto G^{-1} d_{\rm W} G, \tag{8}$$

where

$$G = \prod_{m=1}^{M} (c\mu_m + d)^{-1/4}. \tag{9}$$

The following theorem plays the main role in the construction of the Wirtinger line bundle over Hurwitz space.

THEOREM 1 ([**3**]). *The connection $d_{\rm W}$, defined in the trivial line bundle over $H'_{g,N}(1,\dots,1)$ in terms of the Wirtinger projective connection by formulas* (5), (6), *is flat.*

The flat connection $d_{\rm W}$ determines a character of the fundamental group of $H'_{g,N}(1,\dots,1)$, i.e., the representation

$$\rho\colon \pi_1\big(H'_{g,N}(1,\dots,1)\big) \to \mathbb{C}^*. \tag{10}$$

Denote by $\mathcal{U}$ the universal covering of $H'_{g,N}(1,\dots,1)$; then $\pi_1\big(H'_{g,N}(1,\dots,1)\big)$ acts on the direct product $\mathcal{U}\times\mathbb{C}$ as follows:

$$g(u,z)=(gu,\rho(g)z).$$

The quotient manifold $\mathcal{U}\times\mathbb{C}/\pi_1\big(H'_{g,N}(1,\dots,1)\big)$ has the structure of a holomorphic line bundle over $H'_{g,N}(1,\dots,1)$; we denote this bundle by $\mathcal{T}_{\rm W}$.

DEFINITION 1. The flat holomorphic line bundle $\mathcal{T}_{\rm W}$ equipped with the flat connection $d_{\rm W}$ is called the Wirtinger line bundle over the punctured Hurwitz space $H'_{g,N}(1,\dots,1)$. The (unique up to a multiplicative constant) horizontal holomorphic section of the bundle $\mathcal{T}_{\rm W}$ is called the Wirtinger τ-function of the Hurwitz space $H_{g,N}(1,\dots,1)$ and denoted by $\tau_{\rm W}$.

Taking into account the form (8), (9) of the gauge transformation of connection $d_{\rm W}$ under conformal transformations on the base λ-plane, we see that the Wirtinger tau-function $\tau_{\rm W}$ transforms as follows under conformal transformation (7):

$$\tau_{\rm W}\mapsto\prod_{m=1}^{M}(c\mu_m+d)^{-1/4}\tau_{\rm W}. \tag{11}$$

In genera 0, 1 and 2 the "theta-divisor" $\mathcal{Z}=H_{g,N}(1,\dots,1)\backslash H'_{g,N}(1,\dots,1)$ is empty. Therefore, in this case the bundle $\mathcal{T}_{\rm W}$ is a bundle over the whole Hurwitz space $H_{g,N}(1,\dots,1)$.

REMARK 1. Hyperelliptic coverings ($N=2$) fall within our framework only in genera $g=0$, 1, 2 since for genus $g>2$ one of the theta-constants always vanishes for hyperelliptic curves [**4**].

Let $\mathcal{L}_0$ be a covering of $H'_{g,N}(1,\dots,1)$. In a small neighborhood $U(\mathcal{L}_0)$ of $\mathcal{L}_0$ we may consider the projections $\lambda_1,\dots,\lambda_M$ of branch points as local coordinates on $H'_{g,N}(1,\dots,1)$. The section $\tau_{\rm W}$ can be considered as a holomorphic function on $U(\mathcal{L}_0)$. Its modulus squared, $|\tau_{\rm W}|^2$, is a section of the "real" line bundle $\mathcal{T}\otimes\overline{\mathcal{T}}$.

If the covering $\mathcal{L}$ has genus $g>1$ then it is biholomorphically equivalent to the quotient space $\mathbb{H}/\Gamma$, where $\mathbb{H}=\{z\in\mathbb{C}:\operatorname{Im}z>0\}$; Γ is a strictly hyperbolic Fuchsian group. Denote by $\pi_{\rm F}\colon\mathbb{H}\to\mathcal{L}$ the natural projection. The Fuchsian projective connection $S_{\rm F}$ on $\mathcal{L}$ is given by the Schwarzian derivative $\{z,x\}$, where x is a local coordinate of a point $P\in\mathcal{L}$, $z\in\mathbb{H}$, $\pi_{\rm F}(z)=P$.

Define the determinant of Laplacian operator by the equality

$$\det\Delta=\exp\{-\zeta'_\Delta(0)\},$$

where $\zeta_\Delta(s)$ is the zeta-function of the Laplacian. Consider some covering $\mathcal{L}\in U(\mathcal{L}_0)$; let x be a local parameter on $\mathcal{L}$. Introduce the standard metric $e^{\chi(x,\bar x)}|dx|^2$ of constant curvature -1 on $\mathcal{L}$ by the equality

$$e^{\chi(x,\bar x)}|dx|^2=\frac{|dz|^2}{|\operatorname{Im}z|^2}, \tag{12}$$

where $z\in\mathbb{H}$. Denote by $\zeta=1/\lambda$ the local coordinate in a neighborhood of the infinity of any sheet of covering $\mathcal{L}$.

Introduce functions $\chi^{\rm ext}(\lambda,\bar\lambda)$, $\chi^{\rm int}(x_m,\bar x_m)$, $m=1,\dots,M$ and $\chi_k^\infty(\zeta,\bar\zeta)$, $k=1,\dots,N$ by specifying $x=\lambda$, $x=x_m$ and $x=\zeta$ (in a neighborhood of the point at infinity of kth sheet) in (12) respectively. From now on we identify the branch points P_m with their projections λ_m to $\mathbb{P}$.

Consider the following domain on kth sheet of $\mathcal{L}$: $\mathcal{L}^k_\rho = \{\lambda \in \mathcal{L}^k : \forall m\ |\lambda - \lambda_m| > \rho, |\lambda| < 1/\rho\}$, where λ_m are all the branch points which belong to the k-th sheet $\mathcal{L}^k$ of the covering $\mathcal{L}$. (The sheet $\mathcal{L}^k$ can be considered as a copy of the Riemann sphere $\mathbb{P}$ with appropriate cuts between the branch points; the domain $\mathcal{L}^k_\rho$ is obtained from $\mathcal{L}^k$ by deleting small discs around branch points belonging to this sheet, and the disc around infinity.)

The function $\chi_k^{\rm ext}: \mathcal{L}^k \to \mathbb{R}$ is smooth in the domain $\mathcal{L}^k_\rho$ for any sufficiently small $\rho > 0$. This function has finite limits at the cuts (except the endpoints which are the branch points); at the branch points and at infinity there are the following asymptotics

$$|\partial_\lambda \chi_k^{\rm ext}(\lambda, \bar\lambda)|^2 = \tfrac{1}{4}|\lambda - \lambda_m|^{-2} + O(|\lambda - \lambda_m|^{-3/2}) \tag{13}$$

as $\lambda \to \lambda_m$ and

$$|\partial_\lambda \chi_k^{\rm ext}(\lambda, \bar\lambda)|^2 = 4|\lambda|^{-2} + O(|\lambda|^{-3}) \tag{14}$$

as $\lambda \to \infty$. We define the "truncated" integral T_ρ by

$$T_\rho = \sum_{k=1}^{N} \int_{\mathcal{L}^k_\rho} |\partial_\lambda \chi_k^{\rm ext}|^2\, |d\lambda \wedge d\bar\lambda|/2. \tag{15}$$

Then there exists the finite limit

$$\operatorname{reg} \int_{\mathcal{L}} |\chi_\lambda|^2\, |d\lambda \wedge d\bar\lambda|/2 = \lim_{\rho\to 0} (T_\rho\, |d\lambda \wedge d\bar\lambda|/2 + (8N + 4M)\pi \ln \rho). \tag{16}$$

Define the function $\mathbb{S}_{\rm F}$ by

$$\begin{aligned}\mathbb{S}_{\rm F}(\lambda_1, \dots, \lambda_M) = \frac{1}{12\pi}&\left\{\operatorname{reg}\int_{\mathcal{L}} |\chi_\lambda|^2\, |d\lambda \wedge d\bar\lambda|/2\right\} \\ &+ \frac{1}{6}\sum_{n=1}^{M} \chi^{\rm int}(x_n)|_{x_n=0} - \frac{1}{3}\sum_{k=1}^{N} \chi_k^\infty(\zeta)|_{\zeta=0}.\end{aligned} \tag{17}$$

PROPOSITION 1. *The real-valued function* $\mathbb{S}_{\rm F}$ *defined by* (17) *satisfies the equations*

$$\frac{\partial \mathbb{S}_F}{\partial \lambda_m} = \frac{1}{12}\{z, x_m\}|_{x_m=0}, \qquad m = 1, \dots M.$$

Now we can state the main result of this subsection.

THEOREM 2 ([**3**]). *The modulus squared of the Wirtinger tau-function has the following representation in a neighborhood of a covering* $\mathcal{L}_0$ *from* $H'_{g,N}(1, \dots, 1)$:

$$|\tau_{\rm W}|^2 = e^{-\mathbb{S}_{\rm F}} \frac{\det \Delta}{\det \operatorname{Im} \mathbb{B}} \prod_{\beta\ even} |\Theta[\beta](0 \mid \mathbb{B})|^{-2/(4^{g-1}+2^{g-2})}, \tag{18}$$

where the regularized Dirichlet integral $\mathbb{S}_{\rm F}$ *is defined by* (17).

Here we discuss the Hurwitz spaces $H_{0,N}$, $H_{1,N}$, for which the Wirtinger tau-function can be computed explicitly.

In genus 0 the surface $\mathcal{L}$ is biholomorphically equivalent to the Riemann sphere $\mathbb{P}$. Let γ be the natural coordinate on $\mathbb{P} \setminus \infty$. Then the coefficients of connection $d_{\rm W}$ are simply given by the Schwarzian derivative

$$\mathcal{A}_m = -\frac{1}{12}\{\gamma(x_m), x_m\}|_{x_m=0}. \tag{19}$$

The real potential $\mathbb{S}_{0,N} = \ln|\tau|^2$ satisfying the equations

$$\frac{\partial \mathbb{S}_{0,N}}{\partial \lambda_m} = -\mathcal{A}_m, \qquad m = 1, \dots, M$$

has the form of the Dirichlet integral corresponding to a flat metric on $\mathcal{L}$. To describe this integral we choose one sheet of the covering $\mathcal{L}$ and call it the first sheet. Let us choose the uniformization map $\gamma(P)$ such that $\gamma(\infty^{(1)}) = \infty$, where $\infty^{(1)}$ is the point at infinity of the first sheet. Recall that $\zeta = 1/\lambda$ is the local coordinate in a neighborhood of the infinity of any sheet. Introduce the flat metric $e^{\phi(\lambda,\bar{\lambda})}|d\lambda|^2$ on $\mathcal{L}$ (this metric is singular at $P = \infty^{(1)}$, i.e., $\gamma = \infty$) by the equality

$$e^{\phi(\lambda,\bar{\lambda})}|d\lambda|^2 = |d\gamma|^2. \tag{20}$$

Then potential $\mathbb{S}_{0,N}$ is defined by the following expression:

$$\mathbb{S}_{0,N} = \frac{1}{12\pi} \operatorname{reg} \int_{\mathcal{L}} |\phi_\lambda|^2 \, |d\lambda \wedge d\bar{\lambda}|/2. \tag{21}$$

Here the integral on the right-hand side of (21) is understood in the sense of an appropriate regularization. (It is essentially the same as in the case of higher genus, see (16)). This regularized integral may be explicitly calculated and turns out to be the modulus squared of a holomorphic function of coordinates $(\lambda_1, \dots, \lambda_M)$. This leads to an explicit expression for the Wirtinger tau-function in genus 0 given by the following theorem:

THEOREM 3 ([**3**]). *In genus 0 the Wirtinger tau-function is given by the following formula*

$$\tau_W = \frac{\prod_{k=2}^{N} (d\gamma/d\zeta|_{P=\infty^{(k)}})^{1/6}}{\prod_{k=m}^{M} (d\gamma/dx_m|_{P=P_m})^{1/12}}. \tag{22}$$

Similar ideas also work in the genus 1 case. We only formulate the final result. To this end introduce the notation $f_m \equiv f_m(0)$, $h_k \equiv h_k(0)$, where $v(P) = f_m(x_m)\,dx_m$ near the branch point P_m and $v(P) = h_k(\zeta)\,d\zeta$ near the infinity of the kth sheet.

THEOREM 4 ([**3**]). *The Wirtinger tau-function on Hurwitz space $H_{1,N}$ is given by the formula*

$$\tau_W = \frac{\prod_{k=1}^{N} h_k^{1/6}}{\prod_{m=1}^{M} f_m^{1/12}}. \tag{23}$$

We notice that the result (23) does not depend on normalization of the holomorphic differential $v(P)$: if one makes a transformation $v(P) \to Cv(P)$ with an arbitrary constant C, this constant cancels out in (23) due to the Riemann–Hurwitz formula.

There is a natural covering $H_{g,N}(1, \dots, 1) \to \mathbb{C}^{(M)} \setminus \bigcup_{i \neq j} \{\lambda_i = \lambda_j\}$. Here $\mathbb{C}^{(M)}$ denotes the Mth symmetric power of $\mathbb{C}$. (This implies, in particular, that $H_{g,N}(1, \dots, 1)$ is a Stein manifold.) The number of sheets of this covering is nothing but the Hurwitz number $h_{g,N}$. The Wirtinger tau-function is a multivalued holomorphic function on the manifold $\mathcal{B} = \mathbb{C}^{(M)} \setminus \bigcup_{i \neq j} \{\lambda_i = \lambda_j\}$. (For $g = 0$ and $g = 1$ the 12th power of the Wirtinger tau-function is a single-valued global holomorphic function on $H_{g,N}(1, \dots, 1)$.)

We call two germs of holomorphic functions at one point equivalent if they are proportional. Taking some germ of the Wirtinger tau-function at a point $b \in \mathcal{B}$ and continuing it holomorphically along all closed loops in $\mathcal{B}$ we get a (finite) set of nonequivalent germs at the point b. We conjecture that the number of elements of this set coincides with the Hurwitz number $h_{g,N}$; this might provide an alternative to the $[\mathbf{1},\mathbf{5}]$ procedure for calculation of the Hurwitz numbers.

References

1. T. Ekedahl, S. Lando, M. Shapiro, and A. Vainshtein, *Hurwitz numbers and intersections on moduli spaces of curves*, Invent. Math. **146** (2001), no. 2, 297–327.
2. J. D. Fay, *Theta-functions on Riemann surfaces*, Lecture Notes in Math., vol. 352, Springer-Verlag, Berlin–New York, 1973.
3. A. Kokotov and D. Korotkin, *Tau-functions on Hurwitz spaces*, Math. Phys. Anal. Geom. **7** (2004), no. 1, 47–96. appear).
4. D. Mumford, *Tata lectures on theta. II. Jacobian theta functions and differential equations*, Progr. Math., vol. 43, Birkhäuser, Boston, 1984
5. R. Padharipande, *The Toda equations and the Gromov–Witten theory of the Riemann sphere*, 1999, math.AG/9912166
6. A. N. Tyurin, *Periods of quadratic differentials*, Uspekhi Mat. Nauk **33** (1978), no. 6, 149–195 (Russian); English transl., Russian Math. Surveys **33** (1978), no. 6, 169–221.

Department of Mathematics and Statistics, Concordia University, 7141 Sherbrooke West, Montreal, QC H4B 1R6, Canada

E-mail address, A. Kokotov: `alexey@mathstat.concordia.ca`

E-mail address, D. Korotkin: `korotkin@mathstat.concordia.ca`

Centre de Recherches Mathématiques
CRM Proceedings and Lecture Notes
Volume **37**, 2004

Dyon-Oscillator Duality. Hidden Symmetry of the Yang–Coulomb Monopole

L. G. Mardoyan

ABSTRACT. In this report, in the framework of an analytical approach and with help of a generalized version of the Hurwitz transformation, the five-dimensional bound system composed of the Yang monopole coupled to a particle with an SU(2) isospin and a Coulomb interaction, is constructed from the eight-dimensional quantum oscillator. The generalized Runge–Lenz vector and the SO(6) group of the hidden symmetry are established. It is also shown that group of hidden symmetry makes it possible to calculate the spectrum of the system by a purely algebraic method.

1. Introduction

The objective of the present lecture is to illustrate a property of the Schrödinger equation which is called here the dyon-oscillator duality. The property is the following. The Schödinger equation for an oscillator possesses two parameters—the energy E and the cyclic frequency ω. The quantization leads to the constraint $E = \hbar\omega(N + D/2)$, where $N = 0, 1, 2, \dots$, and D is the dimension of the configuration space of the oscillator. If ω is fixed then E is quantized, and that is the standard situation. Imagine for a moment that now E is fixed. Whence, necessarily ω is quantized, and we are in a nonstandard situation. The question is whether the nonstandard situation corresponds to any physics, i.e., is it possible to find such a transformation that converts the oscillator to some physical system with coupling constant α, being a function of E, and energy ε, depending on ω? If there exists such a transformation, we can confirm that the "nonstandard oscillator" is identical to that physical system. Below will be shown the validity of the described picture for dimensions $D = 1, 2, 4, 8$, and that the final system is a bound system of charge-particle. We recall that a dyon is the hypothetical particle introduced by Schwinger [**17**], which is unlike the Dirac monopole is endowed with not just a magnetic, but an electric charge as well. As the "standard" and "nonstandard" regimes are mutually exclusive, the initial oscillator and the final "charge-dyon"

2000 *Mathematics Subject Classification.* 81S99.

The research reported here was supported partly by the NATO Collaborative Linkage Grant No. 978431 and ANSEF Grant No. PS124-01.

I am very grateful to the organizers who have invited me to this very interesting workshop.

This is the final form of the paper.

system are dual to each other, and that explains the relevancy of the term "dyon-oscillator duality." Note also, that in the initial system the spectrum is discrete only, i.e. the particle has just a finite motion (for such cases it is usual to say that we have a model with confinement). Generally speaking, the spectrum of the final system includes the discrete spectrum as well as the continuous one, i.e., in that model there is no confinement. However, unlike the first model, in the second model we have monopoles. There is some analogy between the dyon-oscillator and the Seiberg–Witten duality [**18**], according to which the gauge theories with strong interactions are equivalent to the theories having weak interaction on one hand and topological nontrivial objects, such as monopoles and dyons, on the other.

2. Coulomb-oscillator duality

Let us consider the equation

$$\frac{d^2R}{du^2} + \frac{D-1}{u}\frac{dR}{du} - \frac{L(L+D-2)}{u^2}R + \frac{2m}{\hbar^2}\left(E - \frac{m\omega^2 u^2}{2}\right)R = 0. \tag{2.1}$$

Here R is the radial part of the wave function for D-dimensional oscillator ($D > 2$), and $L = 0, 1, 2, \ldots,$ are the eigenvalues of the global angular momentum.

After the substitution $r = u^2$ equation (2.1) transforms to the equation

$$\frac{d^2R}{dr^2} + \frac{d-1}{r}\frac{dR}{dr} - \frac{l(l+d-2)}{r^2}R + \frac{2m}{\hbar^2}\left(\epsilon + \frac{e^2}{r}\right)R = 0, \tag{2.2}$$

where $d = D/2+1$, $l = L/2$,

$$\epsilon = -m\omega^2/8, \qquad e^2 = E/4. \tag{2.3}$$

This is quite an unexpected result. If $D = 4, 6, 8, \ldots,$ then $d = 3, 4, 5, \ldots,$ and the equation (2.2) is formally identical to the radial equation for the d-dimensional Coulomb problem (for odd $D > 2$ the value of d is half-integer and can not have the meaning of the dimension of the space in the usual sense). Then, l takes not only integer but half-integer values as well, hence it has the meaning of total momentum and a question arises about the origin of the fermion degree of freedom. The answer to the question will be given later. Finally, as was mentioned in Section 1, the equations (2.1) and (2.2) are dual to each other and the duality transformation is $r = u^2$.

Up to now, just the radial part of the wave function of the oscillator has been considered. For the Schrödinger equation we must take into account the angular part as well. Thus, the duality transformation must also include the transformation of angular variables. If we interpret the change of variables $r = u^2$ as the mechanism of generation of electric charge, (as will be shown later) the transformation of some angular variables is responsible for the generation of magnetic charges.

Therefore, we return to the condition $r = u^2$. In the Cartesian coordinates this condition has the form

$$x_0^2 + x_1^2 + \cdots + x_{d-1}^2 = (u_0^2 + u_1^2 + \cdots + u_{D-1}^2)^2,$$

which is called the Euler identity. According to the Hurwitz theorem [**5**] this equation has a solution bilinear in u_μ only for the following pairs of numbers

$$(D, d) = (1,1);\ (2,2);\ (4,3);\ (8,5).$$

Three remarkable circumstances are connected with the last relation:

- $(D, d) = (2, 2)$: is the Levi-Civita transformation [**10**].
  $(D, d) = (4, 3)$: is the Kustaanheimo–Stiefel transformation [**7**].
  $(D, d) = (8, 5)$: is the Hurwitz transformation [**2, 6, 8**].
- The transformation (3.1) establishes the connection between two fundamental problems of mechanics, the oscillator and Kepler problems.
- The "magic" numbers $D = 1,\ 2,\ 4,\ 8$ have a direct relation to the existence of four basic algebraic structures: real numbers, complex numbers, quaternions and octonions.

Later on we consider only the case $(D, d) = (8, 5)$.

3. Hurwitz transformation

We can write the solution of the Euler's identity as

$$x = H(u; D)u, \tag{3.1}$$

which can be interpreted as a bilinear transformation that maps one Euclidean space into another. Here D is the dimension of the space, H is a $D \times D$ matrix with the elements u_μ, and x, u are D-dimensional columns composed from x_j, u_μ and, possibly, zeroes. So, for the Levi-Civita and Kustaanheimo–Stiefel transformations we have

$$\begin{pmatrix} x_1 \\ x_2 \end{pmatrix} = \begin{pmatrix} u_1 & -u_2 \\ u_2 & u_1 \end{pmatrix} \begin{pmatrix} u_1 \\ u_2 \end{pmatrix},$$

$$\begin{pmatrix} x_1 \\ x_2 \\ x_3 \\ 0 \end{pmatrix} = \begin{pmatrix} u_3 & -u_4 & u_1 & -u_2 \\ u_4 & u_3 & u_2 & u_1 \\ u_1 & u_2 & -u_3 & -u_4 \\ u_2 & -u_1 & -u_4 & u_3 \end{pmatrix} \begin{pmatrix} u_1 \\ u_2 \\ u_3 \\ u_4 \end{pmatrix}.$$

The matrices $H(u; 2)$ and $H(u; 4)$ have the property

$$H(u; 2)H^T(u; 2) = u^2 E(2), \qquad H(u; 4)H^T(u; 2) = u^2 E(4),$$

where the superscript T means transposition, $E(2)$ and $E(4)$ are unit matrices. Due to these properties the Euler identities are fulfilled. Now it is easily deduced, that the transformation $\mathbb{R}^8(\vec{u}) \to \mathbb{R}^5(\vec{x})$ must take the form [**2**]

$$\begin{pmatrix} cx_0 \\ x_1 \\ x_2 \\ x_3 \\ x_4 \\ 0 \\ 0 \\ 0 \end{pmatrix} = \begin{pmatrix} u_0 & u_1 & u_2 & u_3 & -u_4 & -u_5 & -u_6 & -u_7 \\ u_4 & u_5 & -u_6 & -u_7 & u_0 & u_1 & -u_2 & -u_3 \\ u_5 & -u_4 & u_7 & -u_6 & -u_1 & u_0 & -u_3 & u_2 \\ u_6 & u_7 & u_4 & u_5 & u_2 & u_3 & u_0 & u_1 \\ u_7 & -u_6 & -u_5 & u_4 & u_3 & -u_2 & -u_1 & u_0 \\ u_1 & -u_0 & u_3 & -u_2 & u_5 & -u_4 & u_7 & -u_6 \\ u_2 & -u_3 & -u_0 & u_1 & -u_6 & u_7 & u_4 & -u_5 \\ u_3 & u_2 & -u_1 & -u_0 & -u_7 & -u_6 & u_5 & u_4 \end{pmatrix} \begin{pmatrix} u_0 \\ u_1 \\ u_2 \\ u_3 \\ u_4 \\ u_5 \\ u_6 \\ u_7 \end{pmatrix}.$$

Hence it follows that

$$\begin{aligned} x_0 &= u_0^2 + u_1^2 + u_2^2 + u_3^2 - u_4^2 - u_5^2 - u_6^2 - u_7^2, \\ x_1 &= 2(u_0u_4 - u_1u_5 - u_2u_6 - u_3u_7), \\ x_2 &= 2(u_0u_5 + u_1u_4 - u_2u_7 + u_3u_6), \\ x_3 &= 2(u_0u_6 + u_1u_7 + u_2u_4 - u_3u_5), \\ x_4 &= 2(u_0u_7 - u_1u_6 + u_2u_5 + u_3u_4). \end{aligned} \tag{3.2}$$

It is easy to prove that for the matrix $H(u;8)$ there is a condition

$$H(u;8)H^T(u;8) = u^2E(8)$$

that guarantees the validity of the Euler identity.

Adding to (3.2) the transformations [**4**]

$$(3.3)\qquad \begin{aligned} \alpha_T &= \frac{i}{2}\ln\frac{(u_0+iu_1)(u_2-iu_3)}{(u_0-iu_1)(u_2+iu_3)} \in [0,2\pi),\\ \beta_T &= 2\arctan\left(\frac{u_0^2+u_1^2}{u_2^2+u_3^2}\right)^{1/2} \in [0,\pi],\\ \gamma_T &= \frac{i}{2}\ln\frac{(u_0-iu_1)(u_2-iu_3)}{(u_0+iu_1)(u_2+iu_3)} \in [0,4\pi), \end{aligned}$$

we obtain a transformation converting $\mathbb{R}^8$ to the direct product $\mathbb{R}^5\otimes\mathbf{S}^3$ of the space $\mathbb{R}^5(\vec{x})$ and a three-dimensional sphere $\mathbf{S}^3(\alpha_T,\beta_T,\gamma_T)$.

4. Dyon-oscillator duality

In the coordinates (3.2), (3.3) we can transform the Schrödinger equation for the eight-dimensional isotropic oscillator

$$\frac{\partial^2\psi}{\partial u_\mu^2}+\frac{2m}{\hbar^2}\left(E-\frac{m\omega^2u^2}{2}\right)\psi=0,\quad u_\mu\in\mathbb{R}^8$$

into the equation [**12**]

$$(4.1)\qquad \frac{1}{2m}\left(-i\hbar\frac{\partial}{\partial x_j}-\hbar A_j^a\widehat{T}_a\right)^2\psi+\frac{\hbar^2}{2mr^2}\widehat{T}^2\psi-\frac{e^2}{r}\psi=\epsilon\psi,$$

where ϵ and e^2 are defined by the relation (2.3). The operators $\widehat{T}_a$ are the generators of the SU(2) group. In coordinates $(\alpha_T,\beta_T,\gamma_T)$ they are parametrized as follows

$$\begin{aligned} \widehat{T}_1 &= i\left(\cos\alpha_T\cos\beta_T\frac{\partial}{\partial\alpha_T}+\sin\alpha_T\frac{\partial}{\partial\beta_T}-\frac{\cos\alpha_T}{\sin\beta_T}\frac{\partial}{\partial\gamma_T}\right),\\ \widehat{T}_2 &= i\left(\sin\alpha_T\cot\beta_T\frac{\partial}{\partial\alpha_T}-\cos\alpha_T\frac{\partial}{\partial\beta_T}-\frac{\sin\alpha_T}{\sin\beta_T}\frac{\partial}{\partial\gamma_T}\right),\\ \widehat{T}_3 &= -i\frac{\partial}{\partial\alpha_T}. \end{aligned}$$

Recall that the operators $\widehat{T}_a$ satisfy the following commutation relations

$$\left[\widehat{T}_a,\widehat{T}_b\right]=i\varepsilon_{abc}\widehat{T}_c.$$

The five-dimensional vectors $\vec{A}^a$ are given by the expressions

$$\begin{aligned} \vec{A}^1 &= \frac{1}{r(r+x_0)}(0,x_4,x_3,-x_2,-x_1),\\ \vec{A}^2 &= \frac{1}{r(r+x_0)}(0,-x_3,x_4,x_1,-x_2),\\ \vec{A}^3 &= \frac{1}{r(r+x_0)}(0,x_2,-x_1,x_4,-x_3). \end{aligned}$$

The vectors $\vec{A}^a$ are orthogonal to each other,

$$A_j^a A_j^b = \frac{1}{r^2}\frac{r - x_0}{r + x_0}\delta_{ab}$$

and to the vector $\vec{x} = (x_0, x_1, x_2, x_3, x_4)$.

Equation (4.1) is identical to the Pauli equation and therefore we can give to the triplet of five-dimensional vectors $\vec{A}^a$ the meaning of the vector potentials with the line of singularity along the nonpositive x_0 semiaxis.

The five-dimensional vector potentials $\overrightarrow{B}^a$

$$\begin{aligned}
\overrightarrow{B}^1 &= \frac{1}{r(r - x_0)}(0, -x_4, x_3, -x_2, x_1),\\
\overrightarrow{B}^2 &= \frac{1}{r(r - x_0)}(0, -x_3, -x_4, x_1, x_2),\\
\overrightarrow{B}^3 &= \frac{1}{r(r - x_0)}(0, x_2, -x_1, -x_4, x_3)
\end{aligned}$$

with the singularity axis, directed along the nonnegative x_0 semiaxis, are obtained from the vectors $\vec{A}^a$ by the following gauge transformation

$$B_j = \widehat{S} A_j \widehat{S}^{-1} + i\widehat{S}\frac{\partial}{\partial x_j}\widehat{S}^{-1}.$$

Here $A_j = A_j^a \widehat{T}_a$, $B_j = B_j^a \widehat{T}_a$, and

$$\widehat{S} = e^{-i\gamma\widehat{T}_3} e^{-i\beta\widehat{T}_2} e^{-i\alpha\widehat{T}_3}.$$

The hyperspherical angles α, β and γ are defined as

$$\begin{aligned}
\alpha &= \frac{i}{2}\ln\frac{(x_2 - ix_1)(x_4 - ix_3)}{(x_2 + ix_1)(x_4 + ix_3)} \in [0, 2\pi)\\
\beta &= 2\arctan\left(\frac{x_1^2 + x_2^2}{x_3^2 + x_4^2}\right)^{1/2} \in [0, \pi]\\
\gamma &= \frac{i}{2}\ln\frac{(x_2 + ix_1)(x_4 - ix_3)}{(x_2 - ix_1)(x_4 + ix_3)} \in [0, 4\pi).
\end{aligned}$$

Now, it is necessary to explain what physical system equation (4.1) describes.

5. Field tensor

As a first step, we rewrite the five-dimensional vector potentials $\vec{A}^a$ in the following form

$$A_i^a = \frac{2ig}{r(r + x_0)}\tau_{ij}^a x_j. \tag{5.1}$$

Here τ_{ij}^a are the 5×5 matrices having the following explicit form

$$\tau^1 = \frac{1}{2}\begin{pmatrix} 0 & 0 & 0\\ 0 & 0 & -i\sigma^1\\ 0 & i\sigma^1 & 0\end{pmatrix}, \quad \tau^2 = \frac{1}{2}\begin{pmatrix} 0 & 0 & 0\\ 0 & 0 & i\sigma^3\\ 0 & -i\sigma^3 & 0\end{pmatrix}, \quad \tau^3 = \frac{1}{2}\begin{pmatrix} 0 & 0 & 0\\ 0 & \sigma^2 & 0\\ 0 & 0 & \sigma^2\end{pmatrix},$$

where σ^a are the Pauli matrices and the τ^a matrices satisfying $[\tau^a, \tau^b] = i\epsilon_{abc}\tau^c$. It is obvious that $\tau^a_{ij} = -\tau^a_{ji}$ and, therefore, the vectors $\vec{A}^a$ are orthogonal to $\vec{x}$. Moreover, for the τ^a matrices the following relations occur

(5.2) $$4\tau^a_{ij}\tau^b_{jk} = \delta_{ab}(\delta_{ik} - \delta_{i0}\delta_{k0}) + 2i\epsilon_{abc}\tau^c_{ik},$$

(5.3) $$\begin{aligned}\epsilon_{abc}\tau^b_{ij}\tau^c_{km} = \frac{i}{2}[&(\delta_{i0}\delta_{k0} - \delta_{ik})\tau^a_{jm} - (\delta_{i0}\delta_{m0} - \delta_{im})\tau^a_{jk} \\ &+ (\delta_{j0}\delta_{m0} - \delta_{jm})\tau^a_{ik} - (\delta_{j0}\delta_{k0} - \delta_{jk})\tau^a_{im}].\end{aligned}$$

Using the definition of the Yang–Mills field tensor

$$F^a_{ij} = \frac{\partial A^a_j}{\partial x_i} - \frac{\partial A^a_i}{\partial x_j} + \epsilon_{abc}A^{b(+)}_i A^{c(+)}_k$$

and the expressions (5.1) and (5.3) we can write the field tensor F^a_{ij} in a more explicit form

$$F^a_{ij} = \frac{1}{r^2}[(x_j + r\delta_{j0})A^a_i - (x_i + r\delta_{i0})A^a_j - 2i\tau^a_{ij}].$$

A straightforward computation gives

$$F^a_{ij}F^b_{jk} = \frac{1}{r^6}(x_i x_k - r^2\delta_{ik})\delta_{ab} + \frac{1}{r^2}\epsilon_{abc}F^c_{ik}.$$

or

(5.4) $$F^a_{ij}F^b_{ij} = \frac{4}{r^4}\delta_{ab}.$$

6. Topological charge

It is convenient to perform the computation of the topological charge in the five-dimensional hyperspherical coordinates, which we define as

$$\begin{aligned} x_0 &= r\cos\theta, \\ x_2 + ix_1 &= r\sin\theta\sin\frac{\beta}{2}e^{i(\alpha-\gamma)/2}, \\ x_4 + ix_3 &= r\sin\theta\cos\frac{\beta}{2}e^{i(\alpha+\gamma)/2}, \end{aligned}$$

where $r \in [0, \infty)$, $\theta \in [0, \pi]$, $\alpha \in [0, 2\pi)$, $\beta \in [0, \pi]$, $\gamma \in [0, 4\pi)$.

The components of the second rank tensor in the two coordinate systems are interrelated by the formula

$$\bar{f}_{ik} = \frac{\partial x_m}{\partial \bar{x}_i}\frac{\partial x_n}{\partial \bar{x}_k}f_{mn}$$

in our case $\bar{x}_0 = r$, $\bar{x}_1 = \theta$, $\bar{x}_2 = \beta$, $\bar{x}_3 = \alpha$, $\bar{x}_1 = \gamma$. A direct computation gives that $F^a_{rk} \equiv 0$, where $k = r, \theta, \beta, \alpha, \gamma$ (for the other components see the appendix).

Using the explicit forms of the hyperspherical components F^a_{ij} it is possible to verify that the field tensor F^a_{ij} is self-dual

(6.1) $${}^*F^{a(+)\mu\nu} = F^{a(+)\mu\nu}, \quad \mu, \nu = 1, 2, 3, 4.$$

Further, using the definition of the topological charge

$$q = \frac{1}{32\pi^2}\sum_{a=1}^{3}\oint {}^*F^{a(+)\mu\nu}F^{a(+)}_{\mu\nu}\,dS$$

where $dS = (r^4/8)\sin^3\theta\sin\beta\,d\theta\,d\beta\,d\alpha\,d\gamma$, and taking into account the self-duality equations (6.1) and orthogonality conditions (5.4) we obtain that in our case $q = +1$.

We see that equation (4.1) describes the charge-dyon system with an SU(2) Yang monopole [**19**]. Moreover, the Schrödinger equation for the eight-dimensional isotropic oscillator and equation (4.1) are dual to each other.

It is important to note the following fact

$$q^a = \frac{1}{32\pi^2} \oint {}^*F^{a(+)\mu\nu} F^{a(+)}_{\mu\nu} \, dS = \frac{1}{3},$$

i.e., $\frac{1}{3}$ of the topological charge corresponds to each component of the gauge field F^a_{ij}.

7. Hidden symmetry

Since the obtained system is a non-Abelian extension of the five-dimensional Coulomb problem, it is natural to try to construct an analog of the Runge–Lenz vector for the Yang–Coulomb monopole by passing from $\mathbb{R}^3(\vec{x})$ to $\mathbb{R}^5(\vec{x})$ and taking account of the gauge field [**14**]. The first step was made many years ago [**3**]

$$\widehat{M}_k = \frac{1}{2\sqrt{\mu_0}} \left(\hat{p}_i \hat{l}_{ik} + \hat{l}_{ik} \hat{p}_i + \frac{2\mu_0 e^2}{\hbar} \frac{x_k}{r} \right),$$

where $\hat{p}_i = -i\hbar\partial/\partial x_i$, $\hat{l}_{ij} = 1/\hbar(x_i\hat{p}_j - x_j\hat{p}_i)$. The second step can be realized by the substitution [**19**]:

$$\hat{p}_i \to \hat{\pi}_i = \hat{p}_i - \hbar A^a_i \widehat{T}_a,$$
$$\hat{l}_{ij} \to \hat{L}_{ij} = \frac{1}{\hbar}(x_i\hat{\pi}_j - x_j\hat{\pi}_i) - r^2 F^a_{ij} \widehat{T}_a.$$

The following fundamental commutation relations are valid

$$[\hat{\pi}_i, x_j] = -i\hbar\delta_{ij}, \qquad [\hat{\pi}_i, \hat{\pi}_j] = i\hbar^2 F^a_{ij} \widehat{T}_a.$$

It is possible to verify that

$$[\hat{L}_{ik}, x_j] = i\delta_{ij}x_k - i\delta_{kj}x_i, \qquad [\hat{L}_{ik}, \hat{\pi}_j] = i\delta_{ij}\hat{\pi}_k - i\delta_{kj}\hat{\pi}_i,$$
$$[\hat{L}_{ij}, \hat{L}_{mn}] = i\delta_{im}\hat{L}_{jn} - i\delta_{jm}\hat{L}_{in} - i\delta_{in}\hat{L}_{jm} + i\delta_{jn}\hat{L}_{im},$$

i.e., $\hat{L}_{ij}$ are indeed the generators of the group SO(5) and $[\widehat{H}, \hat{L}_{ij}] = 0$.

After some calculations, we have $[\widehat{H}, \widehat{M}_i] = 0$ which means that $\widehat{M}_i$ is in fact the analog of the Runge–Lenz vector for the Yang–Coulomb monopole. It can also be shown that

$$[\hat{L}_{ij}, \widehat{M}_k] = i\delta_{ik}\widehat{M}_j - i\delta_{jk}\widehat{M}_i, \qquad [\widehat{M}_i, \widehat{M}_k] = -2i\widehat{H}\hat{L}_{ik}.$$

These commutation rules generalize relations known from the theory of the Coulomb problem [**9**].

Finally, let us introduce the 6×6 matrix

$$\widehat{D} = \begin{pmatrix} \hat{L}_{ij} & -\widehat{M}'_i \\ \widehat{M}'_j & 0 \end{pmatrix},$$

where $\widehat{M}'_i = (-2\widehat{H})^{-1/2}\widehat{M}_i$. The components $\widehat{D}_{\mu\nu}$ (where μ, $\nu = 0, 1, 2, 3, 4, 5$) satisfy the commutation relations

$$[\widehat{D}_{\mu\nu}, \widehat{D}_{\lambda\rho}] = i\delta_{\mu\lambda}\widehat{D}_{\nu\rho} - i\delta_{\nu\lambda}\widehat{D}_{\mu\rho} - i\delta_{\mu\rho}\widehat{D}_{\nu\lambda} + i\delta_{\nu\rho}\widehat{D}_{\mu\lambda},$$

i.e., $\widehat{D}_{\mu\nu}$ are the generators of the group SO(6). Since $[\widehat{H}, \widehat{D}_{\mu\nu}] = 0$, one concludes that the Yang–Coulomb monopole is provided by the SO(6) hidden symmetry group.

For the continuous spectrum ($\epsilon > 0$) we have

$$[\widehat{\widetilde{M}}_i, \widehat{\widetilde{M}}_k] = -i\hat{L}_{ik},$$

where $\widehat{\widetilde{M}}_i = (2\widehat{H})^{-1/2}\widehat{M}_i$, and the hidden symmetry group is SO(5, 1).

8. Energy levels

The Casimir operators for SO(6) are [**1**]

$$\widehat{C}_2 = \tfrac{1}{2}\widehat{D}_{\mu\nu}\widehat{D}_{\mu\nu},$$
$$\widehat{C}_3 = \epsilon_{\mu\nu\rho\sigma\tau\lambda}\widehat{D}_{\mu\nu}\widehat{D}_{\rho\sigma}\widehat{D}_{\tau\lambda},$$
$$\widehat{C}_4 = \tfrac{1}{2}\widehat{D}_{\mu\nu}\widehat{D}_{\nu\rho}\widehat{D}_{\rho\tau}\widehat{D}_{\tau\mu}.$$

According to [**16**], the eigenvalues of these operators can be taken as

$$C_2 = \mu_1(\mu_1 + 4) + \mu_2(\mu_2 + 2) + \mu_3^2,$$
$$C_3 = 48(\mu_1 + 2)(\mu_2 + 1)\mu_3,$$
$$C_4 = \mu_1^2(\mu_1 + 4)^2 + 6\mu_1(\mu_1 + 4) + \mu_2^2(\mu_2 + 2)^2 + \mu_3^4 - 2\mu_3^2,$$

where μ_1, μ_2 and μ_3 are positive integers, or half-integers and $\mu_1 \geq \mu_2 \geq \mu_3$.

Direct calculation leads to the representation

$$\begin{aligned} \widehat{C}_2 &= -\frac{e^4\mu_0}{2\hbar^2\widehat{H}} + 2\widehat{T}^2 - 4, \\ \widehat{C}_3 &= 48\left(-\frac{\mu_0 e^4}{2\hbar^2\widehat{H}}\right)^{1/2}\widehat{T}^2, \\ \widehat{C}_4 &= \widehat{C}_2^2 + 6\widehat{C}_2 - 4\widehat{C}_2\widehat{T}^2 - 12\widehat{T}^2 + 6\widehat{T}^4. \end{aligned} \tag{8.1}$$

From these equations we can obtain another expression for the eigenvalue C_4

$$C_4 = [C_2 - 2T(T + 1)]^2 + 6[C_2 - 2T(T + 1)] + 2T^2(T + 1)^2$$

and calculate that

$$C_2 - 2T(T + 1) = \mu_1(\mu_1 + 4), \tag{8.2}$$

$$\mu_2^2(\mu_2 + 2)^2 + \mu_3^4 - 2\mu_3^2 = 2T^2(T + 1)^2. \tag{8.3}$$

The energy levels of the Yang–Coulomb monopole can be derived from (8.1) and (8.2)

$$\epsilon_N^T = -\frac{\mu_0 e^4}{2\hbar^2(N/2 + 2)^2}, \tag{8.4}$$

where $\mu_1 = N/2$ and N is a nonnegative integer number. The substitution of the eigenvalues of $\widehat{H}$ and $\widehat{T}^2$ in the equation for $\widehat{C}_3$ gives one more formula for C_3

$$C_3 = 48(\mu_1 + 2)T(T + 1).$$

Now, we have two expressions for C_3 and the comparison leads to the relation

$$T(T + 1) = (\mu_2 + 1)\mu_3. \tag{8.5}$$

Comparing (8.5) with (8.2) we will get the equation

$$(\mu_2^2 - \mu_3^2)[(\mu_2 + 2)^2 - \mu_3^2] = 0.$$

Since $\mu_3 \leq \mu_2$, one concludes that $\mu_3 = \mu_2$. Then, from (8.5) it follows that $\mu_2 = T$. Therefore, N in the formula (8.4) takes only values $N/2 = T, T+1, T+2, \ldots$—the result known from our paper [**13**].

9. Conclusions

Formulas (3.2) and (3.3) together with the ansatz (2.3) form the duality transformation mapping of the eight-dimensional quantum oscillator into the charge-dyon system with the SU(2) Yang monopole. This type of duality is valid not only for the 8D, 4D and 2D oscillators, but also for the oscillator-like systems with the potentials

$$V(u^2) = c_0 + c_1 u^2 + W(u^2),$$

where $W(u^2)$ has a polynomial form

$$W(u^2) = \sum_{n=2}^{\infty} c_n u^{2n}.$$

For such modified potentials, the ansatz (2.3) can be rewritten as

$$\epsilon = -\frac{c_1}{4}, \qquad e^2 = \frac{E - c_0}{4}.$$

The hidden symmetry of the SU(2) Yang–Coulomb monopole makes it possible to solve the equation (4.1) in the five-dimensional hyperspherical, parabolic and elliptic coordinates by the separation of variables method [**11, 13, 15**].

Appendix A

A.1. The Cartesian components of the gauge field tensor F^a_{ij}.

$$F^1_{01} = -\frac{x_4}{r^3}, \qquad F^1_{02} = -\frac{x_3}{r^3}, \qquad F^1_{03} = \frac{x_2}{r^3}, \qquad F^1_{04} = \frac{x_1}{r^3},$$

$$F^1_{12} = \frac{x_2x_4 - x_1x_3}{r^3(r+x_0)}, \qquad F^1_{13} = \frac{x_1x_2 + x_3x_4}{r^3(r+x_0)}, \qquad F^1_{14} = \frac{1}{r^2}\left[\frac{x_1^2 + x_4^2}{r(r+x_0)} - 1\right],$$

$$F^1_{23} = \frac{1}{r^2}\left[\frac{x_2^2 + x_3^2}{r(r+x_0)} - 1\right], \qquad F^1_{24} = \frac{x_1x_2 + x_3x_4}{r^3(r+x_0)}, \qquad F^1_{34} = -\frac{x_1x_2 + x_3x_4}{r^3(r+x_0)}.$$

$$F^2_{01} = -\frac{x_3}{r^3}, \qquad F^2_{02} = \frac{x_4}{r^3}, \qquad F^2_{03} = -\frac{x_1}{r^3}, \qquad F^2_{04} = \frac{x_2}{r^3},$$

$$F^2_{12} = -\frac{x_1x_4 + x_2x_3}{r^3(r+x_0)}, \qquad F^2_{13} = -\frac{1}{r^2}\left[\frac{x_1^2 + x_3^2}{r(r+x_0)} - 1\right], \qquad F^2_{14} = \frac{x_1x_2 - x_3x_4}{r^3(r+x_0)},$$

$$F^2_{23} = \frac{x_3x_4 - x_1x_2}{r^3(r+x_0)}, \qquad F^2_{24} = \frac{1}{r^2}\left[\frac{x_2^2 + x_4^2}{r(r+x_0)} - 1\right], \qquad F^2_{34} = \frac{x_1x_4 + x_2x_3}{r^3(r+x_0)}.$$

$$F^3_{01} = -\frac{x_2}{r^3}, \qquad F^3_{02} = \frac{x_1}{r^3}, \qquad F^3_{03} = -\frac{x_4}{r^3}, \qquad F^3_{04} = \frac{x_3}{r^3},$$

$$F^3_{12} = \frac{1}{r^2}\left[\frac{x_1^2 + x_2^2}{r(r+x_0)} - 1\right], \qquad F^3_{13} = \frac{x_2x_3 - x_1x_4}{r^3(r+x_0)}, \qquad F^3_{14} = \frac{x_1x_3 + x_2x_4}{r^3(r+x_0)},$$

$$F^3_{23} = -\frac{x_1x_3 + x_2x_4}{r^3(r+x_0)}, \qquad F^3_{24} = \frac{x_2x_3 - x_1x_4}{r^3(r+x_0)}, \qquad F^3_{34} = \frac{1}{r^2}\left[\frac{x_3^2 + x_4^2}{r(r+x_0)} - 1\right].$$

A.2. The hyperspherical components of the gauge field tensor F^a_{ij}.

$$F^1_{\theta\beta} = \tfrac{1}{2}\sin\theta\sin\alpha, \qquad F^1_{\theta\alpha} = 0, \qquad F^1_{\theta\gamma} = -\tfrac{1}{2}\sin\theta\sin\beta\cos\alpha,$$

$$F^1_{\beta\alpha} = -\tfrac{1}{4}\sin^2\theta\cos\alpha, \quad F^1_{\beta\gamma} = -\tfrac{1}{4}\sin^2\theta\cos\beta\cos\alpha, \quad F^1_{\alpha\gamma} = \tfrac{1}{4}\sin^2\theta\sin\beta\sin\alpha.$$

$$F^2_{\theta\beta} = \tfrac{1}{2}\sin\theta\cos\alpha, \qquad F^2_{\theta\alpha} = 0, \qquad F^2_{\theta\gamma} = \tfrac{1}{2}\sin\theta\sin\beta\sin\alpha,$$

$$F^2_{\beta\alpha} = \tfrac{1}{4}\sin^2\theta\sin\alpha, \qquad F^2_{\beta\gamma} = \tfrac{1}{4}\sin^2\theta\cos\beta\sin\alpha, \qquad F^2_{\alpha\gamma} = \tfrac{1}{4}\sin^2\theta\sin\beta\cos\alpha.$$

$$F^3_{\theta\beta} = 0, \qquad F^3_{\theta\alpha} = \tfrac{1}{2}\sin\theta, \qquad F^3_{\theta\gamma} = \tfrac{1}{2}\sin\theta\cos\beta,$$

$$F^3_{\beta\alpha} = 0, \qquad F^3_{\beta\gamma} = -\tfrac{1}{4}\sin^2\theta\sin\beta, \qquad F^3_{\alpha\gamma} = 0.$$

References

1. A. O. Barut and R. Rączka, *Theory of group representations and applications*, PWN—Polish Scientific Publishers, Warsaw, 1977.
2. L. S. Davtyan, L. G. Mardoyan, G. S. Pogosyan, A. N. Sissakian, and V. M. Ter-Antonyan, *Generalised KS transformation: from five-dimensional hydrogen atom to eight-dimensional isotrope oscillator*, J. Phys. A **20** (1987), no. 17, 6121–6125.
3. G. Györgi and J. Révai, Hidden symmetry of the Kepler problem, Ž. ksper. Teoret. Fiz. **48** (1965), 1445–1447 (Russian); English transl., Soviet Physics JETP **21** (1965) 967–968.
4. L. V. Hoang, T. J. Viloria, and L. A. Thu, *On the hydrogen-like atom in five-dimensional space*, J. Phys. A **24** (1991), no. 13, 3021–3030.
5. A. Hurwitz, *Über die Zahlentheorie der Quaternionen*, Nachr. Ges. Wiss. Göttingen Math.-Phys. Kl. (1896), 313–340.
6. M. Kibler and P. Winternitz, *Lie algebras under constraints and nonbijective canonical transformations*, J. Phys. A **21** (1988), no. 8, 1787–1803.
7. P. Kustaanheimo and E. Stiefel, *Perturbation theory of Kepler motion based on spinor regularization*, J. Reine Angew. Math. **218** (1965), 204–219.
8. D. Lambert and M. Kibler, *An algebraic and geometric approach to nonbijective quadratic transformations*, J. Phys. A **21** (1988), no. 2, 307–343.
9. L. D. Landau and E. M. Lifshitz, *Quantum mechanics*, Pergamon Press, London, 1977.
10. T. Levi-Civita, *Sur la résolution qualitative du problème restreint des trois corps*, Opere mathematiche. Memorie e note. Vol. II. 1901–1907, Nicola Zanichelli Editore, Bologna, 1956, pp. 411–417.
11. L. G. Mardoyan, *Five-dimensional* SU(2)*-monopole: continuous spectrum*, Phys. Atomic Nuclei **65** (2002), no. 6, 1063–1069.
12. L. G. Mardoyan, A. N.Sissakian, and V. M. Ter-Antonyan, *Oscillator as a hidden non-Abelian monopole*, Preprint JINR no. E2-96-24, Dubna, 1996; hep-th/9601093.
13. ______, *8D oscillator as a hidden* SU(2) *monopole*, Phys. Atomic Nuclei **61** (1998), no. 10, 1746–1750.
14. ______, *Hidden symmetry of the Yang–Coulomb monopole*, Modern Phys. Lett. A **14** (1999), no. 19, 1303–1307.
15. ______, *Bases and interbasis transformations for the* SU(2)*-monopole*, Teoret. Mat. Fiz. **123** (2000), no. 1, 44–56 (Russian); English transl., Theoret. and Math. Phys. **123** (2000), no. 1, 451–462.
16. A. M. Perelomov and V. S. Popov, *The Casimir operators for orthogonal and symplectic groups*, JETP Lett. **2** (1965), 34–37.
17. J. Schwinger, *A magnetic model of matter*, Science **165** (1969), 757–761.
18. N. Seiberg and E. Witten, *Monopoles, duality and chiral symmetry breaking in $N = 2$ supersymmetric QCD*, Nuclear Phys. B **431** (1994), no. 3, 484–550.
19. C. N. Yang, *Generalization of Dirac's monopole to* SU_2 *gauge fields*, J. Math. Phys. **19** (1978), no. 1, 320–328.

Internat. Center for Advanced Studies, Yerevan State University, Alex Manougian Str., 1, 375025 Yerevan, Armenia

E-mail address: mardoyan@icas.ysu.am

Centre de Recherches Mathématiques
CRM Proceedings and Lecture Notes
Volume **37**, 2004

Supersymmetric Calogero–Moser–Sutherland Models: Superintegrability Structure and Eigenfunctions

Patrick Desrosiers, Luc Lapointe, and Pierre Mathieu

Abstract. We first review the construction of the supersymmetric extension of the quantum Calogero–Moser–Sutherland (CMS) models. We stress the remarkable fact that this extension is completely captured by the insertion of a fermionic exchange operator in the Hamiltonian: sCMS models (s for supersymmetric) are nothing but special exchange-type CMS models. Under the appropriate projection, the conserved charges can thus be formulated in terms of the standard Dunkl operators. This is illustrated in the rational case, where the explicit form of the $4N$ (N being the number of bosonic variables) conserved charges is presented, together with their full algebra. The existence of $2N$ commuting bosonic charges settles the question of the integrability of the srCMS model. We then prove its superintegrability by displaying $2N-2$ extra independent charges commuting with the Hamiltonian. In the second part, we consider the supersymmetric version of the trigonometric case (stCMS model) and review the construction of its eigenfunctions, the Jack superpolynomials. This leads to closed-form expressions, as determinants involving supermonomial symmetric functions. Here we focus on the main ideas and the generic aspects of the construction: those applicable to all models whether supersymmetric or not. Finally, the possible Lie superalgebraic structure underlying the stCMS model and its eigenfunctions is briefly considered.

1. Introduction

This presentation has two main objectives. The first is to stress the interpretation of the supersymmetric extension of the quantum Calogero-Moser-Sutherland (CMS) models [**1, 13–15, 21**][1] as special exchange-type CMS models [**12, 16**]. This provides a direct path for establishing their integrability [**3**] and also, as is first

2000 *Mathematics Subject Classification.* 05E05, 05E35, 05A17, 70F10, 35Q40.
The first author would like to thank the Fondation J. A. Vincent for a student fellowship.
The work of the second and third authors is supported by NSERC.
The authors thank F. Lesage for a useful remark.
This is the final form of the paper.

[1]We use the qualitative 'Calogero–Moser–Sutherland' to describe the generic class of models that includes the models studied by Calogero and Sutherland in the quantum case and by Moser in the classical case. In this work, however, we only treat the quantum case. In this context the name of Moser is often omitted. The rationale for its inclusion is due to the fundamental importance of the Lax formulation in the quantum case, which is a direct extension of the classical one that he introduced.

shown here, their superintegrability. For this analysis, which is the content of Section 2, we focus on the rational case. Our second goal is to provide a simple quasi-qualitative presentation of the main ideas underlying our recent construction of the supersymmetric trigonometric CMS (stCMS) eigenfunctions, that is, of the Jack superpolynomials [**3–5**]. This is the content of Section 3. Most technicalities are avoided. We emphasis that our method is a general program for constructing explicitly the eigenfunctions of a CMS-type Hamiltonian, supersymmetric or not. Some subsections are thus formulated in rather general terms.

2. Integrability and superintegrability of the srCMS model

2.1. Supersymmetric quantum mechanics and fermionic-exchange operators. Consider a quantum model, whose Hamiltonian is denoted $\mathcal{H}$, that contains bosonic and fermionic variables. It thus contains in addition to the $2N$ bosonic variables (x_i, p_i), the $2N$ fermionic variables $(\theta_i, \theta_i^\dagger)$ (e.g., $\theta_i\theta_j = -\theta_j\theta_i$, so that $\theta_i^2 = 0$). The bosonic and fermionic variables satisfy respectively a Heisenberg and a Clifford algebra:

$$[x_j, p_k] = i\delta_{jk}, \qquad \{\theta_j, \theta_k^\dagger\} = \delta_{jk}, \tag{2.1}$$

where $\{\ ,\ \}$ stands for the anticommutaror. All other commutators or anticommutators are equal to zero. We usually work with a differential realization of these algebras:

$$p_j = -i\frac{\partial}{\partial x_j}, \qquad \theta_j^\dagger = \frac{\partial}{\partial \theta_j}. \tag{2.2}$$

To construct a supersymmetric Hamiltonian, we will first construct two supersymmetric charges, denoted Q and $Q^\dagger$, and define the Hamiltonian as their anticommutator:

$$\mathcal{H} = \tfrac{1}{2}\{Q, Q^\dagger\}. \tag{2.3}$$

The Hamiltonian is invariant under a supersymmetric transformation if

$$Q^2 = (Q^\dagger)^2 = 0. \tag{2.4}$$

With the charges written under the form

$$Q = \sum_{i=1}^N \theta_i^\dagger A_i(x,p), \qquad Q^\dagger = \sum_{i=1}^N \theta_i A_i^\dagger(x,p), \tag{2.5}$$

(2.4) requires:

$$[A_i, A_j] = 0 = [A_i^\dagger, A_j^\dagger], \quad \forall i, j. \tag{2.6}$$

The generic supersymmetric Hamiltonian thus reads:

$$\mathcal{H} = \frac{1}{2}\Big(\sum_i A_i^\dagger A_i + \sum_{i,j} \theta_i^\dagger \theta_j [A_i, A_j^\dagger]\Big). \tag{2.7}$$

For nonrelativistic models, the Hamiltonian is proportional to the square of the particles' momenta. This forces A_i to be a linear function of the momentum p_i, that is:

$$Q = \sum_j \theta_j^\dagger \big(p_j - i\Phi_j(x)\big), \qquad Q^\dagger = \sum_j \theta_j \big(p_j + i\Phi_j(x)\big). \tag{2.8}$$

Condition (2.6) imposes that the potential $\Phi_j(x)$ be of the form

$$\Phi_j(x) = \partial_{x_j} W(x), \tag{2.9}$$

where $W(x)$ (called the prepotential), is an arbitrary function of the variables $x_1, \ldots, x_N$. The supersymmetric Hamiltonian now takes the form [**9**]:

$$\mathcal{H} = \frac{1}{2} \sum_i (p_i^2 + (\partial_{x_i} W)^2 + \partial_{x_i}^2 W) - \sum_{i,j} \theta_i \theta_j^\dagger \partial_{x_i} \partial_{x_j} W. \tag{2.10}$$

This Hamiltonian is an extension of the purely bosonic model whose potential is $\sum_i [(\partial_{x_i} W)^2 + \partial_{x_i}^2 W]$. We stress that this construction fixes uniquely the supersymmetric extension of the model to be considered. Specializing to a symmetric prepotential that can be broken up into a sum of two-body interactions:

$$W(x) = \sum_{i<j} w(x_{ij}) \quad \text{with } w'(x_{ij}) = X_{ij}, \tag{2.11}$$

where X_{ij} stands for an odd function of $x_{ij} = x_i - x_j$, our Hamiltonian reads:

$$\mathcal{H} = \frac{1}{2} \sum_i p_i^2 + \sum_{i<j} [X_{ij}^2 + X'_{ij}(1 - \theta_{ij}\theta_{ij}^\dagger)] + \sum_{i<j<k} Y_{ijk}, \tag{2.12}$$

with

$$Y_{ijk} = X_{ij}X_{ik} + X_{ji}X_{jk} + X_{ki}X_{kj}. \tag{2.13}$$

This is the supersymmetric Hamiltonian we were looking for.

The main observation at this point is the following: the term

$$\kappa_{ij} \equiv 1 - \theta_{ij}\theta_{ij}^\dagger = 1 - (\theta_i - \theta_j)(\partial_{\theta_i} - \partial_{\theta_j}), \tag{2.14}$$

which captures the whole dependence of the Hamiltonian upon the fermionic variables, is a fermionic-exchange operator [**18**]. In other words, its action on an arbitrary function g of the fermionic variables θ_i and $\theta_i^\dagger$ reads

$$\kappa_{ij} g(\theta_i, \theta_j, \theta_i^\dagger, \theta_j^\dagger) = g(\theta_j, \theta_i, \theta_j^\dagger, \theta_i^\dagger)\, \kappa_{ij}. \tag{2.15}$$

In addition, it satisfies all the properties of an exchange operator:

$$\kappa_{ij} = \kappa_{ji}, \qquad \kappa_{ij}^\dagger = \kappa_{ij}, \qquad \kappa_{ij}\kappa_{jk} = \kappa_{ik}\kappa_{ij} = \kappa_{jk}\kappa_{ki}, \qquad \kappa_{ij}^2 = 1. \tag{2.16}$$

That the supersymmetric extension is fully captured by the introduction of a fermionic exchange operator is thus a generic feature of supersymmetric many-body problems whose interaction is decomposable into a sum of two-body interactions. This is also a key technical tool in our subsequent analysis, as we will shortly explain. But we would like to make the observation that, up to this point, the discussion is quite general and *not restricted to integrable problems*.

Having a Hamiltonian expressed in terms of a fermionic exchange operator implies that under the right projection, we can trade κ_{ij} for the ordinary exchange operator, K_{ij}, that exchanges the variables x_i and x_j:

$$K_{ij} f(x_i, x_j) = f(x_j, x_i) K_{ij}, \tag{2.17}$$

with $f(x_i, x_j)$ standing for a function or an operator. The suitable projection is in our case the one on the space P^{S_N} of functions invariant under the simultaneous exchange of the bosonic and the fermionic variables, that is, invariant under the action of

$$\mathcal{K}_{ij} = K_{ij}\kappa_{ij}. \tag{2.18}$$

On this space, the supersymmetric Hamiltonian reduces to an ordinary exchange-type Hamiltonian [3]. Stated differently: acting on the proper space, we can study many features of the supersymmetric model *without even introducing fermionic degrees of freedom!*

2.2. The integrability of the srCMS model. To work with a concrete example, consider the simple srCMS model [9]:

$$\mathcal{H}_{(r)} = -\frac{1}{2}\sum_{j=1}^{N} \partial_{x_j}^2 + \sum_{\substack{j,k=1\\ j<k}}^{N} \frac{\beta(\beta-\kappa_{jk})}{(x_j-x_k)^2}. \tag{2.19}$$

We will establish the integrability of this model using the Dunkl-operator formalism, thus relying heavily on the projection trick.

Using the rational Dunkl operators,

$$D_j = \partial_{x_j} - \beta\sum_{k\neq j} \frac{1}{x_{jk}} K_{jk}, \tag{2.20}$$

we find

$$-\frac{1}{2}\sum_{j=1}^{N} D_j^2 = -\frac{1}{2}\sum_{j=1}^{N} \partial_{x_j}^2 + \sum_{\substack{j,k=1\\ j<k}}^{N} \frac{\beta(\beta-K_{jk})}{(x_j-x_k)^2}. \tag{2.21}$$

The Dunkl operators are commuting and covariant:

$$[D_j, D_i] = 0, \qquad\qquad K_{ij}D_j = D_iK_{ij}. \tag{2.22}$$

Therefore, the conservation laws of the rCMS model (with exchange terms) are simply $\sum_j D_j^n$, for $n = 1, \ldots, N$. The conserved charges of the rCMS model without exchange terms are obtained by simply projecting these expressions onto the space of symmetric functions, which amounts to replacing every factor K_{ij} by 1 once pushed to the right. On the other hand, the conserved charges of the srCMS model are obtained as follows:

$$\mathcal{H}_n = \sum_j D_j^n|_{P^{S_N}}, \quad n = 1, \ldots, N. \tag{2.23}$$

In particular, $\mathcal{H}_2 = -2\mathcal{H}_{(r)}$. The explicit dependence of the srCMS conserved charges $\mathcal{H}_n$ upon the fermionic variables can be obtained by implementing the projection, that is, replacing every factor K_{ij} by κ_{ij} once shifted to the right. The proof of the commutativity of these charges leans on a simple property of the projection:

$$[A|_{P^{S_N}}, B|_{P^{S_N}}] = [A, B]|_{P^{S_N}} \quad \text{if} \quad [\mathcal{K}_{ij}, A] = [\mathcal{K}_{ij}, B] = 0. \tag{2.24}$$

The operators $\mathcal{H}_n$ meet this requirement since $[\mathcal{K}_{ij}, (\sum_k D_k^n)] = 0$. Therefore, the commutator can be evaluated before doing the projection and its vanishing is an immediate consequence of the commutativity of the Dunkl operators.

We have thus found the supersymmetric extension of all the usual rCMS charges and proved their commutativity. This however does not imply the integrability of the srCMS model because there are more degrees of freedom in the supersymmetric case (these are the $2N$ extra Grassmannian variables). Consequently, one should expect more conserved charges. It is actually rather easy to construct $3N$ extra charges involving explicitly (i.e., even before the projection) some fermionic variables [**3**]:

$$\begin{aligned} \mathcal{Q}_{(n)} &= \sum_i \theta_i D_i^n|_{PS_N}, && n = 0, 1, \ldots, N-1, \\ \mathcal{Q}^\dagger_{(n)} &= \sum_i \theta_i^\dagger D_i^n|_{PS_N}, && n = 0, 1, \ldots, N-1, \\ \mathcal{I}_{(n)} &= \sum_i \theta_i\theta_i^\dagger D_i^n|_{PS_N}, && n = 0, 1, \ldots, N-1. \end{aligned} \tag{2.25}$$

These charges satisfy the following algebra:

$$\begin{gathered} \{\mathcal{Q}_{(n)}, \mathcal{Q}_{(m)}\} = \{\mathcal{Q}^\dagger_{(n)}, \mathcal{Q}^\dagger_{(m)}\} = [\mathcal{I}_{(n)}, \mathcal{I}_{(m)}] = 0, \\ [\mathcal{Q}_{(n)}, \mathcal{H}_{(m)}] = [\mathcal{Q}^\dagger_{(n)}, \mathcal{H}_{(m)}] = [\mathcal{I}_{(n)}, \mathcal{H}_{(m)}] = [\mathcal{H}_{(n)}, \mathcal{H}_{(m)}] = 0. \end{gathered} \tag{2.26}$$

together with

$$\begin{gathered} \{\mathcal{Q}_{(n)}, \mathcal{Q}^\dagger_{(m)}\} = \mathcal{H}_{(n+m)}, \\ [\mathcal{Q}^\dagger_{(n)}, \mathcal{I}_{(m)}] = \mathcal{Q}^\dagger_{(n+m)}, \qquad [\mathcal{Q}_{(n)}, \mathcal{I}_{(m)}] = -\mathcal{Q}_{(n+m)}. \end{gathered} \tag{2.27}$$

There are thus $4N$ conserved charges, $2N$ bosonic and $2N$ fermionic. Only the $2N$ bosonic ones are mutually commuting and independent.

Actually, it would seem at first sight that there are $(N+1)$ $\mathcal{I}_{(n)}$-type charges and thus a total of $2N+1$ mutually commuting conserved charges. However, it can be checked that say $\mathcal{I}_{(N)}$ can be expressed in terms of the lower order $\mathcal{I}_{(n)}$ as well as the $\mathcal{H}_{(n)}$'s. Let us illustrate this in the simple context of two particles ($N = 2$):

$$\begin{aligned} \mathcal{H}_1 &= D_1 + D_2, & \mathcal{H}_2 &= D_1^2 + D_2^2, \\ \mathcal{I}_0 &= \theta_1\theta_1^\dagger + \theta_2\theta_2^\dagger, & \mathcal{I}_1 &= \theta_1\theta_1^\dagger D_1 + \theta_2\theta_2^\dagger D_2. \end{aligned} \tag{2.28}$$

The operator $\mathcal{I}_2$ depends of the preceding four conserved quantities. Indeed, it is easily checked that

$$\mathcal{I}_2 = \mathcal{I}_1\mathcal{H}_1 - \tfrac{1}{2}\mathcal{I}_0(\mathcal{H}_1^2 - \mathcal{H}_2). \tag{2.29}$$

It is clear that there can be no more than $2N$ mutually commuting conserved charges since this is the maximal number of commuting operators in the free case ($\beta = 0$), in which case these quantities are $\{p_i, \theta_i\theta_i^\dagger\}$.

The srCMS model is thus seen to be integrable in the usual sense.[2]

2.3. The superintegrability of the rsCMS model. Consider the following $2N$ quantities:

$$\begin{aligned} \mathcal{L}_n &= \sum_j x_j D_j^{n+1}|_{PS_N}, && n = -1, \ldots, N-2, \\ \mathcal{M}_n &= \sum_j x_j\theta_j\partial_{\theta_j} D_j^{n+1}|_{PS_N}, && n = -1, \ldots, N-2. \end{aligned} \tag{2.30}$$

They satisfy the algebra

$$[\mathcal{L}_n, \mathcal{L}_m] = (n-m)\mathcal{L}_{n+m}, \quad [\mathcal{H}_n, \mathcal{L}_m] = n\mathcal{H}_{n+m}, \quad [\mathcal{H}_n, \mathcal{M}_m] = n\mathcal{I}_{n+m}. \tag{2.31}$$

[2]The integrability can also be established via the Lax formalism [**3**]. The form of the Lax operator is actually the same as in the nonsupersymmetric case, except that every factor X_{ij} (differentiated or not) is replaced by $X_{ij}\kappa_{ij}$.

These relations are easily obtained using:

$$(2.32)\quad [D_i^n, x_j] = \delta_{ij}\Big(nD_i^{n-1} - \beta\sum_{k\neq i}\frac{D_i^n - D_k^n}{D_i - D_k}K_{ij}\Big) - \beta(1-\delta_{ij})\frac{D_i^n - D_j^n}{D_i - D_j}K_{ij},$$

for $n = 0, 1, 2, \ldots,$ and[3]

$$(2.33)\quad [x_iD_i^n, x_jD_j^m] = \delta_{ij}\Big[(n-m)D_i^{n+m-1} + \beta x_i\sum_{k\neq i}\frac{D_i^nD_k^m - D_i^mD_k^n}{D_i - D_k}K_{ij}\Big]$$
$$- \beta(1-\delta_{ij})\Big[\frac{x_iD_i^{n+m} + x_jD_j^{n+m}}{D_i - D_j}K_{ij} - (x_i + x_j)\frac{D_i^mD_j^n}{D_i - D_j}K_{ij}\Big],$$

for $n, m = 1, 2, 3, \ldots$.

We can construct from these, $2N-2$ new and independent conserved charges (i.e., commuting with $\mathcal{H}_2$) :

$$(2.34)\qquad \mathcal{J}_n = \mathcal{H}_{n+1}\mathcal{L}_{-1} - \mathcal{L}_{n-1}\mathcal{H}_1, \qquad \mathcal{K}_n = \mathcal{I}_{n+1}\mathcal{M}_{-1} - \mathcal{M}_{n-1}\mathcal{I}_1.$$

This a direct supersymmetric extension of the argument given in [**10**].[4] The srCMS model is thus not only *super* and *integrable*, but also *superintegrable.*

3. The stCMS model: construction of the eigenfunctions

3.1. The stCMS Hamiltonian. We now move to the second part of this work and discuss the construction of the eigenfunctions of the sCMS models. If the integrability structure of the (s)CMS models is most simply analyzed in the rational case, the study of the eigenfunctions is most naturally done in the trigonometric case.[5] The Hamiltonian of the stCMS model reads [**3, 18**]:

$$(3.1)\qquad \mathcal{H}_{(t)} = -\frac{1}{2}\sum_{i=1}^N \partial_{x_i}^2 + \Big(\frac{\pi}{L}\Big)^2\sum_{i<j}\frac{\beta(\beta - \kappa_{ij})}{\sin^2(\pi x_{ij}/L)} - \Big(\frac{\pi\beta}{L}\Big)^2\frac{N(N^2-1)}{6},$$

where L is the circumference of the circle on which the particles are confined. Removing the contribution of the ground-state wave function,

$$(3.2)\qquad \psi_0(x) = \Delta^\beta(x) \equiv \prod_{j<k}\sin^\beta\Big(\frac{\pi x_{jk}}{L}\Big),$$

the transformed Hamiltonian then becomes

$$(3.3)\qquad \overline{\mathcal{H}} \equiv \frac{1}{2}\Big(\frac{L}{\pi}\Big)^2\Delta^{-\beta}\mathcal{H}_{(t)}\Delta^\beta.$$

---

[3]The expression of $[x_iD_i^n, x_jD_j^m]$ contains many misprints in [**10**].

[4]Furthermore, the algebra (2.31) obviously implies the algebraic linearization of the $4N$ equations of motion:

$$\frac{d\mathcal{H}_n}{dt} = 0, \qquad \frac{d\mathcal{I}_n}{dt} = 0, \qquad \frac{d\mathcal{L}_n}{dt} = -i\mathcal{H}_{n+2}, \qquad \frac{d\mathcal{M}_n}{dt} = -i\mathcal{I}_{n+2}.$$

Such a linearization is also possible for the stCMS model. It follows from a direct generalization of the approach of [**2**] (using the Lax formalism). On the other hand, note that the superintegrability of the classical rCMS model was proved in [**22**].

[5]Of course there are no bound states in the rational case. The presence of bound states necessitates the introduction of an harmonic confinement. But then the eigenfunctions turn out (somewhat surprisingly) to be expressible in terms of the eigenfunctions of the trigonometric case, the Jack polynomials [**20**].

Expressed in terms of the new bosonic variables $z_j = e^{2\pi i x_j/L}$, it finally reads

$$\overline{\mathcal{H}} = \sum_i (z_i\partial_i)^2 + \beta\sum_{i<j}\frac{z_i+z_j}{z_{ij}}(z_i\partial_i - z_j\partial_j) - 2\beta\sum_{i<j}\frac{z_iz_j}{z_{ij}^2}(1-\kappa_{ij}). \tag{3.4}$$

3.2. The $\overline{\mathcal{H}}$ eigenfunctions as symmetric superpolynomials. The eigenfunctions of the Hamiltonian (3.4) are superpolynomials, namely polynomials of the bosonic variables z_i and the fermionic ones θ_i. Looking for the proper generalization of the Jack polynomials, we are interested in eigenfunctions that are symmetric with respect to the simultaneous interchange of both types of variables, i.e., invariant under the action of $\mathcal{K}_{ij}$.

Manifestly, $\overline{\mathcal{H}}$ leaves invariant the space of polynomials of a given degree in z and a given degree in θ, being homogeneous in both sets of variables. We consider eigenfunctions of the form:

$$\mathcal{A}^{(m)}(z,\theta;\beta) = \sum_{\substack{i_1,i_2,\ldots,i_m=1\\ i_1<i_2<\cdots<i_m}}^{N} \theta_{i_1}\cdots\theta_{i_m}A^{(i_1\ldots i_m)}(z;\beta), \qquad m=0,1,2,3,\ldots, \tag{3.5}$$

where $A^{(i_1\ldots i_m)}$ is a homogeneous polynomial in z. Due to the presence of m fermionic variables in its expansion, $\mathcal{A}^{(m)}$ is said to belong to the m-fermion sector. We stress that the simple dependence upon the fermionic variables, which factorizes in monomial prefactors, is a consequence of their anticommuting nature.

We now clarify the symmetry properties, with respect to the z variables, of the eigenfunctions $\mathcal{A}^{(m)}$, assumed to be invariant under the action of the exchange operators $\mathcal{K}_{ij}$. Given that the θ products are antisymmetric, i.e.,

$$\kappa_{i_ai_b}\theta_{i_1}\cdots\theta_{i_m} = -\theta_{i_1}\cdots\theta_{i_m} \quad \text{if } i_a, i_b \in \{i_1,\ldots,i_m\}, \tag{3.6}$$

the superpolynomials $A^{(i_1\ldots i_m)}$ must be partially antisymmetric to ensure the complete symmetry of $\mathcal{A}^{(m)}$. More precisely, the functions $A^{(i_1\ldots i_m)}$ must satisfy the following relations:[6]

$$\begin{aligned} K_{ij}A^{(i_1\ldots i_m)}(z;\beta) &= -A^{(i_1\ldots i_m)}(z;\beta), &&\forall i \text{ and } j\in\{i_1,\ldots,i_m\},\\ K_{ij}A^{(i_1\ldots i_m)}(z;\beta) &= A^{(i_1\ldots i_m)}(z;\beta), &&\forall i \text{ and } j\notin\{i_1,\ldots,i_m\}. \end{aligned} \tag{3.7}$$

We have thus established that any symmetric eigenfunction of the stCMS model contains terms of *mixed symmetry in* z: each polynomial $A^{(i_1\ldots i_m)}$ is completely antisymmetric in the variables $\{z_{i_1},\ldots,z_{i_m}\}$, and totally symmetric in the remaining variables $z/\{z_{i_1},\ldots,z_{i_m}\}$ [**3**].

To proceed further, we need to address the following points: how to label the eigenfunctions; how to define a natural basis for the space of superpolynomials; how to define the eigenfunctions via a triangular expansion in that basis. These points are considered successively in the following subsections.

3.3. Superpartitions. Symmetric polynomials are indexed by partitions. In the same manner, symmetric superpolynomials, i.e., polynomials in P^{S_N}, can be indexed by *superpartitions*. A superpartition in the m-fermion sector is a sequence of nonnegative integers that generates two partitions separated by a semicolon [**3**]:

$$\Lambda = (\Lambda_1,\ldots,\Lambda_m;\Lambda_{m+1},\ldots,\Lambda_N) = (\Lambda^{\rm a};\Lambda^{\rm s}), \tag{3.8}$$

[6] Note that the case $m=1$, with $\mathcal{A}^{(1)} = \sum_i \theta_i A^i(z;1/\beta)$, is special in that regard: $K_{ij}A^k = A^k$ if and only if $i,j\neq k$.

the first one being associated to an antisymmetric function of $\{z_{i_1},\dots,z_{i_m}\}$ (and has thus no repeated parts)

$$\Lambda^{\rm a} = (\Lambda_1,\dots,\Lambda_m), \quad \Lambda_i > \Lambda_{i+1}, \; i=1,\dots,m-1, \tag{3.9}$$

and the second one, to a symmetric function of the variables $\{z_{i_{m+1}},\dots,z_{i_N}\}$

$$\Lambda^{\rm s} = (\Lambda_{m+1},\dots,\Lambda_N), \quad \Lambda_i \geq \Lambda_{i+1}, \; i=m+1,\dots,N. \tag{3.10}$$

In the zero-fermion sector ($m=0$), the semicolon disappears and we recover the partition $\Lambda^{\rm s}$.

The weight (or degree) of a superpartition is simply the sum of its parts. For instance, the only possible superpartitions of weight 3 in the 1-fermion sector are:

$$(3;0), \quad (2;1), \quad (1;2), \quad (1;1,1), \quad (0;3), \quad (0;2,1), \quad (0;1,1,1), \tag{3.11}$$

while in the 2-fermion sector, they are:

$$(3,0;0) \quad (2,1;0) \quad (2,0;1) \quad (1,0;2) \quad (1,0;1,1). \tag{3.12}$$

3.4. The monomial symmetric superpolynomials basis. Given our goal of constructing the superextension of the Jack polynomials, which themselves decompose triangularly in the symmetric monomial basis, the superextension of the latter will provide our natural expansion basis. The monomial symmetric superpolynomials (supermonomials for short) are defined as [**3**]:

$$m_\Lambda(z,\theta) = m_{(\Lambda_1,\dots,\Lambda_m;\Lambda_{m+1},\dots,\Lambda_N)}(z,\theta) = {\sum_{\sigma\in S_N}}' \theta^{\sigma(1,\dots,m)} z^{\sigma(\Lambda)}, \tag{3.13}$$

where the prime indicates that the summation is restricted to distinct terms, and where

$$z^{\sigma(\Lambda)} = z_1^{\Lambda_{\sigma(1)}} \cdots z_m^{\Lambda_{\sigma(m)}} z_{m+1}^{\Lambda_{\sigma(m+1)}} \cdots z_N^{\Lambda_{\sigma(N)}} \quad \text{and} \quad \theta^{\sigma(1,\dots,m)} = \theta_{\sigma(1)}\cdots\theta_{\sigma(m)}. \tag{3.14}$$

Clearly, in the zero-fermion sector, a supermonomial reduces to an ordinary symmetric monomial.[7] Here is an example with $N=4$ that neatly illustrates the mixed symmetry of each component:

$$\begin{aligned} m_{(1,0;1,1)} = {} & \theta_1\theta_2(z_1-z_2)(z_3z_4) + \theta_1\theta_3(z_1-z_3)(z_2z_4) \\ & + \theta_1\theta_4(z_1-z_4)(z_2z_3) + \theta_2\theta_3(z_2-z_3)(z_1z_4) \\ & + \theta_2\theta_4(z_2-z_4)(z_1z_3) + \theta_3\theta_4(z_3-z_4)(z_1z_2) \end{aligned} \tag{3.15}$$

---

[7] There are other simple bases for the space of symmetric superpolynomials. For instance, one can combine the $2N$ algebraically independent symmetric power sums,

$$p_n = \sum_i z_i^n, \qquad q_n = \sum_i \theta_i z_i^n,$$

to form a new basis:

$$p_\Lambda = q_{\Lambda^{\rm a}} p_{\Lambda^{\rm s}} = q_{\Lambda_1}\cdots q_{\Lambda_m} p_{\Lambda_{m+1}} \cdots p_{\Lambda_N}.$$

One can also generate the whole space with the elementary symmetric superfunctions,

$$e_n = m_{(1^n)}, \qquad f_n = m_{(0;1^n)},$$

or with the complete symmetric superfunctions,

$$h_n = \sum_{\substack{\deg \Lambda = n \\ 0-\text{fermion}}} m_\Lambda, \qquad j_n = \sum_{\substack{\deg \Lambda = n \\ 1-\text{fermion}}} (\Lambda_1+1) m_\Lambda.$$

In all cases, $n=1,\dots,N$ for the bosonic functions (p_n, e_n, h_n) while $n=0,\dots,N-1$ for the fermionic ones.

Two other examples (with $N = 3$) will clarify the different role of a zero entry in each sector:

$$\begin{aligned} m_{(3;0)} &= \theta_1 z_1^3 + \theta_2 z_2^3 + \theta_3 z_3^3 \\ m_{(0;3)} &= \theta_1(z_2^3 + z_3^3) + \theta_2(z_1^3 + z_3^3) + \theta_3(z_1^3 + z_2^3). \end{aligned} \tag{3.16}$$

3.5. Jack superpolynomials: a first definition. We now define our first candidates for the role of Jack superpolynomials (in the m-fermion sector) as the unique eigenfunctions of the supersymmetric Hamiltonian $\overline{\mathcal{H}}$,

$$\overline{\mathcal{H}} \mathcal{J}_\Lambda(z, \theta; \beta) = \varepsilon_\Lambda \mathcal{J}_\Lambda(z, \theta; \beta), \tag{3.17}$$

that can be decomposed triangularly in terms of monomial superfunctions:

$$\mathcal{J}_\Lambda(z, \theta; \beta) = m_\Lambda(z, \theta) + \sum_{\Omega;\Omega<\Lambda} c_{\Lambda,\Omega}(\beta) m_\Omega(z, \theta). \tag{3.18}$$

Clearly, for this definition to be complete, we need to specify the ordering ($<$) underlying the triangular decomposition. The simplest and most natural choice at this point is to formulate the ordering in terms of the partitions Λ^* and Ω^* associated respectively to the superpartitions Λ and Ω, by rearranging their parts in decreasing order. (For instance, the rearrangement of $\Lambda = (7, 4, 3, 1; 8, 6, 5, 3, 3, 1)$ is the partition $\Lambda^* = (8, 7, 6, 5, 4, 3, 3, 3, 1, 1)$.) We thus say that $\Omega < \Lambda$ if $\Omega^* < \Lambda^*$ with respect to the dominance ordering on partitions:

$$\Lambda^* \geq \Omega^* \iff \Lambda_1^* + \Lambda_2^* + \cdots + \Lambda_i^* \geq \Omega_1^* + \Omega_2^* + \cdots + \Omega_i^*, \ \forall i, \tag{3.19}$$

with $\Lambda_i^* = (\Lambda^*)_i$. This turns out to be sufficient to calculate explicitly the Jack superpolynomials [**3**]. Here is an example of such a polynomial:

$$\mathcal{J}_{(2;2)} = m_{(2;2)} + \frac{2\beta}{1+\beta} m_{(2;1^2)} + \frac{\beta}{1+\beta} m_{(1;2,1)} + \frac{6\beta^2}{(1+\beta)(1+2\beta)} m_{(1;1^3)}. \tag{3.20}$$

Although complete, the above characterization is not quite precise. Indeed, not all supermonomials m_Ω labeled by a superpartition Ω, dominated by Λ in the above way, do appear in the decomposition. For instance, the terms $m_{(0;2,1,1)}$ and $m_{(0;1^4)}$, although allowed by the ordering condition, are not present in the expression of $\mathcal{J}_{(2;2)}$. This poses the following natural question: can we characterize *precisely* the terms that appear in the supermonomial triangular decomposition of $\mathcal{J}_\Lambda$, that is, can we pinpoint the precise ordering at work? A second (and more ambitious) related question is the following: can we write down an explicit formula for the coefficients $c_{\Lambda,\Omega}$?

The answer to both of these questions is *yes*. The clue to obtain the answer lies in the following observation: the action of $\overline{\mathcal{H}}$ on the supermonomial basis is triangular and this triangularity determines precisely the ordering entering in the definition of the eigenfunctions. Moreover, the action of $\overline{\mathcal{H}}$ can be computed exactly, essentially because it can be reduced to a two-body computation. This turns out to provide all the data required for evaluating the coefficients $c_{\Lambda,\Omega}$. These conclusions are completely general and apply to any CMS model, supersymmetric or not. In the next subsections, we indicate the key steps in reaching these conclusions, keeping the presentation rather general. Technical details can be found in [**4**].

3.6. Generalities: the triangular action of the Hamiltonian, the induced ordering and a determinantal formula for the eigenfunctions. Consider a generic Hamiltonian H and a generic basis m_a. (Typically, the m_a will be monomial functions but another basis could be used.) Suppose that, in a given subspace spanned by three basis elements m_a, m_b, m_c (for some labels a, b, c), the action of H takes the following form:

$$\begin{aligned} Hm_a &= \epsilon_a m_a + v_{ab} m_b + v_{ac} m_c \\ Hm_b &= \epsilon_b m_b + v_{bc} m_c \\ Hm_c &= \epsilon_c m_c, \end{aligned} \tag{3.21}$$

where the eigenvalues are all distinct. The action of H is obviously triangular. It is also clear that the triangularity entails an ordering: $a > b > c$. The key point is that this ordering is precisely the one governing the triangular expansion of the eigenfunctions of H, denoted J_a, in the m_a basis. This follows from the following result: if H acts on the m_a's as in (3.21), its eigenfunction reads [**11**]:

$$J_a \propto \begin{vmatrix} m_c & m_b & m_a \\ \epsilon_c - \epsilon_a & v_{bc} & v_{ac} \\ 0 & \epsilon_b - \epsilon_a & v_{ab} \end{vmatrix}. \tag{3.22}$$

Let us verify that $(H-\epsilon_a)J_a$ does indeed vanish. Since the entries of the determinant are numbers except for the first row, H acts nontrivially only on this row:

$$\begin{aligned} &(H-\epsilon_a)J_a \\ &\propto \begin{vmatrix} (\epsilon_c-\epsilon_a)m_c & (\epsilon_b-\epsilon_a)m_b+v_{bc}m_c & (\epsilon_a-\epsilon_a)m_a+v_{ab}m_b+v_{ac}m_c \\ \epsilon_c-\epsilon_a & v_{bc} & v_{ac} \\ 0 & \epsilon_b-\epsilon_a & v_{ab} \end{vmatrix}. \end{aligned} \tag{3.23}$$

The coefficient of m_a is identically zero. The coefficient of m_b is most simply obtained by setting $m_c = 0$. This leads to

$$(H-\epsilon_a)J_a|_{m_c=0} \propto \begin{vmatrix} 0 & (\epsilon_b-\epsilon_a)m_b & v_{ab}m_b \\ \epsilon_c-\epsilon_a & v_{bc} & v_{ac} \\ 0 & \epsilon_b-\epsilon_a & v_{ab} \end{vmatrix}. \tag{3.24}$$

The first row being proportional to the third one, the determinant is zero. Similarly, by setting $m_b = 0$, we see that the first two rows become proportional, which again enforces the vanishing of the determinant. Since the expansion coefficients of $(H-\epsilon_a)J_a$ in the monomial basis are all zero, we can conclude that $(H-\epsilon_a)J_a = 0$.

The eigenfunction J_a is thus given by the determinant (3.22) up to a multiplicative constant. Note that it is not equal to zero because the eigenvalues being all distinct, the coefficient of m_a cannot vanish. But this shows readily (i.e., by expanding the determinant) that it can be written under the form $J_a \propto (m_a + \cdots)$ where the dots refer to lower order terms in *the ordering induced by the action of H in the m_a basis.* This is precisely the point we wanted to emphasize.

Note that the eigenfunction can be determined uniquely by simply enforcing its leading expansion coefficient to be one (monic condition).

This simple example shows neatly that the ordering governing the decomposition of J_a in the m_a basis is encoded in the triangularity of H on $\{m_a\}$. However, it is somewhat misleading in its simplicity. It suggests that all terms occurring in the decomposition of J_a can be compared with each others and that a single 'chain of ordering' (which refers to the case where all m_i with i comparable to a do appear in the action of Hm_a) is always involved. We will thus consider a second slightly more complicated case that captures the generic features.

Consider a five-dimensional subspace spanned by m_a, m_b, m_c, m_d, m_e and suppose the following action of Hamiltonian:

$$\begin{aligned} Hm_a &= \epsilon_a m_a + v_{ab} m_b + v_{ac} m_c + v_{ad} m_d \\ Hm_b &= \epsilon_b m_b + v_{bc} m_c \\ Hm_c &= \epsilon_c m_c + v_{ce} m_e \\ Hm_d &= \epsilon_d m_d + v_{de} m_e \\ Hm_e &= \epsilon_e m_e. \end{aligned} \tag{3.25}$$

This action implies the following chains of ordering: $a > b > c$ as well as $c > e$ and $d > e$. Hence, we see that the ordering within the subspace is only *partial* because for instance c and d cannot be compared (this is also the case for b and d). The corresponding eigenfunction is[8]

$$J_a \propto \begin{vmatrix} m_e & m_d & m_c & m_b & m_a \\ \epsilon_e - \epsilon_a & v_{de} & v_{ce} & 0 & 0 \\ 0 & \epsilon_d - \epsilon_a & 0 & 0 & v_{ad} \\ 0 & 0 & \epsilon_c - \epsilon_a & v_{bc} & v_{ac} \\ 0 & 0 & 0 & \epsilon_c - \epsilon_a & v_{ab} \end{vmatrix}. \tag{3.26}$$

It thus follows that both m_c and m_d appear in the expression of J_a, even if c and d cannot be compared with each other. The main point is that they are both comparable to a. This example also illustrates our second point: e is comparable to a but m_e does not appear in the expression of $H\,m_a$. In other words, e is compared to a by a sequence of two chains of orderings: $a > b > c$ and $c > e$. We will later relate the number of chains of ordering and the number of applications of a ladder operator on labels.

The construction of determinantal expressions for the eigenfunctions provides a strong motivation for obtaining explicitly the coefficients v_{ab} appearing in the action of H on the monomial basis: they are the building blocks of the expansion coefficients c_{ab} of J_a in that basis. The knowledge of these coefficients leads thus to *closed-form expressions for the eigenfunctions*! Quite remarkably, these coefficients are indeed computable, as we will now show.

3.7. Generalities: the explicit action of H by the universal dressing of a model-dependent $N = 2$ computation. As we just noticed, the specification of the ordering requires the determination of the coefficients v_{ab} that do not vanish in the action of H on m_a. Actually, we will see that it is not much more complicated to compute precisely all these coefficients v_{ab}. And, as pointed out, this yields directly the eigenfunctions.

The strategy is the following: we first compute the action of H in the two-particle sector. Such computations are always very easy. For a CMS-type model, whose Hamiltonian is a sum of two-body interactions, this computation turns out to encode the core value of the coefficient v_{ab}. Indeed, to go from $N = 2$ to a general N simply amounts to dressing the result by a symmetry factor [**4**,**11**,**19**]. And another remarkable fact is that, although the $N = 2$ computation is model dependent, *the*

[8] The validity of this result requires that eigenvalues, corresponding to labels that can be compared, be distinct.

symmetry dressing appears to be universal, that is, the symmetry factor boils down to a symmetry of the labels (partitions or superpartitions) of the eigenfunctions, hence independent of the root structure, or the rational or trigonometric version of the (s)CMS under consideration. This is certainly so for all the cases we have considered so far (including the rational case with confinement [**7**]).

An immediate objection could be formulated with regards to this program: although it is clear that the action of H on m_a is well-defined in the $N = 2$ sector, we know that for a general N-body problem, the action of H on a particular element m_a, that itself does not vanish upon reduction to $N = 2$, may contain terms that disappear upon reduction. Are these terms properly taken into account?

To make the above considerations more concrete and precise, we will now return to the stCMS model. This will also allow us to address the potential objections in a definite context.

3.8. The transposition of the action of $\overline{\mathcal{H}}$ on superpartitions: ladder operators. When the action of the Hamiltonian is a sum of two-body terms $\sum \overline{\mathcal{H}}_{ij}$ (up to a derivative part that acts diagonally), as is the case for the (s)tCMS model, the various terms appearing in the decomposition of $\overline{\mathcal{H}} m_\Lambda$ can be characterized by the action of a ladder operator $R_{ij}^{(\ell)}$ acting on the superpartitions Λ.[9] More precisely, $R_{ij}^{(\ell)}$ acts on two parts Λ_i and Λ_j of a superpartition as [**4**]:

$$
\begin{aligned}
(3.27)\quad & R_{ij}^{(\ell)}(\Lambda_1, \dots, \Lambda_i, \dots, \Lambda_j, \dots, \Lambda_N) \\
& \qquad = \begin{cases} (\Lambda_1, \dots, \Lambda_i - \ell, \dots, \Lambda_j + \ell, \dots, \Lambda_N) & \text{if } \Lambda_i > \Lambda_j, \\ (\Lambda_1, \dots, \Lambda_i + \ell, \dots, \Lambda_j - \ell, \dots, \Lambda_N) & \text{if } \Lambda_j > \Lambda_i. \end{cases}
\end{aligned}
$$

for $i < j$ and $\ell \geq 0$. This action of $R_{ij}^{(\ell)}$ is nonzero only in the following cases:

$$
(3.28)\qquad
\begin{aligned}
&\text{I}: && i, j \in \{1, \dots, m\} \text{ and } \left\lfloor \frac{\Lambda_i - \Lambda_j - 1}{2} \right\rfloor \geq \ell, \\
&\text{II}: && i \in \{1, \dots, m\}, j \in \{m+1, \dots, N\} \text{ and } |\Lambda_i - \Lambda_j| - 1 \geq \ell, \\
&\text{III}: && i, j \in \{m+1, \dots, N\} \text{ and } \left\lfloor \frac{\Lambda_i - \Lambda_j}{2} \right\rfloor \geq \ell.
\end{aligned}
$$

Here $\lfloor x \rfloor$ stands for the largest integer smaller or equal to x. This characterization is obtained by a consideration of the action of $\overline{\mathcal{H}}$ in the $N = 2$ sector. Three cases must then be distinguished, according to their fermion number: $m = 2, 1, 0$, corresponding respectively to pairs (i, j) of type I, II, III. The results pertaining to ordinary Jack polynomials correspond to case III.

Let us make the induced ordering (denoted $\geq^s$) explicit. Given a sequence

$$
(3.29)\qquad \gamma = (\gamma_1, \dots, \gamma_m; \gamma_{m+1}, \dots, \gamma_N),
$$

we will denote by $\overline{\gamma}$ the superpartition whose antisymmetric part is the rearrangement of $(\gamma_1, \dots, \gamma_m)$ and whose symmetric part is the rearrangement of the remaining entries $(\gamma_{m+1}, \dots, \gamma_N)$. For example, we have

$$
(3.30)\qquad \overline{(1, 3, 2; 2, 3, 1, 2)} = (3, 2, 1; 3, 2, 2, 1).
$$

---

[9] If in the Hamiltonian, there are in addition some terms that act nondiagonally (as in the (s)rCMS model [**7**]), their action is taken into account by another ladder operator acting on a single entry of a superpartition.

We say that

$$\Lambda \geq^s \Omega \iff \Omega = \overline{R^{(\ell_k)}_{i_k,j_k} \cdots \overline{R^{(\ell_1)}_{i_1,j_1} \Lambda}} \tag{3.31}$$

for a given sequence of operators $R^{(\ell_1)}_{i_1,j_1}, \dots, R^{(\ell_k)}_{i_k,j_k}$. This is a refinement of the ordering introduced previously at the level of the corresponding partitions Λ^* and a generalization of the dominance ordering, which is recovered in the zero-fermion sector.

All the supermonomials m_Ω appearing in the expansion of $\overline{\mathcal{H}} m_\Lambda$ are precisely those whose labeling superpartition is obtained from Λ by <u>*one application*</u> of the ladder operator R, that is, all Ω that can be written as $\Omega = \overline{R^{(\ell)}_{i,j} \Lambda}$ for some $i < j$, and $\ell > 0$. This implies that the number of nonzero parts of Ω exceeds that of Λ by at most one. On the other hand, the supermonomials m_Γ appearing in the expansion of $\mathcal{J}_\Lambda$ (which is now understood to represent a determinant expansion) are precisely those whose label Γ is obtained from Λ by *any number of applications* of the ladder operator R. We can now make the connection with the loose terminology of the previous sections: one application of R corresponds to a single chain of ordering while multiple chains of ordering are described by the action of a sequence of ladder operators.

The expressions for the coefficients v_{ab}, as well as a detailed description of the symmetry factors, can be found in [**4**] and will not be repeated here. Let us instead address the potential problem pointed out previously: the action of $\overline{\mathcal{H}}$ on a two-part superpartition may contain more than two parts, hence lie beyond the $N = 2$ sector, which was claimed to contain all the relevant information. Consider for instance

$$\overline{\mathcal{H}} m_{(3;1)} = (10 + 4\beta N - 6\beta) m_{(3;1)} + 2\beta m_{(2;2)} + 8\beta m_{(2;1,1)} + 2\beta m_{(1;2,1)} \tag{3.32}$$

which holds for arbitrary values of $N > 2$. However, if we specialize to $N = 2$, the last two terms disappear. How could these be taken into account by a $N = 2$ computation? The point is that the different terms m_Ω in $\overline{\mathcal{H}} m_\Lambda$ are not necessarily linked to the *same* two-body calculation. While the contribution of the first nondiagonal term $m_{(2;2)}$ can be described by the two-body interaction $\overline{\mathcal{H}}_{12}$, that is, $(2;2) = R^{(1)}_{12}(3;1)$, this is not the proper two-body interaction for the description of the other two terms. Indeed, we see that $\underline{(2;1,1)} = R^{(1)}_{13}(3;1,0)$. This shows that the relevant $N = 2$ computation is rather $\overline{\mathcal{H}}_{13} m_{(3;*,0)}$. Similarly, because $(1;1,2) = R^{(2)}_{13}(3;1,0)$ and $\overline{(1;1,2)} = (1;2,1)$, the corresponding two-body problem is again $\overline{\mathcal{H}}_{13} m_{(3;*,0)}$.

It should be heavily stressed that the remarkable fact that the result for general N can be deduced out of the $N = 2$ one is true for the action of $\overline{\mathcal{H}}$ in the supermonomial basis; this would not be possible directly for the eigenfunctions $\mathcal{J}_\Lambda$.

3.9. $\mathcal{J}_\Lambda$ as eigenfunctions of the $\overline{\mathcal{H}}_n$ charges. The method sketched in the previous subsections indicate that we can construct closed-form expressions for the eigenfunctions $\mathcal{J}_\Lambda$ of the stCMS Hamiltonian $\overline{\mathcal{H}}$, our first candidates for the role of Jack superpolynomials. Actually, we can prove that the $\mathcal{J}_\Lambda$'s diagonalize the whole tower of conserved charges $\overline{\mathcal{H}}_n$ [**5**] (recall that the bar indicates that the contribution of the ground-state wave function has been taken away):

$$\overline{\mathcal{H}}_n \mathcal{J}_\Lambda = \epsilon^{(n)}_\Lambda \mathcal{J}_\Lambda. \tag{3.33}$$

Unfortunately these superpolynomials are *not orthogonal.* In other words, the action of the $\overline{\mathcal{H}}_n$'s leaves a residual degeneracy. Indeed, we can check that $\epsilon_\Lambda^{(n)} = \epsilon_\Omega^{(n)}$ for all n if $\Lambda^* = \Omega^*$ even when $\Lambda \neq \Omega$ [**5**]. However, in retrospect, this is not too surprising: we have not constructed the common eigenfunctions of *all* the commuting conserved charges. We still need to diagonalize the $\overline{\mathcal{I}}_n$ charges.

3.10. Constructing orthogonal eigenfunctions: diagonalization of the $\overline{\mathcal{I}}_n$ charges. As just indicated, the eigenfunctions of the Hamiltonian $\overline{\mathcal{H}} = \overline{\mathcal{H}}_2$ happen to be eigenfunctions of all the $\overline{\mathcal{H}}_n$ operators. Similarly, to study the eigenfunctions of the complete set $\{\overline{\mathcal{I}}_n\}$, it suffices to consider only the first nontrivial charge $\overline{\mathcal{I}}_1$ [**5**].

The remaining problem is thus to construct linear combinations of the $\mathcal{J}_\Lambda$ that are eigenfunctions of $\overline{\mathcal{I}}_1$. The strategy is by now clear: we first compute the action of $\overline{\mathcal{I}}_1$ on the $\mathcal{J}_\Lambda$ basis, determine the new ordering induced by the underlying triangularity, compute the expansion coefficients exactly, and then write down the eigenfunctions, now denoted J_Λ, in determinantal form. The self-adjointness property of the charges $\{\overline{\mathcal{H}}_n\}$ and $\{\overline{\mathcal{I}}_n\}$ ensures the orthogonality of the J_Λ. These are thus the *genuine Jack superpolynomials.* The details of this construction can be found in [**5**].

Let us mention that a more direct way to build the orthogonal Jack superpolynomials is also presented in [**5**]. It amounts to simply symmetrizing the product of a nonsymmetric Jack polynomial with a monomial in the fermionic variables. Even though this approach is more direct, we prefer not to include it since it is not as much in the spirit of this presentation.

3.11. A remark on the underlying Lie superalgebraic structure. Let us conclude this section on the stCMS model by addressing the following question: is there a Lie superalgebra structure underlying this problem? Given that the tCMS Hamiltonian is closely linked to the su(N) root structure, the natural guess is that the stCMS Hamiltonian would be related to the superalgebra su($m, N-m$). The root structure of su($m, N-m$) is (see for instance [**8**]):

$$\delta_i - \delta_j, \quad \epsilon_k - \epsilon_\ell, \quad \delta_i - \epsilon_k, \qquad 1 \leq i < j \leq m \text{ and } m+1 \leq k < \ell \leq N-m \tag{3.34}$$

with

$$\delta_i \cdot \delta_j = -\delta_{ij}, \qquad \epsilon_k \cdot \epsilon_\ell = \delta_{k\ell}, \qquad \delta_i \cdot \epsilon_k = 0 \tag{3.35}$$

When $m = 0$, we recover the su(N) roots. The roots $\delta_i - \delta_j$ and $\epsilon_k - \epsilon_\ell$ are said to be bosonic while the remaining roots $\delta_i - \epsilon_k$ are called fermionic. However, it should be clear that there is no genuine notion of statistics at the level of the roots themselves (although the generators do have a definite statistics) since for instance the difference between two fermionic roots could be bosonic (e.g., $\delta_1 - \epsilon_1 - (\delta_2 - \epsilon_1) = \delta_1 - \delta_2$). This, of course, is not a deep observation. However, it readily implies that the Hamiltonian constructed from the root system of a Lie superalgebra, su($m, N-m$) for instance, does not contain anticommuting variables (see [**17**] for example). Hence, the stCMS Hamiltonian does not appear to have an immediate interpretation in terms of the su($m, N-m$) algebra.

However, there does exist a link with the su($m, N-m$) root structure, but at the level of the action of the ladder operator $R_{ij}^{(\ell)}$. Referring to equations (3.27) and (3.28), we see that the action of $R_{ij}^{(\ell)}$ in cases I and III corresponds to the subtraction of positive bosonic roots, while in case II, it amounts to subtract a positive or a negative fermionic root. That both positive and negative fermionic roots are involved is not surprising given the relativity of the positivity requirement for these roots. At the level of superpartitions, the sign of the fermionic roots should not play any role because it is simply related to our choice of relative ordering for the two partitions Λ^a and Λ^s composing the superpartition.

4. Conclusion

The integrability and superintegrability of the sCMS models rely on a fundamental observation: the supersymmetric extension is fully captured by a fermionic exchange operator. This allowed us to infer the (super)integrability by means of a projection argument. Many generalizations of the srCMS model could be shown, in a similar way, to be superintegrable: the srCMS models formulated for general root systems, the extension of the rCMS model with many supersymmetries, the srCMS model with spin degrees of freedom, etc.

On the other hand, we have stressed that the method outlined in Section 3 for constructing eigenfunctions is quite general. It has already been extended to the srCMS case with confinement, that is, to the construction of orthogonal generalized Hermite (super)polynomials [**6, 7**]. Generalized Jacobi and Laguerre polynomials in superspace could also be obtained along this line from the sr/tCMS model with B-type roots. The method can be easily extended to models with more supersymmetries or including spin degrees of freedom. The astonishing power of the method gives us hope that this line of attack could provide a breakthrough in the study of other integrable models.

References

1. F. Calogero, *Solution of the one-dimensional N-body problems with quadratic and/or inversely quadratic pair potentials*, J. Math. Phys. **12** (1971), 419–436.
2. R. Caseiro, J.-P. Françoise, and R. Sasaki, *Algebraic linearization of dynamics of Calogero type for any Coxeter group*, J. Math. Phys. **41** (2000), no. 7, 4679–4686.
3. P. Desrosiers, L. Lapointe, and P. Mathieu, *Supersymmetric Calogero-Moser-Sutherland models and Jack superpolynomials*, Nuclear Phys. B **606** (2001), no. 3, 547–582.
4. ———, *Jack superpolynomials, superpartition ordering and determinantal formulas*, Comm. Math. Phys. **233** (2003), no. 3, 383–402.
5. ———, *Jack polynomials in superspace*, Comm. Math. Phys. 242 (2003), 331–360..
6. ———, *Generalized Hermite polynomials in superspace as eigenfunctions of the supersymmetric rational CMS model*, Nuclear Phys. B **674** (2003), 615–633.
7. ———, *Explicit formulas for the generalized Hermite polynomials in superspace*, J. Phys. A **37** (2004), 1251–1268.
8. L. Frappat, A. Scarrino, and P. Sorba, *Dictionnary on Lie algebras and superalgebras*, Academic Press, San Diego, CA, 2000.
9. D. Z. Freedman and P. F. Mende, *An exactly solvable N-particle system in supersymmetric quantum mechanics*, Nuclear Phys. B **344** (1990), no. 2, 317–343.
10. V. B. Kuznetsov, *Hidden symmetry of the quantum Calogero–Moser system*, Phys. Lett. A **218** (1996), no. 3-6, 212–222.
11. L. Lapointe, A. Lascoux, and J. Morse, *Determinantal formula and recursion for Jack polynomials*, Electron. J. Combin. **7** (2000), no. 1, Note 1 (electronic).
12. J. A. Minahan and A. P. Polychronakos, *Integrable systems for particles with internal degrees of freedom*, Phys. Lett. B **302** (1993), no. 2-3, 265–270.
13. J. Moser, *Three integrable Hamiltonian systems connected with isospectral deformations*, Surveys in Applied Mathematics (Los Alamos, NM, 1974), Academic Press, New York, 1976, pp. 235–258.

14. M. A. Olshanetsky and A. M. Perelomov, *Classical integrable finite-dimensional systems related to Lie algebras*, Phys. Rep. **71** (1981), no. 5, 313–400.
15. ———, *Quantum integrable systems related to Lie algebras*, Phys. Rep. **94** (1983), no. 6, 313–404.
16. A. P. Polychronakos, *Exchange operator formalism for integrable systems of particles*, Phys. Rev. Lett. **69** (1992), no. 5, 703–705.
17. A. Sergeev, *Superanalogs of the Calogero operators and Jack polynomials*, J. Nonlinear Math. Phys. **77** (2001), 59–64.
18. B. S. Shastry and B. Sutherland, *Super Lax pairs and infinite symmetries in the* $1/r^2$ *system*, Phys. Rev. Lett. **70** (1993), no. 26, 4029–4033.
19. K. Sogo, *Eigenstates of Calogero–Sutherland–Moser model and generalized Schur functions*, J. Math. Phys. **35** (1994), no. 5, 2282–2296.
20. ———, *A simple derivation of multivariable Hermite and Legendre polynomials*, J. Phys. Soc. Japan **65** (1996), no. 10, 3097–3099.
21. B. Sutherland, *Exact results for a quantum many-body problem in one dimension*, Phys. Rev. A **4** (1971), no. 5, 2019–2021.
22. S. Wojciechowski, *Superintegrability of the Calogero–Moser system*, Phys. Lett. A **95** (1983), no. 6, 279–281.

Département de physique, de génie physique et d'optique, Université Laval, Québec, QC G1K 7P4, Canada

E-mail address, P. Desrosiers: `pdesrosi@phy.ulaval.ca`

E-mail address, P. Mathieu: `pmathieu@phy.ulaval.ca`

Instituto de Matemática y Física, Universidad de Talca, Casilla 747, Talca, Chile

E-mail address: `lapointe@inst-mat.utalca.cl`

Centre de Recherches Mathématiques
CRM Proceedings and Lecture Notes
Volume **37**, 2004

Complete Sets of Invariants for Classical Systems

Willard Miller, Jr.

Abstract. We consider the general problem of determining exactly when a classical Hamiltonian $H = \sum g^{ij}(\mathbf{x})p_ip_j + V(\mathbf{x})$ in n dimensions admits a constant of the motion that is polynomial in the momenta. If the associated Hamilton–Jacobi equation admits an orthogonal separation of variables, then it is possible to generate algorithmically a canonical basis Q, P where $P_1 = H$, $P_2, \dots, P_n$ are the other second-order constants of the motion associated with the separable coordinates, and $\{Q_i, Q_j\} = \{P_i, P_j\} = 0$, $\{Q_i, P_j\} = \delta_{ij}$. The $2n-1$ functions $Q_2, \dots, Q_n$, $P_1, \dots, P_n$ form a basis for the invariants. We show how to determine for exactly which spaces and potentials the invariant Q_j is a polynomial in the original momenta.

This is a joint work with E. G. Kalnins, J. Kress, and G. Pogosyan.

1. Complete sets of invariants

The basic question that we are considering is: Given a classical Hamiltonian $H = H(x,p) = \sum g^{ij}(\mathbf{x})p_ip_j + V(\mathbf{x})$, where $\mathbf{x} = (x_1, \dots, x_n)$, $\mathbf{p} = (p_1, \dots, p_n)$ how can one find *all* the solutions to the Poisson bracket condition

$$\{H, L\} = \sum_{i=1}^{n} \left(\frac{\partial H}{\partial p_i} \frac{\partial L}{\partial x_i} - \frac{\partial H}{\partial x_i} \frac{\partial L}{\partial p_i} \right) = 0 \tag{1.1}$$

where $L = L(\mathbf{x}, \mathbf{p})$? There is no known general solution to this problem. However, if the Hamilton–Jacobi equation $H(\mathbf{x}, \partial S/\partial \mathbf{x}) = E$ is additively separable in the orthogonal variables $\mathbf{x}$ then a complete integral of the equation can be constructed by quadratures and one can find a basis of $2n-1$ functionally independent solutions to (1.1), [**1**]. Here n of the equations are second-order in the momenta. Indeed there is an explicit canonical change of coordinates from the variables $\mathbf{x}$, $\mathbf{p}$ with $\{x_i, p_j\} = \delta_{ij}$ to variables $\mathbf{Q}$, $\mathbf{P}$ where $P_1 = H$, $P_2, \dots, P_n$ are the other second-order constants of the motion associated with the orthogonal separable x-coordinates, and $\{Q_i, Q_j\} = \{P_i, P_j\} = 0$, $\{Q_i, P_j\} = \delta_{ij}$. Thus the $2n-1$ functions $Q_2, \dots, Q_n$, $P_1, \dots, P_n$ form a basis for the invariants. Each invariant Q_j can be expressed as a sum of the form

$$Q_j = \sum_{k=1}^{n} M_k(x_k, \mathbf{P}). \tag{1.2}$$

2000 *Mathematics Subject Classification.* 37J35, 70H06.
This is a final form of the paper.

Even here, important problems remain, [**3–5**]. For example, when are the constants of the motion L *polynomial* in the canonical momenta p_i, $i = 1, \ldots, n$? Again, there is no known algorithmic way of generating all polynomial solutions.

From the x-based integrals in (1.2) it is difficult to tell if Q_j is a polynomial in the momenta p_i. We adopt a p-based approach to the calculation of the invariants Q_j in which the term M_k take the form $M_k = M(p_k, \mathbf{P})$, and we can say in advance for exactly which separable metrics and potentials Q_j is a polynomial in the momenta. We give, in principle, a complete solution to this problem. Moreover, we show how to characterize each term M_k in (1.2) by the Poisson brackets $\{M_k, P_j\}$.

Note that the system could admit a polynomial invariant

$$L = R(\mathbf{P}, Q_2, \ldots, Q_n)$$

such that L, $\mathbf{P}$ is functionally independent, even if $Q_2, \ldots, Q_n$ are not polynomials. It is a much more difficult problem to classify all such possibilities for polynomial L as functions of possibly nonpolynomial Q_j.

Our approach is to assume orthogonal separation in some coordinate system. This gives n second-order commuting invariants. We can compute a remaining $n-1$ by quadratures. How do we determine if these are polynomials in the momenta? We start with a separable system in two Cartesian coordinates.

2. Cartesian systems in two dimensions

In Cartesian coordinates the Hamiltonian has the form

$$H = p_x^2 + p_y^2 + V(x, y).$$

If we have separation of variables in Cartesian coordinates the potential must take the form [**7**]

$$V(x, y) = X(x) + Y(y). \tag{2.1}$$

We immediately observe that there are already two invariants arising from the separation, namely $L_1 = p_x^2 + X(x)$ and $L_2 = p_y^2 + Y(y)$.

PROBLEM. Calculate a third invariant and determine when it can be chosen to be a polynomial in the canonical momenta.

To do this we compute two functions $M(x, p_x)$ and $N(y, p_y)$ that satisfy the conditions

$$\{H, M\} = 1, \qquad \{H, N\} = 1. \tag{2.2}$$

These equations can be solved in principle if we know the original functions X and Y. Indeed if we write out the first of these conditions, $\{H, M\} = 1$, we obtain

$$2p_x \frac{\partial M}{\partial x} - X' \frac{\partial M}{\partial p_x} = 1.$$

Since we are interested in the dependence of the invariants on the momenta, we take p_x to be the independent variable and consider x to be a function of p_x. Here, L_1, L_2 are treated as constants. Thus the equation takes the form

$$\frac{dM}{dp_x} = -\frac{1}{X'}.$$

This equation can be solved to give

$$M = -\int X'^{-1}\, dp_x$$

where $L_1 = p_x^2 + X$. (We consider $X'^{-1} = dx/dX$ to be a function of $X = L_1 - p_x^2$ to compute the integral.

Once M and N have been determined, we see that $L_3 = N - M$ must be an invariant. It is immediately clear that if $X = x^{1/p}$ where p is an integer then M is a polynomial in p_x. In general, we want dx/dX to be a polynomial in X.

EXAMPLES. For $p = 3$ we have

$$M = -3x^{2/3}p_x - 4x^{1/3}p_x^3 - \tfrac{8}{5}p_x^5.$$

For $p = 4$,

$$M = -4x^{3/4}p_x - 8x^{1/2}p_x^3 - \frac{32}{5}x^{1/4}p_x^5 - \frac{64}{35}p_x^7.$$

It follows from these two examples that the Hamiltonian

$$H = p_x^2 + p_y^2 + x^{1/3} + y^{1/4}$$

has in addition to the obvious invariants

$$L_1 = p_x^2 + x^{1/3}, \qquad L_2 = p_y^2 + y^{1/4},$$

the additional invariant

$$L_3 = 3x^{2/3}p_x + 4x^{1/3}p_x^3 + \frac{8}{5}p_x^5 - 4y^{3/4}p_y - 8y^{1/2}p_y^3 - \frac{32}{5}y^{1/4}p_y^5 - \frac{64}{35}p_y^7. \tag{2.3}$$

It follows from these considerations that all potentials of the form

$$V = \alpha x^{1/p} + \beta y^{1/q} \tag{2.4}$$

have the superintegrability property with three functionally independent invariants which are polynomial in p_x and p_y. This includes the known examples corresponding to $p = 1$, 2.

If $X(x)$ is determined by a polynomial relation of the form

$$\sum_{j=1}^{n} a_j X^j = x$$

we can go even further. Then the function M is always a polynomial in the canonical momentum p_x.

EXAMPLE.

$$X(x) = 2^{-1/3}[\{x + \sqrt{x^2+1}\}^{1/3} - \{x + \sqrt{x^2+1}\}^{-1/3}]. \tag{2.5}$$

The inverse function is

$$x = X^3 + \frac{3}{2^{2/3}}X$$

and the corresponding function $M(x, p_x)$ is given by

$$-M(x, p_x) = \frac{8}{5}p_x^5 + 4Xp_x^3 + 3X^2p_x + \frac{3}{2^{2/3}}p_x.$$

As one further Cartesian case for which polynomial invariants can be generated, consider case when $X(x) = \omega_1^2 x^2$. The corresponding function $M(x, p_x)$ is given by

$$M(x, p_x) = \frac{1}{4\omega_1} \arcsin\left(\frac{\omega_1^2 x^2 - p_x^2}{\omega_1^2 x^2 + p_x^2}\right).$$

If $Y(y) = \omega_2^2 y^2$ this establishes that the Hamiltonian

$$H = p_x^2 + p_y^2 + \omega_1^2 x^2 + \omega_2^2 y^2 \tag{2.6}$$

has the constant of motion

$$L_3 = \frac{1}{4\omega_1} \arcsin\left(\frac{\omega_1^2 x^2 - p_x^2}{\omega_1^2 x^2 + p_x^2}\right) - \frac{1}{4\omega_2} \arcsin\left(\frac{\omega_2^2 y^2 - p_y^2}{\omega_2^2 y^2 + p_y^2}\right), \tag{2.7}$$

in addition to the constants $L_1 = p_x^2 + \omega_1^2 x^2$ and $L_2 = p_y^2 + \omega_2^2 y^2$. In general this invariant is not polynomial in the canonical momenta. However, if ω_1/ω_2 is a fraction p/q for integers p, q then $\omega_1 = ps$, $\omega_2 = qs$ and $L_3' = \sin(4spqL_3)$ will be a rational invariant whose common denominator is a product of powers of L_1 and L_2. The numerator is then an additional polynomial invariant, e.g., consider $\omega_1 = 1$, $\omega_2 = 2$. Then

$$L_3' = \sin(8L_3) = \frac{L_1 L_2^2 - 2(xp_y^2 - 4yp_xp_y - 4xy^2)^2}{L_1 L_2^2},$$

which indicates that $L_3'' = xp_y^2 - 4yp_xp_y - 4xy^2$ is an additional invariant. In general, $L_1^p L_2^q \sin(4spqL_3)$ will be a polynomial invariant, functionally independent of L_1 and L_2.

3. General two-dimensional orthogonal separable systems

On a general Riemannian space a separable potential in the local coordinates x, y has the form

$$H = L_1 = \frac{p_x^2 + p_y^2 + v_1(x) + v_2(y)}{f_1(x) + f_2(y)}. \tag{3.1}$$

and, due to the separability, there is the invariant

$$L_2 = \frac{f_2(y)\big(p_x^2 + v_1(x)\big) - f_1(x)\big(p_y^2 + v_2(y)\big)}{f_1(x) + f_2(y)}.$$

Here, v_1, v_2 are part of the potential, f_1, f_2 are part of the metric. We look for a function $M(H, x, p_x)$ that satisfies

$$\{H, M\} = \frac{1}{f_1(x) + f_2(y)}. \tag{3.2}$$

The condition has the form

$$(-v_1'(x) + f_1'(x)H)\frac{\partial M}{\partial p_x} + 2p_x \frac{\partial M}{\partial x} = 1. \tag{3.3}$$

Note that

$$L_2 = v_1(x) - f_1(x)H + p_x^2.$$

Now we consider x to be a function of the independent variable $U = p_x^2$ and write (3.3) in the form

$$\frac{dM}{dp_x} = \frac{1}{-v_1'(x) + f_1'(x)H} = \frac{dx}{dU}.$$

The solution is

$$M(H, L_2, p_x) = \int U'^{-1}\, dp_x,$$

where

$$U(x) = p_x^2 = -v_1(x) + f_1(x)H + L_2.$$

(We consider $U'^{-1} = dx/dU$ to be a function of $U = p_x^2$.)

Similarly, we compute a function $N(H, y, p_y)$ that satisfies

$$\{H, N\} = \frac{1}{f_1(x) + f_2(y)}. \tag{3.4}$$

The new invariant is $L_3 = N - M$, and also

$$\{L_2, L_3\} = 1. \tag{3.5}$$

Indeed, $\{L_2, M\} = f_2/(f_1 + f_2)$, $\{L_2, N\} = -f_1/(f_1 + f_2)$. This implies that the set L_1, L_2, L_3 is functionally independent.

Similarly we can construct functions $M(H, x, p_x)$, $N(H, y, p_y)$ that satisfy

$$\{H, M\} = \frac{f_1(x)}{f_1(x) + f_2(y)}, \qquad \{H, N\} = \frac{-f_2(y)}{f_1(x) + f_2(y)}, \tag{3.6}$$

Assuming that $|v_i'| + |f_i'| > 0$ for $i = 1, 2$, we see that these equations have the solutions

$$M(H, L_2, p_x) = \int f_1(x) U_1'^{-1}\, dp_x, \qquad N(H, L_2, p_y) = -\int f_2(y) U_2'^{-1}\, dp_y,$$

where

$$U_i = -v_i + f_i H + L_2.$$

Setting $L_4 = N - M$, we see that L_4, not an invariant, satisfies

$$\{H, L_4\} = 1, \qquad \{L_2, L_4\} = 0. \tag{3.7}$$

Thus

$$\{H, L_2\} = \{H, L_3\} = 0, \quad \{L_2, L_3\} = 1, \quad \{H, L_4\} = 1, \quad \{L_2, L_4\} = 0.$$

EXAMPLE. We choose parabolic coordinates in Euclidean space $x' = \frac{1}{2}(\xi^2 - \eta^2)$, $y' = \xi\eta$. Consider first the parabolic-separable Hamiltonian

$$H = L_1 = \frac{p_\xi^2 + p_\eta^2 + \xi}{\xi^2 + \eta^2}. \tag{3.8}$$

We can immediately associate with this the extra invariant

$$L_2 = \frac{\eta^2 p_\xi^2 - \xi^2 p_\eta^2 + \eta^2 \xi}{\xi^2 + \eta^2}.$$

If we look for our functions $M(\xi, p_\xi)$ and $N(\eta, p_\eta)$, we obtain

$$M(\xi, p_\xi) = \frac{1}{4\sqrt{H}} \ln\left(\frac{\sqrt{H} p_\xi + \frac{1}{2} - \xi H}{-\sqrt{H} p_\xi + \frac{1}{2} - \xi H}\right),$$

$$N(\eta, p_\eta) = \frac{1}{4\sqrt{H}} \ln\left(\frac{\sqrt{H}\eta + p_\eta}{\sqrt{H}\eta - p_\eta}\right).$$

If we now consider the constant $\cosh(4(M-N)\sqrt{H})$, we find that it can be written in the form

$$4\cosh(4(M-N)\sqrt{H}) = \frac{L_3^2 H}{(1-4HL_2)L_2},$$

where

$$L_3 = \frac{2\xi\eta}{\xi^2+\eta^2}(p_\xi^2+p_\eta^2) - 2p_\xi p_\eta + \frac{\eta(\xi^2-\eta^2)}{\xi^2+\eta^2} \tag{3.9}$$

is an additional invariant quadratic in the canonical momenta.

What is the scope of this construction so that polynomial invariants are obtained? The solution of (3.2) can be understood in a more general context. We have the dual relations

$$x = F(U-L_2, H), \qquad U(x,H) = -v_1(x) + f_1(x)H + L_2, \qquad U_x \neq 0. \tag{3.10}$$

Here, $H = L_1$, L_2 are parameters. Thus we have

$$1 = F_U U_x, \qquad\qquad F_U U_H + F_H = 0.$$

The condition that $U(x,H)$ is linear in H, i.e., $U_{HH} = 0$, leads to the following necessary and sufficient conditions that the function $x = F(U,H)$ correspond to an invariant M on a Riemannian manifold with potential:

$$F_{HH}F_U^2 - 2F_{UH}F_U F_H + F_{UU}F_H^2 = 0, \quad F_U \neq 0. \tag{3.11}$$

(This equation also appears in the theory of level sets, where it describes those surfaces whose level sets, i.e., intersections with the plane x = constant are straight lines [**8**].) The equation admits an infinite dimensional conformal symmetry group. Indeed if $V = F(U,H)$ is a solution then $G(V)$ is also a solution, for *any* nonconstant function G. Also, this group contains the subgroup of inhomogeneous affine symmetries: if $F(U,H)$ is a solution then so is

$$F([a_{11}U + a_{12}H + a_{13}]/A, [a_{21}U + a_{22}H + a_{23} + a_{23}]/A),$$

where a_{ij} are constants, $\det(a_{ij}) \neq 0$ and

$$A = a_{31}U + a_{32}H + a_{33}.$$

In order to have M as a polynomial in the momentum variable p_x we must have $dx/dU = \partial F/\partial U$ a polynomial in U What is the most general function F that leads to an invariant that is polynomial in p_x?

THEOREM 1. *The function $F(U,H)$ with $F_U \neq 0$ is a solution of* (3.11) *with polynomial dependence on U if and only if it is of the form*

$$F(U,H) = P\left(\frac{U+\alpha H+\beta}{\gamma H+\delta}\right)$$

where P is a (nonconstant) polynomial and α, β, γ, δ are constants with $|\gamma|^2 + |\delta|^2 > 0$.

For the proof, see [**6**]. This theorem generalizes to higher dimensions.

Standard Hamilton–Jacobi theory gives essentially these same constants of the motion, but from a different viewpoint. Our expression for L_3, for example, is

$$L_3 = \int {U'_x}^{-1}\,dp_x - \int {U'_y}^{-1}\,dp_y = M - N,$$

where $U_x = -v_1(x) + f_1(x)H + L_2$, etc. Standard Hamilton-Jacobi theory gives

$$L_3 = \frac{1}{2}\int \frac{dx}{\sqrt{-v_1 + f_1 H + L_2}} - \frac{1}{2}\int \frac{dy}{\sqrt{-v_2 + f_2 H - L_2}} = \widetilde{M} - \widetilde{N}.$$

In the standard theory $\widetilde{M} = \widetilde{M}(H, L_2, x)$, etc., whereas in our approach $M = M(H, L_2, p_x)$, etc. In both cases the condition (3.3) is satisfied. Our approach makes it easier in some cases to determine if polynomial invariants exist. It also points out the bracket relations between M, N and the operators L_j defining the separation.

Examples abound of spaces for which these constructions apply. Consider a family of surfaces in Minkowski space: $ds^2 = dz^2 - dy^2 - dx^2$:

$$\mathbf{X}(t,\xi) = (x, y, z) = (2t\xi, g(t) + (\xi^2 - 1)t, g(t) + (\xi^2 + 1)t). \tag{3.12}$$

The metric on the surface is

$$ds^2 = 4[tg'(t)\,dt^2 - t^2\,d\xi^2] = 4t^2[d\rho^2 - d\xi^2] = (f(\rho) + 1)[d\rho^2 - d\xi^2],$$

where $(d\rho/dt)^2 = g'(t)/t^2$, and we can construct a polynomial invariant for the surface (and for an appropriate added potential) provided that the function $t^2 = F(\rho)$ has a polynomial inverse function, i.e., $\rho = G(t^2)$ where G is a polynomial. Clearly $g'(t) = 4t^4 G'(t^2)^2$ and any polynomial G will determine a surface with a polynomial invariant. For example, choose $G(t^2) = \frac{1}{2}t^4 + t^2$. Then we can take $g(t) = \frac{4}{9}t^9 + \frac{8}{7}t^7 + \frac{4}{5}t^5$ and $\rho(t) = \frac{1}{2}t^4 + t^2$. The resulting M will be third-order polynomial in p_ξ and p_ρ. Similarly, we can determine a potential term $v(\rho)$ with $v' \neq 0$ such that N is a polynomial in p_ξ and p_ρ.

4. Lie form in two dimensions

We know that if a Hamiltonian

$$H = \sum_{i,j=1}^{2} g^{ij} p_i p_j$$

admits a constant of the motion L that is quadratic in the momenta

$$L = \sum_{i,j=1}^{2} a^{ij} p_i p_j, \qquad \{H, L\} = 0 \tag{4.1}$$

and if the roots of the determinant $|a^{ij} - \lambda g^{ij}|$ are distinct, then the eigenforms define new (separable) variables ρ, μ and the Hamiltonian can be written in Liouville form [**2**]

$$H = \frac{p_\rho^2 + p_\mu^2}{f(\rho) + g(\mu)}.$$

However, it may be that the roots of this determinant are equal. In this case H cannot be put into Liouville form, but rather Lie form, which for a suitable choice of variables (nonseparable) is

$$H = \frac{p_x p_y}{x + B(y)}. \tag{4.2}$$

The associated quadratic constant of the motion is

$$L = p_x^2 - 2yH. \tag{4.3}$$

QUESTION. When the roots of L are equal, how can we calculate the third invariant?

We are interested in the the same question when a potential is added to the Hamiltonian.

ANSWER. If we look for a function $N(H, L, y, p_y)$ that is in involution with H, we obtain the equation

$$(x + B(y))N_y + p_y B'(y) N_{p_y} = 0. \tag{4.4}$$

Solving (4.2), (4.3) for x and p_x in terms of the variables H, L, y and p_y, we obtain

$$p_x = \sqrt{L + 2yH}, \qquad x = \frac{p_y}{H}\sqrt{L + 2yH} - B(y).$$

The equation (4.4) for N then has the form

$$\frac{\sqrt{L + 2yH}}{HB'(y)} N_y + N_{p_y} = 0.$$

From this condition get a second invariant

$$L' = H \int \frac{B'(y)}{\sqrt{L + 2yH}}\, dy - p_y. \tag{4.5}$$

We extend these considerations by adding a potential. If we do this and have an extra quadratic constant then H and L have the forms

$$H = \frac{p_x p_y + \frac{1}{2}K(y)}{x + B(y)} + \frac{1}{2}U'(y), \qquad L = p_x^2 - 2yH + U(y). \tag{4.6}$$

Solving (4.6) for p_x and x gives

$$p_x = \sqrt{L - U(y) + 2yH}, \qquad x = \frac{p_y\sqrt{L - U(y) + 2yH + \frac{1}{2}K(y)}}{H - \frac{1}{2}U'(y)} - B(y).$$

Then the equation for N has the form

$$\begin{aligned} &2\sqrt{L - U(y) + 2yH}\,(2H - U'(y))N_y \\ &\quad + [-2U''(y)\sqrt{L - U(y) + 2yH}\,p_y + B'(y)U'(y)^2 + 4B'(y)H^2 \\ &\quad - U''(y)K(y) - 4B'(y)U'(y)H + K'(y)U'(y) - 2K'(y)H]N_{p_y} = 0. \end{aligned} \tag{4.7}$$

This equation can, in principle, be solved directly. For suitable redefinition of the variables $y \to Y$, $p_y \to P_y$, equation (4.7) can be put in the form

$$N_Y + (P_Y + s(Y))N_{P_Y} = 0 \tag{4.8}$$

that can be solved by the further transformation

$$P_{Y'} = P_Y + t(Y), \qquad Y' = Y.$$

Then, provided that

$$t'(Y) - t(Y) + s(Y) = 0,$$

(4.8) reduces to

$$N_{Y'} + P_{Y'} N_{P_{Y'}} = 0.$$

From this we immediately deduce an extra constant of the motion of the form

$$L' = e^{Y'}/P_{Y'}. \tag{4.9}$$

The equation for $t(Y)$ has the solution

$$t(Y) = e^Y \int^Y e^{-u} s(u)\, du.$$

5. Orthogonal separation in three dimensions

If we have a general orthogonal separable coordinate system in three dimensions we could take the Hamiltonian to be

$$H = L_1 = \frac{g_2 - g_3}{\Phi}(p_{x_1}^2 + v_1(x_1)) + \frac{g_3 - g_1}{\Phi}(p_{x_2}^2 + v_2(x_2)) + \frac{g_1 - g_2}{\Phi}(p_{x_3}^2 + v_3(x_3)) \tag{5.1}$$

where $g_i = g_i(x_i), f_i = f(x_i)$ and Φ is the determinant of the Stäckel matrix

$$\begin{pmatrix} 1 & f_1 & g_1 \\ 1 & f_2 & g_2 \\ 1 & f_3 & g_3 \end{pmatrix} \tag{5.2}$$

This automatically gives us two more invariants:

$$L_2 = \frac{f_3 - f_2}{\Phi}(p_{x_1}^2 + v_1(x_1)) + \frac{f_1 - f_3}{\Phi}(p_{x_2}^2 + v_2(x_2)) + \frac{f_2 - f_1}{\Phi}(p_{x_3}^2 + v_3(x_3)), \tag{5.3}$$

$$L_3 = \frac{f_2 g_3 - f_3 g_2}{\Phi}(p_{x_1}^2 + v_1(x_1)) + \frac{f_3 g_1 - f_1 g_3}{\Phi}(p_{x_2}^2 + v_2(x_2)) + \frac{f_1 g_2 - f_2 g_1}{\Phi}(p_{x_3}^2 + v_3(x_3)). \tag{5.4}$$

We need to find an additional two invariants, such that the five form a functionally independent set.

If we look for a function M_1 such that

$$\{H, M_1\} = \frac{g_2 - g_3}{\Phi}, \tag{5.5}$$

then this function satisfies the equation

$$2p_{x_1}\partial_{x_1} M_1 + [-v_1'(x_1) + f_1' H + g_1' L_2]\partial_{p_{x_1}} M_1 = 1, \tag{5.6}$$

which looks like the form we have been using in two dimensions. There are similar equations for the corresponding functions M_i for $i = 2, 3$. For $M_1(H, L_2, L_3, p_{x_1})$ this has the solution

$$M_1 = \int U_1'^{-1}\, dp_{x_1},$$

where $U_1(x_1) = -v_1(x_1) + f_1 H + g_1 L_2 + L_3$ and $L_3 = v_1 - f_1 H - g_1 L_2 + p_{x_1}^2$. (Here, we consider $U_1'^{-1} = dx_1/dU_1$ to be a function of $U_1 = p_{x_1}^2$ to compute the integral. We also assume that $|v_1'| + |f_1'| + |g_1'| > 0$.)

The corresponding invariant that we can calculate from these three functions is $L_3' = M_1 + M_2 + M_3$. This is based on the obvious identity

$$(g_2 - g_3) + (g_3 - g_1) + (g_1 - g_2) = 0.$$

The invariant $L_3' = M_1 + M_2 + M_3$ also commutes with L_2. Finally, from the fact that

$$\partial_{x_1} L_3 = \frac{f_2 g_3 - f_3 g_2}{\Phi}(v'_1 - f_1' H - g_1' L_2)$$

we can verify that (5.6) implies

$$\{L_3, M_1\} = \frac{f_2 g_3 - f_3 g_2}{\Phi}, . \tag{5.7}$$

It follows that $\{L_3, L_3'\} = 1$. Similarly, we can define a new invariant L_2' by requiring that a new function M_1 satisfy

$$\{L_1, M_1\} = \frac{g_1(g_2 - g_3)}{\Phi}, \tag{5.8}$$

with analogous conditions for M_2 and M_3. Then it follows that

$$\{L_2, M_1\} = \frac{g_1(f_3 - f_2)}{\Phi}, \qquad \{L_3, M_1\} = \frac{g_1(f_2 g_3 - f_3 g_2)}{\Phi},$$

with analogous results for M_2, M_3. Thus, from the definition of Φ we see that $\{L_2, L_2'\} = 1$.

Finally we define a function $L_1' = M_1 + M_2 + M_3$ by requiring

$$\{L_1, M_1\} = \frac{f_1(g_2 - g_3)}{\Phi}, \tag{5.9}$$

with similar conditions for M_2 and M_3. Then it follows that

$$\{L_2, M_1\} = \frac{f_1(f_3 - f_2)}{\Phi}, \qquad \{L_3, M_1\} = \frac{f_1(f_2 g_3 - f_3 g_2)}{\Phi},$$

with analogous relations for M_2 and M_3.

In summary, all brackets between the six functions L_i, L_i' are zero except that

$$\{L_3, L_3'\} = \{L_2, L_2'\} = \{L_1, L_1'\} = 1. \tag{5.10}$$

Thus the mapping $(x_1, x_2, x_3, p_{x_1}, p_{x_2}, p_{x_3}) \to (L_1, L_2, L_3, L_1', L_2', L_3')$ is canonical. It is straightforward to generalize this construction to orthogonal separation in n dimensions

Many examples and further results are given in the paper [**6**]. This approach can be extended to compute a set of quantum operators satisfying the canonical commutation relations, given a Schrödinger equation that admits separation in an orthogonal coordinate system. In general, these operators are of infinite order. There are nontrivial obstructions to extending classical polynomial invariants to finite order quantum operator invariants. Research on this question is ongoing.

References

1. V. I. Arnol'd, *Mathematical methods in classical mechanics*, Grad. Texts in Math., vol. 60, Springer-Verlag, New York–Heidelberg, 1978.
2. L. P. Eisenhart, *Riemannian geometry*, 2nd printing, Princeton Univ. Press, Princeton, NJ, 1949.
3. N. W. Evans, *Superintegrability in classical mechanics*, Phys. Rev. A (3) **41** (1990), no. 10, 5666–5676.
4. J. Friš, V. Mandrosov, Ya. A. Smorodinsky, M. Uhlíř, and P. Winternitz, *On higher symmetries in quantum mechanics*, Phys. Lett. **16** (1965), 354–356.
5. J. Hietarinta, *Direct methods for the search of the second invariant*, Phys. Rep. **147** (1987), no. 2, 87–154.

6. E. G.Kalnins, J. Kress, W. Miller, Jr., and G. S. Pogosyan, *Complete sets of invariants for dynamical systems that admit a separation of variables*, J. Math. Phys. **43** (2002), no. 7, 3592–3609.
7. E. G. Kalnins, W. Miller, Jr., and G. S. Pogosyan, *Superintegrability and associated polynomial solutions: Euclidean space and the sphere in two dimensions*, J. Math. Phys. **37** (1996), no. 12, 6439–6467.
8. J. A. Sethian, *Level set methods and fast marching methods. Evolving interfaces in computational geometry, fluid mechanics, computer vision, and materials science*, Cambridge Monogr. Appl. Comput. Math., vol. 3, Cambridge Univ. Press, Cambridge, 1999.

School of Mathematics, University of Minnesota, 206 Church St. SE, Minneapolis, MN 55455, USA

E-mail address: miller@ima.umn.edu

Centre de Recherches Mathématiques
CRM Proceedings and Lecture Notes
Volume **37**, 2004

Higher-Order Symmetry Operators for Schrödinger Equation

A. G. Nikitin

ABSTRACT. We study higher order symmetry operators for the linear Schrödinger equations with arbitrary number of spatial variables. We deduce determining equations for coefficient functions of such operators and consider in detail some cases when these equations can be explicitly solved. In addition, the complete group classification for the nonlinear Schrödinger equation is presented and specific symmetry properties of the Coulomb potential are indicated.

1. Introduction

Higher-order symmetry operators present powerful tools for investigations of partial differential equations. They can be used to construct constants of motion for evolution equations, to search for coordinate systems in which solutions with separated variables exist, and also for investigations of superintegrable systems. The notion of superintegrability was presented by Arnold Sommerfeld in 1923 [**18**] while the systematical inquiry into the superintegrability problem was started by P. Winternitz and co-authors [**6**].

We mention that a nice kind of symmetries called *supersymmetry* is also connected with existence of higher order symmetry operators.

Investigations of symmetries of the Schrödinger equation (SE) started with papers of Niederer [**11**] and Boyer [**5**] who presented a group classification of linear SEs with potential terms. Such a classification is connected with the description of *first-order* symmetry operators for the related equation.

In the fundamental paper [**19**] the complete description of *second-order* symmetry operators for the $1 + 2$-dimensional SE was presented.

A nice statement of the above mentioned results together with their various applications can be found in the Miller's book [**10**].

The problem of the description of *arbitrary-order* symmetry operators for the Schrödinger equation was formulated in papers [**2, 14**]. In paper [**13**] the complete set of such operators was found for the case of a trivial potential. In [**7**] the third-order symmetry operators for $1 + 1$-dimensional SE with an arbitrary potential

2000 *Mathematics Subject Classification.* 35Qxx, 35Kxx.
The author is indebted to the Organizers of the Workshop "Superintegrability in Classical and Quantum Systems" for support of his stay in Montreal.
This is the final form of the paper.

were described and used to construct exact solutions. Moreover, a rather nontrivial connections between the existence of third-order symmetry operators for the *linear* SE and exact solutions of the related *nonlinear* equation was established in [**7**].

The very existence of many of the exact solutions of the nonlinear SE can be understood starting from its symmetries with respect to continuous groups of transformations. The group analysis of the nonlinear SE was carried out by a number of investigators (see, e.g., [**8, 9**] and references cited therein). However, in fact the related results were valid for ad hoc restricted classes of the nonlinear SE. The completed group classification of this equation was presented in the recent paper [**15**], the related problem for systems of reaction-diffusion equations was solved in [**16**].

In the present paper we deduce the determining equations for coefficients of arbitrary order symmetry operators for the $1+m$-dimensional SE with an arbitrary number of spatial variables. We consider in detail the cases when this system can be completely solved, and discuss the related symmetry operators and their possible applications. In addition, we present the complete group classification of the nonlinear SE.

2. Symmetry operators of arbitrary order

Determining equations. Consider the SE with an arbitrary number m of spatial variables

$$L\Psi(t,\mathbf{x}) \equiv \left(i\frac{\partial}{\partial t} - \frac{p^2}{2m} - V(\mathbf{x})\right)\Psi(t,\mathbf{x}) = 0. \tag{2.1}$$

Here $\mathbf{x} = (x_1, x_2, \ldots, x_m)$, $p^2 = -\Delta$ is the Laplace operator in the space of m variables, $\Psi(t,\mathbf{x})$ is a complex wave function.

For simplicity we restrict ourselves to such solutions of (2.1) which are defined on an open set D of the $m+1$-dimensional manifold R consisting of points with coordinates $(t, x_1, x_2, \ldots, x_m)$ and are analytic in the variables t, $x_1, \ldots, x_m$. The set of such solutions forms a complex vector space F_0. Fixing D (e.g., supposing that D coincides with R) we come to the space F_0 of solutions of the SE.

Let us denote by F the vector space of complex valued functions defined and real-analytic on D and by L the linear operator (2.1) defined on F.

Let Q be a n-order differential operator defined on L:

$$\widetilde{Q}_j = h^{a_1,a_2,\ldots,a_j}\frac{\partial^j}{\partial x_{a_1}\partial x_{a_2}\cdots\partial x_{a_j}}. \tag{2.2}$$

We say Q is a symmetry operator for the SE (2.1) provided

$$[L, Q] = \alpha_Q L \tag{2.3}$$

where $[\ ,\ ]$ denotes the commutator and α_Q is a differential operator of order $n-1$.

It is convenient to express $\partial/\partial t$ via p^2 in accordance with (2.1) and represent Q as a sum of j-multiple anticommutators [**7**]

$$Q = \Sigma_{j=0}^{n} Q_j, \quad Q_j = \left[\left[\cdots\left[K^{a_1a_2\cdots a_j}, \frac{\partial}{\partial a_1}\right]_+, \frac{\partial}{\partial a_2}\right]_+, \ldots, \frac{\partial}{\partial a_j}\right]_+, \tag{2.4}$$

where $K^{a_1a_2\cdots a_j}$ are symmetric tensors depending on t and $\mathbf{x}$. Then it is possible to set $\alpha_Q = 0$ in (2.3) in as much as the term $\partial/\partial t$ cannot appear in the l.h.s. of

(2.3). Substituting (2.4) into (2.3) and equating the coefficients for all powers of differential operators we come to the following system of determining equations

$$
\begin{aligned}
&\partial^{(a_{n+1}}K^{a_1a_2\cdots a_n)} = 0,\\
&2\dot{K}^{a_1a_2\cdots a_{2m}} + \frac{1}{2m}\partial^{(a_{2m}}K^{a_1a_2\cdots a_{2m-1})}\\
&\quad + \sum_{k=m}^{[(n-1)/2]}(-1)^{m+k+1}\frac{2(2k+1)!}{(2k-2m+1)!(2m)!}U_k^{a_1a_2\cdots a_{2m}},\\
&2\dot{K}^{a_1a_2\cdots a_{2l+1}} + \frac{1}{2l+1}\partial^{(a_{2l+1}}K^{a_1a_2\cdots a_{2l})}\\
&\quad + \sum_{k=l+1}^{[n/2]}(-1)^{k+l}\frac{2(2k)!}{(2k-2l-1)!(2l+1)!}W_k^{a_1a_2\cdots a_{2l+1}},
\end{aligned}
\tag{2.5}
$$

where

$$
\begin{aligned}
m &= 0,1,\ldots,[n/2], \quad l = 0,1,\ldots,[(n-1)/2],\\
U_k^{a_1a_2\cdots a_{2m}} &= K^{a_1a_2\cdots a_{2m}b_1b_2\cdots b_{2k-2m+1}}\partial_{b_1}\partial_{b_2}\cdots\partial_{b_{2k-2m+1}}V,\\
W_k^{a_1a_2\cdots a_{2l+1}} &= K^{a_1a_2\cdots a_{2l+1}b_1b_2\cdots b_{2k-2l-1}}\partial_{b_1}\partial_{b_2}\cdots\partial_{b_{2k-2l-1}}V
\end{aligned}
$$

and the complete symmetrization is imposed over the indices in brackets.

These equations define both the coefficients of the n-order symmetry (2.4) and potentials V admitting it.

In the following we will consider some cases when the system of determining equations (2.5) can be solved exactly.

Complete set of symmetry operators of arbitrary order. For the case of ad hoc fixed potential $V(\mathbf{x})$ we can try to find exact solutions of (2.5) for *arbitrary n*.

If $V(\mathbf{x}) = 0$ then the system of equations (2.5) is significantly simplified and reduced to the following form

$$
\begin{aligned}
\partial^{(a_{j+1}}K^{a_1a_2\cdots a_j)} &= -2m(j+1)\dot{K}^{a_1a_2\cdots a_{j+1}}, \quad j = 0,1,\ldots,n-1,\\
\partial^{(a_{n+1}}K^{a_1a_2\cdots a_n)} &= 0,\\
\dot{K} &= 0, \qquad j = 0.
\end{aligned}
\tag{2.6}
$$

Equations (2.6) can be solved explicitly for arbitrary n. A differential consequence of (2.6) is

$$
\partial^{(a_{j+1}}\partial^{a_{j+2}}\ldots\partial^{a_{j+s}}K^{a_1a_2\cdots a_j)} = 0, \quad s = n-j+1.
\tag{2.7}
$$

Solutions of (2.7) are the generalized Killing tensors of order s and rank j whose explicit form is presented in [**12**]. The first of relations (2.6) reduces to first-order ordinary equations for arbitrary parameters defining the generalized Killing tensors. They are easily integrated. The number N_n of linearly independent nth-order symmetries is given by the following formula

$$
N_n = \frac{1}{4!}(n+1)(n+2)^3(n+3)
$$

and all these symmetries belong to the enveloping algebra of the Lie algebra of the Schrödinger group.

Third order symmetry operators for 1 + 1-dimensional Schrödinger equation. Another simplified version of the determining equations (2.5) corresponds to a non-trivial potential in the important case of one spatial variable only. Then the first and second order symmetry operators reduce to generators of a Lie group on the set of solutions of the related SE.

Let us consider the case $n = 3$, which corresponds to the simplest non-Lie symmetry, in more detail. For simplicity we set $m = 1$ and $V = U/2$. The related tensors $K^{a_1 a_2 \cdots a_j}$ in (2.4) reduce to scalars and so

$$Q = \left[\left[\left[h_3, \frac{\partial}{\partial x}\right]_+, \frac{\partial}{\partial x}\right]_+, \frac{\partial}{\partial x}\right]_+ + \left[\left[h_2, \frac{\partial}{\partial x}\right]_+, \frac{\partial}{\partial x}\right]_+ + \left[h_1, \frac{\partial}{\partial x}\right]_+ + h_0. \tag{2.8}$$

The corresponding system (2.5) reduces to

$$h_3' = 0, \qquad h_2' + 2\dot{h}_3 = 0, \qquad 2\dot{h}_2 + h_1' - 6h_3U' = 0, \tag{2.9}$$

$$2\dot{h}_1 + h_0' - 4h_2U' = 0, \qquad \dot{h}_0 - h_1U' + h_3U''' = 0, \tag{2.10}$$

where the dots and primes denote derivatives w.r.t. t and x respectively.

Excluding h_0 from (2.10)) and using (2.9) we arrive at the following equation

$$\begin{aligned} aU'''' &- (2\ddot{a}x^2 + 6aU + c - 2\dot{b}x)U'' \\ &- 6(2\ddot{a}x + aU' - \dot{b})U' - 12\ddot{a}U - 2(2\partial_t^4 ax^2 - 2\dddot{b}x + \ddot{c}) = 0 \end{aligned} \tag{2.11}$$

where a, b, c are arbitrary functions of t. Equation (2.11) is nothing but the compatibility condition for system (2.9), (2.10). If the potential U satisfies (2.11) then the corresponding coefficients of the SO have the form

$$\begin{aligned} &h_3 = a, \qquad h_2 = -2\dot{a}x + b, \qquad h_1 = g_1 + 6aU, \\ &h_0 = -\tfrac{4}{3}\dddot{a}x^3 + 2\ddot{b}x^2 - 2\dot{c}x - 4\dot{a}\varphi + 4(b - 2\dot{a}x)U + d \end{aligned} \tag{2.12}$$

where

$$g_1 = 2\ddot{a}x^2 - 2\dot{b}x + c, \qquad \varphi = \int U\,dx, \qquad u = \varphi', \qquad d = d(t).$$

Separating variables in (2.11) we conclude that up to equivalence, this equation can be reduced to one of the following forms:

$$U'' - 3U^2 + 3\omega_1 = 0, \tag{2.13}$$

$$U'' - 3U^2 - 8\omega_2 x = 0, \tag{2.14}$$

$$(U'' - 3U^2)' - 2\omega_3(xU' + 2U) = 0, \tag{2.15}$$

$$\varphi''' - 3(\varphi')^2 - 2\omega_4(x^2\varphi)' = \tfrac{1}{3}\omega_4^2 x^4 + \omega_5, \qquad U = \varphi' \tag{2.16}$$

where $\omega_1, \ldots, \omega_5$ are arbitrary constants.

From (2.8), (2.13)–(2.16) we find the corresponding symmetry operators

$$Q = p^3 + \frac{3}{4}\{U, p\} \equiv 2pH + \frac{1}{2}Up + \frac{i}{4}U', \tag{2.17}$$

$$Q = p^3 + \tfrac{3}{4}\{U, p\} - \omega_2 t, \tag{2.18}$$

$$Q = p^3 + \tfrac{3}{4}\{U, p\} + \omega_3(tH - \tfrac{1}{4}\{x, p\}), \tag{2.19}$$

$$Q_{\pm} = \frac{1}{\sqrt{24}}\left[p^3 \pm \frac{i}{4}\omega\{\{x,p\},p\} + \frac{1}{4}\{3\varphi' - \omega^2 x^2, p\}\right. \\ \left. \pm \frac{i}{2}\omega\left(\varphi + 2x\varphi' - \frac{\omega^2}{3}x^3\right)\right] \exp(\pm i\omega t), \quad \omega = \sqrt{-\omega_4} \tag{2.20}$$

where U are solutions of (2.13)–(2.16) and $H = \frac{1}{2}(-\partial/\partial_x^2 + U(x))$.

Thus, *the Schrödinger equation* (2.1) *admits a third-order* SO *if potential U satisfies one of the equations* (2.13)–(2.16). The explicit form of the corresponding SOs is present in (2.17)–(2.20).

Formula (2.13) presents the Weierstrass equation while relation (2.14) defines the first Painlevé transcendent. Using the generalized Miura ansatz equations (2.15) and (2.16) can be reduced to the second and fourth Painlevé transcendents respectively or to special Riccati equations [**7**].

Operators (2.17)–(2.20) together with Hamiltonian H form rather interesting algebras whose analysis lies out of frames of the present paper.

Symmetries for spherical potentials. Let us consider now the SE in $1+3$-dimensional space and concentrate on the physically important case of spherical potentials $V = V(r^2)$, $r^2 = x_1^2 + x_2^2 + x_3^2$. The related determining equations (2.5) reduce to ordinary ones.

A subclass of higher-order symmetry operators for the SE is formed by polynomials in the first order symmetry operators which form a Lie algebra. However, for some classes of potentials there also exist higher-order symmetries which do not belong to the enveloping algebra of a Lie algebra. We shall call such higher symmetries essential ones.

Examples of essential higher symmetries for the case of $1+1$-dimensional SE are presented in formulas (2.16)–(2.20) (the related SE does not admit Lie symmetries at all). For the case of spherical potentials such symmetries are much more rare as is stated in the following assertion.

THEOREM 2.1. *Essential higher order symmetry operators for the SE with a spherical potential exist only for $V = 1/r$.*

We will not present a proof here. An example of such essential higher order symmetry is the Runge–Lentz vector.

We see that the Coulomb potential is selected from the other possible spherical potentials by existence of essential symmetry for the related SE.

3. Group classification of the nonlinear SE

It was shown in [**7**] that if the linear one dimensional SE (2.1) admits the third order symmetry operator (2.17) than there exists a wide class of solutions of (2.17) which solve also the nonlinear SE with a cubic nonlinearity. Thus there exist rather nontrivial connections between the third-order symmetries for the linear SE and symmetries of the nonlinear SE which cause the existence of exact solutions.

Here we present a complete description of classical Lie symmetries of the nonlinear Schrödinger equation

$$\left(\frac{\partial}{\partial t} + \Delta\right)\Psi + F(\Psi, \Psi^*) = 0, \tag{3.1}$$

where $\Psi = \Psi(t, \mathbf{x})$, $\mathbf{x} \in R_m$, F is an arbitrary function of two variables: Ψ and complex conjugated function Ψ^*.

Classical Lie symmetries of some special classes of equations (3.1) were investigated in numerous papers, e.g., [**7–9, 12, 15, 16**] and references cited therein. Here we present the results of a complete group classification for all nonequivalent nonlinearities F, based on results of papers [**15, 16**].

To describe symmetries of the SE with a nonfixed nonlinearity in the l.h.s., we can use either the classical Lie algorithm or its specific simplified version [**16**] which presupposes the solution of the following operator equation (compare with (2.1)

$$[L, Q] = \alpha_Q L + \varphi(t, \mathbf{x}),$$

where Q is a first order differential operator, φ is an unknown function. We will not present details of calculations, but only the final result of the group classification of (3.1).

For arbitrary F equation (3.1) is invariant with respect the Euclidean group $E(1, m)$ and admits $1 + m(m+1)/2$ symmetry operators of first order

$$P_0 = \frac{\partial}{\partial t}, \qquad P_a = \frac{\partial}{\partial x_a}, \tag{3.2}$$
$$J_{ab} = x_a p_b - x_b p_a, \quad a, b = 1, 2, \ldots, m.$$

Al cases when the basic invariance group $E(1, m)$ can be extended are enumerated in the following tables.

We will present our results in terms of the amplitude $\rho = |\psi|$ and phase $\varphi = (i/2)\ln(\psi^*/\psi)$ of the function Ψ. We will use the following notations

$$I := \Psi\partial_\Psi + \Psi^*\partial_{\Psi^*} = \rho\partial_\rho, \qquad M := i(\Psi\partial_\Psi - \Psi^*\partial_{\Psi^*}) = \partial_\varphi,$$
$$D := t\partial_t + \tfrac{1}{2}x_a\partial_a, \qquad G_a := t\partial_a + \tfrac{1}{2}x_a M,$$
$$\Pi := t^2\partial_t + tx_a\partial_a - \frac{n}{2}tI + \frac{1}{4}x_a x_a M.$$

In the following page we present nonlinearities which are defined up to arbitrary parameters.

We have presented the results of a group classification of the nonlinear Schrödinger equation. We will not discuss possible applications of the obtained results. They consist in the description of all possible models with a nontrivial extension of the basic invariance with respect to the group $E(1, n)$.

We stress that our classification is valid for nonlinear SE with an *arbitrary* number of spatial variables. Rather surprisingly, increasing of the number of these variables does not lead to increasing difficulties in the calculations of symmetries.

TABLE 1. Nonlinearities depending on arbitrary function $f(\Omega)$.

	F	Ω	Additional symmetries
1.1	$f(\Omega)\|\psi\|^{\gamma_1}e^{\gamma_2\varphi}\psi$, $\gamma_1^2 + \gamma_2^2 \neq 0$	$\|\psi\|^{\gamma_2}e^{-\gamma_1\varphi}$	$(\gamma_1^2 + \gamma_2^2)D - \gamma_1 I - \gamma_2 M$
1.2	$(f(\Omega) + (\gamma - i)\delta\ln\|\psi\|)\psi$	$\|\psi\|^\gamma e^{-\varphi}$	$e^{\delta t}(I + \gamma M)$
1.3	$(f(\Omega) + \delta\varphi)\psi$, $\delta \neq 0$	$\|\psi\|$	$e^{\delta t}M$, $e^{\delta t}(\partial_a + \frac{1}{2}\delta x_a M)$
1.4	$f(\Omega)\psi$	$\|\psi\|$	M, G_a
1.5	$f(\Omega)e^{i\psi}$	$\Re\psi$	$D + i(\partial_\psi - \partial_{\psi^*})$
1.6	$f(\Omega) + i(\delta_1 + i\delta_2)\psi$	$\Re\psi$	$ie^{-\delta_1 t}\theta(x)(\partial_\psi - \partial_{\psi^*})$

Here γ, γ_1, γ_2, δ, δ_1, δ_2 are real numbers, $\theta = \theta(x) \in R$ is a solution of equation $\Delta\theta = \delta_2\theta$.

TABLE 2. Nonlinearities depending on arbitrary parameters

	F	Additional symmetries
2.1	0	G_a, I, M, D, Π, $\eta^0\partial_\psi + \eta^{0*}\partial_{\psi^*}$
2.2	$\gamma\psi + \psi^*$	I, $\eta^0\partial_\psi + \eta^{0*}\partial_{\psi^*}$
2.3	$\sigma\|\Re\psi\|^\gamma$, $\gamma \neq 0, 1$	$I + (1-\gamma)D$, $i\theta(x)(\partial_\psi - \partial_{\psi^*})$
2.4	$\sigma \ln \|\Re\psi\|$	$I + D - i(t\Re\sigma + 1/(2n)x_a x_a \Im\sigma)(\partial_\psi - \partial_{\psi^*})$, $i\theta(x)(\partial_\psi - \partial_{\psi^*})$
2.5	$\sigma e^{\Re\psi}$	$D - \partial_\psi - \partial_{\psi^*}$, $i\theta(x)(\partial_\psi - \partial_{\psi^*})$
2.6	$\sigma\|\psi\|^{\gamma_1} e^{\gamma_2\varphi}\psi$, $\gamma_2 \neq 0$	$M - \gamma_2 D$, $\gamma_2 I - \gamma_1 M$
2.7	$\sigma\|\psi\|^\gamma\psi$, $\gamma \neq 0, 4/n$	G_a, M, $I - \gamma D$
2.8	$\sigma\|\psi\|^{4/n}\psi$,	G_a, M, $I - (4/n)D$, Π

In the following $F = (-(\delta_1 + i\delta_2)\ln|\psi| + (\delta_3 - i\delta_4)\varphi)\psi$, $\Delta = (\delta_2 - \delta_3)^2 - 4\delta_1\delta_4$.

2.9	$\delta_4 = 0$, $\delta_3 \neq 0$, $\delta_2 \neq \delta_3$	$e^{\delta_3 t}M$, $e^{\delta_3 t}(\partial_a + \frac{1}{2}\delta_3 x_a M)$, $e^{\delta_2 t}(I - \delta_1/(\delta_2 - \delta_3)M)$
2.10	$\delta_4 = 0$, $\delta_3 \neq 0$, $\delta_2 = \delta_3$	$e^{\delta_3 t}M$, $e^{\delta_3 t}(\partial_a + \frac{1}{2}\delta_3 x_a M)$, $e^{\delta_2 t}(I - \delta_1 t M)$
2.11	$\delta_4 = 0$, $\delta_3 = 0$, $\delta_2 \neq 0$	M, G_a, $e^{\delta_2 t}(\delta_2 I - \delta_1 M)$
2.12	$\delta_4 = 0$, $\delta_3 = 0$, $\delta_2 = 0$, $\delta_1 \neq 0$	M, G_a, $I - \delta_1 t M$
2.13	$\delta_4 \neq 0$, $\Delta > 0$	$e^{\lambda_i t}(\delta_4 I + (\lambda_i - \delta_2)M)$, $i = 1, 2$, $\lambda_1 = \frac{1}{2}(\delta_2 + \delta_3 - \sqrt{\Delta})$, $\lambda_2 = \frac{1}{2}(\delta_2 + \delta_3 + \sqrt{\Delta})$
2.14	$\delta_4 \neq 0$, $\Delta < 0$	$e^{\mu t}(\delta_4 \cos\nu t I + ((\mu - \delta_2)\cos\nu t - \nu\sin\nu t)M)$, $e^{\mu t}(\delta_4 \sin\nu t\, I + ((\mu - \delta_2)\sin\nu t + \nu\cos\nu t)M)$, $\mu = \frac{1}{2}(\delta_2 + \delta_3)$, $\nu = \frac{1}{2}\sqrt{-\Delta}$
2.15	$\delta_4 \neq 0$, $\Delta = 0$	$e^{\mu t}(\delta_4 t I + \frac{1}{2}(\delta_3 - \delta_2)tM + M)$, $e^{\mu t}(\delta_4 I + \frac{1}{2}(\delta_3 - \delta_2)M)$, $\mu = \frac{1}{2}(\delta_2 + \delta_3)$

Here γ, γ_1, γ_2, δ_1, δ_2, δ_3, δ_4 are real numbers, $\sigma \neq 0$; η^0 is an arbitrary solution of the initial equation, $\theta = \theta(x) \in R$ is a solution of the Laplace equation $\Delta\theta = 0$.

4. Concluding remarks

We present symmetry classification of linear and nonlinear Schrödinger equations. Moreover, the linear SEs are classified with respect to higher order symmetries while the nonlinear ones are classified with respect to continues Lie symmetries.

There exist interesting relations between higher order symmetries of the linear SE and supersymmerties. Namely, if the SE admits symmetries (2.20)) (and so the potential satisfies equations (2.16)) it is possible to find the spectrum of the related Hamiltonian using the higher order derivative SUSY approach [**1**]. In addition, if SE admits symmetry (2.17) (and so the potential satisfies relation (2.13)) it admits hidden SUSY with shape invariance of special type discussed in [**17**]. It would be interesting to investigate SUSY aspects of the other SE admitting higher order symmetries, i.e., the symmetries (2.18) and (2.19).

Higher order symmetry operators for equations of supersymmetric and parasupersymmetric quantum mechanics where discussed in papers [**3, 4**].

References

1. A. Andrianov, F. Cannata, M. Ioffe, and D. Nishnianidze, *Systems with higher-order shape invariance: spectral and algebraic properties*, Phys. Lett. A **266** (2000), no. 4-6, 341–349.
2. J. Beckers, N. Debergh, and A. G. Nikitin, *More on symmetries of the Schrödinger equation*, J. Phys. A **24** (1992), no. 21, L1269–L1275.
3. ______, *On supersymmetries in nonrelativistic quantum mechanics,* J. Math. Phys. **33** (1992), no. 1, 152–160.
4. ______, *More on parasupersymmetries of the Schrödinger equation*, Modern Phys. Lett. A **8** (1993), no. 5, 435–444.
5. C. Boyer, *The maximal 'kinematical' invariance group for an arbitrary potential*, Helv. Phys. Acta **47** (1974), 589–605.
6. J. Friš, V. Mandrosov, Ya. A. Smorodinsky, M. Uhlíř, and P. Winternitz, *On higher symmetries in quantum mechanics*, Phys. Lett. **16** (1965), 354–356.
7. W. I. Fushchich and A. G. Nikitin, *Higher symmetries and exact solutions of linear and nonlinear Schrödinger equation*, J. Math. Phys. **38** (1997), no. 11, 5944–5959.
8. W. I. Fushchich and N. I. Serov, *On some exact solutions of the three-dimensional non-linear Schrödinger equation*, J. Phys. A **20** (1987), no. 15, L929–L933.
9. W. I. Fushchich, W. M. Shtelen, and N. I. Serov, *Symmetry analysis and exact solutions of nonlinear equations of mathematical physics*, Math. Appl., vol. 246, Kluwer, Dordreht, 1993.
10. W. Miller, Jr., *Symmetry and separation of variables*, Encyclopedia Math. Appl., vol. 4, Addison-Wesley, Reading, MA–London–Amsterdam, 1977.
11. U. Niederer, *The maximal kinematical invariance group of the free Schrödinger equation*, Helv. Phys. Acta, **45** (1972/73), no. 5, 802–810.
12. A. G. Nikitin, *Generalized Killing tensors of arbitrary rank and order*, Ukrain. Mat. Zh. **43** (1991), no. 6, 786–795 (Russian); English transl., Ukrainian Math. J. **43** (1991), no. 6,734–743.
13. ______, *The complete set of symmetry operators of the Schrödinger equation*, Ukrain. Mat. Zh. **43** (1991), no. 11, 1521–1526 (Russian); English transl., Ukrainian Math. J. **43** (1991), no. 11, 1413–1418.
14. A. G. Nikitin, S. P. Onufryichuk, and W. I. Fushchich, *Higher symmetries of the Schrödinger equation*, Teoret. Mat. Fiz. **91** (1992), no. 2, 268–278 (Russian); English transl., Theoret. and Math. Phys. **91** (1992), no. 2, 514–521.
15. A. G. Nikitin and R. O. Popovich, *Group classification of nonlinear Schrödinger equations*, Ukraïn. Mat. Zh. **53** (2001), no. 8, 1053–1060 (Ukrainian); English transl., Ukrainian Math. J. **53** (2001), no. 8, 1255–1265.
16. A. G. Nikitin and R. J. Wiltshire, *Systems of reaction diffusion equations and their symmetry properties*, J. Math. Phys. **42** (2001), no. 4, 1666–1688.
17. B. F. Samsonov, M. L. Glasser, J. Negro, and L. M. Nieto, *Second order Darboux displacements*, J. Phys. A **36** (2003), no. 39, 100053–100069.
18. A. Sommerfeld, *Atomic structures and and spectral lines*, Methuen, London, 1923, p. 118.
19. P. Winternitz, Ya. A. Smorodinsky, M. Uhlíř, and I. Friš, *Symmetry groups in classical and quantum mechanics*, Jadernaja Fiz. **4** (1967), 625–635 (Russian); English transl., Soviet Nuclear Phys. **4** (1967), 444–450.

INST. OF MATHEMATICS, UKRAINIAN NATIONAL ACADEMY OF SCIENCES, 3 TERESHCHENKIVSKA ST., 01601, KYIV-4, UKRAINE

E-mail address: `nikitin@imath.kiev.ua`

Centre de Recherches Mathématiques
CRM Proceedings and Lecture Notes
Volume **37**, 2004

Symmetries and Lagrangian Time-Discretizations of Euler Equations

Alexei V. Penskoi

ABSTRACT. In the late 80s and early 90s J. Moser and A. P. Veselov considered discrete Lagrangian systems on Lie groups with additional symmetry conditions imposed on the Lagrangians. They observed that such systems are often integrable time-discretizations of integrable Euler equations on these Lie groups. In recent papers we studied Lagrangian discrete systems with additional symmetry requirements on certain infinite-dimensional Lie groups. We will discuss some interesting properties of these systems.

1. Introduction

The purpose of this talk is to review some recent results on discrete Lagrangian systems [**11–14**].

Let M be a manifold and L be a function on $M \times M$. The *discrete Lagrangian system* with the Lagrangian L is the system of difference equations

$$\delta S = 0,$$

which describes the stationary points of the functional $S = S(X)$, defined on the space of sequences $X = (x_k)$, $x_k \in M$, $k \in \mathbb{Z}$ by a formal sum

$$S = \sum_{k \in \mathbb{Z}} L(x_k, x_{k+1}).$$

Sometimes this system has a continuous limit; in that case it is called a discrete version of the corresponding system of differential equations.

For integrable equations it is natural to ask for discretizations which are also integrable. The theory of integrable Lagrangian discretizations of classical integrable systems was initiated by Moser and Veselov [**8, 16**].

Let us assume that M is a Lie group G. It was observed by Veselov and Moser [**8, 9, 16, 17**] that Lagrangians that are symmetric

$$L(x, y) = L(y, x)$$

2000 *Mathematics Subject Classification.* 34K99, 22E65, 70H99.

The author is very grateful to the Organizing Committee of the Workshop on Superintegrability in Classical and Quantum Systems for giving him the possibility to present this talk.

This is the final form of the paper.

and right-invariant (or left-invariant)

$$\forall g \in G \quad L(xg, yg) = L(x, y),$$

often correspond to integrable systems which are discretizations of the Euler equations corresponding to some right-invariant (or left-invariant) metrics on G.

In particular, it was shown that the discrete Lagrangian system on the orthogonal group $\mathrm{O}(n)$ with the Lagrangian $L(X, Y) = \mathrm{tr}(XJY^T)$, where X, $Y \in \mathrm{O}(n)$ and J is a symmetric positive-defined matrix, can be considered as an integrable discrete version of the Euler–Arnol′d top [**1**].

The first attempt to generalize this approach to the infinite-dimensional situation was done in [**9**], where the case of the group of area-preserving plane diffeomorphisms $\mathrm{SDiff}(\mathbb{R}^2)$ was considered.

This paper was followed by recent papers [**11–14**] where discrete Lagrangian systems on the Virasoro group have been discussed. The interest in the Virasoro group was motivated by the important observation due to Khesin and Ovsienko [**10**] that the Korteweg–de Vries equation can be interpreted as an Euler equation on the Virasoro algebra. This result was later generalized by Misiołek [**7**] and Khesin [**5**]. They remarked that the Euler equation corresponding to the right-invariant $H^1_{\alpha,\beta}$-metric on the Virasoro group has the form

$$\alpha(v_t + 3vv_x) - \beta(v_{xxt} + 2v_x v_{xx} + vv_{xxx}) - bv_{xxx} = 0.$$

This is a three-parameter family of integrable equation including the Camassa–Holm, the Hunter–Saxton and the Korteweg–de Vries equations.

The plan of the talk is following. We shall start by recalling the basic facts about the Euler equation in Section 2 and the Virasoro group in Section 3. Then we will expose in Section 4 results of Ovsienko, Khesin and Misiołek [**5, 7, 10**] concerning Euler equations on the Virasoro group. Section 5 is devoted to discrete Lagrangian systems, and the final Section 6 contains results on discrete Lagrangian systems on the Virasoro group.

2. The Euler equation

Let G be a Lie group, $\mathfrak{g}$ its Lie algebra. A right-invariant metric on G is completely defined by its restriction to $\mathfrak{g}$

$$(\ ,\ ) \colon \mathfrak{g} \times \mathfrak{g} \to \mathbb{R}.$$

This scalar product on $\mathfrak{g}$ defines a linear map $A \colon \mathfrak{g} \to \mathfrak{g}^*$ such that

$$(\xi, \eta) = \langle A\xi, \eta \rangle,$$

where ξ, $\eta \in \mathfrak{g}$ and $\langle\ ,\ \rangle$ denotes the natural pairing between $\mathfrak{g}$ and $\mathfrak{g}^*$. The operator A is called the inertia operator.

To describe a geodesic $g(t)$ on G corresponding to $(\ ,\ )$, we transport its velocity vector to the identity by the right translation

$$v(t) = R_{g^{-1}(t)*} \frac{d}{dt} g(t).$$

Since $v(t)$ is an element of $\mathfrak{g}$, we can consider $m = Av$ which is an element of the dual space $\mathfrak{g}^*$. Then m satisfies the Euler equation given by the following explicit formula:

$$\frac{dm}{dt} = -ad^*_{A^{-1}m} m. \tag{2.1}$$

This is a standard result which can be found in [**1**]. If we start with a left-invariant metric, the sign in (2.1) is reversed. Since A is non-degenerate, one can rewrite the Euler equation in terms of v on the Lie algebra $\mathfrak{g}$.

EXAMPLE 1 ([**1**]). Let us consider $\mathrm{SO}(N)$. We identify the dual space $\mathfrak{so}(N)^*$ with the algebra $\mathfrak{so}(N)$ using the standard product $\operatorname{tr} XY^T$ in the space of $N \times N$ matrices. Let J be a symmetric positive-definite matrix. The inertia operator defined as

$$A\Omega = J\Omega + \Omega J, \quad \Omega \in \mathfrak{so}(N)$$

gives the Euler equation on $\mathrm{SO}(N)$. It can be written in a more familiar form as a system

$$\begin{cases} \dot{M} = [M, \Omega], \\ M = J\Omega + \Omega J. \end{cases}$$

This is the Euler–Arnol′d equation for the N-dimensional rigid body.

EXAMPLE 2 ([**1**]). Let us consider the group $\mathrm{SDiff}(\mathbb{R}^2)$ of volume-preserving diffeomorphisms of $\mathbb{R}^2$. The corresponding Lie algebra is the algebra of all vector fields with divergence 0 on $\mathbb{R}^2$. Let us consider the scalar product of two such vector fields defined as

$$(v_1, v_2) = \iint_{\mathbb{R}^2} v_1 \cdot v_2 \, dx_1 \, dx_2,$$

where $\cdot$ denotes the standard scalar product of vectors in $\mathbb{R}^2$. We can find the Euler equation corresponding to this scalar product. We will write it on the Lie algebra in the following form:

$$\begin{cases} \dfrac{\partial v}{\partial t} + v\nabla v = -\operatorname{grad} p, \\ \operatorname{div} v = 0. \end{cases}$$

This is the two-dimensional Euler equation for the ideal fluid. Here p is a pressure, it is some unknown function which plays the role of a Lagrangian multiplier defined by the constraint $\operatorname{div} v = 0$.

3. The Virasoro group and the Virasoro algebra

Let $\mathrm{Diff}_+(S^1)$ be the group of diffeomorphisms of S^1 preserving the orientation. We shall represent an element of $\mathrm{Diff}_+(S^1)$ as a diffeomorphism $f\colon \mathbb{R} \to \mathbb{R}$ such that

(1) $f \in C^\infty(\mathbb{R})$,
(2) $f'(x) > 0$,
(3) $f(x + 2\pi) = f(x) + 2\pi$.

Such a representation is not unique. Indeed, the functions $f + 2\pi k$, $k \in \mathbb{Z}$ represent one element of $\mathrm{Diff}_+(S^1)$.

There exists a nontrivial central extension of $\mathrm{Diff}_+(S^1)$ which is unique up to an isomorphism

This extension is called the Virasoro group (or the Bott–Virasoro group) and is denoted by Vir. Elements of Vir are pairs (f, F), where $f \in \mathrm{Diff}_+(S^1)$, $F \in \mathbb{R}$. The product of two elements is defined with the help of the Bott cocycle as

$$(f, F) \circ (g, G) = \left(f \circ g, F + G + \int_0^{2\pi} \log(f \circ g)' \, d \log g' \right).$$

The unit element of Vir is $(\mathrm{id}, 0)$. The inverse element of (f, F) is $(f^{-1}, -F)$.

The Virasoro algebra vir is a Lie algebra corresponding to the Virasoro group. It is the central extension of the algebra $\mathrm{vect}(S^1)$ of vector fields on the circle S^1

$$\mathrm{vir} = \mathrm{vect}(S^1) \oplus \mathbb{R}.$$

We represent an element of $\mathrm{vect}(S^1)$ as $v(x)\partial_x$, where v is a 2π-periodic function. Thus an element of the Virasoro algebra is a pair $(v(x)\partial_x, a)$. The algebra commutator in vir is defined with the help of the Gelfand–Fuchs cocycle as

$$[v(x)\partial_x, a], [w(x)\partial_x, b]) = \left((-vw_x + v_x w)(x)\partial_x, \int_0^{2\pi} v_{xxx} w\, dx\right).$$

4. The Euler equation for the $H^1_{\alpha,\beta}$-metric on the Virasoro group

In 1987 Khesin and Ovsienko remarked [**10**] that the Korteweg–de Vries equation can be interpreted as an Euler equation on the Virasoro group. This result was later generalized by Misiołek [**7**] and Khesin [**5**] in the following way.

Let α and β be two non-negative real numbers such that $\alpha^2 + \beta^2 \neq 0$. Let us define the $H^1_{\alpha,\beta}$-metric on the Virasoro algebra by the formula

$$\big((v(x)\partial_x, a), (w(x)\partial_x, b)\big)_{H^1_{\alpha,\beta}} = \int_0^{2\pi} \big(\alpha v(x) w(x) + \beta v_x(x) w_x(x)\big)\, dx + ab.$$

THEOREM 1. *The Euler equation corresponding to the right-invariant $H^1_{\alpha,\beta}$-metric on the Virasoro group has the form*

$$\alpha(v_t + 3vv_x) - \beta(v_{xxt} + 2v_x v_{xx} + vv_{xxx}) - bv_{xxx} = 0, \qquad b_t = 0. \tag{4.1}$$

Since b is a constant, we can consider (4.1) as an equation for v depending on three constants α, β, b. We obtain a three-parametric family of integrable equations which we call the Camassa–Holm family. The reason for this name is the following. We have a freedom of multiplication of the equation by a nonzero constant, the two-dimensional scaling symmetry group $v \to \lambda v$, $t \to \mu t$, $x \to \lambda\mu x$ and the Galilean group $v \to v + c$, $x \to x + dt$, $t \to t$. Modulo these symmetries we have just one generic orbit, containing the equation with $\alpha = 1$, $\beta = 1$, $b = 0$:

$$v_t - v_{xxt} + 3vv_x - 2v_x v_{xx} - vv_{xxx} = 0,$$

which is one of the canonical forms of the Camassa–Holm shallow-water equation [**2**]

$$v_t + 2\kappa v_x + \gamma v_{xxx} - v_{xxt} + 3vv_x - 2v_x v_{xx} - vv_{xxx} = 0.$$

We have also four degenerate orbits. When $\alpha \neq 0$, $\beta = 0$, $b \neq 0$ the equation (4.1) is equivalent to the KdV equation:

$$v_t + 3vv_x + v_{xxx} = 0.$$

Further degeneration $\alpha \neq 0$, $\beta = 0$, $b = 0$ leads to the dispersionless KdV equation (sometimes also called the Hopf equation):

$$v_t + 3vv_x = 0.$$

When $\alpha = 0$, $\beta \neq 0$ we have the Hunter–Saxton equation [**4**]

$$v_{xxt} + 2v_x v_{xx} + vv_{xxx} = 0.$$

Finally if both α and β are zero (but b is not) we simply have

$$v_{xxx} = 0.$$

5. Discrete Lagrangian systems

The definition of a discrete Lagrangian system was already given in the introduction. We have already mentioned in the introduction the observation by Veselov and Moser that symmetric and right-invariant (or left-invariant) Lagrangians often correspond to integrable systems which are discretizations of the Euler equations corresponding to some right-invariant (or left-invariant) metrics on G. Let us consider some examples.

EXAMPLE 3 ([**8**]). Let us consider the group $\mathrm{SO}(N)$. Let J be a symmetric positive-definite matrix. The Lagrangian

$$L(X,Y) = \operatorname{tr}(XJY^T)$$

is left-invariant and symmetric. Let us introduce the discrete angular velocity $\omega_k = X_k^T X_{k-1}$ and the discrete angular momentum $M_k = \omega_k^T J - J\omega_k$. The discrete Euler–Lagrange equation $\delta S/\delta X_k = 0$ can be written in the following form (discrete Euler–Arnol′d equation).

$$\begin{cases} M_{k+1} = \omega_k M_k \omega_k^{-1}, \\ M_k = \omega_k^T J - J\omega_k. \end{cases}$$

This equation is integrable and its continuous limit is the Euler–Arnol′d equation of N-dimensional rigid body considered in the Example 1.

EXAMPLE 4 ([**3, 9**]). Let us consider the group $\mathrm{SDiff}(\mathbb{R}^2)$. The Lagrangian

$$L(f,g) = \iint_{\mathbb{R}^2} \operatorname{tr}(J(f)J(g)^{-1})\, dx_1\, dx_2,$$

where $f,\ g \in \mathrm{SDiff}(\mathbb{R}^2)$ and J is the Jacobian, is right-invariant and symmetric. Let $\varphi = f_k \circ f_{k-1}^{-1}$, $\psi = f_{k+1} \circ f_k^{-1}$ and $\chi = \varphi^{-1}$. The discrete Euler–Lagrange equation $\delta S/\delta f_k = 0$ can be written in the following form.

$$\begin{cases} \psi_1 = \tau x_1 + a_1 - \chi_1 \\ \psi_2 = \tau x_2 + a_2 - \chi_2 \\ \operatorname{tr} J(\varphi) = \tau. \end{cases}$$

In these equations τ, a_1, a_2 are constants, ψ_i, $i = 1,\ 2$ are components of $\psi = (\psi_1, \psi_2)\colon \mathbb{R}^2 \to \mathbb{R}^2$, and χ_i, $i = 1\ ,2$ are defined in an analogous way. This discrete system is integrable. If $\tau = 2$ one can find the continuous limit

$$\begin{cases} v_t + v\nabla v = 0, \\ \operatorname{div} v = 0. \end{cases}$$

This is an equation of the isobaric flow, a particular case of the two-dimensional Euler equation for the ideal fluid when the pressure is equal to zero.

6. Discrete Lagrangian systems on the Virasoro group

In this section we present some results of papers [**11–14**]. Let us consider discrete Lagrangian systems on the Virasoro group. It was explained in Section 5 that it is natural to consider right-invariant symmetric Lagrangians.

We shall use the following simple observation to construct Lagrangians. Let us consider a Lagrangian $L(x,y)$ on a Lie group G. Let $H(x) = L(x,e)$, where

e is the identity element of G. A right-invariant Lagrangian $L(x,y)$ is completely determined by H. Indeed, $L(x,y) = L(xy^{-1}, e) = H(xy^{-1})$.

Let us now consider a right-invariant symmetric Lagrangian. It is easy to see that a symmetric Lagrangian corresponds to an inverse-invariant function H:

$$H(x^{-1}) = H(x).$$

Let us start by considering the discrete Lagrangian systems on Vir corresponding to the functions H of the following form [**14**]

$$H\big((f,F)\big) = F^2 + \int_0^{2\pi} V\big(f(x) - x, f'(x)\big)\,dx, \tag{6.1}$$

where f is a diffeomorphism, $F \in \mathbb{R}$, and $V(x_1, x_2)$ is an arbitrary, 2π-periodic in x_1 function of two variables, which satisfies the condition:

$$V_1(0,1) = 0.$$

Let us explain a motivation for such a choice of H. We are interested in Lagrangian time-discretizations of the Camassa–Holm family. This family corresponds to $H^1_{\alpha,\beta}$-metrics, and these metrics depend on elements of the algebra and their first derivatives. So it is natural to consider a function H of the form

$$H\big((f,F)\big) = F^2 + \int_0^{2\pi} U\big(f(x), f'(x), x\big)\,dx, \tag{6.2}$$

where $U(x_1, x_2, x_3)$ is an arbitrary function that is 2π-periodic with respect to the first argument. Periodicity of V or U is related to the fact that $f(x)$ and $f(x) + 2\pi$ represent the same diffeomorphism of S^1. However, it turns out that the continuous limit of the system corresponding to (6.2) is, in general, some quite boring ODE having nothing in common with Euler equations. To obtain something interesting, it is necessary to impose the additional conditions that U has the form $U(x_1, x_2, x_3) = V(x_1 - x_3, x_2)$ and $V_1(0,1) = 0$. In this case the continuous limit is some equation from the Camassa–Holm family.

The form (6.1) is very general since the difference $f(x) - x$, which is a 2π-periodic function, is as natural as $f(x)$ itself. The only property which might look artificial is the last condition on the partial derivative of V. See [**14**] for a detailed discussion.

In general, the Lagrangian corresponding to (6.1) is not symmetric. Hence, it is not expected to give us an integrable system. Nevertheless, it is interesting to look at such Lagrangians because of the following interesting phenomenon. In general, if we consider a Lie group G, a continuous limit of a right-invariant discrete Lagrangian system is a geodesic flow corresponding to some right-invariant metric. This geodesic flow is, in general, not integrable. Let us for example look at the case of $\mathrm{SO}(N)$. It is known that for $N > 3$ a geodesic flow of a general right-invariant metric is nonintegrable (see, e.g., [**15**]). Hence, the result on the Virasoro group turns out to be surprising: in spite of the fact that the class of discrete systems we consider (6.1) is quite general, in the continuous limit we have a family of integrable equations. There are integrable cases of the Euler equations on $\mathrm{SO}(N)$ (for example the Manakov metrics [**6**]) but no analogues of the result for the Virasoro group are known for them.

The heuristic explanation of this phenomenon is that the equations of the Camassa–Holm family can be considered as nonlinear analogues of the harmonic oscillators on the Virasoro group: in the first approximation all Hamiltonian systems near equilibriums behave like harmonic oscillators. Thus, in some sense this demonstrates a universal nature of the Camassa-Holm family of equations. See Discussion in [**14**] for more details.

Let us now consider the question of the integrability of discrete Lagrangian systems on the Virasoro group. Unfortunately we do not know examples of such systems, but we have good candidates for integrability. A good candidate to be an integrable discretization of the KdV equation was considered in [**12**], but it is the case of the discretization of the Hunter-Saxton equation [**13**] which is most promising. Let us consider this case.

The Hunter–Saxton equation corresponds to the Euler equation for the $H^1_{0,1}$-metric. This metric depends only on first derivatives of algebra elements. Hence, it is natural to look at such functions H (6.1) that V depends only on f'. As was explained before, it is natural to study right-invariant symmetric Lagrangians when looking for integrable systems, and symmetric Lagrangians correspond to inverse-invariant functions H.

It can be easily verified that functions H of the form

$$H\big((f,F)\big) = F^2 + \int_0^{2\pi} V\big(f'(x)\big)\,dx \tag{6.3}$$

are inverse-invariant when the function V satisfies the condition

$$xV\left(\frac{1}{x}\right) = V(x). \tag{6.4}$$

The simplest function V satisfying the property (6.4) is $V(x) = \sqrt{x}$. Let us consider this case.

We have $V(x) = \sqrt{x}$. The function V defines H as described in (6.3). The function H defines a Lagrangian L as described above:

$$L\big((f_k, F_k), (f_{k+1}, F_{k+1})\big) = H\big((f_k, F_k) \circ (f_{k+1}, F_{k+1})^{-1}\big).$$

Hence we are considering a functional

$$S = \sum_{k\in\mathbb{Z}} L\big((f_k, F_k), (f_{k+1}, F_{k+1})\big),$$

where $\{(f_k, F_k)\}$ is a sequence of points on Vir. We can find the Euler–Lagrange equations $\delta S/\delta(f_k, F_k) = 0$. They are [**13**]:

$$-\Omega_k + \Omega_{k+1} = 0, \tag{6.5}$$

$$\left[-2\Omega_k(\log((\omega_k)'))' - \tfrac{1}{2}\sqrt{(\omega_k)'} + 2\Omega_{k+1}(\log((\omega_{k+1}^{-1})'))' - \tfrac{1}{2}\sqrt{(\omega_{k+1}^{-1})'}\right]' = 0, \tag{6.6}$$

where (ω_k, Ω_k) and $(\omega_{k+1}, \Omega_{k+1})$ are discrete analogues of angular velocities,

$$(\omega_l, \Omega_l) = (f_{l-1}, F_{l-1}) \circ (f_l, F_l)^{-1}, \quad l \in \mathbb{Z}.$$

The equation (6.5) is just saying that Ω_k is an integral, $\Omega_k = \Omega$. As for (6.6), we can integrate this equation once and put

$$\Phi = \frac{1}{\sqrt{(\omega_k)'}}, \qquad \Psi = \frac{1}{\sqrt{(\omega_{k+1}^{-1})'}}.$$

We obtain the equation

$$8\Omega\left(-\frac{\Phi'}{\Phi}+\frac{\Psi'}{\Psi}\right)+\frac{1}{\Phi}+\frac{1}{\Psi}+C=0,$$

where C is a constant of integration. This equation is equivalent to the equation

$$\Psi'+\Psi\left(\frac{C}{8\Omega}+\frac{1}{8\Omega\Phi}-\frac{\Phi'}{\Phi}\right)+\frac{1}{8\Omega}=0.$$

This is a linear first-order differential equation for Ψ with periodic coefficients depending on Φ. For generic Φ it has only one solution, so Ψ is determined by Φ up to a constant C. Reconstructing ω_{k+1}^{-1} from Ψ, we obtain another constant, so we have a following result: ω_{k+1} is obtained from ω_k by a two-parametric correspondence. To find ω_{k+1} starting from ω_k, we must solve a first-order linear differential equation to find ω_{k+1}^{-1}, and then reconstruct ω_{k+1} from ω_{k+1}^{-1} by inversion. The analogous situation was observed in the case of $\mathrm{SDiff}(\mathbb{R}^2)$ [**9**] which is integrable [**3**]. For this reason we consider system (6.5), (6.6) to be a good candidate for an integrable Lagrangian discretization of the Hunter–Saxton equation. See [**13**] for more detailed discussion.

It should be remarked, that one can obtain the Hunter–Saxton equation as an Euler equation not only on the Virasoro group, but also on the group of orientation-preserving diffeomorphisms of the circle. This leads to a particularly simple discretization

$$\left[\sqrt{(\omega_k)'}+\sqrt{(\omega_{k+1}^{-1})'}\right]'=0,$$

which also has some nice properties, see [**13**] for details.

References

1. V. I. Arnol′d, *Mathematical methods of classical mechanics*, Grad. Texts in Math., vol. 60, Springer-Verlag, New York, 1998.
2. R. Camassa and D. D. Holm, *An integrable shallow water equation with peaked solitons*, Phys. Rev. Lett. **71** (1993), no. 11, 1661–1664.
3. E. V. Ferapontov and A. P. Veselov, appendix to [**9**].
4. J. K. Hunter and R. Saxton *Dynamics of director fields*, SIAM J. Appl. Math. **51** (1991), no. 6, 1498–1521.
5. B. Khesin, G. Misiołek, *Euler equations on homogeneous spaces and Virasoro orbits*, Adv. Math. **176** (2003), no. 1, 116–144.
6. S. V. Manakov, *Note on the integration of Euler equations of the dynamics of an n-dimensional rigid body*, Funktsional. Anal. i Prilozhen. **10** (1976), no. 4, 93–94 (Russian); English transl., Functional Anal. Appl. **10** (1976), no. 4, 328–329.
7. G. Misiołek *A shallow water equation as a geodesic flow on the Bott–Virasoro group*, J. Geom. Phys. **24** (1998), no. 3, 203–208.
8. J. Moser and A. P. Veselov, *Discrete versions of some classical integrable systems and factorization of matrix polynomials*, Comm. Math. Phys. **139** (1991), no. 2, 217–243.
9. ———, *Two-dimensional 'discrete hydrodynamics' and Monge-Ampère equation*, ETH, Zürich, 1993 (preprint); Ergodic Theory Dynam. Systems **22** (2002), no. 5, 1575–1583.
10. V. Yu. Ovsienko and B. A. Khesin, *The super Korteweg–de Vries equation as an Euler equation*, Funktsional. Anal. i Prilozhen. **21** (1987), no. 4, 81–82 (Russian); English transl., Functional Anal. Appl. **21** (1988), no. 4, 329–331.
11. A. V. Penskoi, *Discrete Lagrangian systems on the Virasoro group.* Vestnik Moskov. Univ. Ser. I Math. Mekh. (1996), no 4, 99-102 (Russian); English transl., Moscow Univ. Math. Bull. **51** (1996), no 4, 52–54.
12. ——— *Lagrangian time-discretization of the Korteweg-de Vries equation* Phys. Lett. A **269** (2000), no. 4, 224–229.

13. ———, *Lagrangian time-discretization of the Hunter–Saxton equation*, Phys. Lett. A **304** (2002), no. 5-6, 157–167.
14. A. V. Penskoi and A. P. Veselov, *Discrete Lagrangian systems on the Virasoro group and Camassa–Holm family*, Nonlinearity **16** (2003), no. 2, 683–688.
15. A. P. Veselov, *Conditions for the integrability of Euler equations on* so(4), Dokl. Akad. Nauk SSSR **270** (1983), no. 6, 1298-1300 (Russian); English transl., Soviet Math. Dokl. **27** (1983), no. 3, 740–742.
16. ———, *Integrable systems with discrete time, and difference operators*, Funktsional. Anal. i Prilozhen. **22** (1988), no. 2, 1–13 (Russian); English transl., Funct. Anal. Appl. **22** (1988), no. 2, 83–93.
17. ———, Integrable Lagrangian relations and factorization of matrix polynomials, Funktsional. Anal. i Prilozhen. **25** (1991), no. 2, 38-49 (Russian); English transl., Funct. Anal. Appl. **25** (1991), no. 2, 112–122.

Centre de recherches mathématiques, Université de Montréal, C. P. 6128, succ. Centre-ville, Montréal, Québec, H3C 3J7, Canada

E-mail address: penskoi@crm.umontreal.ca

Centre de Recherches Mathématiques
CRM Proceedings and Lecture Notes
Volume **37**, 2004

Two Exactly-Solvable Problems in One-Dimensional Quantum Mechanics on Circle

L. G. Mardoyan, G. S. Pogosyan, and A. N. Sissakian

Abstract. In this note we establish a relation between two exactly-solvable problems on a circle, namely singular Coulomb and singular oscillator systems.

1. A series of complex mappings $S_{2C} \to S_2$, $S_{4C} \to S_3$ and $S_{8C} \to S_5$ was constructed in a recent paper [**5**]. They extend to spherical geometry the Levi-Civita, Kustaanheimo–Stiefel and Hurwitz transformations, well known for Euclidean space. It was shown that these transformations establish a correspondence between Coulomb and oscillator problems in classical and quantum mechanics for dimensions $(2,2)$, $(3,4)$ and $(5,8)$ on spheres. A detailed analysis of the real mapping on the curved space has been done in [**7**]. It was remarked that in the stereographic projection the relation between Coulomb and oscillator problems functionally coincide with the flat space Levi-Civita and Kustaanheimo–Stiefel transformations. The relation between the quasiradial Schrödinger equations for Coulomb and oscillator problems on the n-dimensional spheres and one- and two-sheeted hyperboloids for $n \geq 2$ was found in the article [**6**].

The present note is devoted to two singular one-dimensional exactly-solvable potentials on circle S_1: $s_0^2 + s_1^2 = R^2$

$$V^{\mathrm{so}}(\vec{s}) = \frac{\omega^2 R^2}{2} \frac{s_1^2}{s_0^2} + \frac{1}{2} \frac{k_1^2 - \frac{1}{4}}{s_1^2}, \qquad V^{\mathrm{c}}(\vec{s}) = -\frac{\mu}{R} \frac{s_0}{|s_1|} + \frac{1}{2} \frac{p^2 - \frac{1}{4}}{s_1^2}, \tag{0.1}$$

(we consider that k_1, $p > 0$) where s_0, s_1 are Cartesian coordinates in the ambient Euclidean space E_2.

The potentials in (0.1) are the well-known analogs of superintegrable [**4**] and exact-solvable [**8**] potentials restricted to one spatial dimension. Below we will prove that these two systems are connected to each other by a one-dimensional

2000 *Mathematics Subject Classification.* 35C05, 35Q40.

The research of the first two authors was partially supported by the NATO Collaborative Linkage Grant No. 978431 and ANSEF Grant No. PS81.

G. S. P. also thanks the support of the Direcciob General de Asuntos del Personal Académico, Universidad Autonoma de Mexico (DGAPA-UNAM) by the grant IN102603-3 *Óptica Matemática*.

We thank our collaborators E. G. Kalnins, W. Miller, Jr., and P. Winternitz for their interest in this work and fruitful discussions.

This is the final form of the paper.

complex duality transformation [**5**], which is related to the more general class of "gauge transformations" introduced in the paper [**3**].

2. The Schrödinger equation describing the nonrelativistic quantum motion on the circle S_1 in the polar coordinate $\varphi \in [-\pi, \pi]$

$$s_0 = R\cos\varphi, \qquad s_1 = R\sin\varphi$$

has the following form ($\hbar = m = 1$)

$$\frac{d^2\Psi}{d\varphi^2} + 2R^2[E - V(\varphi)]\Psi = 0. \tag{0.2}$$

Substituting the singular oscillator potential $V^{so}(\varphi)$ in (0.2), we obtain a Pöschl–Teller-type equation

$$\frac{d^2\Psi}{d\varphi^2} + \left[\epsilon - \frac{k_0^2 - \frac{1}{4}}{\cos^2\varphi} - \frac{k_1^2 - \frac{1}{4}}{\sin^2\varphi}\right]\Psi = 0 \tag{0.3}$$

where $\epsilon = 2R^2E + \omega^2R^4$ and $k_0^2 = \omega^2R^4 + \frac{1}{4}$. The regular (at points $\varphi = 0$ and $\pi/2$) solution of the above equation maybe chosen in following form [**2**]

$$\Psi_n(\varphi) = \sqrt{\frac{2(2n + k_0 \pm k_1 + 1)\Gamma(n + k_0 + k_1 + 1)\Gamma(n \pm k_1 + 1)}{R[\Gamma(1 \pm k_1)]^2\Gamma(n + k_0 + 1)(n)!}} \tag{0.4}$$
$$\times (\sin\varphi)^{1/2 \pm k_1}(\cos\varphi)^{1/2 + k_0}\,{}_2F_1(-n, n + k_0 \pm k_1 + 1; 1 \pm k_1; \sin^2\varphi),$$

with

$$\epsilon = (2n \pm k_1 + k_0 + 1)^2, \qquad n = 0, 1, 2, \dots . \tag{0.5}$$

The energy spectrum of the one-dimensional singular oscillator is given by

$$E_n(R) = \frac{1}{2R^2}[(2n \pm k_1 + \tfrac{1}{2})^2 + (2k_0 + 1)(2n \pm k_1 + 1)]. \tag{0.6}$$

Let us remark that the wave-functions have been normalized in the domain $[0, \pi/2]$. The positive sign in front of k_1 has to taken whenever $k_1 > \frac{1}{2}$, i.e., the additional term to the oscillator potential is repulsive at the origin and the motion takes place only in the domain $\varphi \in [0, \pi/2]$. If $0 < k_1 \le \frac{1}{2}$, i.e., the additional term is attractive at the origin, both the positive and negative signs must be taken into account in the solution. The motion in this case takes place in $\varphi \in [-\pi/2, \pi/2]$. This is indicated by the notation $\pm k_1$, in the formulas.

3. Let us write the Schrödinger equation (0.2) for the singular Coulomb potential $V^{c}(\vec{s})$

$$\frac{d^2\Psi}{d\varphi^2} + \left(2R^2E + 2\mu R\cot|\varphi| - \frac{p^2 - \frac{1}{4}}{\sin^2\varphi}\right)\Psi = 0. \tag{0.7}$$

First we will consider the region $\varphi \in [0, \pi]$. We make now a transformation to the new variable $\theta \in [0, \pi/2]$

$$e^{i\varphi} = \cos\theta, \tag{0.8}$$

which is possible if we continue the variable φ into the complex domain: $\operatorname{Re}\varphi = 0$, $0 \le \operatorname{Im}\varphi < \infty$ (see Figure 1). We also complexify the coupling constant μ by putting $k = i\mu$ so that

$$\mu\cot\varphi = k(1 - 2\sin^{-2}\theta). \tag{0.9}$$

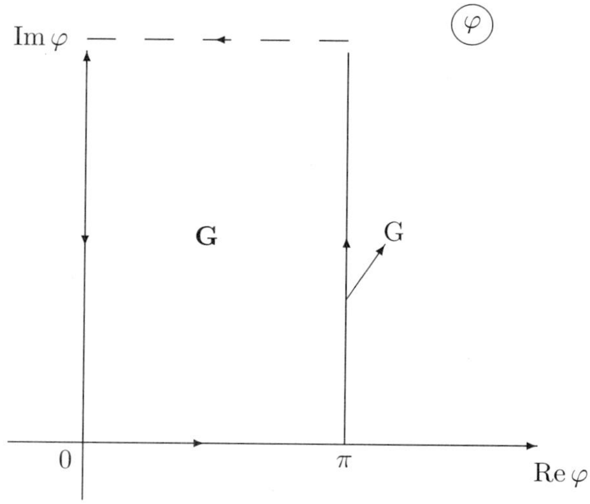

FIGURE 1. Domain $\mathbf{G} = \{0 \leq \operatorname{Re}\varphi \leq \pi; 0 \leq \operatorname{Im}\varphi < \infty\}$ on the complex φ plane.

As a result we obtain the equation

$$\frac{d^2W}{d\theta^2} + \left[\epsilon - \frac{k_0^2 - \frac{1}{4}}{\cos^2\theta} - \frac{k_1^2 - \frac{1}{4}}{\sin^2\theta}\right] W = 0, \tag{0.10}$$

where $W(\theta) = (\cot\theta)^{1/2}\Psi(\theta)$ and

$$\epsilon = 2R^2E + 2kR, \qquad k_0^2 = 2R^2E - 2kR, \qquad k_1^2 = 4p^2. \tag{0.11}$$

From the above equation we see that, up to the substitution (0.11) the equation (0.10) for the one-dimensional singular Coulomb problem coincides with the one-dimensional singular oscillator equation (0.3).

The regular, for $\theta \in [0, \pi/2]$ and $k_0 \geq \frac{1}{4}$, solution of this equation according to (0.4) is

(0.12)

$$\Psi(\theta) = \frac{W(\theta)}{\sqrt{\cot\theta}} = C_n (\sin\theta)^{1\pm k_1} (\cos\theta)^{k_0} {}_2F_1(-n, n + k_0 \pm k_1 + 1; 1 \pm k_1; \sin^2\theta),$$

where C_n is a normalization constant. To compute the constant C_n we require that the wave function (0.12) satisfy the condition

$$R\int_0^\pi \Psi_n \Psi_n^\diamond \, d\varphi = \frac{1}{2}, \tag{0.13}$$

where the symbol "$\diamond$" means the complex conjugate together with the inversion $\varphi \to -\varphi$, i.e., $\Psi^\diamond(\varphi) = \Psi^*(-\varphi)$. [We choose the scalar product involving $\Psi^\diamond$ because for $\varphi \in \mathbf{G}$ and real μ, and ϵ, the function $\Psi^\diamond(\varphi)$ also belongs to the solution space of (0.7)).] By analogy with Refs. [**5**, **6**], we consider the integral over the contour G in the complex plane of the variable φ (see Figure 1)

$$(0.14)\qquad \oint \Psi_n(\varphi)\Psi_n^\diamond(\varphi)\,d\varphi = \int_0^\pi \Psi_n(\varphi)\Psi_n^\diamond(\varphi)\,d\varphi + \int_\pi^{\pi+i\infty} \Psi_n(\varphi)\Psi_n^\diamond(\varphi)\,d\varphi + \int_{\pi+i\infty}^{i\infty} \Psi_n(\varphi)\Psi_n^\diamond(\varphi)\,d\varphi + \int_{i\infty}^{0} \Psi_n(\varphi)\Psi_n^\diamond(\varphi)\,d\varphi.$$

Using the facts that the integrand vanishes as $e^{2ik_0\varphi}$ and that $\Psi_n(\varphi)$ is regular in the domain $\mathbf{G}$ (see Figure 1), according to the Cauchy theorem we have

$$(0.15)\qquad \int_0^\pi \Psi_n(\varphi)\Psi_n^\diamond(\varphi)\,d\varphi = (1-e^{2i\pi k_0})\int_0^{i\infty} \Psi_n(\varphi)\Psi_n^\diamond(\varphi)\,d\varphi.$$

Making the substitution (0.8) in the right integral of (0.15), we find

$$(0.16)\qquad \int_0^\pi \Psi_n(\varphi)\Psi_n^\diamond(\varphi)\,d\varphi = i(1-e^{2i\pi k_0})\int_0^{\pi/2} [\Psi_n]\tan\theta\,d\theta.$$

and after integration over the angle θ we finally get [**1**]

$$(0.17)\qquad C_n = \sqrt{\frac{(-ik_0)(2n+k_0\pm k_1+1)\Gamma(n+1\pm k_1)\Gamma(n+k_0\pm k_1+1)}{R[1-e^{2i\pi k_0}](2n\pm k_1+1)n![\Gamma(1\pm k_1)]^2\Gamma(n+k_0+1)}}.$$

Let us now consider the two most interesting cases.

3.1. *The case when* $p^2 = \frac{1}{4}$. Then the duality transformation (0.8) establishes the connection between the pure Coulomb problem and the singular oscillator with $k_1^2 = 1$. Comparing eqs. (0.5) and (0.11) and putting $k = i\mu$, we get

$$(0.18)\qquad k_0 = -(n+1)+i\sigma, \qquad \sigma = \frac{\mu R}{n+1}$$

and for the energy spectrum

$$(0.19)\qquad E_n(R) = \frac{(n+1)^2}{2R^2} - \frac{\mu^2}{2(n+1)^2}, \qquad n = 0,1,2,\dots.$$

Returning to the variable φ, we obtain that wave function for $0 \le \varphi \le \pi$ has the form

$$(0.20)\qquad \Psi_{n\sigma}(\varphi) = C_n(\sigma)\sin\varphi e^{-i\varphi(n-i\sigma)}{}_2F_1(-n, 1+i\sigma; 2; 1-e^{2i\varphi}),$$

where the normalization constant $C_n(\sigma)$ is

$$(0.21)\qquad C_n(\sigma) = e^{\sigma\pi/2}|\Gamma(1+i\sigma)|\sqrt{\frac{(n+1)^2+\sigma^2}{\pi R}}.$$

The wave function in the region $-\pi \le \varphi < 0$ ($s_1 < 0$) may be determined from (0.20) by the reflection $\varphi \to -\varphi$. Therefore the general solution of the Schrödinger equation for $\varphi \in [-\pi, \pi]$ can be presented in the form of even and odd functions

$$\Psi_{n\sigma}^{(+)}(\varphi) = C_n(\sigma)\sin|\varphi|e^{-i(n-i\sigma)|\varphi|}F(-n, 1+i\sigma; 2; 1-e^{2i|\varphi|}),$$
$$\Psi_{n\sigma}^{(-)}(\varphi) = C_n(\sigma)\sin\varphi e^{-i(n-i\sigma)|\varphi|}F(-n, 1+i\sigma; 2; 1-e^{2i|\varphi|}).$$

Thus by using the relation between the Coulomb and singular oscillator systems we have constructed the wave functions and energy spectrum for a Coulomb system on the one-dimensional sphere.

3.2. Let us now choose $k_1^2 = \frac{1}{4}$ or equivalently $p^2 - \frac{1}{4} = -3/16$. In this case the centrifugal potential term is attractive at the origin for singular Coulomb systems and the motion take place in the domains $\varphi \in (-\infty, \infty)$. For oscillator system this term equal zero and therefore the duality transformation (0.8) connect singular Coulomb and pure oscillator systems.

Let us introduce the quantity ν which takes two values $\nu = \frac{1}{4}$ and $\nu = \frac{3}{4}$. Making all calculations by analogy to previous case, it is easy to obtain the energy spectrum

$$E_n^\nu(R) = \frac{(n+\nu)^2}{2R^2} - \frac{\mu^2}{2(n+\nu)^2}, \quad n = 0, 1, 2, \dots . \tag{0.22}$$

and wave functions

$$\Psi_{n\sigma}^\nu(\varphi) = e^{\sigma\pi/2} 2^\nu \frac{|\Gamma(\nu + i\sigma)|}{\Gamma(2\nu)} \sqrt{\frac{[(n+\nu)^2 + \sigma^2]\Gamma(n+2\nu)}{4\pi R(n+\nu)n!}} \times (\sin\varphi)^\nu e^{-i\varphi(n-i\sigma)} {}_2F_1(-n, \nu + i\sigma; 2\nu; 1 - e^{2i\varphi}),$$

where $\sigma = \mu R/(n+\nu)$.

In the contraction limit $R \to \infty$, $\varphi \to 0$ and $R\varphi \sim x$ - fixed, we see that

$$\epsilon_n = \lim_{R\to\infty} E_n^\nu(R) = -\frac{\mu^2}{2(n+\nu)^2}, \quad n = 0, 1, \dots, \tag{0.23}$$

and

$$\Phi_n^\nu(y) = \lim_{\substack{R\to\infty \\ \varphi\to 0}} \Psi_{n\sigma}^\nu(\varphi) = \frac{\sqrt{\mu}}{\Gamma(2\nu)} \frac{1}{(n+\nu)} \sqrt{\frac{\Gamma(n+2\nu)}{2n!}} y^\nu e^{-|y|/2} {}_1F_1(-n; 2\nu; y), \tag{0.24}$$

where $y = 2\mu x/(n+\nu)$. Formulas (0.23) and (0.24) coincides with the formulas for energy levels ϵ_n and up to the factor $\sqrt{2}$ for wave functions $\Phi_n^\nu(y)$ for two type one-dimensional Coulomb anyons with $\nu = \frac{1}{4}$ and $\nu = \frac{3}{4}$ respectively [**9**].

References

1. A. Erdélyi, W. Magnus, F. Oberhettinger, and F. G. Tricomi, *Tables of integral transforms.* II, McGraw-Hill, New York, 1954.
2. S. Flügge, *Practical quantum mechanics.* I, Grundlehren Math. Wiss., vol. 177, Springer-Verlag, Berlin–New York, 1971.
3. A. González-López, N. Kamran, and P. J. Olver, *Normalizability of one-dimensional quasi-exactly solvable Schrödinger operators*, Comm. Math. Phys. **153** (1993), no. 1, 117–146.
4. C. Grosche, G. S. Pogosyan, and A. N. Sissakian, *Path integral discussion for Smorodinsky–Winternitz potentials.* II. *Two- and three-dimensional sphere*, Fortschr. Phys. **43** (1995), no. 6, 523–563.
5. E. G. Kalnins, W. Miller, Jr., and G. S. Pogosyan, *Coulomb-oscillator duality in spaces of constant curvature*, J. Math. Phys. **41** (2000), no. 5, 2629–2657.
6. ———, *The Coulomb-oscillator relation on n-dimensional spheres and hyperboloids*, Phys. Atomic Nuclei **65** (2002), no. 6, 1086–1094.
7. A. Nersessian and G. Pogosyan, *On the relation of the oscillator and Coulomb systems on (pseudo)spheres*, Phys. Rev. A **63** (2000), no. 2, 020103.
8. M. A. Rodriguez and P. Winternitz, *Quantum superintegrability and exact solvability in n dimensions*, J. Math. Phys. **43** (2002), no. 3, 1309–1322.
9. V. M. Ter-Antonyan, *Dyon-oscillator duality*, quant-ph/0003106.

International Center for Advanced Studies, Yerevan State University, Yerevan, Armenia

E-mail address: mardoyan@icas.ysu.am

International Center for Advanced Studies, Yerevan State University, Yerevan, Armenia; Centro de Ciencias Físicas Universidad Nacional Autónoma de México, Apartado Postal 48–3, 62251 Cuernavaca, Morelos, México, and Laboratory of Theoretical Physics, Joint Institute for Nuclear Research, Dubna, 141980, Moscow Region, Russia

E-mail address: pogosyan@icas.ysu.am, pogosyan@fis.unam.mx and pogosyan@thsun1.jinr.ru

Laboratory of Theoretical Physics, Joint Institute for Nuclear Research, Dubna, Moscow Region, 141980, Russia

E-mail address: sisakian@jinr.ru

Centre de Recherches Mathématiques
CRM Proceedings and Lecture Notes
Volume **37**, 2004

Higher-Order Superintegrability of a Rational Oscillator with Inversely Quadratic Nonlinearities: Euclidean and Non-Euclidean Cases

Manuel F. Rañada and Mariano Santander

Abstract. In the first part, the superintegrability of a rational harmonic oscillator (noncentral harmonic oscillator with rational ratio of frequencies) with nonlinear ("centrifugal") terms is studied in the Euclidean plane. It is proved that inversely quadratic nonlinearities modify the solutions but preserve superintegrability. The constants of motion of the nonlinear system are explicitly obtained.

The second part is devoted to the study of the curvature dependent versions of these systems. We first study the $n = 2$ harmonic oscillator, and then, the oscillator with nonlinear ("centrifugal") terms, on the two-dimensional constant curvature spaces (sphere $\mathbb{S}^2$ and hyperbolic plane $\mathbb{H}^2$). All the mathematical expressions are presented using the curvature κ as a parameter, in such a way that particularizing for $\kappa > 0$, $\kappa = 0$, or $\kappa < 0$, the corresponding properties are obtained for the system on the sphere $\mathbb{S}^2$, the Euclidean plane $\mathbb{E}^2$, or the hyperbolic plane $\mathbb{H}^2$, respectively.

1. Introduction

Friš et al. [**4**] studied the Euclidean $n = 2$ systems which admit separability in two different coordinate systems, and obtained four families V_r, r = a, b, c, d, of superintegrable potentials with constants of motion linear or quadratic in the velocities (momenta). In fact, if we call superseparable a system that admits Hamilton–Jacobi separation of variables (Schrödinger in the quantum case) in more than one coordinate system, then quadratic superintegrability (i.e., superintegrability with linear or quadratic constants of motion) can be considered as a property arising from superseparability. The two first families

$$V_{\rm a} = \frac{1}{2}\omega_0^2(x^2 + y^2) + \frac{k_2}{x^2} + \frac{k_3}{y^2}, \qquad V_{\rm b} = \frac{1}{2}\omega_0^2(4x^2 + y^2) + k_2 x + \frac{k_3}{y^2},$$

can be considered as the most general deformations (with strengths k_2, k_3) of the 1:1 and 2:1 harmonic oscillators preserving quadratic superintegrability (the other

2000 *Mathematics Subject Classification.* 37J35, 70H06.

The first author is supported by projects FPA–2000–1252 and BFM–2000–1066.
The second author is supported by project BFM–2002–03773.
This is the final form of the paper.

two families, V_c and V_d, are related to the Kepler problem). The superintegrability of V_a, which is known as the "Smorodinsky–Winternitz" (SW) potential, has been studied by Evans [**2, 3**] for the general case of n degrees of freedom.

What happens, however, to the SW potential V_a if the first central isotropic function is replaced by a more general noncentral oscillator? Our objective is to give an answer to this question and study this more general potential, not only in the Euclidean plane, but also in spaces of constant curvature. We will see that, even in this more general case, the system is superintegrable as well but with nonquadratic integrals of motion.

2. SW potential with a noncentral oscillator

The two-dimensional harmonic oscillator

$$H_{\mathrm{HO}} = \tfrac{1}{2}(p_1^2 + p_2^2) + \tfrac{1}{2}({\omega_1}^2 q_1^2 + {\omega_2}^2 q_2^2)$$

has the two one-degree of freedom energies, $I_1 = E_1$ and $I_2 = E_2$, as fundamental integrals. The superintegrability of the rational case, $\omega_1 = n_1\omega_0$, $\omega_2 = n_2\omega_0$, with integers n_1, n_2, can be proved by making use of a complex formalism. Let J_j, $j = 1, 2$, be the following two functions $J_j = p_j + i n_j \omega_0 q_j$, $j = 1, 2$; then the time-evolution of these functions is given by

$$\frac{d}{dt} J_1 = i n_1 \omega_0 J_1, \qquad \frac{d}{dt} J_2 = i n_2 \omega_0 J_2,$$

and therefore, the function I_{12} defined as $I_{12} = J_1^{n_2}(J_2^*)^{n_1}$, is a constant of motion. It involves a relation between the two degrees of freedom when the quotient ω_2/ω_1 is rational and, since it is a complex function, it determines not one but two real first integrals, $I_3 = \mathrm{Im}(I_{12})$ and $I_4 = \mathrm{Re}(I_{12})$. Only I_3 must be considered as a fundamental constant of motion (I_4 is a dependent function of I_1, I_2, and I_3).

As an example, for the isotropic case, $\omega_1 = \omega_2 = \omega_0$, we obtain

$$I_3 \equiv \mathrm{Im}(I_{12}) = q_1 p_2 - q_2 p_1, \qquad I_4 \equiv \mathrm{Re}(I_{12}) = p_1 p_2 + {\omega_0}^2 q_1 q_2,$$

and for the first nonisotropic case, $\omega_1 = 2\omega_0$, $\omega_2 = \omega_0$, we arrive to

$$I_3 \equiv \mathrm{Im}(I_{12}) = (q_1 p_2 - q_2 p_1) p_2 - {\omega_0}^2 q_1 q_2^2,$$
$$I_4 \equiv \mathrm{Re}(I_{12}) = p_1 p_2^2 + {\omega_0}^2 (4 q_1 p_2 - q_2 p_1) q_2.$$

The integral I_3 of the 3:1 oscillator will be cubic and, in the general $n_1{:}n_2$ case, the function I_3 will be a polynomial in the velocities (momenta) of degree $n_1 + n_2 - 1$.

2.1. Nonlinear $n = 1$ harmonic oscillator. Consider the following one-dimensional Lagrangian

$$L = \frac{1}{2} v^2 - \left[\frac{1}{2}\omega^2 q^2 + \frac{k}{q^2}\right]$$

representing an oscillator with a inversely quadratic nonlinear term. The associated equation

$$\ddot{q} + \omega^2 q + \frac{c}{q^3} = 0, \qquad c = -2k, \tag{2.1}$$

is a particular case of the so-called Pinney–Ermakov equation [**15**], and therefore it belongs to the restricted subfamily of nonlinear differential equations whose exact solution can be obtained. The solution of (2.1) is given by

$$q = \sqrt{Q}, \qquad Q = u^2 - \frac{c}{W^2}v^2,$$

where u, v, are two independent solutions of the associated linear equation and W is their Wronskian. If $u = A\sin(\omega t + \phi)$ and $v = B\cos(\omega t + \phi)$, then we arrive at

$$q = \frac{1}{\omega A}\sqrt{(\omega^2 A^4 + c)\sin^2(\omega t + \phi) - c}.$$

Notice that this expression satisfies the appropriate linear limit, $\lim_{k\to 0} q = A\sin(\omega t + \phi)$.

Equations similar to (2.1) also arise in some other areas as the study of the so-called "Lewis invariant" for time-dependent harmonic oscillators [**1**], or the theory of Ermakov systems [**6, 14**]. Moreover, this particular potential is also important at the quantum level since it is solvable by the factorization Schrödinger method.

Now we will move to the Hamiltonian formalism. In the following we will make use of the functions $M = M(q,p)$ and $N = N(q,p)$ defined as

$$M = \omega A\sqrt{A^2 - q^2}, \qquad N = \sqrt{\omega^2 A^2 q^2 - 2k}.$$

These two functions, M and N, satisfy the relations

$$MN = \omega A^2 qp, \qquad M^2 + N^2 = \omega^2 A^4 - 2k,$$

and from this we obtain that the time-evolution of the functions M and N is given by

$$\frac{d}{dt}M = -\omega N, \qquad \frac{d}{dt}N = \omega M.$$

Therefore, the complex function K defined as $K = M + \mathrm{i}N$ plays, in this nonlinear case, a similar role to the complex function J of the linear case.

2.2. Nonlinear $n = 2$ harmonic oscillator. The Hamiltonian of the $n = 2$ harmonic oscillator with inversely quadratic nonlinearities ("centrifugal" terms) is given by

$$H = \frac{1}{2}(p_1^2 + p_2^2) + \frac{1}{2}({\omega_1}^2 q_1^2 + {\omega_2}^2 q_2^2) + \frac{k_1}{q_1^2} + \frac{k_2}{q_2^2}. \tag{2.2}$$

We will make use of the notation introduced in the previous subsection. That is, let A_i, $i = 1, 2$, be the two constants functions that, when written as functions of the energies E_i, $i = 1, 2$, read as follows

$$A_i^2 = \frac{1}{{\omega_i}^2}\left[E_i + \sqrt{E_i^2 - 2k_i\omega_i^2}\right], \quad i = 1, 2,$$

and by M_i, N_i, $i = 1, 2$, the following expressions

$$M_i = \omega_i A_i\sqrt{A_i^2 - q_i^2}, \qquad N_i = \sqrt{\omega_i^2 A_i^2 q_i^2 - 2k_i}.$$

Next we consider the rational case of ω_1/ω_2 but with the two other parameters, k_1 and k_2, remaining with arbitrary values.

PROPOSITION 1. *Consider the $n = 2$ nonlinear Hamiltonian (2.2) and let $K_j = K_j(q,p)$, $j = 1, 2$, be the following two functions $K_j = M_j + \mathrm{i}N_j$, $j = 1, 2$. Then, if the quotient ω_2/ω_1 of the linear frequencies is a rational number, $\omega_1 = n_1\omega_0$, $\omega_2 = n_2\omega_0$, the complex function K_{12} defined as*

$$K_{12} = K_1^{n_2}(K_2^*)^{n_1}$$

is a constant of motion.

The time-evolution of the functions K_1, K_2, is given by

$$\frac{d}{dt}K_1 = \mathrm{i}\omega_1 K_1, \qquad \frac{d}{dt}K_2 = \mathrm{i}\omega_2 K_2.$$

Hence we have

$$\begin{aligned}\frac{d}{dt}K_{12} &= n_2 K_1^{(n_2-1)}(K_2^*)^{n_1}\dot{K}_1 + n_1 K_1^{n_2}(K_2^*)^{(n_1-1)}\dot{K}_2{}^* \\ &= (\mathrm{i}\omega_0)(n_2\omega_1 - n_1\omega_2)K_1^{n_2}(K_2^*)^{n_1} = 0.\end{aligned}$$

Thus, if ω_1 and ω_2 are given by $\omega_1 = n_1\omega_0$, $\omega_2 = n_2\omega_0$, then K_{12} is a constant of motion. As in the linear case, K_{12} can be considered as coupling the two degrees of freedom.

Hence the Hamiltonian (2.2) is superintegrable for any rational value of the quotient ω_2/ω_1. In the particular 1:1 case, $\omega_1 = \omega_2 = \omega_0$, the potential reduces to the SW potential

$$V_{\mathrm{a}}(1,1) \equiv V_{\mathrm{a}} = \frac{1}{2}{\omega_0}^2(q_1^2 + q_2^2) + \frac{k_1}{q_1^2} + \frac{k_2}{q_2^2}$$

and in the general case, $\omega_1 = n_1\omega_0$, $\omega_2 = n_2\omega_0$, the potential

$$V_{\mathrm{a}}(n_1, n_2) = \frac{1}{2}{\omega_0}^2(n_1^2 q_1^2 + n_2^2 q_2^2) + \frac{k_1}{q_1^2} + \frac{k_2}{q_2^2}$$

represents a generalized SW potential with a nonisotropic $n_1{:}n_2$ oscillator (notice that in the 2:1 case, this *is not* the family V_{b}).

In this nonlinear $n_1{:}n_2$ case the complex function K_{12} has a rather complicated expression with cross-products of powers of roots as well as some factors depending explicitly of the two energies (the two first integrals). After some calculus we have obtained that, if we denote by I_3 the constant of motion of the associated linear system, the additional third integral J_3 can be written as follows

$$J_3 = I_3^2 + k_1 J_3^{(1)}(q,p) + k_2 J_3^{(2)}(q,p) + k_1 k_2 J_3^{(1,2)}(q,p).$$

That is, the integral J_3 of the nonlinear system ($k_1 \neq 0$, $k_2 \neq 0$), appears as a deformation, not of the function I_3 itself, but of its square I_3^2. Next, we give the expressions of the constant J_3 for the three first cases:

2.2.1. *Smorodinsky–Winternitz potential* $\omega_1 = \omega_2 = \omega_0$. In this case the integral of motion I_3 of the associated linear system is just the angular momentum, $I_3 = \mathrm{Im}(I_{12}) = q_1 p_2 - q_2 p_1$. Then J_3 is given by

$$J_3 = I_3^2 + 2k_1\left(\frac{q_2}{q_1}\right)^2 + 2k_2\left(\frac{q_1}{q_2}\right)^2.$$

2.2.2. *Nonisotropic oscillator with* $\omega_1 = \omega_0$, $\omega_2 = 2\omega_0$. If we denote by I_3 the integral of motion of the associated linear system

$$I_3 = \Im(I_{12}) = (q_1 p_2 - q_2 p_1)p_1 + \omega_0^2 q_1^2 q_2,$$

then we obtain

$$J_3 = \left[I_3 - 2k_1\frac{q_2}{q_1^2}\right]^2 + 2k_2\left(\frac{q_1}{q_2}\right)^2 p_1^2.$$

2.2.3. *Nonisotropic oscillator with* $\omega_1 = \omega_0$, $\omega_2 = 3\omega_0$. If we denote by I_3 the constant of motion of the associated linear system

$$I_3 = \Im(I_{12}) = 3(q_1p_2 - q_2p_1)p_1^2 - \omega_0^2(q_1p_2 - 9q_2p_1)q_1^2,$$

then

$$J_3 = I_3^2 + k_1J_3^{(1)}(\omega_0, k_1) + k_2J_3^{(2)}(\omega_0, k_2) + k_1k_2J_3^{(1,2)}(\omega_0, k_1, k_2),$$

with $J_3^{(1)}(\omega_0, k_1)$, $J_3^{(2)}(\omega_0, k_2)$, and $J_3^{(1,2)}(\omega_0, k_1, k_2)$, given by

$$\begin{aligned} J_3^{(1)} &= \frac{18}{q_1^2}\left[(3q_2p_1 - 4q_1p_2)q_2p_1^3 - \left(\frac{2k_1}{q_1^2}\right)(q_1p_2 - q_2p_1)(q_1p_2 + 3q_2p_1) + 4k_1^2\left(\frac{q_2^2}{q_1^4}\right)\right] \\ &\quad - 6\omega_0^2\left[2(2q_1^2p_2^2 - 6q_1q_2p_1p_2 + 27q_2^2p_1^2) + 9\left(\frac{q_2}{q_1}\right)^2(8k_1 + 3\omega_0^2q_1^4)\right], \\ J_3^{(2)} &= 2\left(\frac{q_1}{q_2}\right)^2(3p_1^2 - \omega_0^2q_1^2)^2, \\ J_3^{(1,2)} &= -\frac{24}{q_1^2q_2^2}(3k_1 + 2\omega_0^2q_1^4). \end{aligned}$$

Summarizing, within the $V_a(n_1, n_2)$ family, only in the particular isotropic 1:1 case the function J_3 is quadratic and, because of this, only in this case the superintegrability arises from separability in two different coordinate systems. In all the remaining cases, $V_a(n_1, n_2)$ is a superintegrable but not superseparable potential.

3. The harmonic oscillator on spaces of constant curvature

It is well known that the central harmonic oscillator and the Kepler problem remain superintegrable [**9, 21**] when we move from the Euclidean plane to the sphere. Moreover, in these last years a certain number of non-Euclidean superintegrable systems have been investigated by different authors (see, e.g., [**5, 7, 10–12, 16–20, 22**]). Our objective is to study, first the harmonic oscillator [**19, 20**], and then the potential $V_a(n_1, n_2)$ on constant curvature spaces.

In the following, all the mathematical expressions will depend on the curvature κ as a parameter, in such a way that for $\kappa > 0$, $\kappa = 0$, or $\kappa < 0$, we will obtain the corresponding property particularized for the dynamical system on the sphere $\mathbb{S}^2$, on the Euclidean plane $\mathbb{E}^2$, or on the hyperbolic plane $\mathbb{H}^2$. In order to present these expressions in a form which holds simultaneously for any value of κ, we will make use [**8, 19**] of the following "tagged" trigonometric functions

$$\mathrm{C}_\kappa(x) = \begin{cases} \cos\sqrt{\kappa}x & \text{if } \kappa > 0, \\ 1 & \text{if } \kappa = 0, \\ \cosh\sqrt{-\kappa}x & \text{if } \kappa < 0, \end{cases} \qquad \mathrm{S}_\kappa(x) = \begin{cases} (1/\sqrt{\kappa})\sin\sqrt{\kappa}x & \text{if } \kappa > 0, \\ x & \text{if } \kappa = 0, \\ (1/\sqrt{-\kappa})\sinh\sqrt{-\kappa}x & \text{if } \kappa < 0, \end{cases}$$

and the κ-tagged tangent $\mathrm{T}_\kappa(x)$ defined in the natural way $\mathrm{T}_\kappa(x) = \mathrm{S}_\kappa(x)/\mathrm{C}_\kappa(x)$.

3.1. Polar and parallel coordinates in $\mathbb{S}^2$, $\mathbb{E}^2$, and $\mathbb{H}^2$. The differential element of distance on the constant curvature spaces $(\mathbb{S}^2, \mathbb{E}^2, \mathbb{H}^2)$ is given by

$$ds^2 = dr^2 + \mathrm{S}_\kappa^2(r)\,d\phi^2$$

when written in polar coordinates (r, ϕ). It reduces to

$$ds_1^2 = dr^2 + (\sin^2 r)\,d\phi^2, \quad ds_0^2 = dr^2 + r^2\,d\phi^2, \quad ds_{-1}^2 = dr^2 + (\sinh^2 r)\,d\phi^2,$$

for the three particular cases of unit sphere, Euclidean plane, and unit Lobachevsky plane. A general standard Lagrangian (Riemannian metric minus a potential) has the following form

$$L(\kappa) = T(\kappa) - U(r,\phi,\kappa), \qquad T(\kappa) = \tfrac{1}{2}(v_r^2 + \mathrm{S}_\kappa^2(r) v_\phi^2),$$

in such a way that for $\kappa = 0$ we recover a standard Euclidean system

$$\lim_{\kappa\to 0} L(\kappa) = \tfrac{1}{2}(v_r^2 + r^2 v_\phi^2) - V(r,\phi), \qquad V(r,\phi) = U(r,\phi,0).$$

Notice the kinetic term $T(\kappa)$, that is determined by the metric, is invariant under the action of the following three κ-dependent vector fields

$$Y_1(\kappa) = (\cos\phi)\frac{\partial}{\partial r} - \left(\frac{\mathrm{C}_\kappa(r)}{\mathrm{S}_\kappa(r)}\sin\phi\right)\frac{\partial}{\partial\phi}, \quad Y_2(\kappa) = (\sin\phi)\frac{\partial}{\partial r} + \left(\frac{\mathrm{C}_\kappa(r)}{\mathrm{S}_\kappa(r)}\cos\phi\right)\frac{\partial}{\partial\phi},$$
$$Y_J = \frac{\partial}{\partial\phi},$$

that close the following Lie algebra

$$[Y_1(\kappa), Y_2(\kappa)] = -\kappa Y_J, \qquad [Y_1(\kappa), Y_J] = -Y_2(\kappa), \qquad [Y_2(\kappa), Y_J] = Y_1(\kappa),$$

corresponding to the algebra of isometries of the spherical (Euclidean, hyperbolic) space.

In three particular cases this Lagrangian system possesses the vector fields, Y_1, Y_2, or Y_J, as exact Noether symmetries

(1) If the potential U is of the form $U = U(z_y)$, with $z_y = \mathrm{S}_\kappa(r)\sin\phi$, then
$$P_1(\kappa) = (\cos\phi)v_r - (\mathrm{C}_\kappa(r)\,\mathrm{S}_\kappa(r)\sin\phi)v_\phi$$
is a constant of motion.

(2) If the potential U is of the form $U = U(z_x)$, $z_x = \mathrm{S}_\kappa(r)\cos\phi$, then
$$P_2(\kappa) = (\sin\phi)v_r + (\mathrm{C}_\kappa(r)\,\mathrm{S}_\kappa(r)\cos\phi)v_\phi$$
is a constant of motion.

(3) If the potential U is of the form $U = U(r)$, then the κ-dependent angular momentum defined as $J(\kappa) = \mathrm{S}_\kappa^2(r)v_\phi$, is a constant of motion.

In differential geometric terms, a 'geodesic parallel' system of coordinates represents the Riemannian version of the Euclidean (x, y) Cartesian system [**13**]. But one important difference is that, when we introduce the curvature κ, then we obtain not just one but two systems that we denote by (u, y) and (x, v) respectively (a more detailed discussion of these properties is given in the appendix, see also [**17, 19**]).

The following expression written in (u, y) 'parallel' coordinates

$$ds^2 = \mathrm{C}_\kappa^2(y)\,du^2 + dy^2$$

represents the differential element of distance on the spaces $(\mathbb{S}^2, \mathbb{E}^2, \mathbb{H}^2)$ with constant curvature κ. So a standard Lagrangian has the following form

$$L(\kappa) = \tfrac{1}{2}(\mathrm{C}_\kappa^2(y)v_u^2 + v_y^2) - U(u,y,\kappa)$$

and the Euclidean system is just given by

$$\lim_{\kappa\to 0} L(\kappa) = \tfrac{1}{2}(v_x^2 + v_y^2) - V(x,y), \qquad V(x,y) = U(x,y,0).$$

Concerning the three κ-dependent linear functions, P_1, P_2, and J, they have now the following expressions in 'parallel' coordinates

$$P_1(\kappa) = \mathrm{C}_\kappa^2(y) v_u, \qquad P_2(\kappa) = \kappa\, \mathrm{S}_\kappa(u)\, \mathrm{C}_\kappa(y)\, \mathrm{S}_\kappa(y) v_u + \mathrm{C}_\kappa(u) v_y,$$
$$J(\kappa) = \mathrm{C}_\kappa(u)\, \mathrm{C}_\kappa(y)\, \mathrm{S}_\kappa(y) v_u - \mathrm{S}_\kappa(u) v_y,$$

Notice that in contrast to the polar coordinates formalism, now the κ-dependence is present in both coordinates, u and y.

3.2. Separability. Next we consider the κ-dependent separability in 'polar' and 'parallel' coordinates:

(i) A κ-dependent potential U that in the 'polar' coordinate system (r, ϕ) has an expression of the form

$$U(\kappa) = F(r) + \frac{G(\phi)}{\mathrm{S}_\kappa^2(r)}, \tag{3.1}$$

is separable. Then it is integrable with the following two quadratic integrals of motion

$$I_1(\kappa) = P_1^2(\kappa) + P_2^2(\kappa) + 2F(r) + \frac{2G(\phi)}{\mathrm{T}_\kappa^2(r)}, \qquad I_2(\kappa) = J^2(\kappa) + 2G(\phi).$$

Integrable potentials in this family can be characterized by the pair $(P_1^2 + P_2^2, J^2)$.

(ii) A κ-dependent potential U that in (u, y) 'parallel' coordinates has an expression of the form

$$U(\kappa) = F(y) + \frac{G(u)}{\mathrm{C}_\kappa^2(y)}, \tag{3.2}$$

is separable. It is therefore integrable with the following two quadratic integrals of motion

$$I_1(\kappa) = P_1^2(\kappa) + 2G(u), \quad I_2(\kappa) = P_2^2(\kappa) + \kappa J^2(\kappa) + 2F(y) + 2\kappa G(u)\, \mathrm{T}_\kappa^2(y).$$

Integrable potentials in this family are characterized by the pair $(P_1^2, P_2^2 + \kappa J^2)$.

(iii) A potential U separable in (x, v) 'parallel' coordinates is of the form

$$U(\kappa) = F(x) + \frac{G(v)}{\mathrm{C}_\kappa^2(x)}, \tag{3.3}$$

and, when written in (u, y) 'parallel' coordinates, it becomes

$$U(\kappa) = F(z_u) + \frac{G(z_y)}{1 - \kappa z_u^2}, \qquad z_u = \mathrm{S}_\kappa(u)\, \mathrm{C}_\kappa(y), \qquad z_y = \mathrm{T}_\kappa(y) / \mathrm{C}_\kappa(u).$$

It is integrable with the following two integrals of motion

$$I_1(\kappa) = P_1^2(\kappa) + \kappa J^2(\kappa) + 2F(z_u) + 2\kappa G(z_y)\left(\frac{z_u^2}{1 - \kappa z_u^2}\right),$$
$$I_2(\kappa) = P_2^2(\kappa) + 2G(z_y).$$

Integrable potentials in this family are characterized by the pair $(P_1^2 + \kappa J^2, P_2^2)$.

In these three separable cases the total energy $I_0(\kappa)$ splits as a sum of two quadratic constants of motion, $I_0(\kappa) = I_1(\kappa) + I_2(\kappa)$. Notice that these three different ways of separability are closely related with the three different possibilities for the contribution of the angular momentum term to the total energy. In (i) the term $\kappa J^2(\kappa)$ determines a constant of motion by itself, in (ii) it is inside $I_2(\kappa)$ together with $P_2^2(\kappa)$, and in (iii) it is inside $I_1(\kappa)$ together with $P_1^2(\kappa)$. In any

TABLE 1.

Separability	Pair of quadratic constants	Euclidean limit
Polar (r,ϕ)	$P_1^2+P_2^2,\ J^2$	$p_x^2+p_y^2,\ J^2$
Parallel (u,y)	$P_1^2,\ P_2^2+\kappa J^2$	$p_x^2,\ p_y^2$
Parallel (x,v)	$P_1^2+\kappa J^2,\ P_2^2$	$p_x^2,\ p_y^2$

case the contribution of the κ-dependent angular momentum has the curvature κ as coefficient and vanishes in the limit $\kappa\to 0$.

If we characterize every constant of motion $I(\kappa)$ by its higher order term in the functions P_1, P_2, J, then Table 1 summarizes this κ-dependent situation.

3.3. The m:1 harmonic oscillator in curved spaces. The Lagrangian of the central harmonic oscillator with curvature κ is given by

$$L=\tfrac{1}{2}(v_r^2+\mathrm{S}_\kappa^2(r)v_\phi^2)-\tfrac{1}{2}\omega_0^2U_{11}(r,\phi,\kappa),\qquad U_{11}=\mathrm{T}_\kappa^2(r),$$

so that the standard oscillator potential on the unit sphere (Higgs oscillator), on the Euclidean plane, or on the unit Lobachevsky plane, arise as the following three particular cases $U_{11}(1)=\tan^2 r$, $V_{11}=U_{11}(0)=r^2$, and $U_{11}(-1)=\tanh^2 r$ (notice that $U_{11}(1)$ is singular on the circle $r=\pi/2$ that separates the sphere $\mathbb{S}^2$ in two halves, and that $U_{11}(-1)$ is well defined and even bounded).

It can be alternatively written in the following three equivalent ways

$$U_{11}(\kappa)=\mathrm{T}_\kappa^2(r)=\mathrm{T}_\kappa^2(y)+\frac{\mathrm{T}_\kappa^2(u)}{\mathrm{C}_\kappa^2(y)}=\mathrm{T}_\kappa^2(x)+\frac{\mathrm{T}_\kappa^2(v)}{\mathrm{C}_\kappa^2(x)} \tag{3.4}$$

displaying the fact that this oscillator is separable simultaneously in these three coordinate systems (r,ϕ), (u,y), and (x,v). In fact, it is a quadratic superintegrable potential with two quadratic constants, with quadratic parts given by P_1^2 and P_2^2, and the angular momentum J as additional integral. Nevertheless this property can also be obtained from a complex factorization. In fact, if we introduce the following two complex functions

$$K_1=P_1(\kappa)+\mathrm{i}\omega_0(\mathrm{T}_\kappa(r)\cos\phi),\qquad K_2=P_2(\kappa)+\mathrm{i}\omega_0(\mathrm{T}_\kappa(r)\sin\phi),$$

then we have

$$\frac{d}{dt}K_1=\left(\frac{\mathrm{i}\omega_0}{\mathrm{C}_\kappa^2(r)}\right)K_1,\qquad \frac{d}{dt}K_2=\left(\frac{\mathrm{i}\omega_0}{\mathrm{C}_\kappa^2(r)}\right)K_2,$$

and hence we obtain that the three functions, I_1, I_2, and K_{12}, defined as $I_1=K_1K_1^*$, $I_2=K_2K_2^*$, and $K_{12}=K_1K_2^*$, are integrals of motion.

The polar (r,ϕ) coordinate system is appropriate for central potentials as U_{11}, but for the study of nonisotropic oscillators is better to make use of parallel coordinates. The results obtained in [**19**] suggest that the appropriate potential $U_{m1}(\kappa)$ for representing the nonisotropic m: 1 oscillator, with m integer or half-integer, on a 2-D manifold of constant curvature κ, is given by

$$U_{m1}(u,y,\kappa)=\frac{\mathrm{T}_\kappa^2(mu)}{\mathrm{C}_\kappa^2(y)}+\mathrm{T}_\kappa^2(y)=\mathrm{T}_\kappa^2(mu)+\frac{\mathrm{T}_\kappa^2(y)}{\mathrm{C}_\kappa^2(mu)}. \tag{3.5}$$

TABLE 2.

Potential	Triple of constants of motion	Euclidean limit
U_{21}	$P_1^2,\ P_2^2+\kappa J^2,\ JP_2$	$4x^2+y^2$
$U_{1/2,1}$	$P_1^2,\ P_2^2+\kappa J^2,\ JP_1$	$\frac{1}{4}x^2+y^2$
U_{31}	$P_1^2,\ P_2^2+\kappa J^2,\ J(3P_2^2-\kappa J^2)$	$9x^2+y^2$

It satisfies the appropriate Euclidean limit, i.e., $\lim_{\kappa\to 0} U_{m1}(\kappa) = (mx)^2 + y^2$, and is separable in (u, y) parallel coordinates with the following two constants of motion

$$I_1(\kappa) = P_1^2(\kappa) + \omega_0^2[\mathrm{T}_\kappa(mu)]^2, \quad I_2(\kappa) = [P_2^2(\kappa) + \kappa J^2(\kappa)] + \omega_0^2\left(\frac{\mathrm{T}_\kappa(y)}{\mathrm{C}_\kappa(mu)}\right)^2,$$

Concerning the superintegrability, it has been proved that also in this κ-dependent case the additional integral $I_3(\kappa)$ can be obtained from a complex factorization; but the situation is not so simple as in the Euclidean plane [**19**]. Table 2 summarizes the situation for three simplest cases.

The associated functions U_{1m} are separable in the (x, v) system. Notice that U_{12} is characterized by the triple $(P_1^2 + \kappa J^2,\ P_2^2,\ JP_1)$; therefore, $U_{1/2,1}$ and U_{12} are different potentials ($U_{1/2,1} \neq U_{12}$, $\kappa \neq 0$) but with the same Euclidean limit V_{12}. Similarly, U_{21} and $U_{1,1/2}$ are also different potentials ($U_{21} \neq U_{1,1/2}$, $\kappa \neq 0$) but with the same Euclidean limit V_{21}.

4. SW potential with a noncentral oscillator in spaces of constant curvature

The κ-dependent version $U_{\rm a}$ of the SW potential $V_{\rm a}$ is given by

$$U_{\rm a}(\kappa) = \frac{1}{2}\omega_0^2 U_{11} + \frac{k_2}{\big(\mathrm{S}_\kappa(u)\,\mathrm{C}_\kappa(y)\big)^2} + \frac{k_3}{\big(\mathrm{S}_\kappa(y)\big)^2}. \tag{4.1}$$

It satisfies the appropriate Euclidean limit, i.e., $\lim_{\kappa\to 0} U_{\rm a}(\kappa) = V_{\rm a}$, and is separable in polar (r, ϕ) and in parallel (u, y) and (x, v) coordinates with the following three quadratic constants of motion

$$\begin{aligned}
I_1(\kappa) &= P_1^2(\kappa) + \omega_0^2\,\mathrm{T}_\kappa^2(u) + \frac{2k_2}{\mathrm{S}_\kappa^2(u)},\\
I_2(\kappa) &= P_2^2(\kappa) + \omega_0^2\left(\frac{\mathrm{T}_\kappa(y)}{\mathrm{C}_\kappa(u)}\right)^2 + 2k_3\left(\frac{\mathrm{C}_\kappa(u)}{\mathrm{T}_\kappa(y)}\right)^2,\\
J_3(\kappa) &= J^2(\kappa) + 2k_2\left(\frac{\mathrm{T}_\kappa(y)}{\mathrm{S}_\kappa(u)}\right)^2 + 2k_3\left(\frac{\mathrm{S}_\kappa(u)}{\mathrm{T}_\kappa(y)}\right)^2,
\end{aligned}$$

In order to study the nonisotropic generalizations of $U_{\rm a}$, as they must be endowed with nonquadratic integrals, the multiple separability approach is not appropriate. Of course the ideal method would be to find two κ-dependent functions generalizing the functions M and N of the Euclidean case. But this means the analysis of the κ-dependent version of the Pinney-Ermakov equation and this still remains as a question to be studied. So we must use a different form of analysis. We will assume that the property observed in the Euclidean case, where the higher term of the constant $J_3(\kappa)$ of the the nonlinear system was just the square of the integral I_3 of the linear system, remains also true in the spaces of constant curvature. We have studied, by making use of this ansatz, the nonlinear deformations of the two first oscillators U_{21} and $U_{1/2,1}$.

(i) The κ-dependent version $U_{\rm a}^{21}(\kappa)$ of the potential $V_{\rm a}(2,1)$ is given by

$$U_{\rm a}^{21}(\kappa) = \frac{1}{2}\omega_0^2 U_{21} + \frac{k_2}{\big(\mathrm{S}_\kappa(2u)\,\mathrm{C}_\kappa(y)\big)^2} + \frac{k_3}{\big(\mathrm{S}_\kappa(y)\big)^2} \tag{4.2}$$

and it is also separable in parallel (u,y) coordinates. The two quadratic constants of motion are

$$I_1(\kappa) = P_1^2(\kappa) + \omega_0^2\,\mathrm{T}_\kappa^2(2u) + \frac{2k_2}{\mathrm{S}_\kappa^2(2u)}$$

$$I_2(\kappa) = P_2^2(\kappa) + \kappa J^2(\kappa) + \omega_0^2\Big(\frac{\mathrm{T}_\kappa(y)}{\mathrm{C}_\kappa(2u)}\Big)^2 + 2\kappa k_2\Big(\frac{\mathrm{T}_\kappa(y)}{\mathrm{S}_\kappa(2u)}\Big)^2 + \frac{2k_3}{\mathrm{T}_\kappa^2(y)}$$

Assuming the existence of an additional constant, and making use of the above mentioned method, we have explicitly obtained $J_3(\kappa)$ that is given by

$$I_3(\kappa) = JP_2 + \frac{\omega_0^2}{2}\left(\frac{\mathrm{T}_\kappa(2u)}{\mathrm{C}_\kappa^2(2u)}\right)\mathrm{T}_\kappa^2(y) - k_3\Big(\frac{\mathrm{S}_\kappa(2u)}{\mathrm{T}_\kappa^2(y)}\Big)$$

$$J_3(\kappa) = [I_3(\kappa)]^2 + \left[\left(\kappa\,\mathrm{S}_\kappa(y)\,\mathrm{C}_\kappa(y)v_u + \frac{v_y}{\mathrm{T}_\kappa(2u)}\right)^2 + \left(\frac{v_y}{\mathrm{S}_\kappa(2u)}\right)^2\right]\mathrm{T}_\kappa^2(y)$$
$$+ \kappa k_2\big(\kappa k_2 + w_0^2\,\mathrm{T}_\kappa^2(2u)\big)\frac{\mathrm{T}_\kappa^4(y)}{\mathrm{S}_\kappa^2(2u)}$$

(ii) The κ-dependent version $U_{\rm a}^{1/2,1}(\kappa)$ of the potential $V_{\rm a}(\frac{1}{2},1)$ is given by

$$U_{\rm a}^{1/2,1}(\kappa) = \frac{1}{2}{\omega_0}^2 U_{1/2,1} + \frac{k_2}{\big(\mathrm{S}_\kappa(u/2)\,\mathrm{C}_\kappa(y)\big)^2} + \frac{k_3}{\big(\mathrm{S}_\kappa(y)\big)^2} \tag{4.3}$$

and is also separable in parallel (u,y) coordinates with the following two quadratic constants of motion

$$I_1(\kappa) = P_1^2(\kappa) + \omega_0^2\,\mathrm{T}_\kappa^2(u/2) + \frac{2k_2}{\mathrm{S}_\kappa^2(u/2)}$$

$$I_2(\kappa) = P_2^2(\kappa) + \kappa J^2(\kappa) + \omega_0^2\left(\frac{\mathrm{T}_\kappa(y)}{\mathrm{C}_\kappa(u/2)}\right)^2 + 2\kappa k_2\left(\frac{\mathrm{T}_\kappa(y)}{\mathrm{S}_\kappa(u/2)}\right)^2 + \frac{2k_3}{\mathrm{T}_\kappa^2(y)}$$

It is also superintegrable with a *fourth-order* constant of motion $J_3(\kappa)$ given by

$$I_3(\kappa) = JP_1 - \omega_0^2\big(\mathrm{T}_\kappa^2(u/2)\,\mathrm{T}_\kappa(y)\big) + 2k_2\left(\frac{\mathrm{T}_\kappa(y)}{\mathrm{T}_\kappa^2(u/2)}\right)$$

$$J_3(\kappa) = [I_3(\kappa)]^2 + k_3\big(\mathrm{S}_\kappa(y)\,\mathrm{C}_\kappa(y)v_u\big)^2\frac{\mathrm{S}_\kappa^2(u)}{\mathrm{T}_\kappa^4(y)}$$

The results obtained for these three particular cases, $U_{\rm a}$, $U_{\rm a}^{21}$, and $U_{\rm a}^{1/2,1}$, suggest that the appropriate potential $U_{\rm a}^{m1}(\kappa)$ for representing the nonisotropic generalization of the SW potential on a 2-D manifold of constant curvature κ, is given by

$$U_{\rm a}^{m1}(\kappa) = \tfrac{1}{2}{\omega_0}^2 U_{m1} + k_2 U_2 + k_3 U_3 \tag{4.4}$$

with

$$U_{m1} = \mathrm{T}_\kappa^2(mu) + \left(\frac{\mathrm{T}_\kappa(y)}{\mathrm{C}_\kappa(mu)}\right)^2, \qquad U_2 = \frac{1}{\big(\mathrm{S}_\kappa(mu)\,\mathrm{C}_\kappa(y)\big)^2}, \qquad U_3 = \frac{1}{\mathrm{S}_\kappa^2(y)}.$$

This potential satisfies the Euclidean limit

$$\lim_{\kappa\to 0} U_{\mathrm{a}}^{m1}(\kappa) = \frac{1}{2}{\omega_0}^2\big((mx)^2 + y^2\big) + \frac{k_2}{x^2} + \frac{k_3}{y^2}$$

and is separable in (u, y) parallel coordinates

$$I_1(\kappa) = P_1^2(\kappa) + \omega_0^2\,\mathrm{T}_\kappa^2(mu) + \frac{2k_2}{\mathrm{S}_\kappa^2(mu)}$$

$$I_2(\kappa) = P_2^2(\kappa) + \kappa J^2(\kappa) + \omega_0^2\left(\frac{\mathrm{T}_\kappa(y)}{\mathrm{C}_\kappa(mu)}\right)^2 + 2\kappa k_2\left(\frac{\mathrm{T}_\kappa(y)}{\mathrm{S}_\kappa(mu)}\right)^2 + \frac{2k_3}{\mathrm{T}_\kappa^2(y)}.$$

It seems natural to conjecture that $U_{\mathrm{a}}^{m1}(\kappa)$ is superintegrable as well, and that the corresponding additional (nonquadratic) integral $J_3(\kappa)$ can also be obtained assuming that it is given by an expression of the form "$J_3(\kappa) = [I_3(\kappa)]^2$ + other terms of lower order," with $I_3(\kappa)$ the constant determined by the linear system (κ-dependent harmonic oscillator U_{m1}). In any case this is a conjecture that must be considered as a final open question.

Appendix A. Geodesic parallel coordinates

Let M be a 2-D Riemannian manifold, O a point on M and g_1, g_2, two orthogonal geodesics by O. Let P be an arbitrary point (in some suitable neighborhood of O) and denote by P_1, P_2, the orthogonal projections of P on g_1, g_2. Then we can characterize the point P by

- the two distances (u, y) of O to P_1 (measured along g_1) and of P_1 to P (measured along the geodesic by P and P_1) respectively.
- the two distances (x, v) of P_2 to P (measured along the geodesic by P and P_2) and of O to P_2 (measured along g_2) respectively.

In the first case we have the parallel coordinates of P relative to (O, g_1) and in the second case relative to (O, g_2). In the (u, y) system the curves 'u = constant' are geodesics and the curves 'y = constant' meet these geodesics orthogonally. In the (x, v) system the geodesics are the curves 'v = constant.' Notice that in the general case we have $u \neq x$ and $v \neq y$.

If M is a space of constant curvature, the (u, y) and (x, v) expressions for the differential element of distance ds^2 are given by **[13]**

$$ds^2 = \mathrm{C}_\kappa^2(y)\,du^2 + dy^2 \qquad \text{and} \qquad ds^2 = dx^2 + \mathrm{C}_\kappa^2(x)\,dv^2,$$

so that in both cases we get $ds^2 = dx^2 + dy^2$ for the particular $\kappa = 0$ case. These two κ-dependent formalisms can be related by the use of the spherical ($\kappa > 0$) or hyperbolic ($\kappa < 0$) trigonometry. We have made use of the (u, y) notation because it is in this notation in which the potentials U_{m1} are separable and have simpler expressions, in contrast with the (x, v) notation that becomes appropriate for the associated $\widetilde{U}_{m1} = U_{1m}$ potentials.

References

1. R. K. Colegrave and M. S. Abdalla, *Invariants for the time-dependent harmonic oscillator*, J. Phys. A **16** (1983), no. 16, 3805–3815.
2. N. W. Evans, *Superintegrability of the Smorodinsky–Winternitz system*, Phys. Lett. A **147** (1990), 483–486.
3. ———, *Group theory of the Smorodinsky-Winternitz system*, J. Math. Phys. **32** (1991), no. 12, 3369–3375.
4. J. Friš, V. Mandrosov, Ya. A. Smorodinsky, M. Uhlíř, and P. Winternitz, *On higher symmetries in quantum mechanics*, Phys. Lett. **16** (1965), 354–356.
5. C. Grosche, G. S. Pogosyan, and A. N. Sissakian, *Path integral discussion for Smorodinsky–Winternitz potentials. II. Two- and three-dimensional sphere*, Fortschr. Phys. **43** (1995), no. 6, 523–563.
6. F. Haas and J. Goedert, *On the linearization of the generalized Ermakov systems*, J. Phys. A **32** (1999), no. 15, 2835–2844.
7. F. Herranz, Á. Ballesteros, M. Santander, and T. Sanz-Gil, *Maximally superintegrable Smorodinsky–Winternitz systems on the N-dimensional sphere and hyperbolic spaces*, in this volume.
8. F. J. Herranz, R. Ortega, and M. Santander, *Trigonometry of spacetimes: a new self-dual approach to a curvature/signature (in)dependent trigonometry*, J. Phys. A **33** (2000), no. 24, 4525–4551.
9. P. W. Higgs, *Dynamical symmetries in a spherical geometry*. I, J. Phys. A **12** (1979), no. 3, 309–323.
10. E. G. Kalnins, J. M. Kress, G. S. Pogosyan, and W. Miller, Jr., *Completeness of superintegrability in two-dimensional constant-curvature spaces*, J. Phys. A **34** (2001), no. 22, 4705–4720.
11. E. G. Kalnins, J. M. Kress, and P. Winternitz, *Superintegrability in a two-dimensional space of nonconstant curvature*, J. Math. Phys. **43** (2002), no. 2, 970–983.
12. E. G. Kalnins, W. Miller, Jr., and G. S. Pogosyan, *Coulomb-oscillator duality in spaces of constant curvature*, J. Math. Phys. **41** (2000), no. 5, 2629–2657.
13. W. Klingenberg, *A course in differential geometry*, Grad. Texts in Math., vol. 51, Springer-Verlag, New York–Heidelberg, 1978.
14. S. Moyo and P. G. L. Leach, *A note on the construction of the Ermakov–Lewis invariant*, J. Phys. A **35** (2002), no. 25, 5333–5345.
15. E. Pinney, *The nonlinear differential equation $y'' + p(x)y + cy^{-3} = 0$*, Proc. Amer. Math. Soc. **1** (1950), 681.
16. M. F. Rañada, *Superintegrabe $n = 2$ systems, quadratic constants of motion, and potentials of Drach*, J. Math. Phys. **38** (1997), no. 8, 4165–4178.
17. M. F. Rañada and M. Santander, *Superintegrable systems on the two-dimensional sphere S^2 and the hyperbolic plane H^2*, J. Math. Phys. **40** (1999), no. 10, 5026–5057.
18. ———, *On some properties of the harmonic oscillator on spaces of constant curvature*, Rep. Math. Phys. **49** (2002), no. 2-3, 335–343.
19. ———, *On harmonic oscillators on the two-dimensional sphere S^2 and the hyperbolic plane H^2*, J. Math. Phys. **43** (2002), no. 1, 431–451.
20. M. F. Rañada, M. Santander, and T. Sanz-Gil, *Superintegrable potentials and the superposition of Higgs oscillators on the sphere S^2*, Classical and Quantum Integrability (Warsaw, 2001), Banach Center Publ., vol. 59, Polish Acad. Sci., Warsaw, 2003, pp. 243–255.
21. E. Schrödinger, *A method of determining quantum mechanical eigenvalues and eigenfunctions*, Proc. Roy. Irish Acad. Sect. A **46** (1940), 9–16.
22. J. J. Sławianowski, *Bertrand systems on spaces of constant sectional curvature*, Rep. Math. Phys. **46** (2000), no. 3, 429–460.

Departamento de Física Teórica, Universidad de Zaragoza, 50009 Zaragoza, Spain
E-mail address: mfran@posta.unizar.es

Departamento de Física Teórica, Universidad de Valladolid, 47011 Valladolid, Spain
E-mail address: santander@fta.uva.es

Centre de Recherches Mathématiques
CRM Proceedings and Lecture Notes
Volume **37**, 2004

A Survey of Quasi-Exactly Solvable Systems and Spin Calogero–Sutherland Models

F. Finkel, D. Gómez-Ullate, A. González-López, M. A. Rodríguez, and R. Zhdanov

ABSTRACT. We present a brief survey of quasi-exactly solvable systems and Calogero–Sutherland models. We describe a general method, based on the simultaneous use of multiple families of Dunkl operators, yielding several new families of quasi-exactly solvable Calogero–Sutherland models with spin.

1. Introduction

The concept of quasi-exact solvability was introduced 15 years ago by Turbiner and Ushveridze [**81,82**]. It was closely related to previous work on "hidden symmetry algebras" by Alhassid, Gürsey, Iachello and coworkers [**1–3**], and motivated by some particular examples, such as the sextic potential studied by Singh et al. [**73**] in 1978. Quasi-exactly solvable (QES) systems have been applied in many areas of Physics and have been studied extensively in the last decade (see [**37, 60, 72, 84**] and references therein). We will present a short description of these systems in the one-dimensional case in Section 2.

In 1971, Calogero [**11, 12**] introduced an integrable and exactly solvable many-particle Hamiltonian on the line with two-body interactions decreasing as the square of the distance between the particles. At the same time, Sutherland [**75, 76**] constructed a similar model on a circle. Later on, this type of models were completely classified by Olshanetsky and Perelomov [**57, 58**], and their properties have attracted the attention of numerous researchers ever since. Using Lie-algebraic techniques, Brink et al. showed in [**10**] that most of the Hamiltonians classified by Olshanetsky and Perelomov are also exactly solvable (in a sense that will be defined in Section 2). In Section 3 we briefly mention some of the main developments in this area.

In Section 4 we discuss the extension of Calogero–Sutherland models to particles with internal degrees of freedom (spin), introduced independently by several authors in the early 90's [**8,42,44,55,77**]. Although several techniques have been applied in

---

2000 *Mathematics Subject Classification.* Primary: 81U15; Secondary: 81R12, 81Q05.
This work was supported by the DGI (Spain) under grant BFM2002-02646. R. Zhdanov would like to acknowledge the financial support of the Spanish Ministry of Education and Culture during his stay at the Universidad Complutense de Madrid.

the study of these models, we shall focus here on the Dunkl operator approach [**23**]. We shall review in this section a generalization of this method introduced in [**25,26**], which can be used to construct several new families of exactly and quasi-exactly solvable spin Calogero–Sutherland models related to the A_N and BC_N root systems.

Finally, in Section 5 we state our conclusions and mention some recent developments on the relation between Calogero–Sutherland models and integrable spin chains with long-range interactions of Haldane–Shastry type [**43, 71**].

2. Quasi-exactly solvable systems

Quasi-exactly solvable (QES) systems occupy an intermediate position between exactly-solvable systems (such as the harmonic oscillator or the Coulomb potential) and the majority of systems, for which the eigenfunctions and spectrum cannot be computed in closed form. In fact, we will limit ourselves in this short introduction to those QES systems called *algebraic*, whose Hamiltonian can be expressed as a polynomial in the generators of a finite-dimensional Lie algebra admitting a finite-dimensional invariant module. The restriction of the Hamiltonian to this module is therefore a finite-dimensional matrix, so that part of the spectrum can be computed algebraically by diagonalizing this matrix. We shall also say (following Turbiner) that a system is *exactly solvable* (ES) if the Hamiltonian leaves invariant an infinite increasing sequence (or flag) of finite-dimensional subspaces.

Though the above ideas apply in any dimension, we will restrict ourselves in this review to one-dimensional problems. In the one-dimensional case, there is essentially one maximal finite-dimensional Lie algebra of first-order differential operators, namely $\mathfrak{sl}(2)$, which in an appropriate coordinate z is spanned by

$$J^- = \partial_z, \qquad J^0 = z\partial_z - \frac{m}{2}, \qquad J^+ = z^2\partial_z - mz. \tag{2.1}$$

If m is a nonnegative integer, this Lie algebra leaves invariant the space $\mathcal{P}_m(z)$ of polynomials in z of degree at most m. A second-order element in the enveloping algebra of this realization of $\mathfrak{sl}(2)$ is of the form

$$-\overline{H}(z) = P(z)\partial_z^2 + \widetilde{Q}(z)\partial_z + \widetilde{R}(z), \tag{2.2}$$

where

$$\widetilde{Q}(z) = Q(z) - \frac{m-1}{2}P'(z), \quad \widetilde{R}(z) = R - \frac{m}{2}Q'(z) + \frac{m}{12}(m-1)P''(z), \tag{2.3}$$

and $P(z)$, $Q(z)$, R are polynomials in z of degrees 4,2, 0, respectively. The operator $\overline{H}(z)$ can be mapped to a Schrödinger operator

$$H = -\partial_x^2 + V(x), \tag{2.4}$$

by the change of variable

$$x = \zeta^{-1}(z) = \int^z \frac{dy}{\sqrt{P(y)}} \tag{2.5}$$

and the gauge transformation with gauge factor

$$\mu(z) = P(z)^{-1/4} \exp\left[\int^z \frac{\widetilde{Q}(y)}{2P(y)}\, dy\right], \tag{2.6}$$

that is,

$$H = \mu(z)\overline{H}(z)\mu^{-1}(z)\big|_{z=\zeta(x)}. \tag{2.7}$$

The explicit form of the potential in terms of the polynomials P, Q and R is

$$V(x) = -R - \frac{1}{12P}\big[m(m+2)(PP'' - \tfrac{3}{4}P'^2) + 3(m+1)(QP' - 2PQ') - 3Q^2\big]. \tag{2.8}$$

Since $\overline{H}(z)$ leaves invariant the polynomial module $\mathcal{P}_m(z)$, the Hamiltonian H preserves the finite-dimensional space

$$\mu(z)\mathcal{P}_m(z)\big|_{z=\zeta(x)}, \tag{2.9}$$

and is therefore quasi-exactly solvable (indeed, algebraic).

The corresponding QES potentials (2.8) can be classified using a residual group of symmetries of the canonical form (2.2), namely the projective group [**35**]. In this way one obtains five types of rational and hyperbolic potentials and another five types of periodic potentials (see [**37**] for more details). Most of the known solvable models, such as the harmonic, Morse or Pöschl–Teller potentials, appear as particular cases in this classification.

3. Calogero–Sutherland Models

Calogero–Sutherland (CS) Hamiltonians represent many-particle systems on the line interacting via a two-body potential. In the simplest (A_N) case, the Hamiltonians can be written as

$$H = -\sum_{i=1}^{N} \partial_{x_k}^2 + g\sum_{i\neq j} V(x_i - x_j). \tag{3.1}$$

In the original Calogero [**11**] and Sutherland [**75**] models, the interaction potentials V are respectively given by

$$V(x) = \frac{1}{x^2}, \qquad V(x) = \frac{1}{\sin^2 x}. \tag{3.2}$$

These models have been applied in many different areas of physics, such as fractional statistics and anyons [**9, 14, 40, 41, 61**], random matrix models [**79**], quantum Hall liquids [**4**], Yang–Mills theories [**19,38**], and soliton wave propagation [**52, 64**].

Olshanetsky and Perelomov [**57,58**] showed that both the Calogero and Sutherland models, based on the A_N root system, can be generalized to an arbitrary root system. They further showed that all these models are completely integrable (both in the classical and quantum cases) provided that the interaction potential is the Weierstrass $\wp$ function or a degenerate version thereof, in which one or both of its periods tends to infinity. More precisely, the integrals of motion of these models are given by the radial parts of the Laplace–Beltrami operators defined in the homogeneous spaces associated to the corresponding root systems.

An important related question is the exact solvability of CS models. Turbiner and collaborators [**10, 70**] have shown that the rational, trigonometric and hyperbolic CS models associated with the A_N and BC_N root systems are in fact exactly solvable in the sense defined in Section 2. The basic idea for proving the exact-solvability of these models is to relate the Hamiltonian to a quadratic combination of the generators of a realization of $\mathfrak{sl}(N)$ preserving a finite-dimensional polynomial module. For the general elliptic CS model, however, only a few explicit solutions are known in the A_N case and for a small number of particles [**21, 30**].

Another approach to the integrability and solvability of CS models relies on expressing the Hamiltonian as a sum of squares of so-called Dunkl operators, originally introduced by Dunkl in his study of families of orthogonal polynomials in several variables [**22, 23**]. The relation of the exact solutions of CS models to these families of orthogonal polynomials (such as the Jack polynomials) has been the subject of extensive research; see, for instance, the papers by Stanley [**74**], Lapointe and Vinet [**53**], Baker and Forrester [**5**], Dunkl [**24**], van Diejen [**20**], or Rösler [**69**].

Inozemtsev and Meshcheryakov found several deformations of CS models with an external potential term not associated with the corresponding root system [**47, 50, 51**] which preserve integrability (in some models, however, this was only established rigorously in the classical case). On the other hand, several QES deformations of CS models have been discussed in the literature [**46, 56**]. A systematic search of QES many-body systems was developed in [**31**] using the classification of QES one-dimensional potentials in [**35**] and Calogero's results on the motion of zeros of solutions of partial differential equations [**13**]. Apart from the rational and hyperbolic CS models of A_N and BC_N type, one obtains in this way several new families of QES deformations, including some of the classically integrable models of Inozemtsev and Meshcheryakov. Using the same method, Gómez-Ullate et al. [**32**] proved that the classically integrable BC_N elliptic model of Inozemtsev [**48**] is QES.

Calogero–Sutherland models have also been studied in the context of superintegrability [**7, 15, 33, 34, 66, 67**]. An N-dimensional Hamiltonian is *superintegrable* if it possesses more than N constants of motion, though not all of them are necessarily in involution. If the Hamiltonian has $2N-1$ constant of motion, it is called *maximally superintegrable.* Superintegrability is related to separation of variables in more than one coordinate system and, at least in some particular cases, to exact solvability [**68, 80**].

4. Calogero–Sutherland spin models

One of the most recent extensions of Calogero–Sutherland models is the introduction of spin interactions. Spin CS models have been constructed using several different techniques, such as the exchange operator method [**62**], the Dunkl operator formalism [**8, 16, 17, 23**], the supersymmetric approach [**10**], and reduction by discrete symmetries [**65**]. We shall discuss separately the A_N case, which is much better understood, and the comparatively less well-known BC_N case.

4.1. The A_N case. The integrability of the A_N rational and trigonometric/hyperbolic spin CS models has been established independently by several authors [**6, 44, 45, 49, 78, 83**]. This was essentially proved by expressing the Hamiltonian as a quadratic combination involving either the original Dunkl operators [**23**],

$$T_i = \frac{\partial}{\partial z_i} + a\sum_{j\neq i}\frac{1}{z_i - z_j}(1-K_{ij}), \quad i = 1,\dots,N, \tag{4.1}$$

for the rational case, or the Dunkl–Cherednik operators [**17**],

$$\widetilde{T}_i = z_i\frac{\partial}{\partial z_i} + a\sum_{j<i}\frac{z_i}{z_i - z_j}(1-K_{ij}) + a\sum_{j>i}\frac{z_j}{z_i - z_j}(1-K_{ij}) + 1 - i, \quad i = 1,\dots,N, \tag{4.2}$$

for the trigonometric/hyperbolic case. In these formulas the operators K_{ij} are coordinate permutation operators, whose action on a function $f(\mathbf{z})$, with $\mathbf{z} =$

$(z_1,\dots,z_N)$, is given by

$$(K_{ij}f)(z_1,\dots,z_i,\dots,z_j,\dots,z_N)=f(z_1,\dots,z_j,\dots,z_i,\dots,z_N), \tag{4.3}$$

and a is a real parameter. The key property ensuring the complete integrability of the A_N rational and trigonometric/hyperbolic spin CS models is the commutativity of the Dunkl and Dunkl–Cherednik operators, namely,

$$[T_i,T_j]=0,\quad [\widetilde{T}_i,\widetilde{T}_j]=0. \tag{4.4}$$

In our recent paper [**25**], we have constructed several new families of spin CS models of A_N type which are QES or, in some cases, ES. In particular, we have obtained three families of QES spin CS models with a potential depending on elliptic functions. The basic idea in our approach is the introduction of a third commuting family of A_N-type Dunkl operators, given by

$$J_i^+=z_i^2\frac{\partial}{\partial z_i}-mz_i+a\sum_{j\neq i}\frac{z_iz_j}{z_i-z_j}(1-K_{ij}), \tag{4.5}$$

where m is a nonnegative integer. These operators, along with the Dunkl (4.1) and the Dunkl–Cherednik (4.2) operators leave invariant the finite-dimensional polynomial module

$$\mathcal{R}_m(\mathbf{z})=\operatorname{span}\left\{\prod_{i=1}^N z_i^{l_i}:0\le l_i\le m,\ i=1,\dots,N\right\}. \tag{4.6}$$

In addition, the operators (4.1) and (4.2) preserve the space $\mathcal{R}_n(\mathbf{z})$ for all values of n. From now on, we shall denote the operators (4.1) by J_i^- and shall use a slightly more symmetric form of the operators (4.2), namely

$$J_i^0=z_i\frac{\partial}{\partial z_i}-\frac{m}{2}+\frac{a}{2}\sum_{j\neq i}\frac{z_i+z_j}{z_i-z_j}(1-K_{ij}). \tag{4.7}$$

Note that, although the operators J_i^0 do not form a commuting family, they still preserve $\mathcal{R}_n(\mathbf{z})$ for all n. Hence any polynomial in $J_i^\epsilon(\epsilon=\pm,0)$ leaves $\mathcal{R}_m(\mathbf{z})$ invariant, while a polynomial in J_i^- and J_i^0 preserves $\mathcal{R}_n(\mathbf{z})$ for all values of n.

The starting point in our construction is the following quadratic combination involving *all* three families of Dunkl operators:

$$-\overline{H}=\sum_i\bigg(c_{++}(J_i^+)^2+c_{00}(J_i^0)^2+c_{--}(J_i^-)^2+\frac{c_{0+}}{2}\{J_i^0,J_i^+\}+\frac{c_{0-}}{2}\{J_i^0,J_i^-\}+c_+J_i^++c_0J_i^0+c_-J_i^-\bigg). \tag{4.8}$$

This operator commutes with all the permutation operators K_{ij}, and preserves the space $\mathcal{R}_m(\mathbf{z})$ (or the spaces $\mathcal{R}_n(\mathbf{z})$ for arbitrary n, if $c_{++}=c_{0+}=c_+=0$).

Let $\mathfrak{S}$ denote the space of spin states $|s_1,\dots,s_N\rangle$, where $-M\le s_i\le M$ and $M\in\frac{1}{2}\mathbb{N}$, and consider the spin permutation operators S_{ij}, $i,j=1,\dots,N$, whose action is given by

$$S_{ij}|s_1,\dots,s_i,\dots,s_j,\dots,s_N\rangle=|s_1,\dots,s_j,\dots,s_i,\dots,s_N\rangle. \tag{4.9}$$

Let Λ_0 be the total antisymmetrizer with respect to both spin and spatial coordinates, which by definition satisfies $K_{ij}S_{ij}\Lambda_0=-\Lambda_0$, or equivalently,

$$K_{ij}\Lambda_0=-S_{ij}\Lambda_0, \tag{4.10}$$

for all i, j. If D is any scalar differential operator we define a linear mapping * by the requirement that

$$(DK_{i_1j_1}\cdots K_{i_rj_r})^* = (-1)^r DS_{i_rj_r}\cdots S_{i_1j_1}.$$

It follows from Eq. (4.10) that

$$\overline{H}\Lambda_0 = \overline{H}^*\Lambda_0.$$

Since $\overline{H}$ commutes with Λ_0 and acts trivially on the spin space $\mathfrak{S}$, the operator $\overline{H}^*$ preserves the finite-dimensional space $\Lambda_0(\mathcal{R}_m(\mathbf{z})\otimes\mathfrak{S})$ (or $\Lambda_0(\mathcal{R}_n(\mathbf{z})\otimes\mathfrak{S})$ for arbitrary n, if $c_{++} = c_{0+} = c_+ = 0$). It can be shown that there exists a change of variables $\mathbf{z} = \boldsymbol{\zeta}(\mathbf{x})$ and a gauge factor $\mu(\mathbf{z})$ such that

$$\mu(\mathbf{z})\overline{H}^*(\mathbf{z})\mu^{-1}(\mathbf{z})\big|_{\mathbf{z}=\boldsymbol{\zeta}(\mathbf{x})} \equiv H = -\sum_i \partial_{x_i}^2 + V(\mathbf{x}), \tag{4.11}$$

where the potential V depends on the spin permutation operators; see [**25**] for more details. We emphasize that in general a second order differential operator cannot be transformed into a Schrödinger operator by a change of variable and a gauge transformation [**18, 28, 36, 72**]. From the properties of $\overline{H}^*$ and the definition of H, it immediately follows that the finite-dimensional space

$$\mathcal{M}_m = \mu(\boldsymbol{\zeta}(\mathbf{x}))\Lambda_0(\mathcal{R}_m(\boldsymbol{\zeta}(\mathbf{x}))\otimes\mathfrak{S}) \tag{4.12}$$

(or $\mathcal{M}_n$ for arbitrary n, if $c_{++} = c_{0+} = c_+ = 0$) is invariant under the action of H. Therefore H is QES or, if $c_{++} = c_{0+} = c_+ = 0$, ES.

The (Q)ES potentials V obtained in this way can be completely classified by exploiting the invariance of the families of Dunkl operators $\{J_i^+, J_i^0, J_i^-\}$ under the action of the projective group on the coordinate z_i, for all $i = 1, \ldots, N$. To this end, let $P(z_i)$ denote the coefficient of $\partial_{z_i}^2$ in $-\overline{H}$. Then $P(z)$ is a polynomial of degree at most four, which transforms under the irreducible multiplier representation $\rho_{4,-2}$ of the projective group [**25, 35, 39, 59**]. The classification is performed by reducing the polynomial P to canonical form. One obtains in this way nine families of (Q)ES spin potentials, all of which can be written as

$$V(\mathbf{x}) = \sum_i U(x_i) + V_{\text{int}}(\mathbf{x}) \tag{4.13}$$

where U plays the role of an external field potential, and the interaction potential V_{int} is of the form

$$V_{\text{int}}(\mathbf{x}) = \sum_{i<j}\big(V^-(x_{ij}^-) + V^+(x_{ij}^+)\big)a(a+S_{ij}), \quad x_{ij}^\pm = x_i \pm x_j, \tag{4.14}$$

with either $V^+ = 0$ or $V^+ = V^-$.

CASE 1. $P(z) = \frac{1}{4}$.

$$U(x) = Ax^4 + Bx^3 + Cx^2 + Dx, \qquad V_{\text{int}}(\mathbf{x}) = \sum_{i\neq j}(x_{ij}^-)^{-2}a(a+S_{ij}).$$

CASE 2. $P(z) = 4z$.

$$U(x) = Ax^6 + Bx^4 + Cx^2 + \frac{D}{x^2}, \quad V_{\text{int}}(\mathbf{x}) = \sum_{i\neq j}\big((x_{ij}^-)^{-2} + (x_{ij}^+)^{-2}\big)a(a+S_{ij})$$

CASE 3. $P(z) = 4(z^2 - 1)$.

$$U(x) = A\cosh^2 2x + B\cosh 2x + C\cosh 2x\sinh^{-2} 2x + D\sinh^{-2} 2x,$$

$$V_{\rm int}(\mathbf{x}) = \sum_{i\neq j}(\sinh^{-2} x_{ij}^- + \sinh^{-2} x_{ij}^+)a(a+S_{ij}).$$

CASE 4. $P(z) = 4(1 - z^2)$.

$$U(x) = A\cos^2 2x + B\cos 2x + C\cos 2x\sin^{-2} 2x + D\sin^{-2} 2x,$$

$$V_{\rm int}(\mathbf{x}) = \sum_{i\neq j}(\sin^{-2} x_{ij}^- + \sin^{-2} x_{ij}^+)a(a+S_{ij}).$$

CASE 5. $P(z) = 4z^2$.

$$U(x) = A\mathrm{e}^{4x} + B\mathrm{e}^{2x} + C\mathrm{e}^{-2x} + D\mathrm{e}^{-4x}, \qquad V_{\rm int}(\mathbf{x}) = \sum_{i\neq j}\sinh^{-2} x_{ij}^-\, a(a+S_{ij}).$$

CASE 6. $P(z) = (1 + z^2)^2$.

$$U(x) = A\cos 4x + B\cos 2x + C\sin 4x + D\sin 2x,$$

$$V_{\rm int}(\mathbf{x}) = \sum_{i\neq j}\sin^{-2} x_{ij}^-\, a(a+S_{ij}).$$

CASE 7. $P(z) = 4(1 - z^2)(1 - k^2 z^2)$.

$$U(x) = A\,\mathrm{sn}^2\, 2x + B\,\mathrm{cn}\, 2x\,\mathrm{dn}\, 2x + \mathrm{sn}^{-2}\, 2x(C + D\,\mathrm{cn}\, 2x\,\mathrm{dn}\, 2x),$$

$$V_{\rm int}(\mathbf{x}) = \sum_{i\neq j}\left(\frac{\mathrm{cn}^2 x_{ij}^-\,\mathrm{dn}^2 x_{ij}^-}{\mathrm{sn}^2 x_{ij}^-} + \frac{\mathrm{cn}^2 x_{ij}^+\,\mathrm{dn}^2 x_{ij}^+}{\mathrm{sn}^2 x_{ij}^+}\right)a(a+S_{ij}).$$

CASE 8. $P(z) = 4(z^2 - 1)(1 - k'^2 z^2)$.

$$U(x) = \mathrm{sn}^{-2}\, 2x(A + B\,\mathrm{dn}\, 2x) + \mathrm{cn}^{-2}\, 2x(C + D\,\mathrm{dn}\, 2x),$$

$$V_{\rm int}(\mathbf{x}) = \sum_{i\neq j}\left(\frac{\mathrm{dn}^2 x_{ij}^-}{\mathrm{sn}^2 x_{ij}^-\,\mathrm{cn}^2 x_{ij}^-} + \frac{\mathrm{dn}^2 x_{ij}^+}{\mathrm{sn}^2 x_{ij}^+\,\mathrm{cn}^2 x_{ij}^+}\right)a(a+S_{ij}).$$

CASE 9. $P(z) = 4(1 - z^2)(k'^2 + k^2 z^2)$.

$$U(x) = \mathrm{dn}^{-2}\, 2x(A + B\,\mathrm{cn}\, 2x) + \mathrm{sn}^{-2}\, 2x(C + D\,\mathrm{cn}\, 2x),$$

$$V_{\rm int}(\mathbf{x}) = \sum_{i\neq j}\left(\frac{\mathrm{cn}^2 x_{ij}^-}{\mathrm{sn}^2 x_{ij}^-\,\mathrm{dn}^2 x_{ij}^-} + \frac{\mathrm{cn}^2 x_{ij}^+}{\mathrm{sn}^2 x_{ij}^+\,\mathrm{dn}^2 x_{ij}^+}\right)a(a+S_{ij}).$$

In the elliptic Cases 7–9, k is the modulus of the elliptic functions and $k' = \sqrt{1-k^2}$ is the complementary modulus. The parameters A, B, C, D that define the external potential are not independent; see **[25]** for their explicit expressions in each case. Note also that in Cases 7 and 8 we could have used the alternative canonical form

$$P(z) = 4z^3 - g_2 z - g_3, \qquad g_2^3 > 27g_3^2 > 0, \tag{4.15}$$

leading to the external and interaction potentials respectively given by

$$U(x) = A\wp(x+\omega_3) + \big(B\wp^2(x+\omega_3) + C\wp(x+\omega_3) + D\big)\big(\wp'(x+\omega_3)\big)^{-2} + E\wp(2x),$$

$$V_{\rm int}(\mathbf{x}) = \sum_{i\neq j}\big(\wp(x_{ij}^-) + \wp(x_{ij}^+)\big)a(a+S_{ij}),$$

where $\wp(x) = \wp(x \mid g_2, g_3)$ is the Weierstrass function with invariants g_2, g_3 and purely imaginary fundamental period $2\omega_3$, and the expressions of the parameters $A, \dots, E$ are given in [**25**].

REMARK 1. Case 1 with $A = B = 0$ yields the rational Calogero A_N spin model. Case 5 with $A = B = 0$ (or $C = D = 0$) is the model studied by Inozemtsev in [**49**], while for $A = B = C = D = 0$ the hyperbolic Sutherland A_N spin model is obtained. The remaining potentials are new. In Cases 1–5, the potential is ES if $A = B = 0$ (in Case 5, the potential is also ES for $C = D = 0$). The only ES potential in Case 6 is the trigonometric Sutherland potential ($A = B = C = D = 0$). The remaining potentials, including all the elliptic potentials in Cases 7–9, are QES.

REMARK 2. The above spin potentials reduce to (Q)ES scalar many-body potentials by the substitution $S_{ij} \to -1$. This is an easy consequence of the fact that the operator $\overline{H}$ preserves the space $\mathcal{S}_m(\mathbf{z})$ of symmetric polynomials in $\mathbf{z}$ of degree at most m, where the permutation operators K_{ij} acts as the identity. It follows that the space $\mu\Lambda_0(\mathcal{S}_m \otimes \mathfrak{S}) = (\mu\mathcal{S}_m) \otimes (\Lambda_s\mathfrak{S})$, where Λ_s is the spin antisymmetrizer, is invariant under H. Hence $H|_{S_{ij}\to -1}$ preserves the space $\mu\mathcal{S}_m$. The list of (Q)ES scalar potentials thus obtained includes all known examples of scalar (Q)ES CS models.

4.2. The $\mathbf{B}_N$ case. In the B_N case, the rational and trigonometric spin CS models were formulated by Yamamoto in [**85**]. In the latter paper, the spectrum of the rational B_N spin model was explicitly determined, and its integrability was shown by means of the Lax pair approach. In [**86**], Yamamoto and Tsuchiya presented an alternative proof of the integrability of this model using Dunkl operators of B_N type. The same operators were later employed by Dunkl to construct a complete basis of eigenfunctions [**24**]. In contrast, the trigonometric/hyperbolic BC_N spin model has received remarkably little attention.

In our recent paper [**26**] we proved that the hyperbolic model is ES. In fact, in the same paper we extended the construction of (Q)ES spin CS models described in Section 4.1 to the BC_N case. To this end, we now consider the following three sets of Dunkl operators:

$$
\begin{aligned}
J_i^- &= \frac{\partial}{\partial z_i} + a\Big(\sum_{j\neq i}\frac{1}{z_i - z_j}(1 - K_{ij}) + \sum_{j\neq i}\frac{1}{z_i + z_j}(1 - \widetilde{K}_{ij})\Big) + \frac{b}{z_i}(1 - K_i),\\
J_i^0 &= z_i\frac{\partial}{\partial z_i} - \frac{m}{2} + \frac{a}{2}\Big(\sum_{j\neq i}\frac{z_i + z_j}{z_i - z_j}(1 - K_{ij}) + \sum_{j\neq i}\frac{z_i - z_j}{z_i + z_j}(1 - \widetilde{K}_{ij})\Big),\\
J_i^+ &= z_i^2\frac{\partial}{\partial z_i} - mz_i + a\Big(\sum_{j\neq i}\frac{z_i z_j}{z_i - z_j}(1 - K_{ij}) - \sum_{j\neq i}\frac{z_i z_j}{z_i + z_j}(1 - \widetilde{K}_{ij})\Big)\\
&\qquad - b' z_i(1 - (-1)^m K_i),
\end{aligned}
\tag{4.16}
$$

where a, b, b' are nonzero real parameters, m is a nonnegative integer, K_i denotes the sign-reversal operator

$$
(K_i f)(z_1, \dots, z_i, \dots, z_N) = f(z_1, \dots, -z_i, \dots, z_N),
\tag{4.17}
$$

and $\widetilde{K}_{ij} = K_i K_j K_{ij}$. The first two families (4.16) were introduced by Dunkl in connection with the rational B_N spin model [**24**], while the third one first appeared in our paper [**26**]. As before, the operator J_i^+ preserves the space $\mathcal{R}_m(\mathbf{z})$, while J_i^- and J_i^0 preserve the spaces $\mathcal{R}_n(\mathbf{z})$ for arbitrary n.

In the BC_N case, the appropriate form of the operator $\overline{H}$ ensuring its commutativity with the permutation and sign-reversal operators is

$$-\overline{H} = \sum_i \big(c_{++}(J_i^+)^2 + c_{00}(J_i^0)^2 + c_{--}(J_i^-)^2 + c_0 J_i^0\big). \tag{4.18}$$

This operator preserves the space $\mathcal{R}_m(\mathbf{z})$ (or the spaces $\mathcal{R}_n(\mathbf{z})$ for arbitrary n, if $c_{++} = 0$).

In addition to the spin permutation operators S_{ij}, we shall also consider the spin reversing operators S_i, defined by

$$S_i|s_1, \dots, s_i, \dots, s_N\rangle = |s_1, \dots, -s_i, \dots, s_N\rangle, \quad i = 1, \dots, N, \tag{4.19}$$

and the operators $\tilde{S}_{ij} = S_i S_j S_{ij}$. Let Λ be the total antisymmetrizer with respect to sign reversals and permutation of both spin and spatial coordinates, which satisfies

$$K_{ij}\Lambda = -S_{ij}\Lambda, \qquad K_i\Lambda = -S_i\Lambda, \tag{4.20}$$

for all i, j. We now define the linear mapping * by the condition

$$(DK_{\alpha_1} \cdots K_{\alpha_r})^* = (-1)^r D S_{\alpha_r} \cdots S_{\alpha_1},$$

where D is a scalar differential operator and the subindex α stands either for ij or i. As before, Eq. (4.20) implies that

$$\overline{H}\Lambda = \overline{H}^*\Lambda.$$

Since $\overline{H}$ commutes with Λ and acts trivially on the spin space $\mathfrak{S}$, it follows that $\overline{H}^*$ leaves invariant the finite-dimensional space $\Lambda(\mathcal{R}_m(\mathbf{z}) \otimes \mathfrak{S})$ (or $\Lambda(\mathcal{R}_n(\mathbf{z}) \otimes \mathfrak{S})$ for arbitrary n, if $c_{++} = 0$). It can again be shown that there exists a change of variables and a gauge transformation mapping $\overline{H}^*$ into a spin Schrödinger operator (4.11); see **[26]** for details. It follows that the finite-dimensional space

$$\mathcal{N}_m = \mu(\zeta(\mathbf{x}))\Lambda\big(\mathcal{R}_m(\zeta(\mathbf{x})) \otimes \mathfrak{S}\big) \tag{4.21}$$

(or $\mathcal{N}_n$ for arbitrary n, if $c_{++} = 0$) is invariant under the action of H. Therefore H is QES or, if $c_{++} = 0$, ES.

In this case, the families of Dunkl operators (4.16) are form-invariant only under inversions and scale transformations. This yields eleven canonical forms for the polynomial P, nine of which lead to QES Hamiltonians of BC_N type **[26]**. The corresponding potentials can be expressed in a unified way in terms of an elliptic function v of primitive half-periods P_1 and P_2 as

$$V(\mathbf{x}) = V_{\mathrm{spin}}(\mathbf{x}) + \sum_i U(x_i), \tag{4.22}$$

where the spin and the scalar external potentials are respectively given by

$$\begin{aligned} V_{\mathrm{spin}}(\mathbf{x}) = a\sum_{i\neq j}\big[\big(v(x_{ij}^-) + v(x_{ij}^+ + P_1)\big)(a + S_{ij}) + \big(v(x_{ij}^+) + v(x_{ij}^- + P_1)\big)(a + \tilde{S}_{ij})\big] \\ + b\sum_i \big(v(x_i) + v(x_i + P_1)\big)(b + S_i), \end{aligned}$$

$$\begin{aligned} U(x) = \lambda(\lambda - 1)[v(x + \tfrac{1}{2}P_1) + v(x - \tfrac{1}{2}P_1)] \\ + \lambda'(\lambda' - 1)[v(x + \tfrac{1}{2}P_1 + P_2) + v(x - \tfrac{1}{2}P_1 + P_2)], \end{aligned}$$

and λ, λ' are constants. The explicit form of the functions v and its half-periods P_1 and P_2 for each case are listed in Table 1. (The potentials in Case 2 have been omitted, since they are not of BC_N type.)

TABLE 1. Function $v(x)$ and its primitive half-periods P_i for each of the BC_N-type potentials (4.22). The constant $K \equiv K(k)$ is the complete elliptic integral of the first kind, $K' = K(k')$, k is the modulus of the elliptic functions, and $k' = \sqrt{1-k^2}$.

CASE	$v(x)$	P_1	P_2
1	x^{-2}	∞	∞
3a	$\sinh^{-2} x$	$\mathrm{i}\pi/2$	∞
3b	$\sin^{-2} x$	$\pi/2$	∞
4a	$\sinh^{-2} x$	∞	$\mathrm{i}\pi/2$
4b	$\sin^{-2} x$	∞	$\pi/2$
5	$\mathrm{dn}^2 x/\mathrm{sn}^2 x$	$K + \mathrm{i}K'$	K
6a	$\mathrm{cn}^2 x\, \mathrm{dn}^2 x/\mathrm{sn}^2 x$	K	$\mathrm{i}K'/2$
6b	$\mathrm{dn}^2 x/(\mathrm{sn}^2 x\, \mathrm{cn}^2 x)$	$\mathrm{i}K'$	$K/2$
7	$\mathrm{cn}^2 x/(\mathrm{sn}^2 x\, \mathrm{dn}^2 x)$	K	$\frac{1}{2}(K + \mathrm{i}K')$

In the elliptic Cases 5–7, the function v can be expressed in a unified way as $v(x) = \wp(x; P_1, P_2)$, where $\wp$ is the Weierstrass function with primitive half-periods P_1 and P_2 listed in Table 1.

The potentials in Cases 1–3 are ES for all values of the parameters. In Case 4a and 4b, the potential is ES only for $U = 0$. The elliptic potentials in Cases 5–7 are always QES. All the potentials presented in Table 1 are new, except for Cases 1 and 4. Case 1 is the rational B_N-type model introduced by Yamamoto [**85**] and studied by Dunkl [**24**], while Case 4b for $U = 0$ is Yamamoto's BC_N-type trigonometric potential [**85**].

5. Conclusions

In this article we have given an overview of a systematic method for constructing of (Q)ES scalar or spin Calogero–Sutherland models. Our approach is close in spirit to the construction of algebraic QES one-dimensional models described in Section 2, the role of the $\mathfrak{sl}(2)$ generators now being played by the families of Dunkl operators presented in the previous section. The (quasi) exact solvability of the Hamiltonian follows from the invariance of the polynomial space (4.6) under the Dunkl operators. All the previously known ES spin Calogero–Sutherland models are obtained as particular cases, as well as several new families of (Q)ES spin CS models, some of them of elliptic type.

Although in this short review we have focused on the description of new QES spin CS models, the spectrum and the integrability of these models can also be established using the properties of Dunkl operators. For instance, in a recent paper [**27**] we have studied the spectrum and proved the integrability of the hyperbolic Sutherland spin model of BC_N type (Case 4a in Table 1).

Another important development not covered in this review is the connection between Calogero–Sutherland models and integrable spin chains with long-range interactions of Haldane–Shastry type [**43, 71**]. In fact, every spin Calogero–Sutherland model can be related to a spin chain with long-range interactions by taking the strong coupling limit, in which the particles are "frozen" in their classical equilibrium positions, and their interaction are restricted to the spin degrees of freedom

[**63, 77**]. In this limit, the constants of motion of the dynamical CS model yield those of the spin chain, thereby proving its integrability [**27, 86**]. An interesting open problem in this respect is the generalization of the above results to the new QES spin CS models presented in the previous section.

References

1. Y. Alhassid, J. Engel, and J. Wu, *Algebraic approach to the scattering matrix*, Phys. Rev. Lett. **53** (1984), no. 1, 17–20.
2. Y. Alhassid, J. Gürsey, and F. Iachello, *Group theory approach to scattering*, Ann. Physics **148** (1983), no. 2, 346–380.
3. ———, *Group theory approach to scattering.* II. *The Euclidean connection*, Ann. Physics **167** (1986), no. 1, 181–200.
4. H. Azuma and S. Iso, *Explicit relation of the quantum Hall effect and the Calogero–Sutherland model*, Phys. Lett. B **331** (1994), no. 1-2, 107–113.
5. T. H. Baker and P. J. Forrester, *The Calogero–Sutherland model and generalized classical polynomials*, Comm. Math. Phys. **188** (1997), no. 1, 175–216.
6. B. Basu-Mallick, *Spin-dependent extension of Calogero–Sutherland model through anyon-like representations of permutation operators*, Nuclear Phys. B **482** (1996), no. 3, 713–730.
7. S. Benenti, C. Chanu, and G. Rastelli, *The super-separability of the three-body inverse-square Calogero system*, J. Math. Phys. **41** (2000), no. 7, 4654–4678.
8. D. Bernard, M. Gaudin, F. D. M. Haldane, and V. Pasquier, *Yang–Baxter equation in long-range interacting systems*, J. Phys. A **26** (1993), no. 20, 5219–5236.
9. L. Brink, T. H. Hansson, S. Konstein, and M. A. Vasiliev, *The Calogero model—anyonic representation, fermionic extension and supersymmetry*, Nuclear Phys. B **401** (1993), no. 3, 591–612.
10. L. Brink, A. Turbiner, and N. Wyllard, *Hidden algebras of the (super) Calogero and Sutherland models*, J. Math. Phys. **39** (1998), no. 3, 1285–1315.
11. F. Calogero, *Solution of the one-dimensional N-body problems with quadratic and/or inversely quadratic pair potentials*, J. Mathematical Phys. **12** (1971), 419–436.
12. ———, *One-dimensional many-body problems with pair interactions whose exact ground-state wave function is of product type*, Lett. Nuovo Cimento **13** (1975), 507–511.
13. ———, *Motion of poles and zeros of special solutions of nonlinear and linear partial differential equations and related "solvable" many-body problems*, Nuovo Cimento B (11) **43** (1978), no. 2, 177–241.
14. A. L. Carey and E. Langmann, *Loop groups, anyons and the Calogero–Sutherland model*, Comm. Math. Phys. **201** (1999), no. 1, 1–34.
15. R. Caseiro, J.-P. Françoise, and R. Sasaki, *Quadratic algebra associated with rational Calogero–Moser models*, J. Math. Phys. **42** (2001), no. 11, 5329–5340.
16. I. Cherednik, *A unification of Knizhnik–Zamolodchikov and Dunkl operators via affine Hecke algebras*, Invent. Math. **106** (1991), no. 2, 411–431.
17. ———, *Integration of quantum many-body problems by affine Knizhnik–Zamolodchikov equations*, Adv. Math. **106** (1994), no. 1, 65–95.
18. É. Cotton, *Sur les invariants différentiels de quelques équations linéaires aux dérivées partielles du second ordre*, Ann. École Norm. **17** (1900), 211–244.
19. E. D'Hoker and D. H. Phong, *Calogero–Moser systems in* SU(N) *Seiberg–Witten theory*, Nuclear Phys. B **513** (1998), no. 1-2, 405–444.
20. J. F. van Diejen, *Confluent hypergeometric orthogonal polynomials related to the rational quantum Calogero system with harmonic confinement*, Comm. Math. Phys. **188** (1997), no. 2, 467–497.
21. J. Dittrich and V. I. Inozemtsev, *On the structure of eigenvectors of the multidimensional Lamé operator*, J. Phys. A **26** (1993), no. 16, L753–L756.
22. C. F. Dunkl, *Reflection groups and orthogonal polynomials on the sphere*, Math. Z. **197** (1988), no. 1, 33–60.
23. ———, *Differential-difference operators associated to reflection groups*, Trans. Amer. Math. Soc. **311** (1989), no. 1, 167–183.
24. ———, *Orthogonal polynomials of types* A *and* B *and related Calogero models*, Comm. Math. Phys. **197** (1998), no. 2, 451–487.

25. F. Finkel, D. Gómez-Ullate, A. González-López, M. A. Rodríguez, and R. Zhdanov, A_N*-type Dunkl operators and new spin Calogero–Sutherland models*, Comm. Math. Phys. **221** (2001), no. 3, 477–497.
26. ———, *New spin Calogero–Sutherland models related to* B_N*-type Dunkl operators*, Nuclear Phys. B **613** (2001), no. 3, 472–496.
27. ———, *On the Sutherland spin model of* B_N *type and its associated spin chain*, Comm. Math. Phys. **233** (2003), no. 2, 191–209.
28. F. Finkel and N. Kamran, *On the equivalence of matrix differential operators to Schrödinger form*, J. Nonlinear Math. Phys. **4** (1997), no. 3–4, 278–286.
29. ———, *The Lie algebraic structure of differential operators admitting invariant spaces of polynomials*, Adv. in Appl. Math. **20** (1998), no. 3, 300–322.
30. L. Gavrilov and A. M. Perelomov, *On the explicit solutions of the elliptic Calogero system*, J. Math. Phys. **40** (1999), no. 12, 6339–6352.
31. D. Gómez-Ullate, A. González-López, and M. A. Rodríguez, *New algebraic quantum many body problems*, J. Phys. A **33** (2000), no. 41, 7305–7335.
32. ———, *Exact solutions of a new elliptic Calogero–Sutherland model*, Phys. Lett. B **511** (2001), no. 1, 112–118.
33. C. Gonera, *On the superintegrability of Calogero–Moser–Sutherland model*, J. Phys. A **31** (1998), no. 19, 4465–4472.
34. C. Gonera and Y. Nutku, *Super-integrable Calogero-type systems admit maximal number of Poisson structures*, Phys. Lett. A **285** (2001), no. 5-6, 301–306.
35. A. González-López, N. Kamran, and P. J. Olver, *Normalizability of one-dimensional quasi-exactly solvable Schrödinger operators*, Comm. Math. Phys. **153** (1993), no. 1, 117–146.
36. ———, *New quasi-exactly solvable Hamiltonians in two dimensions*, Comm. Math. Phys. **159** (1994), 503–537.
37. ———, *Quasi-exact solvability*, Lie Algebras, Cohomology, and New Applications to Quantum Mechanics (Springfield, MO, 1992), Contemp. Math., vol. 160, Amer. Math. Soc., Providence, RI, 1994, pp. 113–140.
38. A. Gorsky and N. Nekrasov, *Hamiltonian systems of Calogero type, and two-dimensional Yang–Mills theory*, Nuclear Phys. B **414** (1994), no. 1-2, 213–238.
39. G. B. Gurevich, *Foundations of the theory of algebraic invariants*, P. Noordhoff Ltd., Groningen, 1964.
40. Z. N. C. Ha, *Exact dynamical correlation functions of Calogero–Sutherland model and one-dimensional fractional statistics*, Phys. Rev. Lett. **73** (1994), no. 12, 1574–1577.
41. ———, *Fractional statistics in one dimension: view from an exactly solvable model*, Nuclear Phys. B **435** (1995), no. 3, 604–636.
42. Z. N. C. Ha and F. D. M. Haldane, *Models with inverse-square exchange*, Phys. Rev. B **46** (1992), no. 15, 9359–9368.
43. F. D. M. Haldane, *Exact Jastrow–Gutzwiller resonating-valence-bond ground state of the spin-*$\frac{1}{2}$ *antiferromagnetic Heisenberg chain with* $1/r^2$ *exchange*, Phys. Rev. Lett. **60** (1988), no. 7, 635–638.
44. K. Hikami and M. Wadati, *Integrability of Calogero–Moser spin system*, J. Phys. Soc. Japan **62** (1993), no. 2, 469–472.
45. ———, *Integrable spin-*$\frac{1}{2}$ *particle systems with long-range interactions*, Phys. Lett. A **173** (1993), no. 3, 263–266.
46. X. Hou and M. Shifman, *A quasi-exactly solvable* N*-body problem with the* $\mathrm{sl}(N+1)$ *algebraic structure*, Internat. J. Modern Phys. A **14** (1999), no. 19, 2993–3003.
47. V. I. Inozemtsev, *New completely integrable multiparticle dynamical systems*, Phys. Scripta **29** (1984), no. 6, 518–520.
48. ———, *Lax representation with spectral parameter on a torus for integrable particle systems*, Lett. Math. Phys. **17** (1989), no. 1, 11–17.
49. ———, *Integrable model of interacting fermions confined by the Morse potential*, Internat. J. Modern Phys. A **12** (1997), no. 1, 195–200.
50. V. I. Inozemtsev and D. V. Meshcheryakov, *Integrable particle systems in an external field and singular solutions of the inhomogeneous Burgers–Hopf equation*, Phys. Lett. A **106** (1984), no. 3, 105–108.
51. ———, *The discrete spectrum states of finite-dimensional quantum systems connected with Lie algebras*, Phys. Scripta **33** (1986), no. 2, 99–104.

52. A. Kasman, *Bispectral* KP *solutions and linearization of Calogero–Moser particle systems*, Comm. Math. Phys. **172** (1995), no. 2, 427–448.
53. L. Lapointe and L. Vinet, *Exact operator solution of the Calogero–Sutherland model*, Comm. Math. Phys. **178** (1996), no. 2, 425–452.
54. D. F. Lawden, *Elliptic functions and applications*, Appl. Math. Sci., vol. 80, Springer-Verlag, New York, 1989.
55. J. A. Minahan and A. P. Polychronakos, *Integrable systems for particles with internal degrees of freedom*, Phys. Lett. B **302** (1993), no 2-3, 265–270.
56. A. Minzoni, M. Rosenbaum, and A. Turbiner, *Quasi-exactly solvable many-body problems*, Modern Phys. Lett. A **11** (1996), no. 24, 1977–1984.
57. M. A. Olshanetsky and A. M. Perelomov, *Classical integrable finite-dimensional systems related to Lie algebras*, Phys. Rep. **71** (1981), no. 5, 313–400.
58. ———, *Quantum integrable systems related to Lie algebras*, Phys. Rep. **94** (1983), no. 6, 313–404.
59. P. J. Olver, *Equivalence, invariants, and symmetry*, Cambridge Univ. Press, Cambridge, 1995.
60. ———, *A quasi-exactly solvable travel guide*, GROUP21: Physical Applications and Mathematical Aspects of Geometry, Groups, and Algebras, vol. 1, World Scientific, Singapore, 1997, pp. 285–295.
61. A. P. Polychronakos, *Nonrelativistic bosonization and fractional statistics*, Nuclear Phys. B **324** (1989), no. 3, 597–622.
62. ———, *Exchange operator formalism for integrable systems of particles*, Phys. Rev. Lett. **69** (1992), no. 5, 703–705.
63. ———, *Lattice integrable systems of Haldane–Shastry type*, Phys. Rev. Lett. **70** (1993), no. 15, 2329–2331.
64. ———, *Waves and solitons in the continuum limit of the Calogero–Sutherland model*, Phys. Rev. Lett. **74** (1995), no. 26, 5153–5157.
65. ———, *Generalized Calogero models through reductions by discrete symmetries*, Nuclear Phys. B **543** (1999), no. 1-2, 485–498.
66. M. F. Rañada, *Superintegrability of the Calogero–Moser system: constants of motion, master symmetries and time-dependent symmetries*, J. Math. Phys. **40** (1999), no. 1, 236–247.
67. ———, *On the second Lax structure for the Calogero–Moser system: time-dependent constants and superintegrability*, Phys. Lett. A **277** (2000), no. 4-5, 219–222.
68. M. A. Rodríguez and P. Winternitz, *Quantum integrability and exact solvability in n dimensions*, J. Math. Phys. **43** (2002), no. 3, 1309–1322.
69. M. A. Rösler, *Generalized Hermite polynomials and the heat equation for Dunkl operators*, Comm. Math. Phys. **192** (1998), no. 3, 519–542.
70. W. Rühl and A. Turbiner, *Exact solvability of the Calogero and Sutherland models*, Modern Phys. Lett. A **10** (1995), no. 29, 2213–2221.
71. B. S. Shastry, *Exact solution of an $S = \frac{1}{2}$ Heisenberg antiferromagnetic chain with long-ranged interactions*, Phys. Rev. Lett. **60** (1988), no. 7, 639–642.
72. M. A. Shifman, *New findings in quantum mechanics (partial algebraization of the spectral problem)*, Internat. J. Modern Phys. A **4** (1989), no. 12, 2897–2952.
73. V. Singh, S. N. Biswas, and K. Datta, *Anharmonic oscillator and the analytic theory of continued fractions*, Phys. Rev. D (3) **18** (1978), no. 6, 1901–1908.
74. R. P. Stanley, *Some combinatorial properties of Jack symmetric functions*, Adv. Math. **77** (1989), no. 1, 76–115.
75. B. Sutherland, *Exact results for a quantum many-body problem in one dimension*, Phys. Rev. A **4** (1971), no. 5, 2019–2021.
76. ———, *Exact results for a quantum many-body problem in one dimension.* II, Phys. Rev. A **5** (1972), no. 3, 1372–1376.
77. B. Sutherland and B. S. Shastry, *Solution of some integrable one-dimensional quantum systems*, Phys. Rev. Lett. **71** (1993), no. 1, 5–8.
78. K. Takemura and D. Uglov, *The orthogonal eigenbasis and norms of eigenvectors in the spin Calogero–Sutherland model*, J. Phys. A **30** (1997), no. 10, 3685–3717.
79. N. Taniguchi, B. S. Shastry, and B. L. Altshuler, *Random matrix model and the Calogero–Sutherland model: a novel current-density mapping*, Phys. Rev. Lett. **75** (1995), no. 20, 3724–3727.

80. P. Tempesta, A. V. Turbiner, and P. Winternitz, *Exact solvability of superintegrable systems*, J. Math. Phys. **42** (2001), no. 9, 4248–4257.
81. A. V. Turbiner, *Quasi-exactly solvable problems and* $\mathfrak{sl}(2)$ *algebra*, Comm. Math. Phys. **118** (1988), no. 3, 467–474.
82. A. V. Turbiner and A. G. Ushveridze, *Spectral singularities and quasi-exactly solvable quantal problem*, Phys. Lett. A **126** (1987), no. 3, 181–183.
83. H. Ujino, A. Nishino, and M. Wadati, *New nonsymmetric orthogonal basis for the Calogero model with distinguishable particles*, Phys. Lett. A **249** (1998), no. 5-6, 459–464.
84. A. G. Ushveridze, *Quasi-exactly solvable models in quantum mechanics*, Institute of Physics Publishing, Bristol, 1994.
85. T. Yamamoto, *Multicomponent Calogero model of* B_N*-type confined in a harmonic potential*, Phys. Lett. A **208** (1995), no. 4-6, 293–302.
86. T. Yamamoto and O. Tsuchiya, *Integrable* $1/r^2$ *spin chain with reflecting end*, J. Phys. A **29** (1996), no. 14, 3977–3984.

Departamento de Física Teórica II, Universidad Complutense, 28040 Madrid, Spain
E-mail address, F. Finkel: `federico@ciruelo.fis.ucm.es`
E-mail address, A. González-López: `artemio@fis.ucm.es`
E-mail address, M. A. Rodríguez: `rodrigue@fis.ucm.es`

Centre de recherches mathématiques, Université de Montréal, C.P. 6128, succ. Centre-Ville, Montréal, QC H3C 3J7, Canada
Current address: Dipartimento de Matematica, Universit di Bologna, Piazza di Porta S. Donato, 5, 40127 Bologna, Italy
E-mail address: `ullate@crm.umontreal.ca`

Institute of Mathematics, Ukrainian Academy of Sciences, 3 Tereshchenkivska St., 01601, Kyiv-4, Ukraine
E-mail address: `renat@imath.kiev.ua`

Centre de Recherches Mathématiques
CRM Proceedings and Lecture Notes
Volume **37**, 2004

On the Classification of Third-Order Integrals of Motion in Two-Dimensional Quantum Mechanics

Mikhail Sheftel

Abstract. We study potentials possessing third-order integrals of motion for the two-dimensional stationary Schrödinger equation. Such an equation must admit at least one third-order Lie symmetry. We derive the determining equations for third-order integrals and for the potentials which possess such integrals. After imposing a special set of constraints on the form of third-order integrals we arrive at four different classes of such potentials and study in detail the simplest one. We study also the superintegrable system which have two third-order integrals of the types 3 and 4.

1. Lie symmetries and integrals of motion in quantum mechanics

We consider the stationary Schrödinger equation in the matrix form in characteristic variables $z = x + iy$, $\bar{z} = x - iy$

$$\hbar^2 \begin{pmatrix} u \\ v \end{pmatrix}_{z\bar{z}} = \frac{1}{2}(V(z,\bar{z}) - E)\begin{pmatrix} u \\ v \end{pmatrix}, \tag{1.1}$$

where u, v are components of the wave function $\psi = u + iv$ and we put the mass $m = 1$. The *Hamiltonian* of the system has the form

$$H = -2\hbar^2 D_z D_{\bar{z}} + V(z,\bar{z}). \tag{1.2}$$

Introducing the differential operator

$$\Delta = -\tfrac{1}{2}(H - E) = \hbar^2 D_z D_{\bar{z}} - \tfrac{1}{2}(V - E) \tag{1.3}$$

we can write the Schrödinger equation in the form

$$\Delta \begin{pmatrix} u \\ v \end{pmatrix} = \begin{pmatrix} 0 \\ 0 \end{pmatrix}. \tag{1.4}$$

Lie symmetries of the Schrödinger equation are generated by the flows (Lie equations)

$$\begin{pmatrix} u \\ v \end{pmatrix}_\tau = \begin{pmatrix} f \\ g \end{pmatrix} \tag{1.5}$$

2000 *Mathematics Subject Classification.* 35Q40, 81R12.

The author thanks Pavel Winternitz for many helpful advices and Simon Gravel for useful discussions.

This is the final form of the paper.

which leave the equation (1.4) invariant. Here τ is the parameter of Lie group. The *determining equation* for symmetries

$$\Delta \begin{pmatrix} f \\ g \end{pmatrix} = \begin{pmatrix} 0 \\ 0 \end{pmatrix} \tag{1.6}$$

coincides with the Schrödinger equation in the form (1.4).

The *recursion operator* L is defined by the commutator relation

$$[\Delta, L] = \lambda\Delta \iff [H, L] = \lambda(H - E). \tag{1.7}$$

For energy independent symmetries and recursion operators we split this commutation relation with respect to the energy E and obtain $\lambda = 0$ and $[H, L] = 0$.

Hence, the recursion operator coincides with the integral of motion X.

The order N of the integral is defined as the order of the highest derivative of the wave function which it depends on. If $N \geq 2$, then by Lie symmetry we mean a higher-order generalized symmetry.

2. Existence conditions and general form for a third-order integral

The classification of second-order integrals was made by Ya. A. Smorodinsky, P. Winternitz and coauthors in sixties [**1, 4, 6**] and was revised from a new point of view in the recent paper [**5**]. Here we are interested in the classification of quantum mechanical systems possessing third-order integrals of motion and in obtaining all such integrals. The problem was posed by Gravel and Winternitz in [**3**] where it was solved for a class of superintegrable systems admitting one first-order and one third-order integral and all such potentials and their integrals were found explicitly. Gravel obtained a complete classification of potentials that permit the separation of variables in Cartesian coordinates and allow a third-order integral of motion [**2**]. By construction, such potentials also allow a second-order integral. We make an attempt to attack a more difficult problem of finding all potentials with a third-order integral of motion without an additional requirement that they should also admit a lower-order one. To simplify the problem, we did not study it here in its full generality but instead imposed a set of constraints simplifying the form of a third-order integral.

We know [**1, 4–6**] that the existence of a second-order integral leads to a separation of variables in the Schrödinger equation. The important question is what an additional property of the Schrödinger equation follows from the existence of a third-order integral. It is related to the question whether potentials possessing a third-order integral, but no lower order integrals exist at all.

The conditions for the existence of third-order integrals and corresponding potentials were formulated in [**3**] using the real coordinates of the physical phase space x, y, p_x, p_y. Here we use instead the complex characteristic variables z, $\bar{z}$ and the operators of total derivatives D_z, $D_{\bar{z}}$.

We start with a third-order integral X of a general form

$$\begin{aligned} X = AD_z^3 + \bar{A}D_{\bar{z}}^3 + BD_z^2 D_{\bar{z}} + \overline{B}D_z D_{\bar{z}}^2 + CD_z^2 + \overline{C}D_{\bar{z}}^2 + FD_z D_{\bar{z}} \\ + GD_z + \overline{G}D_{\bar{z}} + H \end{aligned} \tag{2.1}$$

with the reality condition $\bar{X} = X$ implying $\overline{F} = F$ and $\bar{H} = H$. Here A, B, C, F, G, H and their complex conjugates are functions of z, $\bar{z}$.

The commutation relation $[H, X] = 0$ gives a set of equations, most of which are immediately integrated to give

$$
\begin{aligned}
&A(z) = \tfrac{1}{6}\Lambda z^3 + \tfrac{1}{2}A_2 z^2 + A_1 z + A_0, \qquad C(z) = \tfrac{1}{2}(\Lambda + M)z^2 + C_1 z + C_0 \\
&B(z,\bar z) = -\tfrac{1}{2}\Lambda z^2 \bar z - A_2 z\bar z + \tfrac{1}{2}\bar A_2 z^2 + B_1 z - A_1 \bar z + B_0 \\
&F(z,\bar z) = -M z\bar z + (\bar A_2 - \overline{C}_1)z + (A_2 - C_1)\bar z + F_0
\end{aligned} \tag{2.2}
$$

together with complex conjugate formulas. The constants Λ, A_2, A_1, A_0, B_1, B_0, C_1, C_0, M, F_0 satisfy the conditions $\bar\Lambda = -\Lambda$, $\overline{B}_1 = -B_1$, $\bar M = M$, $\overline{F}_0 = F_0$ following from the reality of X. The six remaining equations are

$$
\begin{aligned}
\hbar^2 G_{\bar z} = {}& \tfrac{1}{4}\Lambda(z^2\bar z V_{\bar z} - z^3 V_z) + \tfrac{1}{4}A_2(2z\bar z V_{\bar z} - 3z^2 V_z) - \tfrac{1}{4}\bar A_2 z^2 V_{\bar z} \\
& + \tfrac{1}{2}A_1(\bar z V_{\bar z} - 3zV_z) - \tfrac{1}{2}B_1 z V_{\bar z} - \tfrac{3}{2}A_0 V_z - \tfrac{1}{2}B_0 V_{\bar z}
\end{aligned} \tag{2.3}
$$

$$
\begin{aligned}
\hbar^2(G_z + \overline{G}_{\bar z}) = {}& \tfrac{1}{2}\Lambda(z^2\bar z V_z - z\bar z^2 V_{\bar z}) + \tfrac{1}{2}A_2(2z\bar z V_z - \bar z^2 V_{\bar z}) \\
& + \tfrac{1}{2}\bar A_2(2z\bar z V_{\bar z} - z^2 V_z) + B_1(\bar z V_{\bar z} - zV_z) + A_1 \bar z V_z + \bar A_1 z V_{\bar z} \\
& - B_0 z - \overline{B}_0 V_{\bar z} + \hbar^2 M
\end{aligned} \tag{2.4}
$$

$$
\begin{aligned}
\hbar^2 H_z = {}& -\tfrac{1}{4}\Lambda(z\bar z^2 V_{z\bar z} - z^2\bar z V_{zz} + \bar z^2 V_{\bar z} - 2z\bar z V_z) \\
& + \tfrac{1}{2}A_2(z\bar z V_{zz} - \tfrac{1}{2}\bar z^2 V_{z\bar z}) + \tfrac{1}{2}\bar A_2(3\bar z V_{\bar z} - 2zV_z + z\bar z V_{z\bar z} - \tfrac{1}{2}z^2 V_{zz}) \\
& + \tfrac{1}{2}A_1 \bar z V_{zz} + \tfrac{1}{2}\bar A_1(zV_{z\bar z} + 3V_{\bar z}) - \tfrac{1}{2}B_1(zV_{zz} - \bar z V_{z\bar z} + V_z) \\
& - \tfrac{1}{2}B_0 V_{zz} - \tfrac{1}{2}\overline{B}_0 V_{z\bar z} + \tfrac{1}{2}M(z\bar z V_z - \bar z^2 V_{\bar z}) + \tfrac{1}{2}C_1 \bar z V_z \\
& + \tfrac{1}{2}\overline{C}_1(zV_z - 2\bar z V_{\bar z}) - \overline{C}_0 V_{\bar z} - \tfrac{1}{2}F_0 V_z
\end{aligned} \tag{2.5}
$$

$$
\begin{aligned}
\hbar^2 H_{z\bar z} = {}& \tfrac{1}{4}\Lambda[\tfrac{1}{3}(\bar z^3 V_{\bar z\bar z\bar z} - z^3 V_{zzz}) + z^2\bar z V_{zz\bar z} - z\bar z^2 V_{z\bar z\bar z} + \bar z^2 V_{\bar z\bar z} - z^2 V_{zz}] \\
& + \tfrac{1}{2}A_2[z\bar z V_{zz\bar z} - \tfrac{1}{2}(z^2 V_{zzz} + \bar z^2 V_{z\bar z\bar z} - \bar z V_{z\bar z})] \\
& + \tfrac{1}{2}\bar A_2[z\bar z V_{z\bar z\bar z} - \tfrac{1}{2}(z^2 V_{zz\bar z} + \bar z^2 V_{\bar z\bar z\bar z} - zV_{z\bar z})] \\
& + \tfrac{1}{2}A_1(\bar z V_{zz\bar z} - zV_{zzz}) + \tfrac{1}{2}\bar A_1(zV_{z\bar z\bar z} - \bar z V_{\bar z\bar z\bar z}) \\
& - \tfrac{1}{2}A_0 V_{zzz} - \tfrac{1}{2}\bar A_0 V_{\bar z\bar z\bar z} + \tfrac{1}{2}B_1(\bar z V_{z\bar z\bar z} - zV_{zz\bar z}) - \tfrac{1}{2}B_0 V_{zz\bar z} \\
& - \tfrac{1}{2}\overline{B}_0 V_{z\bar z\bar z} + \tfrac{1}{2}C_1(\bar z V_{z\bar z} - zV_{zz}) + \tfrac{1}{2}\overline{C}_1(zV_{z\bar z} - \bar z V_{\bar z\bar z}) - \tfrac{1}{2}C_0 V_{zz} \\
& - \tfrac{1}{2}\overline{C}_0 V_{\bar z\bar z} - \tfrac{1}{2}F_0 V_{z\bar z} - \tfrac{1}{2}G(z,\bar z)V_z - \tfrac{1}{2}\overline{G}(z,\bar z)V_{\bar z}
\end{aligned} \tag{2.6}
$$

and two equations that are complex conjugate to (2.3)) and (2.5)).

The compatibility condition $(\overline{G}_z)_{\bar z} = (\overline{G}_{\bar z})_z$ of the equation complex conjugate to (2.3) and of (2.4) can be obtained in the form

$$
\hbar^2 G_{zz} = \tfrac{3}{2}\bar A(\bar z)V_{\bar z\bar z} - BV_{zz} - \tfrac{1}{2}\overline{B}V_{z\bar z} + \tfrac{1}{2}(\overline{B}_{\bar z} - 2B_z)V_z + \tfrac{1}{2}[3\bar A'(\bar z) - 2\overline{B}_z]V_{\bar z}. \tag{2.7}
$$

The compatibility condition $(G_{zz})_{\bar z} = (G_{\bar z})_{zz}$ of equations (2.7) and (2.3) is a linear third-order partial differential equation for the potential V possessing a third-order integral

$$(2.8)\quad \tfrac{1}{2}\Lambda[z^3V_{zzz}-\bar{z}^3V_{\bar{z}\bar{z}\bar{z}}+z^2\bar{z}V_{zz\bar{z}}-z\bar{z}^2V_{z\bar{z}\bar{z}}+8(z^2V_{zz}-\bar{z}^2V_{\bar{z}\bar{z}})+12(zV_z-\bar{z}V_{\bar{z}})]$$
$$+\tfrac{1}{2}A_2(3z^2V_{zzz}+2z\bar{z}V_{zz\bar{z}}-\bar{z}^2V_{z\bar{z}\bar{z}}+16zV_{zz}+12V_z)$$
$$+\tfrac{1}{2}\bar{A}_2(3\bar{z}^2V_{\bar{z}\bar{z}\bar{z}}+2z\bar{z}V_{z\bar{z}\bar{z}}-z^2V_{zz\bar{z}}+16\bar{z}V_{\bar{z}\bar{z}}+12V_{\bar{z}})$$
$$+A_1(3zV_{zzz}+\bar{z}V_{zz\bar{z}}+8V_{zz})+\bar{A}_1(3\bar{z}V_{\bar{z}\bar{z}\bar{z}}+zV_{z\bar{z}\bar{z}}+8V_{\bar{z}\bar{z}})$$
$$-B_1(zV_{zz\bar{z}}-\bar{z}V_{z\bar{z}\bar{z}})+3A_0V_{zzz}+3\bar{A}_0V_{\bar{z}\bar{z}\bar{z}}-B_0V_{zz\bar{z}}-\overline{B}_0V_{z\bar{z}\bar{z}}=0.$$

It is a necessary but not sufficient third-order existence condition for a third-order integral.

The compatibility condition $(H_{\bar{z}})_z=(H_z)_{\bar{z}}$ of the equation (2.5) and its complex conjugate has the form of the following linear second-order existence condition for a third-order integral

$$(2.9)\quad M[z^2V_{zz}-\bar{z}^2V_{\bar{z}\bar{z}}+3(zV_z-\bar{z}V_{\bar{z}})]+(C_1-A_2)(2zV_{zz}+3V_z)$$
$$-(\overline{C}_1-\bar{A}_2)(2\bar{z}V_{\bar{z}\bar{z}}+3V_{\bar{z}})+2(C_0-A_1)V_{zz}-2(\overline{C}_0-\bar{A}_1)V_{\bar{z}\bar{z}}.$$

This equation coincides with the existence condition for a second-order integral [**5**]. Since we need only a third-order integral, not accompanied by a second-order integral, condition (2.9) should be trivially satisfied by $M=0$, $C_1=A_2$, $C_0=A_1$ and hence $F=F_0=\text{const}$.

The compatibility condition $(H_z)_{\bar{z}}=H_{z\bar{z}}$ of equations (2.5) and (2.6) is obtained in the form

$$(2.10)\quad \tfrac{1}{6}\Lambda[z^3V_{zzz}-\bar{z}^3V_{\bar{z}\bar{z}\bar{z}}+6(z^2V_{zz}-\bar{z}^2V_{\bar{z}\bar{z}}+zV_z-\bar{z}V_{\bar{z}})]$$
$$+\tfrac{1}{2}A_2(z^2V_{zzz}+4zV_{zz}+3V_z)+\tfrac{1}{2}\bar{A}_2(\bar{z}^2V_{\bar{z}\bar{z}\bar{z}}+4\bar{z}V_{\bar{z}\bar{z}}+3V_z)$$
$$+A_1(zV_{zzz}+2V_{zz})+\bar{A}_1(\bar{z}V_{\bar{z}\bar{z}\bar{z}}+2V_{\bar{z}\bar{z}})+A_0V_{zzz}+\bar{A}_0V_{\bar{z}\bar{z}\bar{z}}$$
$$-\tfrac{1}{2}M(zV_z+\bar{z}V_{\bar{z}})-\tfrac{1}{2}C_1V_z-\tfrac{1}{2}\overline{C}_1V_{\bar{z}}+GV_z+\overline{G}V_{\bar{z}}=0.$$

Next we could eliminate G and $\overline{G}$ using the equation (2.10) and equations obtained previously and obtain *non-linear existence conditions* for the potential V possessing a third-order integral. These conditions were obtained, though in different variables, in [**3**]. We do not present the nonlinear existence conditions here, since we shall not use them explicitly. Our strategy is to solve the linear existence conditions first and then account for non-linear conditions implicitly in the process of determining G and $\overline{G}$ from the remaining equations.

3. Third-order integrals with definite parity

Following [**3**] without any loss of generality we impose on the integral X in (2.1) with coefficients determined by (2.2) the requirement that it should have a symmetrized form with definite parity

$$(3.1)\quad X=A_{300}\{L_3^3,1\}+A_{210}\{L_3^2,p_1\}+A_{201}\{L_3^2,p_2\}$$
$$+A_{120}\{L_3,p_1^2\}+A_{102}\{L_3,p_2^2\}+A_{111}\{L_3,p_1p_2\}$$
$$+A_{030}\{1,p_1^3\}+A_{021}\{1,p_1^2p_2\}+A_{012}\{1,p_1p_2^2\}+A_{003}\{1,p_2^3\}$$
$$+\{g_1,p_1\}+\{g_2,p_2\}$$

where $\{a,b\}=ab+ba$ denotes the anticommutator, A_{ijk} are arbitrary real constants and g_1, g_2 are real functions of z, $\bar{z}$. We adopt here the notation used in [**3**].

This symmetric form of the integral X implies the following constraints on the coefficients of (2.1)

$$M = 0,\ C_1 = A_2,\ C_0 = A_1,\ F_0 = 0 \implies F = 0.$$

The first three of these equations mean that the existence condition (2.9) for a second-order integral is satisfied trivially.

The relation between the notations of coefficients in (2.2) and (3.1) has the form

$$\begin{gathered}\Lambda = 12i\hbar^3 A_{300}, \quad A_2 = 4\hbar^3(A_{210} + iA_{201}), \quad A_1 = 2\hbar^3\big[A_{111} + i(A_{102} - A_{120})\big], \\ A_0 = 2\hbar^3\big[A_{012} - A_{030} + i(A_{003} - A_{021})\big], \\ B_1 = -4i\hbar^3(A_{120} + A_{102}), \quad B_0 = -2\hbar^3\big[3(A_{030} + iA_{003}) + A_{012} + iA_{021}\big]\end{gathered} \tag{3.2}$$

where we have used the usual definitions of linear and angular momenta in (3.1)

$$p_1 = -i\hbar\partial_x, \qquad p_2 = -i\hbar\partial_y, \qquad L_3 = xp_2 - yp_1 = -i\hbar(x\partial_y - y\partial_x).$$

The functional coefficients G and $\overline{G}$ of X are related to g_1 and g_2 as follows

$$G = 2\hbar(g_1 + ig_2) + \tfrac{1}{6}\Lambda z + \tfrac{1}{4}A_2, \qquad \overline{G} = 2\hbar(g_1 - ig_2) - \tfrac{1}{6}\Lambda\bar{z} + \tfrac{1}{4}\bar{A}_2. \tag{3.3}$$

Due to relations (3.3)) the symmetric form (3.1) of X obviously implies a constraint for the coefficient H in (2.1), namely

$$H(z,\bar{z}) = \tfrac{1}{2}(G_z + \overline{G}_{\bar{z}}). \tag{3.4}$$

As a consequence the equation (2.5) and its complex conjugate follow from the equation (2.4) differentiated in z, $\bar{z}$ respectively and the equation (2.6) is reduced to (2.10).

4. Simplification of integrals by using the Schrödinger equation

An additional simplification of the form of X could follow from the use of the Schrödinger equation

$$H\psi = E\psi \implies (p_1^2 + p_2^2)\psi = 2(E - V)\psi.$$

Thus, the terms containing $p_1^2 + p_2^2$ are reduced to first-order terms and hence they are included in g_1 and g_2. This implies the following replacements in X of (3.1)

$$\begin{aligned} A_{120}\{L_3, p_1^2\} + A_{102}\{L_3, p_2^2\} &\mapsto \tilde{A}_{120}\{L_3, p_1^2 - p_2^2\} \\ A_{030}\{1, p_1^3\} + A_{012}\{1, p_1p_2^2\} &\mapsto \tilde{A}_{030}\{p_1, p_1^2 - p_2^2\} \\ A_{021}\{1, p_1^2p_2\} + A_{003}\{1, p_2^3\} &\mapsto \tilde{A}_{021}\{p_2, p_1^2 - p_2^2\} \end{aligned}$$

with the result

$$\begin{aligned} X = {} & A_{300}\{L_3^3, 1\} + A_{210}\{L_3^2, p_1\} + A_{201}\{L_3^2, p_2\} + A_{120}\{L_3, p_1^2 - p_2^2\} \\ & + A_{111}\{L_3, p_1p_2\} + A_{030}\{p_1, p_1^2 - p_2^2\} + A_{021}\{p_2, p_1^2 - p_2^2\} \\ & + \{g_1, p_1\} + \{g_2, p_2\} \end{aligned} \tag{4.1}$$

where we skip the tildes. However, the coefficients g_1, g_2 of the integral X in (4.1) will acquire a dependence on energy E. Since we are interested here only in energy independent symmetries, this is not allowed. For simplicity we still restrict ourselves in this paper to the special form (4.1) of X by imposing the constraints $A_{102} = -A_{120}$, $A_{012} = -A_{030}$, $A_{003} = -A_{021}$ in the formula (3.1). We plan to study the general case without these constraints in a future publication.

The relations (3.2) are simplified to

$$A_1 = 2\hbar^3(A_{111} - 2iA_{120}), \quad A_0 = -4\hbar^3(A_{030} + iA_{021}), \quad B_1 = 0, \quad B_0 = \bar{A}_0$$

with the same formulas for Λ and A_2 as in (3.2).

To simplify the form of equations remaining unsolved, we introduce the notation

$$a = A_{210} + iA_{201}, \qquad b = A_{111} + 2iA_{120}, \qquad c = A_{030} + iA_{021}$$

$$f_1(z) = \hbar(-iA_{300}z^3 - az^2 - \bar{b}z + 2c) \tag{4.2}$$

$$f_2(z,\bar{z}) = \hbar(3iA_{300}z^2\bar{z} + 2az\bar{z} - \bar{a}z^2 + \bar{b}\bar{z} + 2\bar{c}) \tag{4.3}$$

which is similar but not identical to the notation in [**3**]. We note the following relations for the future use

$$f_{2,\bar{z}} = -f_1'(z), \qquad \bar{f}_{2,\bar{z}} = -f_{2,z}, \qquad f_{2,\bar{z}\bar{z}} = 0, \qquad f_{2,zz} = \bar{f}_1''(\bar{z})$$

with primes denoting ordinary derivatives; plus the complex conjugate relations.

The remaining independent equations become

$$G_{\bar{z}} = 3f_1(z)V_z + f_2(z,\bar{z})V_{\bar{z}} \tag{4.4}$$

$$\overline{G}_z = 3\bar{f}_1(\bar{z})V_{\bar{z}} + \bar{f}_2(z,\bar{z})V_z \tag{4.5}$$

$$G_z + \overline{G}_{\bar{z}} = 2\big[f_2(z,\bar{z})V_z + \bar{f}_2(z,\bar{z})V_{\bar{z}}\big] \tag{4.6}$$

$$G_{zz} = 2f_2V_{zz} + \bar{f}_2V_{z\bar{z}} - 3\bar{f}_1V_{\bar{z}\bar{z}} + 3f_{2,z}V_z - 5\bar{f}_1'(\bar{z})V_{\bar{z}} \tag{4.7}$$

$$\begin{aligned} 3(f_1V_{zzz} + \bar{f}_1V_{\bar{z}\bar{z}\bar{z}}) - f_2V_{zz\bar{z}} - \bar{f}_2V_{z\bar{z}\bar{z}} + 8(f_1'V_{zz} + \bar{f}_1'V_{\bar{z}\bar{z}}) \\ + 6(f_1''V_z + \bar{f}_1''V_{\bar{z}}) = 0 \end{aligned} \tag{4.8}$$

$$V_zG + V_{\bar{z}}\overline{G} = \hbar^2(f_1V_{zzz} + \bar{f}_1V_{\bar{z}\bar{z}\bar{z}} + 2f_1'V_{zz} + 2\bar{f}_1'V_{\bar{z}\bar{z}} + f_1''V_z + \bar{f}_1''V_{\bar{z}}). \tag{4.9}$$

5. Euclidean group of transformations and classification of third-order integrals

Further simplification of the form (4.1) of a third-order integral and of the third-order linear existence condition (4.8) is achieved by the Euclidean group E(2) of transformations of coordinate systems on the plane

$$\begin{aligned} x &= x'\cos\alpha + y'\sin\alpha + x_0, & p_1 &= p_1'\cos\alpha + p_2'\sin\alpha \\ y &= -x'\sin\alpha + y'\cos\alpha + y_0, & p_2 &= -p_1'\sin\alpha + p_2'\cos\alpha \end{aligned}$$

$$L_3 = L_3' - (x_0\sin\alpha + y_0\cos\alpha)p_1' + (x_0\cos\alpha - y_0\sin\alpha)p_2'.$$

As a result we have to distinguish *four basically different cases* corresponding to the choice of simplest representatives of four different orbits of the group E(2).

CASE 1. $A_{300} = 1$, $A_{210} = A_{201} = A_{111} = 0$.

$$\begin{aligned} X_1 = \{L_3^3, 1\} + A_{120}\{L_3, p_1^2 - p_2^2\} + A_{030}\{p_1, p_1^2 - p_2^2\} + A_{021}\{p_2, p_1^2 - p_2^2\} \\ + \{g_1, p_1\} + \{g_2, p_2\} \end{aligned} \tag{5.1}$$

with the linear existence condition

$$\begin{aligned} 3\big[z^3V_{zzz} - \bar{z}^3V_{\bar{z}\bar{z}\bar{z}} + z^2\bar{z}V_{zz\bar{z}} - z\bar{z}^2V_{z\bar{z}\bar{z}} + 8(z^2V_{zz} - \bar{z}^2V_{\bar{z}\bar{z}}) + 12(zV_z - \bar{z}V_{\bar{z}})\big] \\ + 2A_{120}\big[3\bar{z}V_{\bar{z}\bar{z}\bar{z}} + zV_{z\bar{z}\bar{z}} + 8V_{\bar{z}\bar{z}} - (3zV_{zzz} + \bar{z}V_{zz\bar{z}} + 8V_{zz})\big] \\ + 2i\big[c(3V_{zzz} - V_{z\bar{z}\bar{z}}) + \bar{c}(3V_{\bar{z}\bar{z}\bar{z}} - V_{zz\bar{z}})\big] = 0 \end{aligned} \tag{5.2}$$

where c is defined in (4.2).

CASE 2. $A_{300} = 0$, $A_{210} = 1$, $A_{201} = A_{120} = A_{111} = 0$.

$$X_2 = \{L_3^2, p_1\} + A_{030}\{p_1, p_1^2 - p_2^2\} + A_{021}\{p_2, p_1^2 - p_2^2\} + \{g_1, p_1\} + \{g_2, p_2\} \tag{5.3}$$

with the linear existence condition

$$\begin{aligned}3(z^2V_{zzz} + \bar{z}^2V_{\bar{z}\bar{z}\bar{z}}) + 2z\bar{z}(V_{zz\bar{z}} + V_{z\bar{z}\bar{z}}) - z^2V_{zz\bar{z}} - \bar{z}^2V_{z\bar{z}\bar{z}} + 16(zV_{zz} + \bar{z}V_{\bar{z}\bar{z}})\\ + 12(V_z + V_{\bar{z}}) - 2\big[c(3V_{zzz} - V_{z\bar{z}\bar{z}}) + \bar{c}(3V_{\bar{z}\bar{z}\bar{z}} - V_{zz\bar{z}})\big] = 0\end{aligned} \tag{5.4}$$

with the same definition of c.

CASE 3. $A_{111} = 1$, $A_{300} = A_{210} = A_{201} = A_{120} = A_{030} = A_{021} = 0$.

$$X_3 = \{L_3, p_1p_2\} + \{g_1, p_1\} + \{g_2, p_2\} \tag{5.5}$$

with the linear existence condition

$$3(zV_{zzz} + \bar{z}V_{\bar{z}\bar{z}\bar{z}}) + zV_{z\bar{z}\bar{z}} + \bar{z}V_{\bar{z}zz} + 8(V_{zz} + V_{\bar{z}\bar{z}}) = 0. \tag{5.6}$$

CASE 4. $A_{030} = 1$, $A_{300} = A_{210} = A_{201} = A_{120} = A_{111} = A_{021} = 0$.

$$X_4 = \{p_1, p_1^2 - p_2^2\} + \{g_1, p_1\} + \{g_2, p_2\} \tag{5.7}$$

with the linear existence condition

$$3(V_{zzz} + V_{\bar{z}\bar{z}\bar{z}}) - V_{zz\bar{z}} - V_{z\bar{z}\bar{z}} = 0. \tag{5.8}$$

6. Study of Case 4

The linear existence condition (5.8) can be factorized as

$$(\partial_z + \partial_{\bar{z}})[3(V_{zz} + V_{\bar{z}\bar{z}}) - 4V_{z\bar{z}}] = 0,$$

which is equivalent to

$$3(V_{zz} + V_{\bar{z}\bar{z}}) - 4V_{z\bar{z}} = 10f''(z - \bar{z})$$

where f is an arbitrary function. The substitution $V = W + f(z - \bar{z})$ leads to the equation for W

$$3(W_{zz} + W_{\bar{z}\bar{z}}) - 4W_{z\bar{z}} = 0.$$

In the characteristic variables

$$\xi = \tfrac{1}{2}[(\sqrt{5} - i)z + (\sqrt{5} + i)\bar{z}], \qquad \eta = \tfrac{1}{2}[(\sqrt{5} + i)z + (\sqrt{5} - i)\bar{z}]$$

this equation becomes $W_{\xi\eta} = 0$ with the general solution $W = \alpha(\xi) + \beta(\eta)$. Finally, we obtain the general solution of the equation (5.8)

$$V = f(z - \bar{z}) + \alpha(\xi) + \beta(\eta)$$

where f, α and β are arbitrary functions of one variable. Here $\bar{f} = f$, $\bar{\alpha} = \alpha$, $\bar{\beta} = \beta$, $\bar{\xi} = \xi$ and $\bar{\eta} = \eta$, so that the potential is real $\bar{V} = V$.

Since in Case 4 we have $a = 0$, $b = 0$, $c = 1$, $f_1 = f_2 = 2\hbar$, the equation (4.4) becomes

$$G_{\bar{z}} = 2\hbar(3V_z + V_{\bar{z}}) = 2\hbar[(2\sqrt{5} - i)\alpha'(\xi) + (2\sqrt{5} + i)\beta'(\eta) + 2f'(z - \bar{z})].$$

It is integrated in the form

$$G = 2\hbar[(3 - i\sqrt{5})\alpha(\xi) + (3 + i\sqrt{5})\beta(\eta) - 2f(z - \bar{z}) + 2g(z)].$$

Similarly the equation (4.5) gives the complex conjugate quantity

$$\overline{G} = 2\hbar[(3 + i\sqrt{5})\alpha(\xi) + (3 - i\sqrt{5})\beta(\eta) - 2f(z - \bar{z}) + 2\bar{g}(\bar{z})].$$

Substituting G and $\overline{G}$ into the equation (4.6)

$$G_z + \overline{G}_{\bar{z}} = 4\hbar(V_z + V_{\bar{z}})$$

we obtain after the separation of $z, \bar{z}$

$$g'(z) = -\bar{g}'(\bar{z}) = \lambda = \text{const}$$

so that

$$g(z) = \lambda z + g_0, \qquad \bar{g}(\bar{z}) = -\lambda \bar{z} + \bar{g}_0.$$

We substitute these $g(z)$, $\bar{g}(\bar{z})$ into G, $\overline{G}$ and, in order to simplify all the following, introduce instead of α, β, f new unknowns ρ, σ, F

$$\rho(\xi) = \alpha(\xi) + i\frac{\lambda}{5}\xi + \frac{i}{2\sqrt{5}}(g_0 - \bar{g}_0), \qquad \sigma(\eta) = \beta(\eta) - i\frac{\lambda}{5}\eta - \frac{i}{2\sqrt{5}}(g_0 - \bar{g}_0)$$

$$F(z - \bar{z}) = f(z - \bar{z}) - \frac{\lambda}{5}(z - \bar{z}) - \frac{1}{2}(g_0 + \bar{g}_0), \qquad \overline{F} = F.$$

Then the expressions for G and $\overline{G}$ become

$$G = 2\hbar(g_1 + ig_2) = 2\hbar[(3 - i\sqrt{5})\rho(\xi) + (3 + i\sqrt{5})\sigma(\eta) - 2F(z - \bar{z})]$$
$$\overline{G} = 2\hbar(g_1 - ig_2) = 2\hbar[(3 + i\sqrt{5})\rho(\xi) + (3 - i\sqrt{5})\sigma(\eta) - 2F(z - \bar{z})]$$

and the potential has the form

$$V = \rho(\xi) + \sigma(\eta) + F(z - \bar{z}) + \tfrac{1}{2}(g_0 + \bar{g}_0), \quad \xi = \sqrt{5}x + y, \ \eta = \sqrt{5}x - y. \tag{6.1}$$

The only remaining equation (4.9) with our expressions for $V, G, \overline{G}$ reduces to

$$\begin{aligned} iF'[i(\xi - \eta)][\sigma(\eta) - \rho(\xi)] - F[i(\xi - \eta)][\sigma'(\eta) + \rho'(\xi)] \\ + \sigma'(\eta)[\sigma(\eta) + 2\rho(\xi)] + \rho'(\xi)[2\sigma(\eta) + \rho(\xi)] = \frac{\hbar^2}{4}[\sigma'''(\eta) + \rho'''(\xi)] \end{aligned} \tag{6.2}$$

where the relation $z - \bar{z} = i(\xi - \eta)$ has been used.

We consider (6.2) as an equation for one function F of one variable $\xi - \eta$ and so its coefficients should also be functions of $\xi - \eta$ only. In general, this gives equations for σ and ρ that lead to a contradiction. The only possible exceptions occur in cases when $F = \text{const}$ or $\rho = \sigma = 0$.[1] Consider first the cases when $F = \text{const}$.

SUBCASE 4.1. $\sigma' \cdot \rho' \neq 0$.

Then differentiating (6.2) with respect to ξ and η we obtain the equation $\sigma''(\eta)\rho'(\xi) + \rho''(\xi)\sigma'(\eta) = 0$ which has the general solution $\sigma(\eta) = C_1 e^{\lambda\eta} + C_2$, $\rho(\xi) = C_3 e^{-\lambda\xi} + C_4$ where λ is a separation constant. Substitution of this solution into (6.2) gives $\lambda = 0$ and hence $\sigma' = 0$, $\rho' = 0$, in contradiction to our assumption.

SUBCASE 4.2. $\rho = 0$, $\sigma'(\eta) \neq 0$.

Equation (6.2) becomes $\sigma'(\eta)[\sigma(\eta) - F] = (\hbar^2/4)\sigma'''(\eta)$. It can be easily integrated with the result $\sigma = 3\hbar^2\mathcal{P}(\eta + C) + F$ where $\mathcal{P}$ is the elliptic Weierstrass function with arbitrary invariants g_2, g_3 and $C = \text{const}$. The potential has the form

[1] If one of unknowns ρ or σ is a nonzero constant then it can be annihilated by a redefinition of the other unknown.

$V = 3\hbar^2\mathcal{P}(\eta + C)$ where we skip a constant additive term. The functions G, g_1, g_2 become

$$G = 2\hbar[(3 + i\sqrt{5})3\hbar^2\mathcal{P}(\eta + C) + (1 + i\sqrt{5})F]$$

$$g_1 = 9\hbar^2\mathcal{P}(\eta + C) + F, \qquad g_2 = \sqrt{5}[3\hbar^2\mathcal{P}(\eta + C) + F].$$

The third-order integral (5.7) takes the form

(6.3) $$X_4 = \{p_1, p_1^2 - p_2^2\} + \{9\hbar^2\mathcal{P}(\eta + C) + F, p_1\} + \sqrt{5}\{3\hbar^2\mathcal{P}(\eta + C) + F, p_2\}.$$

SUBCASE 4.3. $\sigma = 0$, $\rho'(\xi) \neq 0$.

In a similar way we obtain the result $\rho = 3\hbar^2\mathcal{P}(\xi + C) + F$, $V = 3\hbar^2\mathcal{P}(\xi + C)$ and the integral is

(6.4) $$X_4 = \{p_1, p_1^2 - p_2^2\} + \{9\hbar^2\mathcal{P}(\xi + C) + F, p_1\} - \sqrt{5}\{3\hbar^2\mathcal{P}(\xi + C) + F, p_2\}.$$

SUBCASE 4.4. $F' \neq 0$, $\rho = 0$.

Then if $\sigma \neq 0$, the equation (6.2) becomes

$$iF' - \frac{\sigma'(\eta)}{\sigma(\eta)}F = \frac{\hbar^2}{4}\frac{\sigma'''(\eta)}{\sigma(\eta)} - \sigma'(\eta).$$

Since F depends only on $\xi - \eta$, we have $\sigma'(\eta)/\sigma(\eta) = \lambda = \text{const}$ and $\sigma(\eta) = C_1 e^{\lambda\eta}$. The condition that the right-hand side should also be constant gives $C_1 = 0$ and hence $\sigma = \rho = 0$, so that the equation (6.2) is identically satisfied for any F. Thus, we obtain the translationally invariant potential

$$V = F(z - \bar{z}) = V(y), \qquad G = \overline{G} = -4\hbar V(y), \qquad g_1 = -2V(y), \qquad g_2 = 0.$$

A third-order integral exists for any such potential

(6.5) $$X_4 = \{p_1, p_1^2 - p_2^2\} - 2\{V(y), p_1\},$$

but it is a linear combination of the Hamiltonian and the cube of the first-order integral P_1. The alternative assumption leads to the same result.

It is important to note that in the all subcases of Case 4 the potential V depends only on one variable hence the Hamiltonian has also a first order integral, the case already studied before in [**3**]. This integral is p_2 in Subcase 4.4 and in Subcases 4.2 and 4.3 it will be p_1, or p_2 after rotations transforming η and ξ into y and x, respectively.

7. A class of superintegrable systems with two third-order integrals

We consider now superintegrable systems possessing two third-order integrals of the types belonging to the Cases 3 and 4. We solve the system of two third-order linear existence conditions (5.6) and (5.8) for the potential V. The general solution is

(7.1) $$V = \frac{c}{(z - \bar{z})^2} + d \cdot (z - \bar{z}) \qquad \text{or} \qquad V = -\frac{c}{4y^2} + dy,$$

where c, d are constants. The solution for G in X_4 (5.7) becomes

$$G = \overline{G} = 2\hbar g_1 = \hbar\left(\frac{c}{y^2} - 4dy + 2g_0\right), \qquad g_2 = 0$$

and the integral of the type 4 takes the form

(7.2) $$X_4 = \{p_1, p_1^2 - p_2^2\} + \left\{\frac{c}{2y^2} - 2dy + g_0, p_1\right\}.$$

Further restrictions on the potential V and the integral X_4 are found in the process of calculation of the second third-order integral X_3 of the type 3. In Case 3 we have

$$A_{300} = A_{210} = A_{201} = A_{120} = A_{030} = A_{021} = 0, \qquad A_{111} = 1$$
$$a = c = 0, \qquad b = 1, \qquad f_1 = -\bar{f}_2 = -\hbar z, \qquad \bar{f}_1 = -f_2 = -\hbar \bar{z}.$$

The equation (4.4) for G now becomes

$$G_{\bar{z}} = \hbar(\bar{z}V_{\bar{z}} - 3zV_z) = \hbar(3z + \bar{z})\left[\frac{2c}{(z-\bar{z})^3} - d\right]$$

together with the complex conjugate equation for $\overline{G}$, from which we obtain

$$G = \hbar\left[\frac{2c(z+\bar{z})}{(z-\bar{z})^2} - \frac{d}{2}(z-\bar{z})^2 - 4dz\bar{z} + g(z)\right], \qquad \bar{c} = c, \qquad \bar{d} = -d \tag{7.3}$$

and $\overline{G}$ is equal to the complex conjugate expression. We substitute these G, $\overline{G}$ into the equation (4.6)

$$G_z + \overline{G}_{\bar{z}} = 2\hbar(\bar{z}V_z + zV_{\bar{z}})$$

with the result

$$g'(z) + 4dz = -[\bar{g}'(\bar{z}) - 4d\bar{z}] = \lambda = \text{const}, \qquad \bar{\lambda} = -\lambda.$$

Substitute the solution $g(z) = -2dz^2 + \lambda z + 2g_0$ and the complex conjugate $\bar{g}(\bar{z})$ in (7.3)

$$G = \hbar\left[2c\frac{(z+\bar{z})}{(z-\bar{z})^2} - \frac{d}{2}(z-\bar{z})^2 - 4dz\bar{z} - 2dz^2 + \lambda z + 2g_0\right]$$

and the complex conjugate expression for $\overline{G}$ with $\bar{c} = c$, $\bar{d} = -d$, $\bar{\lambda} = -\lambda$. We substitute these G and $\overline{G}$ in the last equation (4.9)

$$V_z G + V_{\bar{z}}\overline{G} = -\hbar^3(zV_{zzz} + \bar{z}V_{\bar{z}\bar{z}\bar{z}} + 2V_{zz} + 2V_{\bar{z}\bar{z}}) \tag{7.4}$$

and use our previous result $V = (c/(z-\bar{z})^2) + d\cdot(z-\bar{z})$. Equation (7.4) becomes

$$\left[\frac{2c}{(z-\bar{z})^3} - d\right](\overline{G} - G) = 0$$

and hence $\overline{G} = G$, $d = 0$, $\lambda = 0$, $\bar{g}_0 = g_0$, so that

$$\overline{G} = G = 2\hbar\left[\frac{c(z+\bar{z})}{(z-\bar{z})^2} + g_0\right], \qquad g_1 = \frac{c(z+\bar{z})}{(z-\bar{z})^2} + g_0, \qquad g_2 = 0.$$

The potential takes the final form $V = -c/4y^2$ and so we again end up with the Hamiltonian admitting the first-order integral p_1. Two third-order integrals admitted by this V are

$$X_3 = \{L_3, p_1p_2\} + \left\{-\frac{cx}{2y^2} + g_0, p_1\right\}, \quad X_4 = \{p_1, p_1^2 - p_2^2\} + \left\{\frac{c}{2y^2} + \tilde{g}_0, p_1\right\}.$$

This potential also admits two second-order integrals [**5**].

8. Conclusion

We consider the problem of finding all the potentials in two-dimensional quantum mechanics which possess a third-order integral without an additional restriction of having lower-order integrals, and all third-order integrals. We do not treat this problem here in its full generality, but consider a special form of a third-order integral, obtained under some simplifying constraints. Further simplifications are due to the fact that the integral must have a definite parity. We also simplify by using the Euclidean group of transformations of the plane. As a result we arrive at the classification of such potentials and integrals valid only for the considered special case. We distinguish four different classes of the potentials and corresponding third-order integrals. We give explicitly a general form of the integral in each case and a linear existence condition which is a third-order PDE for the potential V. This is a necessary, but by no means sufficient condition. In the simplest fourth case we solved the linear existence condition for V and the equations for the integral and obtained explicitly the potential and the integral. Unfortunately this case turned out to be trivial, since the potentials in this case depend only on one variable and hence the Hamiltonian has also a first-order integral, the case already studied before in [**3**].

In our analysis we avoided solving very complicated nonlinear existence conditions for V. We hope that this strategy will work in the remaining three cases. The only problem is how to solve linear PDEs with variable polynomial coefficients which are the existence conditions in these cases. We also plan to consider a similar classification of third-order integrals and the corresponding potentials, in the general case without any simplifying assumptions.

References

1. J. Friš, V. Mandrosov, Ya. A. Smorodinsky, M. Uhliř, and P. Winternitz, *On higher symmetries in quantum mechanics*, Phys. Lett. **16** (1965), 354–356.
2. S. Gravel, *Hamiltonians separable in Cartesian coordinates and third-order integrals of motion*, J. Math. Phys. (to appear); math-ph/0302028.
3. S. Gravel and P. Winternitz, *Superintegrable systems with third-order integrals in classical and quantum mechanics*, J. Math. Phys, **43** (2002), no. 12, 5902–5912.
4. A. Makarov, Ya. A. Smorodinsky, Kh. Valiev, and P. Winternitz, *A systematic search for non-relativistic systems with dynamical symmetries*, Nuovo Cimento A **52** (1967), 1061–1084.
5. M. B. Sheftel, P. Tempesta, and P. Winternitz, *Superintegrable systems in quantum mechanics and classical Lie theory*, J. Math. Phys. **42** (2001), no. 2, 659–673.
6. P. Winternitz, Ya. A. Smorodinsky, M. Uhliř, and I. Friš, *Symmetry groups in classical and quantum mechanics*, Jadernaja Fiz. **4** (1967), 625–635 (Russian); English transl., Soviet Nuclear Phys. **4** (1967), 444–450.

Department of Physics, Boğaziçi University, 34342 Bebek, Istanbul, Turkey and Feza Gürsey Institute, P.O. Box 6, Çengelköy, 81220 Istanbul, Turkey

E-mail address: mikhail.sheftel@boun.edu.tr and sheftel@gursey.gov.tr

Centre de Recherches Mathématiques
CRM Proceedings and Lecture Notes
Volume **37**, 2004

Towards a Classification of Cubic Integrals of Motion

Raymond G. McLenaghan, Roman G. Smirnov, and Dennis The

Abstract. We employ the isometry group invariants approach introduced recently to study Killing tensors of valence three defined in the Euclidean plane. The corresponding invariants are found to be homogeneous polynomials of the parameters of the linear space of the Killing tensors. The invariants are used to classify the nontrivial first integrals of motion which are cubic in the momenta of Hamiltonian systems defined in the Euclidean plane. The integrable cases isolated by Holt and Fokas–Lagerström are investigated from this viewpoint.

1. Introduction

The study of Hamiltonian systems with two degrees of freedom admitting integrals of motion which are cubic in the momenta has been for many years an active area of research [**2, 3, 6, 11**]. It is an integral part of the well-established theory of superintegrable systems which originated in the pioneering works by Winternitz et al. [**4, 15**] (see, for example, [**14**]).

The main problem for Hamiltonian systems with two degrees of freedom can be characterized as follows. Let $(M, \mathbf{g})$ be a two-dimensional pseudo-Riemannian manifold of constant curvature. Consider a Hamiltonian system on M defined by a natural Hamiltonian of the form

$$H(\mathbf{q}, \mathbf{p}) = \tfrac{1}{2} g^{ij}(\mathbf{q}) p_i p_j + V(\mathbf{q}), \tag{1.1}$$

via the canonical Poisson bi-vector $\mathbf{P}_0 = (\partial/\partial q^i) \wedge (\partial/\partial p_i)$, $i = 1, 2$, where g^{ij} are the components of the metric tensor $\mathbf{g}$ and $(\mathbf{q}, \mathbf{p}) \in T^*M$ are the standard position-momenta coordinates. We wish to describe the Hamiltonian system defined by (1.1), or, more specifically, by its potential function V, whose complete integrability is assured by the existence of an additional first integral of the form:

$$F(\mathbf{q}, \mathbf{p}) = L^{ijk}(\mathbf{q}) p_i p_j p_k + B^k(\mathbf{q}) p_k, \quad i, j, k = 1, 2, \tag{1.2}$$

where L^{ijk} denote the components of a Killing tensor $\mathbf{L}$ of valence three defined on $(M, \mathbf{g})$ (see below). The problem appears to be very difficult, since there are three quantities involved, namely the Killing tensor $\mathbf{L}$, vector field $\mathbf{B}$ and potential

---

2000 *Mathematics Subject Classification.* Primary 70H06, 53B50; Secondary 53A55, 35F05.

The first author was supported in part by NSERC.
The second author was supported in part by the Alexander von Humboldt Foundation.
The third author was supported in part by NSERC.

V related via the partial differential equations stemming from the vanishing of the Poisson bracket $\{H, F\} = \mathbf{P}_0 \, \mathrm{d}H \, \mathrm{d}F = 0$, all of which can be varied. Most of the known Hamiltonian systems with the designated property have been isolated by fixing the form of the potential V in (1.1) and then trying to solve the system of equations determined by $\{H, F\} = 0$ under this condition (see, for instance, [**3, 11**]).

In this paper we consider the problem from a different angle. More specifically, we begin by classifying all admissible, nontrivial Killing tensors defining the leading term in (1.2). This can be done by making use of the new method developed by the authors in [**7–10**] which is based on group invariants of Killing tensors defined on pseudo-Riemannian spaces of constant curvature under the action of the isometry group. The group of isometries of M is the most natural choice for such a classification, since the action of the group preserves the dynamics of the system defined by (1.1)–(1.2). In this paper we are mainly concerned with the case $M = \mathbb{R}^2$, where $\mathbb{R}^2$ denotes the Euclidean plane. Accordingly, in what follows we find the isometry group invariants of Killing tensors of valence three defined in $\mathbb{R}^2$ and split the space of Killing tensors into invariant subspaces whose respective Killing tensors are characterized by the invariants of the isometry group. This procedure entails a classification of first integrals of motion which are cubic in the momenta of Hamiltonian systems defined by (1.1) in $\mathbb{R}^2$.

2. Cubic integrals of motion

Consider a Hamiltonian system defined by (1.1). Assume it admits an additional first integral of the form

$$G(\mathbf{q}, \mathbf{p}) = L^{ijk}(\mathbf{q}) p_i p_j p_k + K^{ij} p_i p_j(\mathbf{q}) + B^{\ell}(\mathbf{q}) p_\ell + U(\mathbf{q}), \tag{2.1}$$

where $i, j, k = 1, 2$. The vanishing of the Poisson bracket $\{H, G\} = 0$ decouples into the following conditions expressed in component and coordinate-free forms respectively:

$$L^{(ijk}_{,f} g^{\ell)f} - \tfrac{3}{2} L^{f(ij} g^{k\ell)}_{,f} = 0 \iff [\mathbf{L}, \mathbf{g}] = 0, \tag{2.2}$$

$$K^{(ij}_{,f} g^{k)f} - K^{f(i} g^{jk)}_{,f} = 0 \iff [\mathbf{K}, \mathbf{g}] = 0, \tag{2.3}$$

$$B^{(i}_{,f} g^{j)f} - \tfrac{1}{2} B^f g^{ij}_{,f} - 3L^{fij} V_{,f} = 0 \iff [\mathbf{B}, \mathbf{g}] = 3\mathbf{L} \, \mathrm{d}V, \tag{2.4}$$

$$U_{,f} g^{fi} - 2K^{fi} V_{,f} = 0 \, (U_{,i} = 2K^j{}_i V_{,j}) \iff \mathrm{d}U \mathbf{g} = 2\mathbf{K} \, \mathrm{d}V, \tag{2.5}$$

$$B^f V_{,f} = 0 \iff \mathbf{B}(V) = 0, \tag{2.6}$$

where $_{,f}$, $(\ ,\ )$ and $[\ ,\ ]$ denote partial differentiation, symmetrization and the Schouten bracket respectively. Note $[\mathbf{B}, \mathbf{g}] = \mathcal{L}_{\mathbf{B}} \mathbf{g}$, where $\mathcal{L}$ denotes the Lie derivative operator. It follows immediately from (2.2) and (2.3) that $\mathbf{L}$ and $\mathbf{K}$ are Killing tensors of valence three and two respectively, while $\mathbf{B}$ is a vector field, since it satisfies the tensor equation (2.4). Equation (2.6) reveals that the potential function V is preserved by the vector field $\mathbf{B}$. We observe next that equations (2.2)–(2.6) separate into two groups, namely (2.2), (2.4), (2.6) and (2.3), (2.5), involving only components of G which are odd and even in the momenta respectively. This fact implies that the first integral can be written as $G = G_{odd} + G_{even}$, where

$$\begin{aligned} G_{\mathrm{odd}}(\mathbf{q}, \mathbf{p}) &= L^{ijk}(\mathbf{q}) p_i p_j p_k + B^k(\mathbf{q}) p_k, \\ G_{\mathrm{even}}(\mathbf{q}, \mathbf{p}) &= K^{ij}(\mathbf{q}) p_i p_j + U(\mathbf{q}) \end{aligned} \tag{2.7}$$

and furthermore that $\{H, G_{\text{odd}}\} = \{H, G_{\text{even}}\} = 0$.

DEFINITION 2.1. We say that a Hamiltonian system defined by (1.1) is *trivial* if it admits a first integral $G = X^i p_i$ linear in the momenta, where $\mathbf{X}$ is a Killing vector, satisfying $\mathbf{X}(V) = 0$. Accordingly, we say that a Killing tensor $\mathbf{L}$ of valence three is *trivial* if it is a symmetric product of the metric $\mathbf{g}$ and a Killing vector $\mathbf{X}$, i.e., $\mathbf{L} = \mathbf{g} \odot \mathbf{X}$.

These arguments confirm that (1.2) is the most general form of a first integral cubic in the momenta and a Hamiltonian system (1.1) admitting a first integral of the form (2.1) is in fact *superintegrable*, provided it is not trivial, for in this case one can combine the metric $\mathbf{g}$ with $\mathbf{X}$ and square of $\mathbf{X}$ to obtain cubic and quadratic terms in the momenta respectively in the representation (2.1). Note the decomposition (2.7) was proven in [**11**] by employing a time reflection symmetry assumption that does not seem to be required. We conclude therefore that for a Hamiltonian system (1.1) admitting a cubic first integral (1.2) determined by its Killing tensor $\mathbf{L}$ and vector field $\mathbf{B}$, the following tensorial equations are essential:

$$[\mathbf{L}, \mathbf{g}] = 0, \qquad \text{(Killing tensor equation for } \mathbf{L}) \tag{2.8}$$

$$[\mathbf{B}, \mathbf{g}] = 3\mathbf{L}\, dV, \qquad \text{(tensor equation for } \mathbf{B}) \tag{2.9}$$

$$\mathbf{B}(V) = 0. \qquad \text{(compatibility condition for } V) \tag{2.10}$$

Let us fix $\mathbf{L}$, $\mathbf{B}$ and V in (1.2) and (1.1) respectively. Since $(M, \mathbf{g})$ is of constant curvature, the dimension of the corresponding isometry group $I(M)$ is three. Therefore the space $\mathcal{K}^1(M)$ of Killing vectors (infinitesimal isometries) of M is spanned by three Killing vectors $\mathbf{X}_1$, $\mathbf{X}_2$ and $\mathbf{X}_3$. Clearly, in view of linearity of the Schouten bracket, the vector field $\widetilde{\mathbf{B}}$ given by

$$\widetilde{\mathbf{B}} = \mathbf{B} + \sum_{i=1}^{3} k_i \mathbf{X}_i, \tag{2.11}$$

is also a solution to the tensor equation (2.9), where $k_1, k_2, k_3 \in \mathbb{R}$ are arbitrary constants. Therefore as far as the equation (2.9) is concerned, each $\mathbf{B}$ generates a space $S_{\mathbf{B}} \subset TM$ of vector fields satisfying (2.9), $S_{\mathbf{B}} = \{\mathbf{B}\} + \mathcal{K}^1(M)$. This is not true however for the equation (2.10).

We also note that the right-hand side of (2.9) measures the deviation of the vector field $\mathbf{B}$ from a Killing vector.

REMARK 2.1. The space of vector fields $S_{\mathbf{B}}$ is not closed under the Lie commutator. Indeed, let $\mathbf{B}_1, \mathbf{B}_2 \in S_{\mathbf{B}}$. Hence, in view of (2.11) $\mathbf{B}_1 = \mathbf{B} + \mathbf{Y}_1$ and $\mathbf{B}_2 = \mathbf{B} + \mathbf{Y}_2$, where $\mathbf{Y}_1, \mathbf{Y}_2 \in \mathcal{K}^1(M)$. Taking into account (2.9) and (2.10) and using the standard properties of Lie derivatives, we get $\big[[\mathbf{B}_1, \mathbf{B}_2], \mathbf{g}\big] = 3\tilde{\mathbf{L}}\, dV + 3\mathbf{L}\, d\widetilde{V}$, where $\tilde{\mathbf{L}} = \mathcal{L}_{\mathbf{Y}_1 - \mathbf{Y}_2}\mathbf{L}$ and $\widetilde{V} = (\mathbf{Y}_1 - \mathbf{Y}_2)(V)$.

REMARK 2.2. Lie derivative deformations of $\mathbf{B}$ with respect to Killing vectors do not preserve the form of $\mathbf{B}$. Indeed, using the same arguments as in Remark 2.1, it is easy to see, that $\big[[\mathbf{B}, \mathbf{X}], \mathbf{g}\big] = 3\tilde{\mathbf{L}}\, dV + 3\mathbf{L}\, d\widetilde{V}$, where $\mathbf{X} \in \mathcal{K}^1(M)$ and $\tilde{\mathbf{L}} = \mathcal{L}_{\mathbf{X}}\mathbf{L}$, $\widetilde{V} = \mathbf{X}(V)$. Note the right hand side of the last expession is different from that of (2.9).

REMARK 2.3. Let $\mathbf{B} \in \mathcal{K}^1(M)$. This immediately entails that the corresponding Hamiltonian system is trivial and superintegrable, enjoying a second additional first integral of the form $G = L^{ijk}p_ip_jp_j$. The right-hand side of (2.9) in this case has changed too.

In what follows, we find and use isometry group invariants of Killing tensors of valence three defined in the Euclidean plane $\mathbb{R}^2$ to classify cubic integrals of motion (1.2) up to the leading term represented by a Killing tensor of valence three for the Hamiltonian systems with two degrees of freedom defined in $\mathbb{R}^2$.

3. Isometry group invariants of Killing tensors of valence three

Let $(M, \mathbf{g})$ be the Euclidean plane $\mathbb{R}^2$. In this section we apply the method developed recently by the authors in [**7–10**] to the problem of classification of Killing tensors belonging to the linear space $\mathcal{K}^3(\mathbb{R}^2)$.

Consider the linear space $\mathcal{K}^3(\mathbb{R}^2)$. The dimension of the space can be computed by employing the Takeuchi–Thompson formula [**12, 13**]:

$$\dim \mathcal{K}^3(\mathbb{R}^2) = \frac{1}{2}\binom{2+3}{3+1}\binom{2+3-1}{3} = 10.$$

Alternatively, we can solve the Killing tensor equation $[\mathbf{L}, \mathbf{g}] = 0$, $\mathbf{L} \in \mathcal{K}^3(\mathbb{R}^2)$ for the metric tensor of $\mathbb{R}^2$ in (say) Cartesian coordinates (x, y), to get

$$\begin{aligned} L^{111} &= a_1 + 3\alpha_1 y + 3\beta_1 y^2 + \gamma y^3, \\ L^{112} &= a_2 + \alpha_2 y - \alpha_1 x - 2\beta_1 xy + \beta_2 y^2 - \gamma x y^2, \\ L^{122} &= a_3 - \alpha_2 x - \alpha_3 y - 2\beta_2 xy + \beta_1 x^2 + \gamma y x^2, \\ L^{222} &= a_4 + 3\alpha_3 x + 3\beta_2 x^2 - \gamma x^3. \end{aligned} \tag{3.1}$$

The ten parameters $a_1, \ldots, \gamma$ are the constants of integration representing also the dimension of $\mathcal{K}^3(\mathbb{R}^2)$. Accordingly, we can view each element of $\mathcal{K}^3(\mathbb{R}^2)$ as an algebraic object determined by its ten parameters $a_1, \ldots, \gamma$. Furthermore, for the linear space $\mathcal{K}^3(\mathbb{R}^2)$, we can formulate an analogue of the main problem of the classical theory of algebraic invariants (see [**5, 10**] for more details). Indeed, recall that the main problem of the classical theory of algebraic invariants developed by Hilbert [**5**] in the modern mathematical language reads:

PROBLEM 3.1. Determine the linear action of a group G on a K-vector space V. Then in the ring of polynomial functions $K[V]$ describe the subring $K[V]^G$ of all polynomial functions on V which are invariant under the action of G (i.e., the *invariants*).

In the present study, we formulate its analogue as follows:

PROBLEM 3.2. Let Σ be the space spanned by the ten parameters $a_1, \ldots, \gamma$ that appear in (3.1). Determine the action induced by the isometry group $I(\mathbb{R}^2)$ on Σ. Then in the space of functions defined in Σ describe the subspace of all functions in Σ which are invariant under the action induced by the isometry group.

Note that the linear spaces Σ and $\mathcal{K}^3(\mathbb{R}^2)$ are isomorphic. The solution of Problem 3.2 consists of two essential parts: first determining the action of $I(\mathbb{R}^2)$ in Σ and second finding the invariants. Unlike in the classical theory of algebraic invariants [**5**], where the group acts naturally in a space of polynomials by linear substitutions, in the present situation determining the action of the isometry group is more complex. To make the first step, we proceed as in the analogous problems considered by the authors in [**7, 9, 10**] for the linear spaces $\mathcal{K}^2(\mathbb{R}^2)$ and $\mathcal{K}^2(\mathbb{R}^2_1)$, where $\mathbb{R}^2_1$ denotes the Minkowski plane. Observe that the generators of the Lie algebra $i(\mathbb{R}^2)$ of the isometry group $I(\mathbb{R}^2)$ with respect to Cartesian coordinates (x, y) are the vector fields $\mathbf{X} = \partial/\partial x$, $\mathbf{Y} = \partial/\partial y$ and $\mathbf{R} = y\partial/\partial x - x\partial\partial y$ whose flows represent translations along x, y and a rotation respectively. They enjoy the following commutator relations:

$$[\mathbf{X}, \mathbf{Y}] = 0, \qquad [\mathbf{X}, \mathbf{R}] = -\mathbf{Y}, \qquad [\mathbf{Y}, \mathbf{R}] = \mathbf{X}. \tag{3.2}$$

Now we introduce the *projection map* $\pi\colon \mathcal{K}^3(\mathbb{R}^2) \to \mathcal{X}\Sigma$ defined for a fixed $\mathbf{L}^0 \in \mathcal{K}^3(\mathbb{R}^2)$ as follows:

$$\pi(\mathbf{L}^0) = \sum_{i=1}^{4} a_i^0 \frac{\partial}{\partial a_i} + \sum_{i=1}^{3} \alpha_i^0 \frac{\partial}{\partial \alpha_i} + \sum_{i=1}^{2} \beta_i^0 \frac{\partial}{\partial \beta_i} + \gamma^0 \frac{\partial}{\partial \gamma}, \tag{3.3}$$

where $a_1^0, \ldots, \gamma^0$ are the ten parameters that determine $\mathbf{L}^0$ via the representation (3.1) with respect to Cartesian coordinates (x, y) and $a_1, \ldots \gamma$ are the ten coordinate functions of the space Σ. In order to determine the induced action of $I(\mathbb{R}^2)$ in Σ, we use the composition map $\pi \circ \mathcal{L}$ of π and the Lie derivative operator $\mathcal{L}$. Define the following vector fields in $\mathcal{X}\Sigma$:

$$\mathbf{V}_1 := \pi\mathcal{L}_{\mathbf{X}}\mathbf{L}, \qquad \mathbf{V}_2 := \pi\mathcal{L}_{\mathbf{Y}}\mathbf{L}, \qquad \mathbf{V}_3 := \pi\mathcal{L}_{\mathbf{R}}\mathbf{L}, \tag{3.4}$$

where $\mathbf{L}$ is the general Killing tensor, whose components with respect to Cartesian coordinates (x, y) are exposed in (3.1). Note that the Lie derivative deformations of $\mathbf{L}$ with respect to Killing vectors are themselves Killing tensors of valence three and as such represent elements of the linear space $\mathcal{K}^3(\mathbb{R}^2)$. This fact is a consequence of the Jacobi identity for the Schouten bracket acting in the space of symmetric contravariant tensorial quantities. We conclude therefore that the formulas (3.4) are well-defined. Employing the formulas (3.4) in conjunction with (3.1), we arrive at the following representations of $\mathbf{V}_1$, $\mathbf{V}_2$, $\mathbf{V}_3 \in \mathcal{X}\Sigma$ with respect to the coordinate functions $a_1, \ldots, \gamma$:

$$\begin{aligned}
\mathbf{V}_1 &= -\alpha_1 \frac{\partial}{\partial a_2} - \alpha_2 \frac{\partial}{\partial a_3} + 3\alpha_3 \frac{\partial}{\partial a_4} - 2\beta_1 \frac{\partial}{\partial \alpha_2} + 2\beta_2 \frac{\partial}{\partial \alpha_3} - \gamma \frac{\partial}{\partial \beta_2} \\
\mathbf{V}_2 &= 3\alpha_1 \frac{\partial}{\partial a_1} + \alpha_2 \frac{\partial}{\partial a_2} - \alpha_3 \frac{\partial}{\partial a_3} + 2\beta_1 \frac{\partial}{\partial \alpha_1} + 2\beta_2 \frac{\partial}{\partial \alpha_2} + \gamma \frac{\partial}{\partial \beta_1} \\
\mathbf{V}_3 &= 3a_2 \frac{\partial}{\partial a_1} + (2a_3 - a_1) \frac{\partial}{\partial a_2} + (a_4 - 2a_2) \frac{\partial}{\partial a_3} - 3a_3 \frac{\partial}{\partial a_4} \\
&\quad + \alpha_2 \frac{\partial}{\partial \alpha_1} - 2(\alpha_1 + \alpha_3) \frac{\partial}{\partial \alpha_2} + \alpha_2 \frac{\partial}{\partial \alpha_3} + \beta_2 \frac{\partial}{\partial \beta_1} - \beta_1 \frac{\partial}{\partial \beta_2}
\end{aligned} \tag{3.5}$$

Our next observation is that $\mathbf{V}_1$, $\mathbf{V}_2$, $\mathbf{V}_3$ satisfy the following commutator relations:

$$[\mathbf{V}_1, \mathbf{V}_2] = 0, \qquad [\mathbf{V}_1, \mathbf{V}_3] = -\mathbf{V}_2, \qquad [\mathbf{V}_2, \mathbf{V}_3] = \mathbf{V}_1. \tag{3.6}$$

Therefore $\mathbf{V}_1$, $\mathbf{V}_2$, $\mathbf{V}_3$ form a basis for a Lie algebra $i_{\mathcal{K}^3(\mathbb{R}^2)}(\Sigma)$, which is a Lie subalgebra of $\mathcal{X}\Sigma$. Furthermore, its generators $\mathbf{V}_1$, $\mathbf{V}_2$ and $\mathbf{V}_3$ satisfy the same commutator relations as the generators $\mathbf{X}$, $\mathbf{Y}$ and $\mathbf{R}$ of $i(\mathbb{R}^2) = \mathcal{K}^1(\mathbb{R}^2)$ in $T\mathbb{R}^2$ (compare (3.2) with (3.5)). Therefore we have proven the following

PROPOSITION 3.1. *The vector space $i_{\mathcal{K}^3(\mathbb{R}^2)}(\Sigma) \subset \mathcal{X}\Sigma$ is a Lie subalgebra of $T\Sigma$ isomorphic to the Lie algebra of Killing vectors $i(\mathbb{R}^2) = \mathcal{K}^1(\mathbb{R}^2)$.*

Proposition 3.1 establishes the induced action of $I(\mathbb{R}^2)$ in $\Sigma \simeq \mathcal{K}^3(\mathbb{R}^2)$. Therefore we have solved the first part of Problem 3.2.

Recall that invariance of an object under entire Lie group is equivalent to infinitesimal invariance under infinitesimal generators of the corresponding Lie algebra. This observation and the result of Proposition 3.1 prompts the following

DEFINITION 3.1. A smooth function $F\colon \Sigma \to \mathbb{R}$ is said to be an $I(\mathbb{R}^2)$-*invariant of* $\mathcal{K}^3(\mathbb{R}^2)$ iff

$$\mathbf{V}_i(F) = 0, \quad i = 1, 2, 3, \tag{3.7}$$

where $\mathbf{V}_i$, $i = 1, 2, 3$ are the generators of the Lie algebra $i_{\mathcal{K}^3(\mathbb{R}^2)}(\Sigma)$ isomorphic to the Lie algebra $i(\mathbb{R}^2) = \mathcal{K}^1(\mathbb{R}^2)$ of Killing vectors defined on $\mathbb{R}^2$.

Solving the second part of Problem 3.2 comes down to solving the system of partial differential equations (3.7) determined by the generators of $i_{\mathcal{K}^3(\mathbb{R}^2)}(\Sigma)$. A simple counting argument reveals that there are $7 = 10$ (dimension of Σ) $-$ 3 (dimension of the Lie group $I(\mathbb{R}^2)$) essential invariants that determine the space of all $I(\mathbb{R}^2)$-invariants of $\mathcal{K}^3(\mathbb{R}^2)$. In what follows we may not need them though, since we wish to consider Problem 3.2 in conjunction with the dynamical problem envisaged in Introduction. More specifically, we will exclude from further consideration all Killing tensors that are trivial, according to Definition 2.1. Ultimately, this will lead to a different (smaller) set of essential $I(\mathbb{R}^2)$-invariants of the linear space $\mathcal{K}^3(\mathbb{R}^2)$.

4. Essential integral submanifolds of the space Σ

In this section we carry over the ideas exhibited in the previous section to the classification problem of the linear space $\mathcal{K}^3(\mathbb{R}^2)$, that is find and use $I(\mathbb{R}^2)$-invariants of $\mathcal{K}^3(\mathbb{R}^2)$ to classify Killing tensors of valence three and thus, the corresponding first integrals of motions of (1.1) up to their leading terms.

First, we wish to exclude from further consideration the Killing tensors corresponding to the trivial first integrals (see Definition 2.1). In Cartesian coordinates the most general Killing tensor $\mathbf{L}_{\rm tr} \in \mathcal{K}^3(\mathbb{R}^2)$ of this type can be obtained via multiplication of the metric $\mathbf{g}$ of $\mathbb{R}^2$ by the most general Killing vector $\mathbf{X}_{\rm gen}$ of $\mathbb{R}^2$ spanned by $\mathbf{X}$, $\mathbf{Y}$ and $\mathbf{R}$:

$$\mathbf{X}_{\rm gen} = (a + cy)\frac{\partial}{\partial x} + (b - cx)\frac{\partial}{\partial y}, \qquad a, b, c \in \mathbb{R}.$$

Thus, the components of $\mathbf{L}_{\rm tr}$ are given by $L_{\rm tr}^{ijk} = g^{(ij}X_{\rm gen}^{k)}$. Comparing the last formula with (3.1) and taking into account $g^{12} = 0$, $g^{11} = g^{22} = 1$, we come to the conclusion that in Σ the subspace $\mathcal{T} \subset \Sigma$ corresponding to the space of trivial Killing tensors in $\mathcal{K}^3(\mathbb{R}^2)$ is determined by the following conditions:

$$a_1 - 3a_3 = 0, \quad 3a_2 - a_4 = 0, \quad \alpha_1 + \alpha_3 = 0, \quad \alpha_2 = 0, \quad \beta_1 = \beta_2 = 0, \quad \gamma = 0. \tag{4.1}$$

It is easy to see that the space $\mathcal{T} \subset \Sigma$ defined by (4.1) is an integral submanifold in Σ whose tangent space $T\mathcal{T}$ is spanned by the vector fields

$$\mathbf{T}_1 = 3\frac{\partial}{\partial a_1} + \frac{\partial}{\partial a_3}, \qquad \mathbf{T}_2 = \frac{\partial}{\partial a_2} + 3\frac{\partial}{\partial a_4}, \qquad \mathbf{T}_3 = \frac{\partial}{\partial \alpha_1} - \frac{\partial}{\partial \alpha_3}. \tag{4.2}$$

Indeed, the system $\{\mathbf{T}_1, \mathbf{T}_2, \mathbf{T}_3\}$ is in involution ($\{\mathbf{T}_1, \mathbf{T}_2, \mathbf{T}_3\}$ is a basis of an abelian Lie algebra in $T\Sigma$), and as such, in view of Frobenius' theorem, is integrable. Interestingly, the system

$$\{\mathbf{V}_1, \mathbf{V}_2, \mathbf{V}_3, \mathbf{T}_1, \mathbf{T}_2, \mathbf{T}_3\} \tag{4.3}$$

is a basis of a six-dimensional Lie algebra in $T\Sigma$ with the following commutator table:

	$\mathbf{V}_1$	$\mathbf{V}_2$	$\mathbf{V}_3$	$\mathbf{T}_1$	$\mathbf{T}_2$	$\mathbf{T}_3$
$\mathbf{V}_1$	0	0	$-\mathbf{V}_2$	0	0	$\mathbf{T}_2$
$\mathbf{V}_2$	0	0	$\mathbf{V}_1$	0	0	$-\mathbf{T}_1$
$\mathbf{V}_3$	$\mathbf{V}_2$	$-\mathbf{V}_1$	0	$\mathbf{T}_2$	$-\mathbf{T}_1$	0
$\mathbf{T}_1$	0	0	$-\mathbf{T}_2$	0	0	0
$\mathbf{T}_2$	0	0	$\mathbf{T}_1$	0	0	0
$\mathbf{T}_3$	$-\mathbf{T}_2$	$\mathbf{T}_1$	0	0	0	0

Therefore the system of vector fields (4.3) is also integrable. According to Definition 3.1, in order to find essential $I(\mathbb{R}^2)$-invariants of $\mathcal{K}^3(\mathbb{R}^2)$ of *nontrivial* Killing tensors, we have to solve the following system of PDE's:

$$\mathbf{T}_i(F) = 0, \quad \mathbf{V}_i(F) = 0, \qquad i = 1, 2, 3 \tag{4.4}$$

for a smooth function $F\colon \Sigma \to \mathbb{R}$. The equations $F = c$, where c is a constant, will determine integral submanifolds of the integrable distribution of vector fields (4.3).

Although in principle the system (4.4) can be solved by the method of characteristics, the actual execution of the procedure is an arduous computational task. To alleviate the difficulties, we solve (4.4) in two steps. More specifically, let us first find all essential $I(\mathbb{R}^2)$-invariants of $\mathcal{K}^3(\mathbb{R}^2)$ for a subsystem of PDE's, namely the following.

$$\mathbf{T}_i(F) = 0, \quad \mathbf{V}_j(F) = 0, \qquad i = 1, 2, 3 \quad j = 1, 2. \tag{4.5}$$

Employing the method of characteristics, we find the following set of five essential $I(\mathbb{R}^2)$-invariants of $\mathcal{K}^3(\mathbb{R}^2)$ for the system (4.5) as functions of the ten parameters $a_1, \ldots, \gamma$:

$$\begin{aligned}
\tilde{I}_1 &= \gamma\alpha_2 - 2\beta_1\beta_2,\\
\tilde{I}_2 &= \gamma(\alpha_1 + \alpha_3) + \beta_2^2 - \beta_1^2,\\
\tilde{I}_3 &= \gamma^2(a_1 - 3a_3) + 3\gamma(\beta_2\alpha_2 - \beta_1(\alpha_1 + \alpha_3)) + 2\beta_1(\beta_1^2 - 3\beta_2^2),\\
\tilde{I}_4 &= \gamma^2(3a_2 - a_4) - 3\gamma(\beta_1\alpha_2 + \beta_2(\alpha_1 + \alpha_3)) + 2\beta_2(3\beta_1^2 - \beta_2^2),\\
\tilde{I}_5 &= \gamma.
\end{aligned} \tag{4.6}$$

Note that all of the invariants (4.6) are *homogeneous polynomials*. Therefore we conclude that the most general $I(\mathbb{R}^2)$-invariant $\widetilde{F}\colon \Sigma \to \mathbb{R}$ satisfying (4.5) is given by

$$\widetilde{F} = \widetilde{F}(\tilde{I}_1, \tilde{I}_2, \tilde{I}_3, \tilde{I}_4, \tilde{I}_5). \tag{4.7}$$

Now let us treat the functions $\tilde{I}_i$, $i = 1, \ldots, 5$ as new coordinates and apply the last vector field $\mathbf{V}_3$ to $\widetilde{F}$. This leads to the first order linear partial differential equation

$$0 = -2\tilde{I}_2\frac{\partial \widetilde{F}}{\partial \tilde{I}_1} + 2\tilde{I}_1\frac{\partial \widetilde{F}}{\partial \tilde{I}_2} + 3\tilde{I}_4\frac{\partial \widetilde{F}}{\partial \tilde{I}_3} - 3\tilde{I}_3\frac{\partial \widetilde{F}}{\partial \tilde{I}_4} \tag{4.8}$$

and four essential $I(\mathbb{R}^2)$-invariants of $\mathcal{K}^3(\mathbb{R}^2)$ as functions of $\tilde{I}_i$, $i = 1, \ldots, 5$ that are found to be:

$$
\begin{aligned}
I_1 &= 2(\tilde{I}_1^2 - 3\tilde{I}_2^2)\tilde{I}_1\tilde{I}_3\tilde{I}_4 + (3\tilde{I}_1^2 - \tilde{I}_2^2)\tilde{I}_2(\tilde{I}_3^2 - \tilde{I}_4^2), \\
I_2 &= \tilde{I}_1^2 + \tilde{I}_2^2, \\
I_3 &= \tilde{I}_3^2 + \tilde{I}_4^2, \\
I_4 &= \tilde{I}_5.
\end{aligned} \tag{4.9}
$$

This result puts in evidence that the function

$$\widetilde{F} = \widetilde{F}(I_1, I_2, I_3, I_4) \tag{4.10}$$

is the solution to Problem 3.2 for the *nontrivial* elements of $\mathcal{K}^3(\mathbb{R}^2)$, that is the Killing tensors that determine nontrivial, according to Definition 2.1, first integrals of (1.1) cubic in the momenta. Now we can use the *essential $I(\mathbb{R}^2)$-invariants of $\mathcal{K}^3(\mathbb{R}^2)$* I_i, $i = 1, \ldots, 4$ (4.9) to classify nontrivial Killing tensors of valence three up to the action of the isometry group $I(\mathbb{R}^2)$ according to whether I_i, $i = 1, \ldots, 4$ are equal to zero or not. Therefore there are at most $2^4 = 16$ distinct classes of nontrivial Killing tensors characterized by I_i, $i = 1, \ldots, 4$ (4.9). The elements of each class are equivalent up to the action of the isometry group $I(\mathbb{R}^2)$. We note however that the Killing tensors of valence three of the first integrals isolated by Holt [**6**] and Fokas–Lagerström [**3**] are *nontrivial*, yet the essential $I(\mathbb{R}^2)$-invariants I_i, $i = 1, \ldots, 4$ all vanish identically (see the next section). This happens due to the fact that for the Killing tensors of both Holt and Fokas-Lagerström cubic in the momenta first integrals of motion the parameters γ, β_1 and β_2 vanish identically, which, in turn, kills all of the essential $I(\mathbb{R}^2)$-invariants I_i, $i = 1, \ldots, 4$ given by (4.9). As might be expected, we need extra invariant(s) to distinguish the cases when $\gamma = \beta_1 = \beta_2 = 0$. Indeed, observe that γ is an essential invariant of $\mathcal{K}^3(\mathbb{R}^2)$ for nontrivial Killing tensors of valence three. Therefore we can consider an integral hypersurface of the space $\Sigma\backslash\mathcal{T}$ determined by the condition $\gamma = 0$. Using the same arguments as above, we find that the homogeneous polynomial $\beta_1^2 + \beta_2^2$ is an $I(\mathbb{R}^2)$-invariant of $\mathcal{K}^3(\mathbb{R}^2)$ in this case. Therefore we can consider the integral submanifold $\Sigma_1\backslash\mathcal{T}$ of $\Sigma\backslash\mathcal{T}$ determined by the conditions: $\gamma = 0$, $\beta_1^2 + \beta_2^2 = 0$. It is easy to see by using the same counting argument as that given in the end of Section 3 that there exists only *one* essential $I(\mathbb{R}^2)$-invariant of $\mathcal{K}^3(\mathbb{R}^2)$ in the subspace $\Sigma_1\backslash\mathcal{T}$. Employing the procedure described above, we find it to be the following homogeneous polynomial of the parameters α_1, α_2 and α_3:

$$I_* = (\alpha_1 + \alpha_3)^2 + \alpha_2^2. \tag{4.11}$$

Now we can classify Killing tensors of valence three in the space $\Sigma_1\backslash\mathcal{T}$ up to the action of the isometry group $I(\mathbb{R}^2)$ according to whether I_* is zero or not. In view of homogeneity of the polynomial I_* the Killing tensors of valence three determined by the condition $I_* \neq 0$ are equivalent up to rescaling $\tilde{x} = kx, \tilde{y} = ky$, $k \in \mathbb{R}$ (see (3.1)) that preserves the signature of the metric g. Note also that there exist no other essential $I(\mathbb{R}^2)$-invariants of $\mathcal{K}^3$ involving α_1, α_2 and α_3. Thus, there are only two equivalence classes of Killing tensors in $\Sigma_1\backslash\mathcal{T}$. Let $\Sigma_2 = \Sigma\backslash\Sigma_1$. We summarize our observations in Table 1.

In the next section we apply our classification scheme to study the cubic in the momenta first integrals of motion isolated by Holt [**6**] and Fokas–Lagerström [**3**].

TABLE 1. Group invariant classification of Killing tensors of valence three of nontrivial cubic in the momenta first integrals of motion.

Case	Integral submanifold in Σ	Invariants	# of equivalence classes of elements of $\mathcal{K}^3(\mathbb{R}^2)$
$\gamma^2+\beta_1^2+\beta_2^2 \neq 0$	$\Sigma_2 \backslash \mathcal{T}$	I_1, I_2, I_3, I_4	$2^4 = 16$
$\gamma^2+\beta_1^2+\beta_2^2 = 0$	$\Sigma_1 \backslash \mathcal{T}$	I_*	$< 2^4 = 16$

5. The Holt and Fokas–Lagerström potentials

Consider the Holt and Fokas–Lagerström integrable systems characterized by the existence of the first integrals of motion

$$F_{\rm H}(\mathbf{q},\mathbf{p}) = 2p_1^3+3p_1p_2^2+3p_1(-3(q^2)^2+2(q^1)^2+2\delta)(q^2)^{-2/3}+18p_2q^1(q^2)^{1/3} \tag{5.1}$$

and

$$F_{\rm FL}(\mathbf{q},\mathbf{p}) = (p_1^2-p_2^2)(q^1p_2-q^2p_1)-4(q^2p_1+q^1p_2)((q^1)^2-(q^2)^2)^{-2/3} \tag{5.2}$$

respectively given in the standard position-momenta coordinates. Comparing the cubic terms in $F_{\rm H}$ and $F_{\rm FL}$ with the general formulas (3.1) and taking into account the conditions (4.1), we conclude that the corresponding Killing tensors of valence three $\mathbf{L}_{\rm H}$ and $\mathbf{L}_{\rm FL}$ determining the leading terms of (5.1) and (5.2) respectively are nontrivial, yet they both satisfy the invariant condition $\gamma = \beta_1 = \beta_2 = 0$. Therefore both $\mathbf{L}_{\rm H}$ and $\mathbf{L}_{\rm FL}$ belong to the space $\Sigma_1\backslash\mathcal{T}$ and as such can be characterized by the essential $I(\mathbb{R}^2)$-invariant I_* given by (4.11). Indeed, comparing (5.1) and (5.2) with (3.1), we get for $\mathbf{L}_{\rm H}$: $I_* = 0$ and for $\mathbf{L}_{\rm FL}$: $I_* = \frac{4}{9}$. Hence, each of $\mathbf{L}_{\rm H}$ and $\mathbf{L}_{\rm FL}$ span the two equivalence classes of nontrivial Killing tensors of valence three in $\Sigma_1\backslash\mathcal{T}$. In other words, any Killing tensor belonging to $\Sigma_1\backslash\mathcal{T}$ can be obtained from either $\mathbf{L}_{\rm H}$ or $\mathbf{L}_{\rm FL}$ via the translations and rotations of the coordinates (q^1, q^2) determined by the generators $\mathbf{X}$, $\mathbf{Y}$ and $\mathbf{R}$ respectively or by rescaling.

Therefore any new integrable cases of (1.1) admitting first integrals of motion that are cubic in the momenta within the space $\Sigma_1\backslash\mathcal{T}$ can only be found by varying the respective vectors fields $\mathbf{B}_{\rm H}$ and $\mathbf{B}_{\rm FL}$ in (5.1) and (5.2) and finding the corresponding potential functions V of (1.1) via the formulas (2.9) and (2.10).

6. Concluding remarks

In this paper we have extended the isometry group invariants approach developed in [**7–10**] by the authors to the study of Killing tensors of valence three in the Euclidean plane, or elements of the linear space $\mathcal{K}^3(\mathbb{R}^2)$. More specifically, Problem 3.2, which is an analogue of the main problem of the classical theory of algebraic invariants [**5**], has been solved for the linear space of nontrivial Killing tensors of valence three defined in the Euclidean plane.

The approach allows us to classify first integrals cubic in the momenta for Hamiltonian systems defined in the Euclidean plane up to their leading cubic terms by employing essential $I(\mathbb{R}^2)$-invariants of $\mathcal{K}^3(\mathbb{R}^2)$. We have studied the Killing tensors $\mathbf{L}_{\rm H}$ and $\mathbf{L}_{\rm FL}$ of valence three determined by the first integrals of motion isolated by Holt and Fokas–Lagerström respectively and described the integral submanifold in the space Σ determined by the parameters of the generic Killing tensor of valence three defined in the Euclidean plane, that contains both of them. It has also been demonstrated that all other Killing tensors in this space are related to $\mathbf{L}_{\mathrm{H}}$ and $\mathbf{L}_{\mathrm{FL}}$ via the rigid motions of the Euclidean space $\mathbb{R}^2$.

We are convinced that the method used in this paper to study Hamiltonian systems can be successfully employed in other areas of mathematical physics. Recall, that Delong [**1**] studied the ten-parameter conformal symmetry group of the two-dimensional wave equation. He has proved that there are $(2k+1)(2k+2)(2k+3)/6$ independent k-order symmetries generated by the ten basic elements of this conformal symmetry group. This formula is an analogue of the Takeuchi–Thompson formula [**12, 13**] used in the present study. Thus, one can derive invariants of the ten-parameter conformal symmetry group in the space, say, of second-order symmetries generated by the basic elements and use them in classification-type problems that arise in the study of conformal symmetries of the two-dimensional wave equation.

Acknowledgements. The authors wish to thank the organizers of the workshop "Superintegrability in Classical and Quantum System" for the invitation to participate and acknowledge illuminating discussions on the subject with Willard Miller and Pavel Winternitz. We also thank Willard Miller for bringing to our attention the reference [**1**].

References

1. R. P. Delong, Jr., *Killing tensors and Hamilton–Jacobi equation*, Ph.D. thesis, University of Minnesota, 1982.
2. J. Drach, *Sur l'intégration logique des équations de la dynamique à deux variables. Forces conservatives. Intégrales cubiques. Mouvement dans le plan*, C. R. Acad. Sci. Paris **200** (1935), 22–26.
3. A. S. Fokas and P. A. Lagerstrom, *Quadratic and cubic invariants in classical mechanics*, J. Math. Anal. Appl. **74** (1980), no. 2, 325–341.
4. J. Friš, V. Mandrosov, Ya. A. Smorodinsky, M. Uhlíř, and P. Winternitz, *On higher symmetries in quantum mechanics*, Phys. Lett. **16** (1965), 354–356.
5. D. Hilbert, *Theory of algebraic invariants*, Cambridge Univ. Press, Cambridge, 1993.
6. C. R. Holt, *Construction of new integrable Hamiltonians in two degrees of freedom*, J. Math. Phys. **23** (1982), no. 6, 1037–1046.
7. R. G. McLenaghan, R. G. Smirnov, and D. The, *Group invariant classification of separable Hamiltonian systems in the Euclidean plane and the* O(4)*-symmetric Yang–Mills theories of Yatsun*, J. Math. Phys. **43** (2002), no. 3, 1422–1440.
8. ______, *The* 1881 *problem of Morera revisited*, Differential Geometry and Its Applications (Opava, 2001), Math. Publ., vol. 3, Silesian Univ. Opava, Opava, 2001, pp. 333–341.
9. ______, *Group invariants of Killing tensors in the Minkowski plane*, SPT 2002: Symmetry and Perturbation Theory (Cala Gonone, 2002), World Sci. Publishing, River Edge, NJ, 2002, pp. 153–161.
10. ______, *An extension of the classical theory of algebraic invariants to pseudo-Riemannian geometry and Hamiltonian mechanics*, University of Paderborn, 2002 (preprint).
11. K. Nakagawa and H. Yoshida, *A list of all integrable two-dimensional homogeneous polynomial potentials with a polynomial integral of order at most four in the momenta*, J. Phys. A **34** (2001), no. 41, 8611–8630.
12. M. Takeuchi, *Killing tensor fields on spaces of constant curvature*, Tsukuba J. Math. **7** (1983), no. 2, 233–255.
13. G. Thompson, *Killing tensors in spaces of constant curvature*, J. Math. Phys. **27** (1986), no. 11, 2693–2699.
14. A. V. Tsiganov, *The Drach superintegrable systems*, J. Phys. A. **33** (2000), no. 41, 7407–7422.

15. P. Winternitz, Ya. A. Smorodinsky, M. Uhlíř, and I. Friš, *Symmetry groups in classical and quantum mechanics*, Jadernaja Fiz. **4** (1967), 625–635 (Russian); English transl., Soviet Nuclear Phys. **4** (1967), 444–450.

Department of Applied Mathematics, University of Waterloo, Waterloo, ON N2L 3G1, Canada

E-mail address: `rgmclena@sirius.math.uwaterloo.ca`

Universität Paderborn, Fachbereich 17 Mathematik, Pohlweg 98, 33098 Paderborn, Germany

E-mail address: `rsmirnov@sirius.math.uwaterloo.ca`

Department of Mathematics, University of British Columbia, 121-1984 Mathematics Rd., Vancouver, BC V6T 1Z2, Canada

Current address: Department of Mathematics and Statistics, McGill University, 805 Sherbrooke W., Montréal, QC H3A 2K6, Canada

E-mail address: `dthe@math.mcgill.ca`

Centre de Recherches Mathématiques
CRM Proceedings and Lecture Notes
Volume **37**, 2004

Integrable Systems Whose Spectral Curves Are the Graph of a Function

Kanehisa Takasaki

ABSTRACT. For some integrable systems, such as the open Toda molecule, the spectral curve of the Lax representation becomes the graph $C = \{(\lambda, z) \mid z = A(\lambda)\}$ of a function $A(\lambda)$. Those integrable systems provide an interesting "toy model" of separation of variables. Examples of this type of integrable systems are presented along with generalizations for which $A(\lambda)$ lives on a cylinder, a torus or a Riemann surface of higher genus.

1. Introduction

Separation of variables (SOV) is an old and new subject, which was born in the Hamilton–Jacobi theory of the nineteenth century [**33**], revived by Moser in the light of the modern theory of integrable systems [**25**], and extended by Sklyanin and his collaborators to a wide range of classical and quantum integrable systems [**31**]. Moreover, recent studies have revealed a number of new aspects of SOV in the context of bi-Hamiltonian structures [**3, 20–23**], universal Hamiltonian structures [**15, 16**], algebraic surfaces [**4, 12**], geometric Langlands duality [**10**], string theory [**11**], affine Jacobi varieties [**26, 27, 32**], etc.

This article is a supplement to the joint work with Takebe [**34**] on SOV of integrable systems. This work was motivated by a rather naive question—what would be the simplest nontrivial examples of SOV of integrable systems? The first plan of this research was to use such systems as a "toy model" for better understanding of Sklyanin's work on SOV of quantum integrable systems (in particular, Calogero–Moser and Ruijsenaars–Schneider systems [**18, 19**]). Unfortunately, the program on quantum integrable systems has been unsuccessful due to technical difficulties. However, as we shall see below, searching for the simplest examples of SOV is an interesting issue in itself.

2000 *Mathematics Subject Classification.* 14H70, 37J35, 58F07, 70H06.

This work was partly supported by the Grant-in-Aid for Scientific Research (No. 12640169 and No. 14540172) from the Ministry of Education, Culture, Sports and Technology.

I would like to thank Takashi Takebe for collaboration for years, Evgueni Sklyanin for valuable comments in an early stage of this research, and Dmitry Korotkin and Jean-Pierre Françoise for discussion and comments at the workshop.

This is the final form of the paper.

2. Separation of variables of integrable Hamiltonian systems

Let us consider a Hamiltonian system with canonical variables $q_1, \ldots, q_n$, $p_1, \ldots, p_n$ and a Hamiltonian $H = H(q_1, \ldots, q_n, p_1, \ldots, p_n)$. This system is said to be integrable (or completely integrable) if it has n functionally independent (i.e., $dH_1 \wedge \cdots \wedge dH_n \neq 0$) first integrals $H_1, \ldots, H_n$ in involution (i.e., $\{H_j, H_k\} = 0$). The joint level set of the first integrals,

$$H_\ell(q_1, \ldots, q_n, p_1, \ldots, p_n) = E_\ell, \quad \ell = 1, \ldots, n, \tag{2.1}$$

is a Lagrangian submanifold of the $2n$-dimensional phase space, which (after canonical transformations of coordinates, if necessary) can be expressed as

$$p_j = \partial S/\partial q_j \tag{2.2}$$

in terms of a function $S = S(q_1, \ldots, q_n, E_1, \ldots, E_n)$. In the traditional terminology of Hamilton–Jacobi theory, S is a "complete solution" of the joint Hamilton–Jacobi equations

$$H_\ell(q_1, \ldots, q_n, \partial S/\partial q_1, \ldots, \partial S/\partial q_n) = E_\ell, \quad \ell = 1, \ldots, n. \tag{2.3}$$

The system can be thereby transferred to an action-angle system with the action variables E_ℓ and the angle variables $\phi_\ell = \partial S/\partial E_\ell$.

In this setup, SOV means that the equations of the joint level set take the "separated" form

$$f_j(\lambda_j, \mu_j, E_1, \ldots, E_n) = 0, \quad j = 1, \ldots, n, \tag{2.4}$$

in another set of Darboux coordinates $\lambda_1, \ldots, \lambda_n$, $\mu_1, \ldots, \mu_n$ ("separation variables"). One can then assume the generating function S in the separated form

$$S = \sum_{j=1}^{n} S_j(\lambda_j, E_1, \ldots, E_n) \tag{2.5}$$

and obtain each term S_j by solving the ODE

$$f_j(\lambda_j, dS_j/d\lambda_j, E_1, \ldots, E_n) = 0. \tag{2.6}$$

Note that the existence of first integrals itself tells us nothing about separability; a prescription of SOV has to be sought elsewhere. Sklyanin's "magic recipe" [**31**] is a prescription based on the Lax representation of integrable systems. The idea stems from the work of the Montreal group on generalized Moser systems [**1**], and is also closely related to the algebro-geometric method of the Moscow group [**7**]. Suppose that the integrable system has a Lax representation of the form

$$\dot{L}(z) = [L(z), M(z)] \tag{2.7}$$

with $N \times N$ matrices $L(z)$ and $M(z)$ that depends on a spectral parameter z. The characteristic polynomial

$$f(\lambda, z) = \det(\lambda I - L(z)) = \lambda^N + \sum_{\ell=1}^{N-1} p_\ell(z)\lambda^{N-\ell} \tag{2.8}$$

of $L(z)$ does not depend on the time variable. A set of first integrals in involution can be obtained from the coefficients of expansion of the $p_\ell(z)$'s in a suitable basis of functions of z. The curve

$$C = \{(\lambda, z) \mid f(\lambda, z) = 0\} \tag{2.9}$$

defined by $f(\lambda, z)$ is called the "spectral curve" of the L-matrix $L(z)$. To each point (λ, z) of C is associated the eigenspace of $L(z)$ with eigenvalue λ. For many integrable systems, this eigenspace is one-dimensional at almost all points of C, and the components of a suitably normalized eigenvector $\phi(\lambda, z)$ are meromorphic functions on C with a finite number of poles. It is the coordinates λ_j, z_j of these poles that play the role of separation variables in Sklyanin's magic recipe. These variables are constrained by the equation

$$f(\lambda_j, z_j) = 0 \tag{2.10}$$

of the spectral curve, and these equations can be interpreted as the separated equations (2.4) of the joint level set of first integrals. Moreover, in most cases, the Poisson brackets of these separation variables take the separated form

$$\{\lambda_j, \lambda_k\} = \{z_j, z_k\} = 0, \qquad \{\lambda_j, z_k\} = \delta_{jk} g(\lambda_k, z_k), \tag{2.11}$$

where $g(\lambda_k, z_k)$ is a common function of λ_k and z_k (typically, $1, \lambda_k, z_k$ or $\lambda_k z_k$). In other words, the symplectic form of the system can be written as

$$\Omega = \sum_{j=1}^{N} \frac{dz_j \wedge d\lambda_j}{g(\lambda_j, z_j)}. \tag{2.12}$$

3. If the spectral curve is the graph of a function

Our concern lies in the case where the spectral curve becomes the graph of a function $A(\lambda)$:

$$C = \{(\lambda, z) \mid z = A(\lambda)\}. \tag{3.1}$$

Here is a list of examples of integrable systems of this type:

(1) Open Toda molecule
(2) Open Ruijsenaars–Toda molecule
(3) Rational and trigonometric Calogero-Moser systems
(4) Rational and trigonometric Ruijsenaars-Schneider systems

These integrable systems are usually understood to be associated with no spectral curve, because the usual Lax representation does not contain a spectral parameter. The fact is that they do have a Lax representation with a spectral parameter, and that the spectral curve takes the form mentioned above.

Separation of variables of these examples is almost parallel to the more standard cases. The function $A(\lambda)$ is expanded as a linear combination of a suitable basis $f_0(\lambda), f_1(\lambda), \ldots, f_N(\lambda)$ of functions as

$$A(\lambda) = f_0(\lambda) + \sum_{\ell=1}^{N} E_\ell f_\ell(\lambda), \tag{3.2}$$

and the separation variables comprise N pairs $(\lambda_1, z_1), \ldots, (\lambda_N, z_N)$ of variables that satisfy the equations

$$z_j = A(\lambda_j), \quad j = 1, \ldots, N. \tag{3.3}$$

These equations, in turn, determines the first integrals H_ℓ as a function $E_\ell = H_\ell(\lambda_1, \ldots, \lambda_N, z_1, \ldots, z_N)$ of the separation variables. More precisely, we assume that the non-degeneracy condition

$$\Delta = \det(a_k(\lambda_j) \mid j, k = 1, \ldots, N) \neq 0 \tag{3.4}$$

is satisfied. One can then use Cramer's formula to solve the foregoing equations for the coefficients E_l as

$$E_\ell = \Delta_\ell/\Delta, \tag{3.5}$$

where Δ_ℓ denotes the determinant in which the ℓ-th row of Δ is replaced by $a_\ell(\lambda_j) \to z_j - a_0(\lambda_j)$. This gives an explicit formula of the first integrals.

For illustration, we consider the case of the open Toda molecule in detail below.

4. Example: open Toda molecule

4.1. L-matrix and spectral curve. The open Toda molecule of length N has the Hamiltonian

$$H = \frac{1}{2}\sum_{j=1}^{N} p_j^2 + \sum_{j=2}^{N} e^{q_{j-1}-q_j}. \tag{4.1}$$

A Lax representation with a spectral parameter z is presented by Krichever and Vaninsky [**17**]. The L-matrix reads

$$L(z) = \sum_{j=2}^{N} c_j E_{j,j-1} + \sum_{j=1}^{N} b_j E_{j,j} + \sum_{j=2}^{N} E_{j-1,j} + zE_{N,1}, \tag{4.2}$$

where $E_{j,k}$ stands for the matrix unit with the (j,k) element equal to 1 and the others equal to 0; b_j and c_j are the Flaschka variables, $b_j = p_j$ and $c_j = c^{q_{j-1}-q_j}$. Note that $L = L(0)$ is the usual tridiagonal L-matrix without spectral parameter.

This L-matrix can be derived from the L-matrix $L_{\mathrm{cl}}(z)$ of the closed Toda molecule (namely, the Toda lattice with periodic boundary condition $q_{j+N} = q_j$, $p_{j+N} = p_j$) as follows. The Hamiltonian of the closed Toda molecule is the sum

$$H_{\mathrm{cl}} = H + c_1, \quad c_1 = e^{q_N - q_1} \tag{4.3}$$

of the Hamiltonian of the open Toda molecule and the potential between q_N and q_1. The associated L-matrix is also the sum of two pieces:

$$L_{\mathrm{cl}}(z) = L(z) + c_1 z^{-1} E_{1,N}. \tag{4.4}$$

In the limit as the coupling between q_N and q_1 is turned off ($c_1 \to 0$), this L-matrix becomes the aforementioned matrix $L(z)$.

This correspondence carries over to the spectral curve as well. The characteristic polynomial of $L_{\mathrm{cl}}(z)$ takes the well known form

$$\det(\lambda I - L_{\mathrm{cl}}(z)) = z + c_1 \cdots c_N z^{-1} - P(\lambda), \tag{4.5}$$

where $P(\lambda)$ is a polynomial of degree N. The spectral curve defined by this polynomial is a hyperelliptic curve of genus $N-1$. In the limit as $c_1 \to 0$, the characteristic polynomial reduces to

$$\det(\lambda I - L(z)) = z - A(\lambda), \tag{4.6}$$

where $A(\lambda)$ is nothing but the characteristic polynomial of $L = L(0)$:

$$A(\lambda) = \det(\lambda I - L). \tag{4.7}$$

The spectral curve of the open Toda molecule thus turns out to be the graph of $A(\lambda)$.

4.2. Separation variables. Having the L-matrix $L(z)$, one can obtain separation variables from poles of a suitably normalized eigenvector of $L(z)$. This is indeed achieved in the work of Krichever and Vaninsky [**17**]. One can derive the same set of separation variables from those of the periodic Toda molecule; all necessary data for the latter can be found in the work of Flaschka and McLaughlin [**9**]. We here present yet another approach based on Moser's classical method for solving the open Toda molecule [**24**].

Moser's method uses one of the diagonal elements of the resolvent $(\lambda I - L)^{-1}$:

$$R(\lambda) = {}^{\mathrm{t}}\mathbf{e}_N(\lambda I - L)^{-1}\mathbf{e}_N, \quad {}^{\mathrm{t}}\mathbf{e}_N = (0, \ldots, 0, 1). \tag{4.8}$$

This is a rational function of the form

$$R(\lambda) = B(\lambda)/A(\lambda). \tag{4.9}$$

Since the numerator $B(\lambda)$ is a polynomial of degree $N-1$, $R(\lambda)$ can be expanded in partial fractions as

$$R(\lambda) = \sum_{j=1}^{N} \frac{\rho_j}{\lambda - \alpha_j}, \qquad A(\lambda) = \prod_{j=1}^{N}(\lambda - \alpha_j). \tag{4.10}$$

Moser discovered that the new dynamical variables α_j and ρ_j obey the simple evolution equations

$$\dot{\alpha}_j = 0, \qquad \dot{\rho}_j = \alpha_j \rho_j, \tag{4.11}$$

which can be readily solved as $\alpha_j(t) = \alpha_j(0)$, $\rho_j(t) = e^{\alpha_j(0)t}\rho_j$. The L-matrix can be reproduced from α_j and ρ_j by the Stieltjes method of continued fraction. This is the way that Moser solved the open Toda molecule.

From our point of view, the roots of $B(\lambda)$ are one half of the separation variables:

$$B(\lambda) = \prod_{k=1}^{N-1}(\lambda - \lambda_k). \tag{4.12}$$

The other half $z_1, \ldots, z_{N-1}$ are given by the values $z_k = A(\lambda_k)$ of $A(\lambda)$ at the roots of $B(\lambda)$. The Poisson brackets of these variables turn out to take the form

$$\{\lambda_j, \lambda_k\} = \{z_j, z_k\} = 0, \qquad \{\lambda_j, z_k\} = \delta_{jk} z_k. \tag{4.13}$$

One can derive these Poisson brackets in several different ways. The most direct way is to use the result of Faybusovich and Gekhtman [**8**] on the Poisson brackets of $A(\lambda)$ and $B(\lambda)$.

A comment on the number of variables will be in order. These $2N-2$ separation variables λ_k, z_k amount to $N-1$ of the N degrees of freedom of the open Toda molecule. The rest is the degree of freedom carried by the center of mass $Q = \sum_{j=1}^{N} q_j/N$ and the total momentum $P = \sum_{j=1}^{N} p_j$, which disappear in the center-of-mass frame.

4.3. Many faces of symplectic form. We now have at least three different sets of dynamical variables besides the canonical ones q_j, p_j—Moser's variables α_j, ρ_j, the separation variables λ_k, z_k, and the action-angle variables E_ℓ, ϕ_ℓ. Accordingly, the symplectic form Ω of the system has many different expressions. We present their explicit form below (in the center-of-mass frame, for simplicity).

Let us start from the expression

$$\Omega = \sum_{k=1}^{N-1} d\log A(\lambda_k) \wedge d\lambda_k \tag{4.14}$$

in the separation variables λ_k and $z_k = B(\lambda_k)$. Substituting $A(\lambda_k) = \prod_{j=1}^{N}(\lambda_k - \alpha_j)$ leads to an intermediate expression

$$\Omega = \sum_{k=1}^{N-1}\sum_{j=1}^{N} \frac{d\lambda_k \wedge d\alpha_j}{\lambda_k - \alpha_j}, \tag{4.15}$$

which can be reorganized to the expression

$$\Omega = \sum_{j=1}^{N} d\log B(\alpha_j) \wedge d\alpha_j \tag{4.16}$$

in the new variables α_j and $B(\alpha_j)$. As Faybusovich and Gekhtman point out [**8**] (see also Vaninsky's recent paper [**36**]), this is exactly the symplectic form that Atiyah and Hitchin [**2**] introduced on the moduli space of SU(2) monopoles. In terms of Moser's variables α_j and $\rho_j = B(\alpha_j)/A'(\alpha_j)$, the last expression reads

$$\Omega = \sum_{j=1}^{N} d\log \rho_j \wedge d\alpha_j + \sum_{j\neq k} \frac{d\alpha_j \wedge d\alpha_k}{\alpha_j - \alpha_k}. \tag{4.17}$$

An expression in action-angle variables can be obtained as follows. Let E_ℓ be the coefficients of $A(\lambda)$:

$$A(\lambda) = \lambda^N + \sum_{\ell=2}^{N} E_\ell \lambda^{N-\ell}. \tag{4.18}$$

(E_1 is equal to the total momentum, hence absent in the center-of-mass frame.) The foregoing expression of Ω can now be rewritten as

$$\Omega = \sum_{\ell=2}^{N} dE_\ell \wedge d\phi_\ell, \tag{4.19}$$

where

$$\phi_\ell = \sum_{k=1}^{N-1} \int^{\lambda_k} \frac{\lambda^{N-\ell}}{A(\lambda)}\, d\lambda. \tag{4.20}$$

This shows that ϕ_ℓ is an angle variable conjugate to the action variable E_ℓ. It is easy to see that generating function S of the action-angle variables, too, has a similar simple expression:

$$S = \sum_{k=1}^{N-1} \int^{\lambda_k} \log A(\lambda)\, d\lambda. \tag{4.21}$$

5. Other examples

5.1. Open RT molecule. The open Ruijsenaars–Toda (RT) molecule may be thought of as a limit of the closed RT molecule (namely, the RT lattice with periodic boundary condition $q_{j+N} = q_j$ and $p_{j+N} = p_j$). The passage to the open molecule is parallel to the case of the Toda lattice. The closed molecule has the Hamiltonian

$$H_{\text{cl}} = \sum_{j=1}^{N} e^{p_j}(1 + e^{q_{j-1}-q_j})^{1/2}(1 + e^{q_j - q_{j+1}})^{1/2}, \tag{5.1}$$

which turns into the Hamiltonian H of the open molecule as the potential $e^{q_N - q_1}$ between q_1 and q_N is turned off.

SOV of this case is more delicate than the Toda molecules, because curves arising here are NOT a spectral curve in the usual sense. This is already the case for the closed molecule. The closed RT molecule was solved by Bruschi and Ragnisco [**30**] by an algebro-geometric method, but they had to to use an unusual Lax representation with a matrix $L(\lambda, z)$ that depends on two parameters λ and z. Since $L(\lambda, z)$ amounts to $L(z) - \lambda I$ in the usual Lax representation, they considered the curve defined by the equation

$$f(\lambda, z) = \det L(\lambda, z) = 0, \tag{5.2}$$

and obtained a set of separation variables as Flaschka and McLaughlin [**9**] did for the closed Toda molecule. Therefore, bearing in mind the case of the Toda molecules, it is not difficult to see what appears in the limit to the open RT molecule.

In the limit to the open molecule, as expected, the hyperelliptic curve of Bruschi and Ragnisco degenerates to the graph of a polynomial $A(\lambda)$ of the form

$$A(\lambda) = \lambda^N + \sum_{\ell=1}^{N-1} E_\ell \lambda^{N-\ell} + 1. \tag{5.3}$$

Separation variables $\lambda_1, \ldots, \lambda_{N-1}$ are the roots of a polynomial $B(\lambda)$ of degree $N-1$, and their conjugate variables $z_1, \ldots, z_{N-1}$ are defined by $z_k = A(\lambda_k)$. The Poisson brackets are slightly different from those of the Toda molecules:

$$\{\lambda_j, \lambda_k\} = \{z_j, z_k\} = 0, \qquad \{\lambda_j, z_k\} = \delta_{jk}\lambda_k z_k. \tag{5.4}$$

The symplectic form Ω of the system, too, is modified as

$$\Omega = \sum_{k=1}^{N-1} d\log A(\lambda_k) \wedge d\log\lambda_k = \sum_{j=1}^{N} d\log B(\lambda_j) \wedge d\log\alpha_j = \sum_{\ell=1}^{N-1} dE_\ell \wedge d\phi_\ell, \tag{5.5}$$

where α_j's are the roots of $A(\lambda)$, and ϕ_ℓ's are given by

$$\phi_\ell = \sum_{k=1}^{N-1} \int^{\lambda_k} \frac{\lambda^{N-\ell}}{A(\lambda)} d\log\lambda. \tag{5.6}$$

5.2. Rational and trigonometric CM systems. The rational and trigonometric Calogero–Moser (CM) systems may be thought of as degeneration of the elliptic CM system. This allows us to find a Lax representation with a spectral parameter for these systems.

Let us recall the following L-matrix [**14**] for the elliptic CM system:

$$L(z) = \sum_{j=1}^{N} p_j E_{j,j} + \sum_{j\neq k} gi\phi(q_j - q_k, z) E_{j,k}, \tag{5.7}$$

where $\phi(u,z)$ is written in terms of the Weierstrass σ function as

$$\phi(u,z) = \frac{\sigma(u+z)}{\sigma(u)\sigma(z)}. \tag{5.8}$$

As the elliptic curve degenerates to a rational curve, $\phi(u,z)$ turns into a trigonometric (hyperbolic) function

$$\phi(u,z) \to \frac{\sinh(u+z)}{\sinh(u)\sinh(z)} = \coth(u) + \coth(z)$$

or a rational function

$$\phi(u,z) \to \frac{u+z}{uz} = \frac{1}{u} + \frac{1}{z}.$$

Remakably, the limit of $\phi(u,z)$ splits into two functions of a single variable. Accordingly, the limit of $L(z)$ becomes the sum of two matrices as

$$L(z) \to \begin{cases} L + gi\coth(z)K & \text{(trigonometric)} \\ L + giz^{-1}K & \text{(rational)} \end{cases} \tag{5.9}$$

where

$$L = \sum_{j=1}^{N} p_j E_{j,j} + \sum_{j\neq k} gi\coth(q_j - q_k) E_{j,k} \tag{5.10}$$

in the trigonometric limit,

$$L = \sum_{j=1}^{N} p_j E_{j,j} + \sum_{j\neq k} gi(q_j - q_k)^{-1} E_{j,k} \tag{5.11}$$

in the rational limit, and

$$K = \sum_{j\neq k} E_{j,k} \tag{5.12}$$

for both cases. L is nothing but the usual L-matrix of the trigonometric and rational CM systems; K is the constant matrix that appears in the method of Hamiltonian reduction (or the "projection method") for solving these systems [**13, 28, 29**]. The linear combination of L and K in the limit (5.9) of $L(z)$ gives an L-matrix with a spectral parameter for the rational and trigonometric CM systems

We modify these rational and trigonometric L-matrices slightly. Note that K can be expressed as

$$K = \mathbf{e}{}^{t}\mathbf{e} - I, \quad {}^{t}\mathbf{e} = (1, \ldots, 1). \tag{5.13}$$

Therefore the term proportional to K in (5.9) is a linear combination of a rank-one matrix and a scalar matrix; the latter does not affect to the Lax equation. We can thus drop the scalar term and *redefine* the L-matrix for the degenerate CM systems as

$$L(z) = L + gi\coth(z)\mathbf{e}{}^{t}\mathbf{e} \tag{5.14}$$

for the trigonometric case and

$$L(z) = L + giz^{-1}\mathbf{e}^{\mathrm{t}}\mathbf{e} \tag{5.15}$$

for the rational case.

We now apply the so called Weinstein-Aronszajn formula

$$\det(M + \mathbf{x}^{\mathrm{t}}\mathbf{y}) = (1 + {}^{\mathrm{t}}\mathbf{y}M^{-1}\mathbf{x}) \det M$$

to the characteristic polynomial of $L(z)$. This yields the expression

$$\det(\lambda I - L(z)) = \begin{cases} P_0(\lambda) - gi\coth(z)P_1(\lambda) & \text{(trigonometric)} \\ P_1(\lambda) - giz^{-1}P_1(\lambda) & \text{(rational)} \end{cases} \tag{5.16}$$

where

$$P_0(\lambda) = \det(\lambda I - L), \quad P_1(\lambda) = {}^{\mathrm{t}}\mathbf{e}(\widetilde{\lambda I - L})\mathbf{e}. \tag{5.17}$$

($\widetilde{M}$ denote the cofactor matrix of M.) The equation of the spectral curve of $L(z)$ thus turns out to take the form

$$z = \frac{P_1(\lambda)}{P_0(\lambda)} \tag{5.18}$$

for the rational case and

$$e^{2z} = \frac{P_0(\lambda) + P_1(\lambda)}{P_0(\lambda) - P_1(\lambda)} \tag{5.19}$$

for the trigonometric form. This is what we have sought.

It should be mentioned that the same equation of the spectral curve for the trigonometric case was also derived by Vaninsky [**35**] from a different point of view. We can proceed further to the construction of separation coordinates, though we omit details here.

5.3. Rational and trigonometric RS systems. The rational and trigonometric Ruijsenaars–Schneider (RS) systems can be treated in the same way as the CM systems. We again start from the elliptic RS system, and consider the limit as the elliptic curve degenerates to a rational curve. It is convenient to use the following L-matrix of the elliptic RS system (which is gauge equivalent to the L-matrix of Bruschi and Calogero [**6**]):

$$L(z) = \sum_{j,k=1}^{N} (h_j h_k)^{1/2} \phi(q_j - q_k + \gamma, z) E_{j,k}, \tag{5.20}$$

where γ is a constant and h_j's are fairly complicated functions of the canonical variables. In the limit to the trigonometric or rational models, the L-matrix again splits into the sum of two matrices, one being the L-matrix L without spectral parameter, and the other a rank-one matrix. One can thus use the Weinstein-Aronszajn formula again to show that the spectral curve is the graph of a function $A(\lambda)$. The same curve was derived by Braden and Marshakov [**5**] in the context of supersymmetric gauge theories.

6. Variants and generalizations

The goal of this research is to search for new integrable systems from the point of view of SOV. Let us show a few results below.

6.1. Trigonometric and elliptic analogues. Apart from the CM and RS systems, the foregoing examples are associated with a pair of polynomials $A(\lambda)$ and $B(\lambda)$. The main result of the joint work with Takebe [**34**] is to construct the following analogues defined on a cylinder and a torus:

(1) Trigonometric version

$$A(\lambda) = \prod_{j=1}^{N} \sinh(\lambda - \alpha_j), \qquad B(\lambda) = \prod_{j=1}^{N} \sinh(\lambda - \lambda_j). \tag{6.1}$$

(2) Elliptic version

$$A(\lambda) = \prod_{j=1}^{N} \sigma(\lambda - \alpha_j), \qquad B(\lambda) = \prod_{j=1}^{N} \sigma(\lambda - \lambda_j). \tag{6.2}$$

The construction can be further extended to the case where $A(\lambda)$ and $B(\lambda)$ have a different number of zeroes.

6.2. Elliptic fibration. Another possible elliptic analogue is to modify the z direction of the (λ, z) space to an elliptic curve. A naive way will be to replace

$$\Omega = \sum_{j=1}^{N} d\log z_j \wedge d\lambda_j \to \Omega = \sum_{j=1}^{N} \frac{dz_j}{y_j} \wedge d\lambda_j \tag{6.3}$$

where y_j and z_j are constrained by the equation

$$y_j^2 = 4z_j^3 + g_2 z_j + g_3$$

of an elliptic curve. Actually, one can go further by allowing the coefficients g_2 and g_3 to depend on λ_j as

$$y_j^2 = 4z_j^3 + g_2(\lambda_j) z_j + g_3(\lambda_j), \tag{6.4}$$

where $g_2(\lambda)$ and $g_3(\lambda)$ are functions (e.g., polynomials) of λ. Geometrically, this is the structure called an "elliptic fibration." Integrable systems of this type, too, are considered in the joint work with Takebe [**34**].

6.3. Higher genera. Generalization to a complex algebraic curve C_0 of genus $g > 1$ (and a Riemann surface of more general type) will be achieved in several different ways. Let us propose an idea below.

The idea is to choose the (A, B) pair as nonzero elements of the vector space $\mathcal{L}(R_1 + \cdots + R_N)$ of meromorphic functions $f(P)$ ($P \in C_0$) on C_0 with at most simple poles at N fixed points $R_1, \ldots, R_N$. If $N > 2g - 2$ (or if $N > g$ and $R_1, \ldots, R_N$ are in general position), the Riemann–Roch theorem shows that this vector space is spanned by $N - g + 1$ linearly independent elements $f_0(P), f_1(P), \ldots, f_{N-g}(P)$. We write $A(P)$ and $B(P)$ by this basis as

$$A(P) = \sum_{\ell=0}^{N-g} u_\ell f_\ell(P), \qquad B(P) = \sum_{\ell=0}^{N-g} v_\ell f_\ell(P) \tag{6.5}$$

and view $(u_\ell)_{\ell=0}^{N-g}$ and $(v_\ell)_{\ell=0}^{N-g}$ as homogeneous coordinates on two copies of the projective space $\mathbf{P}^{N-g}$. The (A, B) pair is thus parametrized by $2N - 2g$ "moduli," which play the role of dynamical variables. Now let us choose a holomorphic differential $d\lambda$ on C_0, and consider its (multi-valued) primitive function

$$\lambda(P) = \int^P d\lambda. \tag{6.6}$$

In this quite general setup, one can use a trick based on the residue theorem [**34**] to derive the identity

$$\sum_{j=1}^{N} d\log B(P_j) \wedge d\lambda(P_j) = \sum_{j=1}^{N} d\log A(Q_j) \wedge d\lambda(Q_j), \tag{6.7}$$

where P_j and Q_j are zeroes of $A(P)$ and $B(P)$, respectively; d is understood to be the total differential in the $2N - 2g$ moduli of the (A, B) pair. This defines a symplectic form Ω on the moduli space. The inhomogeneous coordinates $H_\ell = u_\ell/u_0$ give a set of functions in involution with respect to Ω, thereby defining an integrable system. The role of separation variables is played by

$$z_j = A(Q_j), \qquad \lambda_j = \lambda(Q_j), \tag{6.8}$$

though they are not independent variables.

References

1. M. R. Adams, J. Harnad, and J. Hurtubise, *Darboux coordinates and Liouville–Arnold integration in loop algebras*, Comm. Math. Phys. **155** (1993), no. 2, 385–413.
2. M. Atiyah and N. Hitchin, *The geometry and dynamics of magnetic monopoles*, M. B. Porter Lectures, Princeton Univ. Press, Princeton, NJ, 1988.
3. M. Błaszak, *On separability of bi-Hamiltonian chain with degenerate Poisson structures*, J. Math. Phys. **39** (1998), no. 6, 3213–3235.
4. F. Bottacin, *Poisson structure on Hilbert schemes of points of a surface and integrable systems*, Manuscripta Math. **97** (1998), no. 4, 517–527.
5. H. W. Braden and A. Marshakov, *Singular phases of Seiberg–Witten integrable systems: weak and strong coupling*, Nuclear Phys. B **595** (2001), no. 1-2, 417–466.
6. M. Bruschi and F. Calogero, *The Lax representation for an integrable class of relativistic dynamical systems*, Comm. Math. Phys. **109** (1987), no. 3, 481–492.
7. B. A. Dubrovin, I. M. Krichever, and S. P. Novikov, *Integrable systems.* I, Dynamical System. IV, Encyclopaedia of Math. Sci., vol. 4, Springer-Verlag, Berlin, 1990, pp. 173–280.
8. L. Faybusovich and M. Gekhtman, *Poisson brackets on rational functions and multi-Hamiltonian structure for integrable lattices*, Phys. Lett. A **272** (2000), no. 4, 236–244.
9. H. Flaschka and D. W. McLaughlin, *Canonically conjugate variables for the Korteweg–de Vries equation and the Toda lattice with periodic boundary conditions*, Prog. Theoret. Phys. **55** (1976), no. 2, 438–456.
10. E. Frenkel, *Affine algebras, Langlands duality and Bethe Ansatz*, XIth International Congress of Mathematical Physics (Paris 1994), Internat. Press, Cambridge, MA, 1995, pp. 606–642.
11. A. Gorsky, N. Nekrasov and V. Rubtsov, *Hilbert schemes, separated variables, and D-branes*, Comm. Math. Phys. **222** (2001), no. 2, 299–318.
12. J. C. Hurtubise, *Integrable systems and algebraic surfaces*, Duke Math. J. **83** (1996), no. 1, 19–50.
13. D. Kazhdan, B. Kostant and S. Sternberg, *Hamiltonian group actions and dynamical systems of Calogero type*, Comm. Pure. Appl. Math. **31** (1978), no. 4, 481–507.
14. I. M. Krichever, *Elliptic solutions of Kadomtsev–Petviashvili equation and integrable systems of particle*, Funct. Anal. Appl. **14** (1980), 282–290.
15. I. M. Krichever and D. H. Phong, *On the integrable geometry of soliton equations and $N = 2$ supersymmetric gauge theories*, J. Differential Geom. **45** (1998), no. 2, 349–389.
16. ———, *Symplectic forms in the theory of solitons*, Surveys in Differential Geometry: Integrable Systems, Surv. in Differ. Geom., vol. 4, Int. Press, Boston, MA, 1999, pp. 239–313.

17. I. Krichever and K. Vaninsky, *The periodic and open Toda lattice*, Mirror Symmetry. IV (Montreal, 2000), AMS/IP Stud. Adv. Math., vol. 33, Amer. Math. Soc., Providence, RI, 2002, pp. 139–158.
18. V. B. Kuznetsov and E. K. Sklyanin, *Separation of variables in the A_2 type Jack polynomials* 1995, solv-int/9508002.
19. ———, *Separation of variables for A_2 Ruijsenaars model and new integral representation for A_2 Macdonald polynomials*, J. Phys. A **29** (1996), no. 11, 2779–2804.
20. F. Magri G. Falqui, and M. Pedroni, *The method of Poisson pairs in the theory of nonlinear PDEs*, 2000, nlin.SI/0002009.
21. ———, *The bihamiltonian approach to separation of variables in mechanics*, Proceedings of the Workshop on Nonlinearity, Integrability, and All That. Twenty Years after NEEDS '79 (Gallipoli, 1999), World Scientific, Singapore, 2000, pp. 258–266.
22. C. Morosi and G. Tondo, *Quasi-bi-Hamiltonian systems and separability*, J. Phys. A. **30** (1997), no. 8, 2799–2806.
23. ———, *On a class of dynamical systems both quasi-bi-Hamiltonian and bi-Hamiltonian*, Phys. Lett. A **247** (1998), no. 1-2, 59–64.
24. J. Moser, *Finitely many mass points on the line under the influence of an exponential potential—an integrable system*, Dynamical Systems, Theory and Applications (Seattle, WA, 1974), Lecture Notes in Phys., vol. 38, Springer, Berlin, 1975, pp. 467–497.
25. ———, *Geometry of quadrics and spectral theory*, The Chern Symposium 1979 (Berkeley, CA, 1979), Springer, New York–Berlin, 1980, pp. 147–188.
26. A. Nakayashiki and F. A. Smirnov, *Cohomology of affine Jacobi varieties and integrable systems*, Comm. Math. Phys. **217** (2001), no. 3, 623–652.
27. ———, *Euler characteristics of theta divisors of Jacobians for spectral curves*, The Kowalevski Property (Leeds, 2000), CRM Proc. Lecture Notes, vol. 32, Amer. Math. Soc., Providence, RI, 2002, pp. 239–246.
28. M. A. Olshanetsky and A. M. Perelomov, *Explicit solutions of the Calogero models in the classical case and geodesic flows on symmetric spaces of zero curvature*, Lett. Nuovo Cimento (2) **16** (1976), no. 11, 333–339.
29. ———, *Explicit solutions of some completely integrable systems*, Lett. Nuovo Cimento (2) **17** (1976), no. 3, 97–101.
30. O. Ragnisco and M. Bruschi, *The periodic relativistic Toda lattice: direct and inverse problem*, Inverse Problems **5** (1989), no. 3, 389–405.
31. E. K. Sklyanin, *Separation of variables—new trends*, Progr. Theoret. Phys. Suppl. **118** (1995), 35–60.
32. F. A. Smirnov and V. Zeitlin, *Affine Jacobians of spectral curves and integrable models*, 2002, math-ph/0203037.
33. P. Stäckel, *Über die Integration der Hamilton–Jacobishen Differentialgleichung mittelst Separation der Variablen*, Habilitationsschrift, Univ. Halle–Wittenberg, 1981.
34. K. Takasaki and T. Takebe, *An integrable system on the moduli space of rational functions and its variants*, J. Geom. Phys. **47** (2003), no. 1, 1–20.
35. K. L. Vaninsky, *On explicit parametrisation of spectral curves for Moser–Calogero particles and its applications*, Internat. Math. Res. Notices (1999), no. 10, 509–529.
36. ———, *The Atiyah–Hitchin bracket and the open Toda lattice*, J. Geom. Phys. **46** (2003), no. 3-4, 283–307.

Department of Fundamental Sciences, Faculty of Integrate Human Studies, Kyoto University, Yoshida, Sakyo-ku, Kyoto 606-8501, Japan

E-mail address: `takasaki@math.h.kyoto-u.ac.jp`

Centre de Recherches Mathématiques
CRM Proceedings and Lecture Notes
Volume **37**, 2004

On Superintegrable Systems in E_2: Algebraic Properties and Symmetry Preserving Discretization

Piergiulio Tempesta

ABSTRACT. We survey some results concerning a class of superintegrable systems defined on the Euclidean plane. Their algebraic properties as well as their discrete analogs are discussed.

1. Introduction

The purpose of this paper is to review some results, recently obtained in [**46, 47, 63, 64, 68**], regarding a class of superintegrable systems in quantum mechanics. These systems, introduced in [**26, 48, 76**] and known as the *Smorodinsky–Winternitz systems*, are defined on the Euclidean plane and characterized by integrals of motion quadratic in the momenta.

Given a quantum system with Hamiltonian H in the Euclidean space E_n, the stationary Schrödinger equation has the form

$$H\psi = E\psi, \quad H = -\tfrac{1}{2}\Delta + V(x_1, \dots, x_n). \tag{1.1}$$

As is well known, in nonrelativistic quantum mechanics an integral of motion is expressed by a time-independent linear operator X_a which commutes with the Hamiltonian. The usual restriction is to consider operators polynomial in the momenta. A system is said to be completely integrable if

$$[H, X_a] = 0, \quad [X_a, X_b] = 0 \qquad (a, b = 1, \dots, n-1) \tag{1.2}$$

and superintegrable if it admits k additional independent operators Y_l, $l = 1, \dots, k$, $0 < k \leq (n-1)$, commuting with the Hamiltonian. In particular, a system is said to be minimally superintegrable if $k = 1$ and maximally superintegrable if $k = n-1$, $n \geq 2$.

This definition is modeled on the notion of integrability in classical mechanics [**2**]. Let us denote by $\mathcal{A}$ the Lie algebra of first integrals for a Hamiltonian system with n degrees of freedom, defined on a $2n$-dimensional symplectic manifold

---

2000 *Mathematics Subject Classification.* 81-02, 39-06.

The research reported in this paper is based on joint work in collaboration with D. Levi, M. Sheftel, A. Turbiner and P. Winternitz, whom the author thanks gratefully.

This is the final form of the paper.

(M, Ω). If $\dim \mathcal{A} = n$, then the Liouville–Arnol′d theorem ensures that the canonical equations of motion are integrable by quadratures. If $\dim \mathcal{A} > n$, the first integrals cannot be in involution anymore, i.e., the algebra $\mathcal{A}$ is no longer commutative (see, e.g., [**9, 21, 67**] for a discussion of the geometrical picture underlying noncommutative integrability and [**51**] for a generalization of the classical Liouville theorem).

Maximally superintegrable systems possess many interesting physical properties. For instance, at the classical level, all bounded trajectories are closed. At the quantum level, the corresponding spectra are characterized by accidental degeneracy (removed by the additional observables Y_l). This is the case of the hydrogen atom and the harmonic oscillator, whose symmetry operators form an $o(4)$ and an $su(3)$ algebra, respectively [**5, 22, 36, 52**].

The search for superintegrable systems, both in classical and quantum mechanics, has been actively pursued in the last decades. An extensive literature now exists on different aspects of the theory, like the group-theoretical properties of superintegrability (see, e.g., [**18–23, 26, 48, 63, 76, 77**]), its relation with solvability ([**43, 59, 68**]) and separation of variables (see, e.g., [**7, 17, 18, 29, 50, 58, 76**]), the extension of the notion of superintegrability to non-Euclidean spaces (see, e.g., [**4, 35, 37, 39–42, 53, 54, 57**]), the existence of recurrence operators (see, e.g., [**12, 25, 28, 67**]), higher-order integrals of motion (see, e.g., [**6, 16, 24, 27, 33, 34, 38, 49, 56**]), quantum ([**3, 10, 13, 14**]) and discrete ([**46, 47, 74**]) analogs of superintegrable systems, etc. (the list of references, while extensive, is far from being complete).

In this paper, we will focus on the algebraic properties of the Smorodinsky–Winternitz Hamiltonians.

In Section 2, the notion of integrability in quantum mechanics will be related to the existence of higher-order symmetries of the Schrödinger equation and a group theoretical classification of integrable and superintegrable systems in E_2 with first-order and second-order integrals of motion will be presented. In Section 3, the algebraic structure of the superintegrable Hamiltonians will be analyzed. In particular, it will be shown that the corresponding Hilbert space of bound states consists of a flag of finite-dimensional invariant subspaces. In Section 4, the symmetry-preserving discretization of the superintegrable systems on a regular uniform two-dimensional lattice is discussed in the light of Rota's approach to the theory of finite-difference operators.

2. Lie symmetries of superintegrable systems in E_2

In this section, the problem of the integrability and superintegrability of the Schrödinger equation will be studied from the point of view of Lie group theory. As is well known, in the case of Hamiltonian systems with infinitely many degrees of freedom, the notion of integrability is related to the existence of infinitely many "generalized" or higher-order symmetries [**1, 55**]. These symmetries depend on derivatives of the dependent variables with respect to the independent ones. Another basic feature of integrability is the existence of a recursion operator that generates higher symmetries from lower ones. Our aim is to study the two–dimensional stationary Schrödinger equation and to establish conditions on the potential $V(x, y)$ in (1.1) for it to have one, or more symmetries depending on first and second derivatives of the wave function ψ. The same conditions imply the existence of a recursion operator, although this aspect will not be discussed here (see, e.g., [**63, 64**]).

We recall that a symmetry group G of a system F of differential equations is a local group of transformations, acting on the manifold $M = X \times U$ of the independent and dependent variables and leaving the system invariant. The corresponding generators can be calculated algorithmically using Lie's theorem [**55**]. It can be stated as follows. Let G be a local group of transformations on M. If the prolongation of every infinitesimal generator $\mathbf{v}$ of G on the k–th jet space of M annihilates the system F on the set of its solutions, i.e.:

$$\mathrm{pr}^{(k)}\,\mathbf{v}[F]|_{F=0} = 0. \tag{2.1}$$

then G is a symmetry group of F. This condition furnishes a set of *determining equations* which specify completely the coefficients of the vector field $\mathbf{v}$.

The motivation for studying symmetry groups of systems of differential equations is twofold. First, a class of exact solutions can be constructed by applying a symmetry group to a known solution. Second, if G acts regularly, with s–dimensional orbits, it is possible to associate with the original system a reduced one, defined on the quotient manifold of M by G, having s fewer independent variables. If one can solve the reduced system, then solutions of the original problem can be constructed.

In the evolutionary formalism [**55**], a m–th order Lie symmetry is expressed in terms of a vector field of the form

$$\mathbf{w} = Q(x, y, u, u_x, u_y, u_{xx}, u_{xy}, \dots)\partial_u, \tag{2.2}$$

where $u(x, y)$ denotes the real or imaginary part of ψ, and Q is called the characteristic function of $\mathbf{w}$. If the order of the symmetry is m, then Q will contain derivatives of u up to the order m. Condition (2.1) in the case of the two–dimensional Schrödinger equation reads

$$D_{xx}Q + D_{yy}Q - 2(V - E)Q = 0. \tag{2.3}$$

Consequently we obtain a flow

$$\frac{d}{d\lambda}u = Q, \tag{2.4}$$

compatible with the Schrödinger equation (1.1). Solving (2.3) together with (1.1) we construct a one–parameter family of solutions of (1.1). Higher-order Lie symmetries in classical and quantum mechanics were introduced in [**1**] and in [**23–25**]. In this paper, they are explicitly used to characterize the notion of superintegrability in the Euclidean plane. In the subsequent considerations, we will restrict to the case in which the characteristic function Q depends on derivatives up to second order only. Also, we deal with energy-independent symmetries, namely Q will be the same for all energies E. We shall classify the potentials under the action of the Euclidean group $E(2)$. Its Lie algebra $e(2)$ admits the basis

$$L_3 = y\partial_x - x\partial_y, \qquad P_1 = \partial_x, \qquad P_2 = \partial_y.$$

2.1. First-order symmetries. Consider first the case $Q = Q(x, y, u, u_x, u_y)$. From the determining equations one deduces that the potential $V(x, y)$ has to satisfy the equation

$$[\alpha(y\partial_x - x\partial_y) + \beta\partial_x + \gamma\partial_y]V(x, y) = 0 \tag{2.5}$$

where α, β and γ are real constants. The solution of (2.5) is

$$V = V(\xi), \qquad \xi = \tfrac{1}{2}\alpha(x^2+y^2) - \gamma x + \beta y \tag{2.6}$$

For $\alpha \neq 0$ we use a translation to annul β and γ. For $\alpha = 0$ we rotate to put $\beta = 0$. We obtain the following result.

THEOREM 2.1. *First-order energy independent Lie symmetries of the two-dimensional stationary Schrödinger equation exist if and only if the potential allows a geometric symmetry, i.e., satisfies* (2.5). *The flow corresponding to this symmetry is*

$$u_\lambda = (k_1 X + k_2)u + h(x,y), \tag{2.7}$$

with

$$X = \alpha L_3 + \beta P_1 + \gamma P_2, \qquad \psi = u_1 + iu_2. \tag{2.8}$$

Here α, β, $\gamma \in \mathbb{R}$, k_1, $k_2 \in \mathbb{R}$ and h is a real solution of the Schrödinger equation (1.1).

2.2. Second-order Lie symmetries. Let $Q = Q(x, y, u, u_x, u_y, u_{xy}, u_{yy})$ (the dependence on u_{xx} is eliminated using the Schrödinger equation). Solving the determining equations (2.3) we obtain for the characteristic Q the expression

$$\begin{aligned} Q = \{aL_3^2 + b(L_3P_1 + P_1L_3) + c(L_3P_2 + P_2L_3)& \\ + d(P_1^2 - P_2^2) + 2eP_1P_2 + \phi(x,y)\}u + h(x,y),& \end{aligned} \tag{2.9}$$

with

$$\begin{aligned} \phi_x &= -2(ay^2 + 2by + d)V_x + 2(axy + bx - cy - e)V_y, \\ \phi_y &= 2(axy + bx - cy - e)V_x + 2(-ax^2 + 2cx + d)V_y. \end{aligned} \tag{2.10}$$

The compatibility condition for eq. (2.10) gives us the condition on the potential, necessary and sufficient for a second-order symmetry to exist, namely

$$\begin{aligned} (-axy - bx + cy + e)(V_{xx} - V_{yy}) + [a(x^2 - y^2) - 2by - 2cx - 2d]V_{xy}& \\ -3(ay + b)V_x + 3(ax - c)V_y = 0.& \end{aligned} \tag{2.11}$$

The following result holds [**63**].

THEOREM 2.2. *A second-order Lie symmetry of the two-dimensional stationary Schrödinger equation exists if and only if there exists a second-order operator of the form*

$$\begin{aligned} X = aL_3^2 + b(L_3P_1 + P_1L_3) + c(L_3P_2 + P_2L_3) + d(P_1^2 - P_2^2)& \\ + 2eP_1P_2 + \alpha L_3 + \beta P_1 + \gamma P_2 + \phi(x,y).& \end{aligned} \tag{2.12}$$

that commutes with the Hamiltonian.

Commutativity of operator (2.12) with the Hamiltonian implies that the potential satisfies (2.5) and the compatibility condition (2.11) and that the function ϕ satisfies (2.10).

The operator X can be classified into orbits under the action of the Euclidean group $E(2)$. One can prove that there are four classes of operators, each of them corresponding to a potential allowing *separation of variables.* We have a general statement.

THEOREM 2.3. *A second-order Lie symmetry exists if and only if the potential $V(x,y)$ is such that the Schrödinger equation* (1.1) *allows the separation of variables in Cartesian, polar, parabolic or elliptic coordinates.*

From the previous discussion it emerges that there exits a clear correspondence between symmetries of the Schrödinger equation and integrals of motion. Now, since we wish to classify superintegrable systems, we allow an additional symmetry to exist. We have to study two different classes of systems:

(i) Systems admitting one first-order and one second-order symmetry;
(ii) Systems admitting two second-order symmetries.

For the class (i), the results can be summed up as follows.

THEOREM 2.4. *Four $E(2)$ classes of potentials exist, allowing one first-order Lie symmetry and at least one second-order one. They are $\omega^2 r^2$, α/r, αx^{-2} and αx. For $V=\omega^2 r^2$ the second-order symmetries form a u(2) algebra. For $V(r)=\alpha/r$, they form an o(3), o(2,1) or e(2) algebra for a fixed energy $E<0$, $E>0$ and $E=0$, respectively. For $V=\alpha x^{-2}$ and $V=\alpha x$ the second-order Lie symmetries generate infinite-dimensional Lie algebras.*

We notice that, for spherically symmetric potentials, this is in agreement with Bertrand's theorem [**8**], which states that the only rotationally invariant potentials for which all finite trajectories are closed are the Coulomb and the harmonic potentials.

For the class (ii), we have a general result.

THEOREM 2.5. *Four superintegrable systems with two second-order Lie symmetries exist in the Euclidean space E_2. For each of them the Schrödinger equation allows the separation of variables in at least two coordinate systems. They are:*

(2.13)
$$\begin{aligned} H_{\mathrm{I}} &= -\frac{1}{2}(\partial_x^2+\partial_y^2)+\omega^2(x^2+y^2)+\frac{1}{2}\left(\frac{A}{x^2}+\frac{B}{y^2}\right),\\ X_1 &= P_1^2-P_2^2-2\omega^2(x^2-y^2)-\frac{A}{x^2}+\frac{B}{y^2},\\ X_2 &= L_3^2-\left(\frac{A}{\cos^2\phi}+\frac{B}{\sin^2\phi}\right). \end{aligned}$$

The Schrödinger equation separates in Cartesian, polar and elliptic coordinates.

(2.14)
$$\begin{aligned} H_{\mathrm{II}} &= -\frac{1}{2}(\partial_x^2+\partial_y^2)+\omega^2\left(2x^2+\frac{y^2}{2}\right)+\frac{\alpha}{2y^2}+\beta x,\\ X_1 &= P_1^2-P_2^2-\omega^2(4x^2-y^2)-2\beta x+\frac{\alpha}{y^2},\\ X_2 &= L_3P_2+P_2L_3-2\omega^2xy^2+\frac{2\alpha x}{y^2}-\beta y^2. \end{aligned}$$

The Schrödinger equation separates in Cartesian and parabolic coordinates.

(2.15)
$$\begin{aligned} H_{\mathrm{III}} &= -\frac{1}{2}(\partial_x^2+\partial_y^2)+\frac{\alpha}{2r}+\frac{1}{r^2}\left(\frac{\beta+\gamma\cos\vartheta}{\sin^2\vartheta}\right),\\ X_1 &= L_3^2-2\left(\frac{\beta+\gamma\cos\vartheta}{\sin^2\vartheta}\right),\\ X_2 &= L_3P_2+P_2L_3+\alpha\cos\vartheta+2\frac{\gamma(\cos^2\vartheta+1)+2\beta\cos\vartheta}{r\sin^2\vartheta}. \end{aligned}$$

The Schrödinger equation separates in polar and parabolic coordinates (and also elliptic ones).

$$
\begin{aligned}
(2.16)\qquad H_{\mathrm{IV}} &= -\frac{1}{2}\frac{1}{\xi^2+\eta^2}(\partial_\xi^2+\partial_\eta^2)+\frac{2\alpha+\beta\xi+\gamma\eta}{\xi^2+\eta^2},\\
X_1 &= L_3P_1+P_1L_3+\frac{\beta\eta(\eta^2-\xi^2)+\gamma\xi(\xi^2-\eta^2)-4\alpha\eta\xi}{(\xi^2+\eta^2)},\\
X_2 &= L_3P_2+P_2L_3+2\frac{\alpha(\xi^2-\eta^2)+\eta\xi(\gamma\xi-\beta\eta)}{(\xi^2+\eta^2)}.
\end{aligned}
$$

The Schrödinger equation separates in two different parabolic coordinate systems.

The Hamiltonians (2.13)–(2.16) were discovered in [**26**] by means of a direct approach.

We conclude that the superintegrable systems in the Euclidean plane with quadratic integrals of motion are *superseparable*: they admit separation of variables in at least two coordinate systems (see also [**7, 17, 18, 58**] for an interesting study of the connection between the problems of separation of variables in the Hamilton–Jacobi equation and in the Schrödinger equation).

3. Superintegrability and exact solvability

An important property of some quantum mechanical problems is their "exact solvability" (see, e.g., [**72, 73**]). A quantum mechanical exactly solvable system with Hamiltonian H is characterized by the property that its Hilbert space of bound states consists of an infinite flag of finite-dimensional subspaces

$$(3.1)\qquad S_0 \subset S_1 \subset \cdots \subset S_n \subset \cdots$$

preserved by the Hamiltonian

$$(3.2)\qquad HS_i \subset S_i.$$

A basis for the entire space is obtained as a union of bases for the invariant subspaces. Therefore, an exactly solvable system is described by a Hamiltonian which can be transformed into a block diagonal form with finite-dimensional blocks.

The following properties are sufficient for exact solvability.

(i) In appropriate coordinates s_i the Hamiltonian H can be written as

$$(3.3)\qquad H = ghg^{-1}$$

where the function g is called the gauge factor, and

$$(3.4)\qquad h = c_aJ_a + c_{a,b}J_aJ_b,$$

i.e. is a quadratic combination of first order differential operators J_a, generating a parabolic subalgebra of a finite–dimensional semisimple Lie algebra, usually called the *hidden algebra* underlying the considered model. Here c_a and $c_{a,b}$ are constants, subject to some further conditions in order to ensure the normalizability of the wave functions [**32**].

(ii) The bound state wave functions $\Psi_n(\vec{x})$ have the form

$$(3.5)\qquad \Psi_n(\vec{x}) = g(\vec{x})P_n(\vec{s}), \quad s_i = s_i(\vec{x}).$$

where the functions $P_n(\vec{s})$ are polynomials of order n in certain variables s_i. These polynomials define a flag, whose subspaces S_i are labeled by n. In principle, (3.4) ensures that the Hamiltonian H will preserve, or decrease the order of the polynomials P_n. We allow the gauge factor $g(\vec{x})$ to be energy dependent.

As is well known, for one-dimensional problems, the hidden algebra can be only a subalgebra of $\mathrm{sl}(2,\mathbb{R})$. Indeed, this algebra, realized as

$$J_n^+ = x^2\partial_x - nx, \qquad J_n^0 = x\partial_x - \frac{n}{2}, \qquad J_n^- = \partial_x. \tag{3.6}$$

is the only Lie algebra of first order differential operators which acts on the real line and possesses finite-dimensional representations (for a discussion see, for example, [**30**]). The flag of polynomials $\mathcal{P}_n = \langle x^k \mid 0 \le k \le n\rangle$ is preserved by any element of the universal enveloping algebra of the $sl(2,\mathbb{R})$ parabolic subalgebra J_n^0, J_n^- for any n. Therefore, a Hamiltonian written as an element of this enveloping algebra defines an *exactly solvable* system. When a Hamiltonian possesses only one finite-dimensional invariant subspace $\mathcal{P}_n$, it will be said to be *quasi-exactly solvable* [**65,70,75**]. A necessary and sufficient condition for a one-dimensional Hamiltonian to be quasi-exactly solvable is that it belongs to the universal enveloping algebra of the $\mathrm{sl}(2,\mathbb{R})$-algebra taken in realization (3.6) [**71**].

For two-dimensional problems, the following algebras can play the role of hidden Lie algebras: $sl(3,\mathbb{R})$, $sl(2,\mathbb{R})\oplus sl(2,\mathbb{R})$, $o(3,1)$, a parametric family $gl(2,\mathbb{R}) \ltimes \mathbb{R}^{r+1}$ and some of their subalgebras [**30, 31, 65**]. Let us consider the algebra $gl(3,\mathbb{R}) \supset sl(3,\mathbb{R})$, realized as

$$\begin{gathered} J_1 = \partial_t, \qquad J_2 = \partial_u, \qquad J_3 = t\partial_t, \\ J_4 = u\partial_u, \qquad J_5 = u\partial_t, \qquad J_6 = t\partial_u, \\ J_7 = t^2\partial_t + tu\partial_u - nt, \qquad J_8 = tu\partial_t + u^2\partial_u - nu, \\ X = n. \end{gathered} \tag{3.7}$$

The linear space of polynomials $\mathcal{P}_n^{(2)} = \langle t^k u^m \mid 0 \le k+m \le n\rangle$ forms a flag, which is preserved by any element of the universal enveloping algebra of the parabolic subalgebra $\{J_1,\dots,J_6\}$ [**72, 73**].

We will prove that the superintegrable Hamiltonians (2.13)–(2.16) are *exactly solvable*, with hidden algebra given by the maximal parabolic subalgebra of $\mathrm{sl}(3,\mathbb{R})$, formed by the generators $\{J_1,\dots,J_6\}$ [**68**].

Let us consider first the Hamiltonian (2.13). Its eigenfunctions are

$$\Psi_{n,m}(x,y) = x^{p_1}y^{p_2}L_n^{(-1/2+p_1)}(\omega x^2)L_m^{(-1/2+p_2)}(\omega y^2)e^{-\omega x^2/2-\omega y^2/2}, \tag{3.8}$$

where $L_k^{(\alpha)}(z)$ are Laguerre polynomials, n, $m = 0, 1, 2, \dots$, and the parameters p_1, p_2 are defined by $A = p_1(p_1-1)$, $B = p_2(p_2-1)$. Performing a gauge rotation of $H_{\mathrm{I}}(x,y)$ and then a change of variables, we get

$$\begin{aligned} h^{\mathrm{I}} &\equiv \frac{1}{\omega}(\Psi_{0,0}(x,y))^{-1}H_{\mathrm{I}}(x,y)\Psi_{0,0}(x,y)|_{t=\omega x^2,u=\omega y^2} \\ &= -2t\partial_t^2 - 2u\partial_u^2 + 2t\partial_t + 2u\partial_u - (2p_1+1)\partial_t \\ &\qquad - (2p_2+1)\partial_u + 1 + p_1 + p_2, \end{aligned} \tag{3.9}$$

with eigenvalues $E(n,m) = n+m$, n, $m = 0, 1, 2, \dots$.

The same procedure leads to the operators

$$\hat{x}_{\mathrm{C}}^{\mathrm{I}} = 2t\partial_t^2 - 2u\partial_u^2 - 2t\partial_t + 2u\partial_u + (2p_1+1)\partial_t - (2p_2+1)\partial_u - p_1 + p_2, \tag{3.10}$$

$$\hat{x}_{\mathrm{R}}^{\mathrm{I}} = 2\{2tu(\partial_t - \partial_u) + [(2p_1+1)u - (2p_2+1)t]\}(\partial_t - \partial_u) - (p_1+p_2)^2. \tag{3.11}$$

These three operators h^{I}, $\hat{x}^{\mathrm{I}}_{\mathrm{C}}$, $\hat{x}^{\mathrm{I}}_{\mathrm{R}}$ generate an infinite-dimensional algebra. In particular,

$$[h^{\mathrm{I}}, \hat{x}^{\mathrm{I}}_{\mathrm{C}}] = [h^{\mathrm{I}}, \hat{x}^{\mathrm{I}}_{\mathrm{R}}] = 0.$$

If we rewrite the operators h^{I}, $\hat{x}^{\mathrm{I}}_{\mathrm{C}}$, $\hat{x}^{\mathrm{I}}_{\mathrm{R}}$ in terms of the generators $\{J_1, \ldots, J_6\}$, we get

$$h^{\mathrm{I}} = -2J_3J_1 - 2J_4J_2 + 2J_3 + 2J_4 - (2p_1+1)J_1 - (2p_2+1)J_2, \tag{3.12}$$

$$\hat{x}^{\mathrm{I}}_{\mathrm{C}} = 2J_3J_1 - 2J_4J_2 - 2J_3 + 2J_4 + (2p_1+1)J_1 - (2p_2+1)J_2, \tag{3.13}$$

$$\begin{aligned} \hat{x}^{\mathrm{I}}_{\mathrm{R}} = 4J_3J_5 + 4J_4J_6 - 8J_3J_4 + 2(2p_1+1)J_5 - 2(2p_2+1)J_3 \\ - 2(2p_1+1)J_4 + 2(2p_2+1)J_6. \end{aligned} \tag{3.14}$$

In the new variables t, u, the operators (3.12)–(3.14) preserve a triangular flag of polynomials

$$\mathcal{P}_n(t,u) = \langle t^p u^q | 0 \leq p+q \leq n \rangle$$

Therefore, the new Hamiltonian h^{I} (as well as the operators (3.12)–(3.14)) possesses infinitely many finite-dimensional invariant subspaces and polynomial eigenfunctions.

A very similar analysis can be performed for the Hamiltonian (2.14). The algebraic form of this Hamiltonian and of its integrals of motion is given by

$$h^{\mathrm{II}} = -J_1J_1 - 2J_4J_2 + 2J_3 + 2J_4 - (1+2p_2)J_2 + \tfrac{3}{2} + p_2, \tag{3.15}$$

$$\hat{x}^{\mathrm{II}}_{\mathrm{C}} = 2J_1J_1 - 4J_4J_2 - 4J_3 + 4J_4 - 2(1+2p_2)J_2 - 1 + 2p_2, \tag{3.16}$$

$$\hat{x}^{\mathrm{II}}_{\mathrm{P}} = -4J_4J_6 + 4J_1J_4 - 2J_5 + (1+2p_2)J_1 - 2(1+2p_2)J_6. \tag{3.17}$$

The flag of polynomials preserved by operators (3.15)–(3.17) is the same as in the previous case.

The Hamiltonian (2.15) in parabolic coordinates has the form

$$H_{\mathrm{III}}(\xi,\eta) = -\frac{1}{2}\frac{1}{\xi^2+\eta^2}(\partial_\xi^2 + \partial_\eta^2) + \frac{1}{\xi^2+\eta^2}\left(2\alpha + \frac{\beta_1}{\xi^2} + \frac{\beta_2}{\eta^2}\right), \tag{3.18}$$

with

$$x = \tfrac{1}{2}(\xi^2 - \eta^2), \qquad y = \xi\eta.$$

The eigenfunctions corresponding to the energy E are

$$\Psi_{n,m} = \xi^{p_1}\eta^{p_2} L_n^{(-1/2+p_1)}(\sqrt{-2E}\xi^2) L_m^{(-1/2+p_2)}(\sqrt{-2E}\eta^2) e^{-\sqrt{-E/2}(\xi^2+\eta^2)} \tag{3.19}$$

where $2\beta_1 = p_1(p_1-1)$, $2\beta_2 = p_2(p_2-1)$. Writing the Schrödinger equation $H_{\mathrm{III}}\Psi = E\Psi$ in the form

$$\left[-\frac{1}{2}(\partial_\xi^2+\partial_\eta^2) - E(\xi^2+\eta^2) + \frac{\beta_1}{\xi^2} + \frac{\beta_2}{\eta^2}\right]\Psi = -2\alpha\Psi, \tag{3.20}$$

the previous equation becomes

$$Q_{\mathrm{III}}\Psi = -2\alpha\Psi, \tag{3.21}$$

where

$$Q_{\mathrm{III}} \equiv (\xi^2+\eta^2)(H_{\mathrm{III}} - E) - 2\alpha. \tag{3.22}$$

Since

$$Q_{\mathrm{III}} = H_{\mathrm{I}}(\xi,\eta; -E, \beta_1, \beta_2), \tag{3.23}$$

(cf. (2.13)) we can conclude that the first Hamiltonian, written in *Cartesian* coordinates, coincides with a modified third Hamiltonian Q^{III}, written in *parabolic* coordinates. The spectral parameter (-2α) corresponds to the energy E^{I} of the first case. After a gauge rotation of the operator Q_{III}

$$q^{\mathrm{III}} \equiv \frac{1}{\sqrt{-2E}} G_{\mathrm{III}}^{-1} Q_{\mathrm{III}}(x,y) G_{\mathrm{III}}, \tag{3.24}$$

with the gauge factor (now energy dependent)

$$G_{\mathrm{III}} = \xi^{p_1} \eta^{p_2} e^{-\sqrt{-E/2}(\xi^2+\eta^2)}, \tag{3.25}$$

and after a change of coordinates $t = \sqrt{-2E}\xi^2$, $u = \sqrt{-2E}\eta^2$ we get

$$q^{\mathrm{III}} = -2t\partial_t^2 - 2u\partial_u^2 + 2t\partial_t + 2u\partial_u - (2p_1+1)\partial_t - (2p_2+1)\partial_u + 1 + p_1 + p_2. \tag{3.26}$$

The spectrum of this operator, which coincides with the operator h^{I} (3.9), is given by

$$-2\alpha/\sqrt{-E} = 2(n+m) + 1 + p_1 + p_2, \quad n, m = 0, 1, 2, \ldots. \tag{3.27}$$

The operator q^{III} commutes with the operators $\hat{x}_{\mathrm{C}}^{\mathrm{III}} \doteq \hat{x}_{\mathrm{C}}^{\mathrm{I}}$, $\hat{x}_{\mathrm{R}}^{\mathrm{III}} \doteq \hat{x}_{\mathrm{R}}^{\mathrm{I}}$. The Hamiltonian (3.18) is exactly solvable, since q^{III} can be rewritten in terms of the generators $\{J_1, \ldots, J_6\}$ of the maximal parabolic subalgebra of $sl(3,\mathbb{R})$ (see (3.7)).

Now let us investigate the relation between the operators commuting with the original Hamiltonian H_{III} of (3.18) and the operators $\hat{x}_{\mathrm{C}}^{\mathrm{III}}$, $\hat{x}_{\mathrm{R}}^{\mathrm{III}}$.

The operators that commute with H_{III}, also given in parabolic coordinates, are

$$\begin{aligned} X_{\mathrm{P}} &= \frac{1}{\xi^2+\eta^2}(\eta^2\partial_\xi^2 - \xi^2\partial_\eta^2 + 2\alpha(\xi^2-\eta^2) - 2\beta_1\frac{\eta^2}{\xi^2} + 2\beta_2\frac{\xi^2}{\eta^2}), \\ X_{\mathrm{R}} &= (\xi\partial_\eta - \eta\partial_\xi)^2 - 2(\xi^2+\eta^2)(\frac{\beta_1}{\xi^2} + \frac{\beta_2}{\eta^2}). \end{aligned} \tag{3.28}$$

Due to (3.23), the operators that commute with Q_{III} of (3.21) can be obtained from those commuting with H_{I}. It turns out that they are related to the integrals X_{P} and X_{R} by the formulae

$$X_{\mathrm{C}}^{\mathrm{III}} = -X_{\mathrm{P}} + (\xi^2-\eta^2)(H^{\mathrm{III}} - E), \qquad X_{\mathrm{R}}^{\mathrm{III}} = X_{\mathrm{R}}. \tag{3.29}$$

In general, let us consider a Hamiltonian H and an operator Q, defined by the relation

$$H = \frac{Q+K}{\xi^2+\eta^2} + E, \tag{3.30}$$

where K is a constant and E is the energy. Let X be an operator commuting with the Hamiltonian: $[H, X] = 0$. The operator Q then satisfies

$$[Q, X] = (\xi^2+\eta^2)\left[X, \frac{1}{\xi^2+\eta^2}\right](\xi^2+\eta^2)(H-E). \tag{3.31}$$

Thus, if the operator X commutes with H, it will commute "weakly" with Q, i.e., on functions Ψ satisfying $(H-E)\Psi = 0$.

It follows that when relating operators that commute with H and those that commute with Q, we are allowed to consider linear combinations of the type $AX_{\mathrm{P}} + BX_{\mathrm{R}} + f(\xi,\eta)(H-E)$. Here A and B are constants, but $f(\xi,\eta)$ can be any function, since $H - E$ vanishes on the "*energy shell.*"

The fourth superintegrable Hamiltonian, in the standard parabolic coordinates ξ, η is expressed by

$$H_{\mathrm{IV}}(\xi,\eta) = -\frac{1}{2}\frac{1}{\xi^2+\eta^2}(\partial_\xi^2 + \partial_\eta^2) + \frac{1}{\xi^2+\eta^2}(2\alpha + \beta\xi + \gamma\eta). \tag{3.32}$$

The eigenfunctions are products of Laguerre polynomials times an energy dependent multiplier, namely

$$G_{\mathrm{IV}} = (\xi - \beta/2E)^{p_1}(\eta - \gamma/2E)^{p_2} e^{-\sqrt{-E/2}[(\xi-\beta/2E)^2+(\eta-\gamma/2E)^2]}$$

with a condition $\alpha + \beta^2/4E + \gamma^2/4E + \sqrt{-2E} = 0$ and p_1, $p_2 = 0, 1$.

The corresponding Schrödinger equation $H_{\mathrm{IV}}\Psi = E\Psi$ can be transformed into

$$(3.33)\quad \left[-\frac{1}{2}(\partial_\xi^2 + \partial_\eta^2) - E\left(\xi - \frac{\beta}{2E}\right)^2 - E\left(\eta - \frac{\gamma}{2E}\right)^2\right]\Psi = \left(-2\alpha - \frac{\beta^2+\gamma^2}{4E}\right)\Psi.$$

Let us introduce the operator

$$Q_{\mathrm{IV}} \equiv (\xi^2+\eta^2)(H_{\mathrm{IV}} - E) - 2\alpha - \frac{\beta^2+\gamma^2}{4E}.$$

Equation (3.33) becomes

$$Q_{\mathrm{IV}}\Psi = \tilde{\alpha}\Psi,$$

with the new spectral parameter $\tilde{\alpha} = -2\alpha - (\beta^2+\gamma^2)/4E$.

As in the previous case, the operator Q_{IV} can be easily related to H_{I}:

$$(3.34)\quad Q_{\mathrm{IV}} = H_{\mathrm{I}}(\xi, \eta; -E, 0, 0).$$

(cf. (2.13), (3.23)) In other words, the first Hamiltonian, written in *Cartesian* coordinates, coincides with a modified fourth Hamiltonian Q_{IV} written in *parabolic* coordinates. The spectral parameter $\tilde{\alpha}$ corresponds to the energy of the first case, $\tilde{\alpha} \leftrightarrow E^{\mathrm{I}}$.

Following the procedure adopted in the first case, we write the gauge rotated operator Q_{IV} in new coordinates $t = \sqrt{-2E}(\xi - \beta/2E)^2$, $u = \sqrt{-2E}(\eta - \gamma/2E)^2$. The gauge rotated operator

$$q^{\mathrm{IV}} \equiv \frac{1}{\sqrt{-2E}} G_{\mathrm{IV}}^{-1} Q_{\mathrm{IV}}(x, y) G_{\mathrm{IV}}$$

coincides with the operator h^{I} (3.9) and q^{III} (3.26). Its spectrum is equal to $2(n+m)+1+p_1+p_2$, n, $m = 0, 1, 2, \ldots$.

The operator q^{IV} commutes with the operators $\hat{x}_{\mathrm{C}}^{\mathrm{IV}} \doteq \hat{x}_{\mathrm{C}}^{\mathrm{I}}$, $\hat{x}_{\mathrm{R}}^{\mathrm{IV}} \doteq \hat{x}_{\mathrm{R}}^{\mathrm{I}}$ and is exactly solvable: it can be rewritten in terms of the generators $\{J_1, \ldots, J_6\}$ of the maximal parabolic subalgebra of $sl(3, \mathbb{R})$ (see (3.7)).

The operators commuting with the original Hamiltonian (3.32) H_{IV} are

$$(3.35)\quad \begin{aligned} X_1 &= \frac{1}{2(\xi^2+\eta^2)}\{\eta^2\partial_\xi^2 - \xi^2\partial_\eta^2 + 2\alpha(\xi^2-\eta^2) + 2\xi\eta(\gamma\xi - \beta\eta)\}, \\ X_2 &= \frac{1}{(\xi^2+\eta^2)}\{\xi\eta(\partial_\xi^2+\partial_\eta^2) + (-\beta\eta+\gamma\xi)(\xi^2-\eta^2) - 4\alpha\xi\eta - \partial_{\xi\eta}^2\}. \end{aligned}$$

The operators commuting with Q_{IV} (3.34) are related to the previous ones by the formulae

$$(3.36)\quad Q^{(1)} = -2X_1 + (\xi^2-\eta^2)(H_{\mathrm{IV}} - E) + \frac{\gamma^2-\beta^2}{4E},$$

$$(3.37)\quad Q^{(2)} = X_2 + 2\xi\eta(H_{\mathrm{IV}} - E) - \frac{\beta\gamma}{2E}.$$

To conclude this analysis, we observe that the Hamiltonian $H_{\rm IV}(\xi,\, \eta)$ can be easily reduced to a Hamiltonian corresponding to an harmonic oscillator with frequency $\omega = \sqrt{-E/2}$ and a displaced equilibrium point $\xi = \beta/2E$, $\eta = \gamma/2E$:

$$Q_{\rm IV} = -\frac{1}{2}(\partial_\xi^2 + \partial_\eta^2) - E\left[\left(\xi - \frac{\beta}{2E}\right)^2 + \left(\eta - \frac{\gamma}{2E}\right)^2\right]. \tag{3.38}$$

Therefore, the "Hamiltonian" $Q_{\rm IV}$ commutes with $Q^{(1,2)}$ and L_z, where

$$\left(\xi - \frac{\beta}{2E}\right)\partial_\eta - \left(\eta - \frac{\gamma}{2E}\right)\partial_\xi \equiv L_z. \tag{3.39}$$

This is coherent with the fact that the harmonic oscillator is invariant under an SU(2) group. The operators (3.36), (3.37), (3.38) form the basis of a $u(2)$ symmetry algebra with $Q_{\rm IV}$ as its center.

4. Superintegrable systems on a lattice

In this section, it will be shown that many algebraic properties of continuous n–dimensional quantum systems can be preserved in a discrete space-time by means of the finite operator theory, which is a modern formulation of the so called *umbral calculus*. This calculus, which originates in the works of Sylvester (who invented the name), Cayley, Blissard and others, was considered for a long time as a basic tool for constructing identities in combinatorics. Only in the 70's of the twentieth century, with the contributions of G. C. Rota, S. Roman and others, umbral calculus found a rigourous foundation in the finite operator theory, based on the linear algebra of shift-invariant operators [**11, 60–62**]. Using this approach one can study in a rigorous way the analogy existing between the continuous and the finite difference calculus. The physical motivations for considering quantum models on a lattice are manifold (see, e.g. [**47**] for a brief discussion). In particular, in loop quantum gravity the existence of a discrete structure of space-time geometry, realized in terms of a regular lattice, is postulated [**69**]. In [**15**], the umbral calculus has been carefully studied in connection with quantum mechanics. It turns out that there are infinitely many ways to discretize a quantum system preserving its symmetry properties. In particular, Lorentz and Galilei algebras can be preserved on a lattice [**47**]. As a consequence, the notions of integrability and superintegrability can be consistently formulated in lattice quantum mechanics [**46, 47**]. This approach is closely related to that introduced in [**66, 74**]. In these interesting papers, the Lie algebraic discretization of quantum mechanics in terms of operators in Fock spaces has been proposed. It provides a general procedure for discretizing exactly and quasi-exactly solvable systems.

Let us briefly review some basic definitions necessary in the subsequent considerations. We will work essentially on a one–dimensional equally spaced lattice; the extension of our approach to the multidimensional case is straightforward.

Let $\mathbb{K}$ be a field of characteristic 0 and $\mathbb{K}[x]$ the algebra of polynomials in one variable with coefficients in $\mathbb{K}$. By a *polynomial sequence* we denote a sequence of polynomials $p_n(x)$, $n = 0, 1, 2, \ldots$, such that $p_n(x)$ is of degree n for all n. Let us denote by T the shift operator, acting as $T\,f(x) = f(x + \sigma)$, where σ is the (constant) distance between two points in the lattice.

DEFINITION 4.1. A delta operator Q is a shift-invariant (i.e. it commutes with shifts) linear endomorphism of $\mathbb{K}[x]$, such that

$$Q\,x = c, c \neq 0. \tag{4.1}$$

DEFINITION 4.2. A polynomial sequence $p_n(x)$ is called a sequence of basic polynomials for the delta operator Q if, for any $n \in \mathbb{N}$,

$$Q\,p_n(x) = np_{n-1}(x), \qquad p_0(x) = 1, \qquad p_n(0) = 0. \tag{4.2}$$

Standard examples of delta operators are given by the operators ∂_x (the usual derivative), the discrete derivatives $\Delta^+ = (T-1)/\sigma$, $\Delta^- = (1-T^{-1})/\sigma$. The associated basic polynomials are x^n, $x_{(n)} = x(x-\sigma)\cdots(x-(n-1)\sigma)$, and $x^{(n)} = x(x+\sigma)\cdots(x+(n-1)\sigma)$ respectively. Another example is represented by the Laguerre operator, defined by $Q\,f(x) = -\int_0^\infty e^{-t}f'(x+t)\,dt$, whose basic polynomials are the so called basic Laguerre polynomials [**62**]. Generally, we will deal with delta operators of the form [**46, 47**]

$$\Delta\,f(x) = \frac{1}{\sigma}\sum_{k=l}^{m} a_k\,T^k\,f(x), \tag{4.3}$$

where the coefficients $a_k \in \mathbb{R}$ satisfy the requirements

$$\sum_k a_k = 0, \qquad \sum_k ka_k = 1. \tag{4.4}$$

which ensure that the operator $Q = \Delta$ is a delta operator, having ∂_x as its continuum limit.

To fix all constants a_k we must impose $m-l-1$ further conditions, for instance

$$\gamma_q \equiv \sum_{k=l}^{m} a_k k^q = 0, \quad q = 2, 3, \ldots, m-l. \tag{4.5}$$

We will say that the order of the operators of type (4.3)–(4.5) is $m-l$ since they provide an approximation of order σ^{m-l} to the continuous derivative ∂_x. For instance, the symmetric derivative $\Delta^s = (T-T^{-1})/2\sigma$ is a second-order delta operator.

Now, in keeping with the previous point of view, in order to discretize quantum systems let us introduce a pair of conjugate operators: a delta operator Δ and a shift-invariant operator β, such that the Heisenberg commutation relation is preserved [**44, 46, 47, 66**]:

$$[\Delta, x\beta] = 1. \tag{4.6}$$

It follows that $[\Delta, (x\beta)^n] = n(x\beta)^{n-1}$. Therefore, $\{(x\beta)^n\}_{n\geq 0}$ is a basic sequence of operators for Δ under the operation of commutation. The explicit expression of β is provided by

$$\beta = ([\Delta, x])^{-1}, \tag{4.7}$$

since β commutes with Δ. Let us now consider some examples. The continuum case corresponds to $\Delta = \partial_x$, $\beta = 1$. For the right discrete derivative Δ^+, we have $\beta = T^{-1}$. For Δ^-, $\beta = T$. For Δ^s we get $\beta = [(T+T^{-1})/2]^{-1}$.

The basic tool we adopt for constructing linear field theories on a lattice, and in particular discrete quantum models, is the *umbral correspondence*.

Let us denote by $\mathcal{L}$ the algebra of linear operators acting on $\mathbb{K}[[x]]$, the algebra of formal power series in a variable x.

DEFINITION 4.3. An umbral correspondence is a mapping $\mathcal{R}\colon \mathcal{L} \to \mathcal{L}$ defined by

$$(x\beta_1)^n \xrightarrow{\mathcal{R}} (x\beta_2)^n, \tag{4.8}$$

where $P_n^1 = (x\beta_1)^n$ and $P_n^2 = (x\beta_2)^n$ are basic sequences of operators for two delta operators Q_1 and Q_2, respectively.

Due to the uniqueness of the basic sequence associated with a delta operator Q, the umbral correspondence (4.8) naturally induces a correspondence between the two operators Q_1 and Q_2, according to the following scheme:

$$\begin{array}{ccc} (x\beta_1)^n & \xrightarrow{\mathcal{R}} & (x\beta_2)^n \\ \downarrow & & \downarrow \\ Q_1 & \xrightarrow{\mathcal{R}} & Q_2. \end{array} \tag{4.9}$$

We shall also denote the induced correspondence between delta operators by the same symbol $\mathcal{R}$.

Lut us consider the interesting particular case when Q_1 is the standard derivative ∂_x, and Q_2 is a discrete derivative Δ. Then, we have the correspondence (still denoted by $\mathcal{R}$)

$$\begin{array}{ccc} x^n & \xrightarrow{\mathcal{R}} & (x\beta)^n \\ \downarrow & & \downarrow \\ \partial_x & \xrightarrow{\mathcal{R}} & \Delta. \end{array} \tag{4.10}$$

The notion of umbral operator has been introduced in **[62]**: it is an operator $R\colon \mathbb{K}[x] \to \mathbb{K}[x]$ (not necessarily shift-invariant) which maps a basic sequence $p_n(x)$ into another basic sequence $q_n(x)$:

$$p_n(x) \xrightarrow{R} q_n(x). \tag{4.11}$$

We observe that, since $\beta = \beta(T)$ and any constant is invariant under the action of a shift operator, the action of R is deduced from the action of $\mathcal{R}$ simply by applying the sequence of operators $(x\beta)^n$ onto 1:

$$(x\beta_1)^n 1 \xrightarrow{R} (x\beta_2)^n 1. \tag{4.12}$$

The correspondences (4.9) and (4.10) preserve Lie algebras. For instance, let $\mathfrak{g}_1$ be a m-dimensional Lie algebra, generated by vector fields $\{\mathbf{v}_i\}_{i=1,\dots,m}$ of the form

$$\mathbf{v}_i = \sum_j a_j(x_1, \dots, x_p)\partial_{x_j}. \tag{4.13}$$

The correspondence (4.10) maps $\mathfrak{g}_1$ isomorphically into a formal algebra $\mathfrak{g}_2$, spanned by the operators $\{\mathbf{v}_i^U\}_{i=1,\dots,m}$, with

$$\mathbf{v}_i^U = \sum_j a_j(x_1\beta_{x_1}, \dots, x_p\beta_{x_p})\Delta_{x_j}. \tag{4.14}$$

Obviously, the structure constants of the two algebras remain the same.

Using the umbral correspondence, we can preserve on the lattice all the infinitesimal aspects of Lie group theory, namely the symmetry generators. In general, it is not possible to integrate these generators to obtain the corresponding finite symmetry transformations. Nevertheless, one can still use the algebra of generators to perform symmetry reductions.

As was shown in Section 2, higher-order symmetries have a simple quantum mechanical interpretation: they can be considered as integrals of motion of the stationary Schrödinger equation. The operators X_j, as well as the Hamiltonian H, can be viewed as elements of the enveloping algebra of the Weyl–Heisenberg algebra generated by the operators $\{p_j, x_j, \hbar\}$, which satisfy the canonical commutation relations

$$[p_j, x_k] = -i\hbar\delta_{jk}, \tag{4.15}$$

with all other commutators vanishing. This paves the way to a consistent discretization of superintegrable systems. Indeed, the umbral correspondence preserves the commutation relations (4.15) between quantum mechanical operators. Here we report only the results concerning the discretization of the generalized harmonic oscillator. This case has been already considered in [**66, 74**]. In the umbral formalism, analogously we have:

$$H_{\mathrm{I}}^D = -\frac{1}{2}(\Delta_x^2 + \Delta_y^2) + \omega^2[(x\beta_x)^2 + (y\beta_y)^2] + \frac{a}{2}(x\beta_x)^{-2} + \frac{b}{2}(y\beta_y)^{-2}. \tag{4.16}$$

with the integrals of motion

$$X_1 = [-\tfrac{1}{2}\Delta_x^2 + 2\omega^2(x\beta_x)^2 + a(x\beta_x)^{-2}] - [-\tfrac{1}{2}\Delta_y^2 + 2\omega^2(y\beta_y)^2 + b(y\beta_y)^{-2}] \tag{4.17}$$

and

$$\begin{aligned} X_2 = {} & (x\beta_x\Delta_y - y\beta_y\Delta_x)^2 \\ & - [a(1 + (x\beta_x)^{-2}(y\beta_y)^2) + b(1 + (x\beta_x)^2(y\beta_y)^{-2})]. \end{aligned} \tag{4.18}$$

Similarly, the discrete version of the system H_{II} is

$$H_{\mathrm{II}}^D = -\frac{1}{2}(\Delta_x^2 + \Delta_y^2) + \omega^2\left[2(x\beta_x)^2 + \frac{1}{2}(y\beta_y)^2\right] + \frac{a}{2}(y\beta_y)^{-2} + bx\beta_x. \tag{4.19}$$

The second-order operators commuting with the Hamiltonian (4.19) are

$$X_1 = \Delta_x^2 - \Delta_y^2 - \omega^2(4(x\beta_x)^2 - (y\beta_y)^2) - 2b(x\beta_x) + a(y\beta_y)^{-2}, \tag{4.20}$$

and

$$\begin{aligned} X_2 = {} & [(y\beta_y)\Delta_x - (x\beta_x)\Delta_y]\Delta_y + \Delta_y[(y\beta_y)\Delta_x - (x\beta_x)\Delta_y] \\ & - 2\omega^2(x\beta_x)(y\beta_y)^2 + 2a(x\beta_x)(y\beta_y)^{-2} - b(y\beta_y)^2. \end{aligned} \tag{4.21}$$

Many aspects of umbral calculus in quantum theory, such as the analysis of spectral properties of umbral operators and the question of convergence of the formal power series appearing in the theory, deserve to be further investigated. In particular, as pointed out in [**15**], discrete analogs of continuous self-adjoint Hamiltonians may not be essentially self-adjoint. Hence, in principle a purely algebraic discretization procedure may not be sufficient to fix the dynamics on the lattice. A suitable choice of physical constraints is then required to determine unambiguously the spectral properties of a discrete Hamiltonian. This is an open problem.

Finally, we mention that the umbral formalism can be also generalized to provide q–deformations of quantum systems (see, for instance, the discussion in [**11, 45, 60**]).

References

1. R. L. Anderson and N. H. Ibragimov, *Lie–Bäcklund transformations in applications*, SIAM Stud. Appl. Math., vol. 1, Society for Industrial and Applied Mathematics, Philadelphia, PA, 1979.
2. V. I. Arnol′d, *Mathematical methods in classical mechanics*, Grad. Texts in Math., vol. 60, Springer-Verlag, New York–Heidelberg, 1978.
3. Á. Ballesteros and F. J. Herranz, *Integrable deformations of oscillator chains from quantum algebras*, J. Phys. A **32** (1999), no. 50, 8851–8862.
4. Á. Ballesteros, F. Herranz, M. Santander, and T. Sanz-Gil, *Maximal superintegrability on N−dimensional curved spaces*, J. Phys. A **36** (2003), no. 7, L93–L99.
5. V. Bargmann, *Zur Theorie des Wasserstoffatoms. Bemerkungen zur gleichnamigen Arbeit von V. Fock*, Z. Phys. **99** (1936), 576–582.
6. J. Beckers, N. Debergh, and A. Nikitin, *More on symmetries of the Schrödinger equation*, J. Phys. A **24** (1991), no. 21, L1269–L1275.
7. S. Benenti, C. Chanu, and G. Rastelli, *The super-separability of the three-body inverse-square Calogero system*, J. Math. Phys. **41** (2000), no. 7, 4654–4678.
8. J. Bertrand, *Théorème relatif au mouvement d'un point attiré vers un centre fixe*, C. R. Acad. Sci. **77** (1873), 849–853.
9. O. I. Bogoyavlenskij, *Extended integrability and bi–Hamiltonian systems*, Comm. Math. Phys. **196** (1998), no. 1, 19–51.
10. D. Bonatsos, C. Daskaloyannis, and K. Kokkotas, *Deformed oscillator algebras for two-dimensional quantum superintegrable systems*, Phys. Rev. A (3) **50** (1994), no. 5, 3700–3709.
11. A. Di Bucchianico and D. Loeb, *A selected survey of umbral calculus*, Electron. J. Combin. **2** (1995), Dynamic Survey 3 (electronic).
12. R. Caseiro, *Master integrals, superintegrability and quadratic algebras*, Bull. Sci. Math. **126** (2002), no. 8, 617–630.
13. T. L. Curtright and C. K. Zachos, *Deformation quantization of superintegrable systems and Nambu mechanics*, New J. Phys. **4** (2002), 83.1–83.16 (electronic).
14. C. Daskaloyannis, *Quadratic Poisson algebras of two-dimensional classical superintegrable systems and quadratic associative algebras of quantum superintegrable systems*, J. Math. Phys **42** (2001), no. 3, 1100–1119.
15. A. Dimakis, F. Müller-Hoissen, and T. Striker, *Umbral calculus, discretization, and quantum mechanics on a lattice*, J. Phys. A **29** (1996), no. 21, 6861–6876.
16. J. Drach, *Sur l'intégration logique des équations de la dynamique à deux variables. Forces conservatives. Integrales cubiques. Mouvement dans le plan*, C. R. Acad. Sci. **200** (1935), 22–26.
17. L. P. Eisenhart, *Separable systems of Stäckel*, Ann. Math. **35** (1934), 284–305.
18. N. W. Evans, *Superintegrability in classical mechanics*, Phys. Rev. A (3) **41** (1990), no. 10, 5666–5676.
19. ———, *Superintegrability of the Winternitz system*, Phys. Lett. A **147** (1990), no. 8-9, 483–486.
20. ———, *Group theory of Smorodinsky–Winternitz system*, J. Math. Phys. **32** (1991), no. 12, 3369–3375.
21. F. Fassò and T. Ratiu, *Compatibility of symplectic structures adapted to noncommutatively integrable systems*, J. Geom. Phys. **27** (1998), no. 3-4, 199–220.
22. V. Fock, *Zur Theorie des Wasserstoffatoms*, Z. Phys. **98** (1935), 145–154.
23. A. S. Fokas, *Group-theoretical aspects of constants of motion and separable solutions in classical mechanics*, J. Math. Anal. Appl. **68** (1979), no. 2, 347–370.
24. A. S. Fokas and P. A. Lagerstrom, *Quadratic and cubic invariants in classical mechanics*, J. Math. Anal. Appl. **74** (1980), no. 2, 325–341.
25. ———, *On the use of Lie–Bäcklund operators in quantum mechanics*, J. Math. Anal. Appl. **74** (1980), no. 2, 342–358.
26. I. Friš, V. Mandrosov, J. Smorodinsky, M. Uhlíř, and P. Winternitz, *On higher symmetries in quantum mechanics*, Phys. Lett. **16** (1965), 354–356.
27. W. I. Fushchych and A. G. Nikitin, *Higher symmetries and exact solutions of linear and nonlinear Schrödinger equation*, J. Math. Phys. **38** (1997), no. 11, 5944–5959.

28. M. Giordano, G. Marmo, A. Simoni, and F. Ventriglia, *Integrable and syperintegrable systems in classical and quantum mechanics*, Nonlinear Physics: Theory and Experiment II (Gallipoli, 2002) (M. Ablowitz, M. Boiti, F. Pempinelli and B. Prinari, eds.), Word Scientific, Singapore, 2003, pp. 179–185.
29. C. Gonera, *Isochronic potentials and new family of superintegrable systems*, J. Phys. A: Math. Gen. **37** (2004),no.13, 4085–4095.
30. A. González-López, N. Kamran, and P. J. Olver, *Quasi-exactly solvable Lie algebras of differential operators in two complex variables*, J. Phys. A **24** (1991), no. 17, 3995–4008.
31. ______, *Lie algebras of vector fields in the real plane*, Proc. London Math. Soc. (3) **64** (1992), no. 2, 339–368.
32. ______, *Normalizability of one-dimensional quasi-exactly solvable Schrödinger operators*, Comm. Math. Phys. **153** (1993), no. 1, 117–146.
33. S. Gravel, *Hamiltonians separable in Cartesian coordinates and third-order integrals of motion*, J. Math. Phys. **45** (2004), no. 3, 1003–1019.
34. S. Gravel and P. Winternitz, *Superintegrability with third-order integrals in quantum and classical mechanics*, J. Math. Phys. **43** (2002), no. 12, 5902–5912.
35. C. Grosche, G. S. Pogosyan, and A. N. Sissakian, *Path integral discussion for Smorodinskky–Winternitz potentials. II. The two- and three-dimensional sphere*, Fortschr. Phys. **43** (1995), no. 6, 523–563.
36. J. Jauch and E. L. Hill, *On the problem of degeneracy in quantum mechanics*, Phys. Rev. **57** (1940), 641–645.
37. J. Harnad, L. Vinet, O. Yermolayeva, and A. Zhedanov, *Two-dimensional Krall–Sheffer polynomials and integrable systems*, J. Phys. A **34** (2001), no. 48, 10619–10625.
38. C. R. Holt, *Construction of new integrable Hamiltonians in two degrees of freedom*, J. Math. Phys. **23** (1982), no. 6, 1037–1046.
39. E. G. Kalnins, J. M. Kress, W. Miller, Jr., and P. Winternitz, *Superintegrable systems in Darboux spaces*, J. Math. Phys. (44) (2003), n. 12, 5811–5848.
40. E. G. Kalnins, J. M. Kress, and P. Winternitz, *Superintegrability in a two-dimensional space of nonconstant curvature*, J. Math. Phys. **43** (2002), no. 2, 970–983.
41. E. G. Kalnins, W. Miller, Jr., and G. S. Pogosyan, *Superintegrability and associated polynomial solutions: Euclidean space and the sphere in two dimensions*, J. Math. Phys. **37** (1996), no. 12, 6439–6467.
42. ______, *Completeness of superintegrability in two-dimensional constant-curvature space*, J. Phys. A **34** (2001), no. 22, 4705–4720.
43. P. Létourneau and L. Vinet, *Superintegrable systems: polynomial algebras and quasi-exactly solvable Hamiltonians*, Ann. Physics **243** (1995), no. 1, 144–168.
44. D. Levi, J. Negro and M. A. del Olmo, *Discrete derivatives and symmetries of difference equations*, J. Phys A **34** (2001), no. 10, 2023–2030.
45. D. Levi, J. Negro and M. A. del Olmo, *Discrete q-derivatives and symmetries of q-difference equations*, J. Phys A **37** (2004), no. 10, 3459–3473.
46. D. Levi, P. Tempesta, and P. Winternitz, *Umbral calculus, difference equations and the discrete Schrödinger equation*, J. Math. Phys. (to appear).
47. ______, *Lorentz and Galilei invariance on lattices*, Phys. Rev. D 69 (2004), no. 10, 105011.
48. A. A. Makarov, J. A. Smorodinsky, Kh. Valiev and P. Winternitz, *A systematic search for nonrelativistic systems with dynamical symmetries*, Nuovo Cimento A **52** (1967), 1061–1084.
49. R. G. McLenaghan, R. G. Smirnov, and D. The, *Group invariant classification of separable Hamiltonian systems in the Euclidean plane and the* O(4) *symmetric Yang–Mills theories of Yatsun*, J. Math. Phys. **43** (2002), no. 3, 1422–1440.
50. W. Miller, Jr. *Symmetry and Separation of Variables*, Encyclopedia Math. Appl., vol. 4, Addison-Wesley, Reading, MA–London–Amsterdam, 1977.
51. A. S. Miščhenko and A. T. Fomenko, *A generalized Liouville method of integration of Hamiltonian systems*, Funkcional. Anal. i Priložen. 12 (1978), no. 2, 46–56 (Russian); Functional Anal. Appl. **12** (1978), no. 2, 113–121.
52. M. Moshinsky, J. Patera, and P. Winternitz, *Canonical transformation and accidental degeneracy. III. A unified approach to the problem*, J. Math. Phys. **16** (1975), no. 1, 82–92.
53. M. A. del Olmo, M. A. Rodríguez, and P. Winternitz, *Integrable systems based on* $\mathrm{SU}(p,q)$ *homogeneous manifolds*, J. Math. Phys. **34** (1993), no. 11, 5118–5139.

54. ———, *The conformal group* SU(2, 2) *and integrable systems on a Lorentzian hyperboloid*, Fortschr. Phys. **44** (1996), no. 3, 199–233.
55. P. J. Olver, *Applications of Lie groups to differential equations*, Grad. Texts in Math, vol. 107, Springer-Verlag, New York, (second edition) 1993.
56. M. F. Rañada, *Superintegrable $n = 2$ systems, quadratic constants of motion and the potentials of Drach*, J. Math. Phys. **38** (1997), no. 8, 4165–4178.
57. M. F. Rañada and M. Santander, *Superintegrable systems on the two-dimensional sphere* S^2 *and the hyperbolic plane* H^2, J. Math. Phys. **40** (1999), no. 10, 5026–5057.
58. H. P. Robertson, *Bemerkung über separierbare Systeme in der Wellenmechanik*, Math. Ann. **98** (1928), 749–752.
59. M. A. Rodríguez and P. Winternitz, *Quantum superintegrability and exact solvability in n dimensions*, J. Math. Phys. **43** (2002), no. 3, 1309–1322.
60. S. Roman, *The umbral calculus*, Pure Appl. Math., vol. 111, Academic Press, Inc., New York, 1984.
61. S. Roman and G. C. Rota, *The umbral calculus*, Advances in Math. **27** (1978), no. 2, 95–188.
62. G. C. Rota (ed.), *Finite operator calculus*, Academic Press, Inc., New York–London, 1975.
63. M. B. Sheftel, P. Tempesta, and P. Winternitz, *Superintegrable systems in quantum mechanics and classical Lie theory*, J. Math. Phys. **42** (2001), no. 2, 659–673.
64. ———, *Recursion operators, higher-order symmetries and superintegrability in quantum mechanics*, Czechoslovak J. Phys. **51** (2001), no. 4, 392–399.
65. M. A. Shifman and A. V. Turbiner, *Quantal problems with partial algebraization of the spectrum*, Comm. Math. Phys. **126** (1989), no. 2, 347–365.
66. Y. Smirnov and A. Turbiner, *Lie algebraic discretization of differential equations*, Modern Phys. Lett. A **10** (1995), no. 24, 1795–1802.
67. G. G. Sparano and G. Vilasi, *Noncommutative integrability and recursion operators*, J. Geom. Phys. **36** (2000), no. 3-4, 270–284.
68. P. Tempesta, A. V. Turbiner, and P. Winternitz, *Exact solvability of superintegrable systems*, J. Math. Phys. **42** (2001), no. 9, 4248–4257.
69. T. Thiemann, *Lectures on loop quantum gravity*, Lecture Notes in Phys., vol. 631, Springer, Berlin, 2003, pp. 41–135.
70. A. V. Turbiner, *Quasi-exactly-solvable problems and* sl(2) *algebra*, Comm. Math. Phys. **118** (1988), no. 3, 467–474.
71. ———, *Lie algebras and polynomials in one variable*, J. Phys. A **25** (1992), no. 18, L1087–L1093.
72. ———, *Lie algebras and linear operators with invariant subspace*, Lie Algebras, Cohomology, and New Applications in Quantum Mechanics (Springfield, MO, 1992), Contemp. Math., vol. 160, Amer. Math. Soc., Providence, RI, 1994, pp. 263–310.
73. ———, *Quasi-exactly-solvable differential equations*, CRC Handbook of Lie Group Analysis of Differential Equations. Vol. 3. New Trends in Theoretical Developments and Computational Methods, CRC Press, Boca Raton, FL, 1995, Chapter 12, pp. 331–366.
74. ———, *Canonical discretization.* I. *Discrete faces of (an)harmonic oscillator*, Internat. J. Modern Phys. A **16** (2001), no. 9, 1579–1603.
75. A. Ushveridze, *Quasy-exactly-solvable models in quantum mechanics*, Institute of Physics Publishing, Bristol, 1994.
76. P. Winternitz, J. Smorodinsky, M. Uhlíř, and I. Friš, *Symmetry groups in classical and quantum mechanics*, Jadernaja Fiz. **4** (1966), 625–635 (Russian); English transl., Soviet J. Nuclear Phys. **4** (1967), 444–450.
77. S. Wojciechowski, *Superintegrability of the Calogero–Moser system*, Phys. Lett. A **95** (1983), no. 6, 279–281.

Centre de recherches mathématiques, Université de Montréal, C.P. 6128, succ. Centre-Ville, Montréal, QC H3C 3J7, Canada

Current address: SISSA, via Beirut 2–4, 34014 Trieste, Italy

E-mail address: `tempesta@crm.umontreal.ca, tempesta@fm.sissa.it`

Centre de Recherches Mathématiques
CRM Proceedings and Lecture Notes
Volume **37**, 2004

Perturbations of Integrable Systems and Dyson–Mehta Integrals

Alexander V. Turbiner

ABSTRACT. We show that the existence of algebraic forms of quantum, exactly-solvable, completely-integrable $A-B-C-D$ and G_2, F_4, E_6, E_7, E_8 Olshanetsky–Perelomov Hamiltonians allow one to develop the *algebraic* perturbation theory, where corrections are computed by means of pure linear algebra. A Lie-algebraic classification of such perturbations is given. In particular, this scheme allows to an explicit study of anharmonic many-body problems. The approach also allows one to calculate ratios of a certain generalized Dyson–Mehta integrals algebraically, which is of interest by itself.

1. Introduction

This talk is devoted to a study of perturbations of the quantum-mechanical integrable systems. The feature of integrability of these systems attracts a lot of attention which in our opinion is not fully justified. The main problem of quantum mechanics is *finding the spectrum of the Hamiltonian via solving the Schroedinger equation*. However, a knowledge of operators commuting with the Hamiltonian does not help to solve the main problem of quantum mechanics. At least, for the moment such a connection is not known. The simplest illustration of the situation is provided by the one-dimensional quantum dynamics—by definition, any one-dimensional dynamics is completely- and super-integrable—but it does not imply that a general one-dimensional Schroedinger equation with an arbitrary potential can be solved exactly in whatever sense. The problem remains to be transcendental and is equivalent to the diagonalization of a generic matrix of infinite size. Existence of integrals commuting with the Hamiltonian gives a chance to assign definite quantum numbers to the eigenstates, which also can be done *constructively* only for the situation when exact solutions of the original Schrödinger equation are known. However, among known integrable systems in quantum mechanics there exists a class of systems with an outstanding property—these integrable systems are *exactly-solvable* in the sense that their eigenfunctions and eigenvalues can be found exactly. The notion *exactly* needs to be explained and we introduce the notion of *exact-solvability*.

2000 *Mathematics Subject Classification.* 17B10, 17B25, 17B80, 47A55, 81R05, 81R12.
This is the final form of the paper.

Let us take the Hamiltonian

$$\mathcal{H} = -\Delta + V(x), \quad x \in \mathbb{R}^d.$$

The system is called *integrable*, if there exist k functionally-independent operators commuting with $\mathcal{H}$,

$$[\mathcal{H}, I_i] = 0, \quad i = 1, 2 \ldots, k.$$

If $k > d - 1$, the system is called *superintegrable*. If $k = d - 1$ and for all i, j

$$[I_i, I_j] = 0,$$

the system is called *completely-integrable*.

The main problem of quantum mechanics is to solve the Schrödinger equation

$$\mathcal{H}\Psi(x) = E\Psi(x), \quad \Psi(x) \in L^2(\mathbb{R}^d).$$

Integrability in whatever sense does not help to solve the Schroedinger equation. However, among integrable systems there are several *exactly-solvable* systems [**5–7, 9, 10**]. They can be used as zero approximations to study realistic physical systems.

Now we introduce a definition of *exact-solvability*. Let us assume that a linear operator h possesses infinitely-many finite-dimensional invariant subspaces $\mathcal{V}_n$, $n = 0,\ 1, \ldots,$ which can ordered

$$\mathcal{V}_0 \subset \mathcal{V}_1 \subset \mathcal{V}_2 \subset \cdots \subset \mathcal{V}_n \subset \cdots \subset \mathcal{V},$$

thus forming an *infinite flag (filtration)* $\mathcal{V}$. The operator h preserves the flag $\mathcal{V}$.

GENERAL DEFINITION ([**14, 15**]). An operator h which preserves an infinite flag of finite-dimensional spaces $\mathcal{V}$ is called an *exactly-solvable operator with flag* $\mathcal{V}$.

If given h preserves several flags and among them there is a flag for which $\dim \mathcal{V}_n$ is *maximal* for any given n, such a flag is called *minimal*.

Below we deal with certain linear spaces of polynomials in several variables. It leads to a notion of *characteristic vector* [**8**]. Let us consider the triangular linear space of polynomials in k variables

$$\mathcal{P}_n^{(\alpha_1,\ldots,\alpha_k)} = \langle s_1^{p_1} s_2^{p_2} \cdots s_k^{p_k} \mid 0 \leq \alpha_1 p_1 + \alpha_2 p_2 + \cdots + \alpha_k p_k \leq n \rangle, \tag{1.1}$$

where α's are positive numbers and $n = 0,\ 1,\ 2, \ldots$. A characteristic vector is a vector with components which are equal to the coefficients (weights) α_i in front of p_i:

$$\vec{f} = (\alpha_1, \alpha_2, \ldots, \alpha_k). \tag{1.2}$$

Taking the spaces with $n = 0,\ 1,\ 2, \ldots,$ we arrive at the flag which has $\mathcal{P}_n^{(\alpha_1,\ldots,\alpha_k)}$ as *generating* linear space. We call this flag $\mathcal{P}^{(\alpha_1,\ldots,\alpha_k)}$. The flag with $\vec{f}_0 = (1, 1, \ldots, 1)$ is called *basic* and denoted $\mathcal{P}^{(k)} \equiv \mathcal{P}^{(1,\ldots,1)}$. It is clear that is minimal among all possible flags ever. This flag has a Lie-algebraic interpretation.

Let us take the algebra gl_{k+1} in almost degenerate or totally symmetric representation characterized by spins $(n, \underbrace{0, 0, \ldots 0}_{k-1})$. Consider operators

$$
\begin{aligned}
\mathcal{J}_i^- &= \frac{\partial}{\partial x_i}, && i = 1, 2 \ldots k,\\
\mathcal{J}_{ij}^0 &= x_i \frac{\partial}{\partial x_j}, && i, j = 1, 2 \ldots k,\\
\mathcal{J}^0 &= \sum_{i=1}^{d} x_i \frac{\partial}{\partial x_i} - n, && \\
\mathcal{J}_i^+ &= x_i \mathcal{J}^0 = x_i \left(\sum_{j=1}^{d} x_j \frac{\partial}{\partial x_j} - n \right), && i = 1, 2 \ldots, k,
\end{aligned}
\tag{1.3}
$$

where n is any real or complex number. It is easy to check that these operators obey the commutation relations of the algebra gl_{k+1}. The operators $\mathcal{J}_I^-$, $\mathcal{J}_{ij}{}^0$, $\mathcal{J}^)$ form the maximal affine subalgebra $b \subset \mathrm{gl}_{k+1}$. Furthermore, if $n = 0, 1, 2, \ldots$, then the *finite-dimensional* irreducible representations appear with representation spaces

$$\mathcal{P}_n^{(k)} = \left\langle x_1{}^{p_1} x_2{}^{p_2} \cdots x_d{}^{p_k} \;\middle|\; 0 \le \sum p_i \le n \right\rangle$$

REMARK. The flag $\mathcal{P}^{(k)}$ is made out of irreducible finite-dimensional representation spaces of the algebra gl_{k+1} taken in realization (1.3).

There exist other flags associated with irreducible finite-dimensional representation spaces of the Lie algebras of differential (difference) operators.

DEFINITION. The operator h is called *algebraic*, if it preserves a flag of polynomials. It has the form $\sum \mathrm{Pol}_n \cdot \partial^n$. This form is called *algebraic.*

It is evident that the existence of an algebraic form does *not* guarantee that the operator preserves the flag of polynomials.

THEOREM 1. *The linear differential operator h preserves the flag $\mathcal{P}^{(k)}$ iff $h = P(\mathcal{J}(b \subset \mathrm{gl}_{k+1}))$, where P is a polynomial in generators of the maximal affine subalgebra b of the algebra gl_{k+1}, taken in the realization* (1.3).

In particular, an *exactly-solvable* second order differential operator h (preserving the flag $\mathcal{P}^{(k)}$) has the form

$$h = P_2^{(ij)}(x) \partial_i \partial_j + P_1^{(i)}(x) \partial_i \tag{1.4}$$

where $P_2^{(ij)}(x)$ and $P_1^{(i)}(x)$ are 2nd- and 1st-degree polynomials in the coordinates x with arbitrary coefficients. It is the standard multidimensional *hypergeometrical* operator. It was shown in [**4, 12**] that the Calogero and Sutherland models (equivalently, the A_k-rational and trigonometric models), and the BC_k-rational and trigonometric models are of the hypergeometrical type (1.4). In order to get these models in the form (1.4) for the Hamiltonian, the ground state eigenfunction should be taken as a gauge factor and then Weyl-invariant (trigonometric) polynomials should be taken as new variables.

TABLE 1. Characteristic vectors of the Olshanetsky–Perelomov Hamiltonians.

Model	Rational	Trigonometric
A_N	$\underbrace{(1,1,\dots,1)}_{N}$	$\underbrace{(1,1,\dots,1)}_{N}$
BC_N	$\underbrace{(1,1,\dots,1)}_{N}$	$\underbrace{(1,1,\dots,1)}_{N}$
G_2	$(1,2)$	$(1,2)$
F_4	$(1,2,2,3)$	$(1,2,2,3)$
E_6	$(1,1,2,2,2,3)$	$(1,1,2,2,2,3)$
E_7	$(1,2,2,2,3,3,4)$	?
E_8	$(1,3,5,5,7,7,9,11)$	?

Another natural hypergeometrical operator appears in connection with the flag $\mathcal{P}^{(1,2)}$. Let us consider the algebra $g^{(2)}$ generated by

$$
\begin{aligned}
&L^1 = \partial_1, && L^2 = x_1\partial_1 - n/3, \\
&L^3 = x_2\partial_2 - n/6, && L^4 = x_1^2\partial_1 + 2x_1x_2\partial_2 - nx_1, \\
&L^5 = \partial_2, && L^6 = x_1\partial_2, \\
&L^7 = x_1^2\partial_2, && T = x_2\partial_{11}^2,
\end{aligned}
\tag{1.5}
$$

(see [**11**]), where n is any real or complex number. If $n = 0, 1, 2, \dots$, the algebra $g^{(2)}$ has finite-dimensional irreps with representation spaces given by $\mathcal{P}_n^{(1,2)}$. These representation spaces form the flag $\mathcal{P}^{(1,2)}$. It is rather obvious that except for the operator L^4 the remaining operators in (1.5) have the space $\mathcal{P}_n^{(1,2)}$ as a common invariant subspace for any n. Therefore a second-degree differential operator preserving the flag $\mathcal{P}^{(1,2)}$ is constructed by taking a second-degree polynomial in L_1, L_2, L_3, L_5, L_6 plus the generator T. Its explicit form is

$$
\begin{aligned}
h = {} & \big(a_1^{(11)}x_1^2 + a_2^{(11)}x_1 + a_3^{(11)}x_2 + a_4^{(11)}\big)\partial_1\partial_1 \\
& + \big(a_1^{(12)}x_1^3 + a_2^{(12)}x_1^2 + a_3^{(12)}x_1x_2 + a_1^{(12)}x_4 + a_5^{(12)}x_2 + a_6^{(12)}\big)\partial_1\partial_2 \\
& + \big(a_1^{(22)}x_1^4 + a_2^{(22)}x_1^3 + a_3^{(22)}x_1^2x_2 + a_4^{(22)}x_1^2 + a_5^{(22)}x_2^2 \\
& \qquad + a_6^{(22)}x_1x_2 + a_7^{(22)}x_1 + a_8^{(22)}x_2 + a_9^{(22)}\big)\partial_2\partial_2 \\
& + \big(b_1^{(1)}x_1 + b_2^{(1)}\big)\partial_1 + \big(b_1^{(2)}x_1^2 + b_2^{(2)}x_1 + b_3^{(2)}x_2 + b_4^{(2)}\big)\partial_2.
\end{aligned}
\tag{1.6}
$$

It was shown in [**11**] that the algebraic form of both rational and trigonometric G_2 models are a particular type of (1.6). It can be easily checked that the flag $\mathcal{P}^{(1,2)}$ is minimal for these models.

There exist other hypergeometrical operators related with flags of polynomials other than $\mathcal{P}^{(k)}$ or $\mathcal{P}^{(1,2)}$. In Table 1 we list the characteristic vectors of minimal flags of the Olshanetsky–Perelomov Hamiltonians.

2. Perturbation theory

The existence of algebraic forms of exactly-solvable operators allows the construction of an *algebraic* perturbation theory, where the procedure of finding corrections is *linear algebra procedure*.

Consider the spectral problem,

(2.1) $$(h_0 + \lambda h_1)\phi = E\phi,$$

where λ is a formal parameter, the solution of thatwhich we look for in perturbation theory,

(2.2) $$\phi = \sum \lambda^k \phi_k, \qquad E = \sum \lambda^k E_k.$$

The following theorem holds:

THEOREM 2 ([**17**]). *Let h_0 be an exactly-solvable operator with flag $\mathcal{V}$. Let the perturbation $h_1 = v_1\mathbf{1}$ be such that* (i) *v_1 is an element of a finite-dimensional space $\mathcal{V}_n$ from the flag and* (ii) *we look for $\phi \in \mathcal{V}$. Then the perturbation theory is algebraic: $\exists p(k)$ such that kth correction $\phi_k \in \mathcal{V}_{p(k)}$ and hence can be found algebraically.*

The equation to solve to find the kth correction is:

$$(h_0 - E_0)\phi_k = \sum_{i=1}^{k} E_i\phi_{k-i} - v_1\phi_{k-1}$$

As an immediate consequence of this theorem the following can be proved. If h_0 depends on some parameters holomorphically then the coefficients in ϕ_k and E_k are rational functions parameters. In the case of the absence of parameters in h_0 the coefficients in ϕ_k and E_k are rational numbers.

This form of the perturbation theory is characterized by a certain self-similarity property—a calculation of the kth correction is similar to the calculation of the first correction but with a modified perturbation potential, when v_1 is replaced by

$$v_k = -(\sum_{i=1}^{k-1} E_i\phi_{k-i} - v_1\phi_{k-1})/\phi_0.$$

3. Algebraic forms of the Olshanetsky–Perelomov Hamiltonians

As was indicated in the previous section the existence of algebraic forms of exactly-solvable problems allows one to construct perturbation theory algebraically for certain classes of potentials. Thus, the exactly-solvable Olshanetsky–Perelomov Hamiltonians for which there exist algebraic forms can be possible unperturbed problems for developing such a perturbation theory. We present those algebraic forms below.

In order to find the algebraic forms of the Olshanetsky-Perelomov Hamiltonians [**10**] the following strategy is used [**2, 4, 11, 12**],

- Gauging away the ground-state eigenfunction, $(\Psi_0)^{-1}\mathcal{H}\Psi_0$,
- Olshanetsky–Perelomov Hamiltonians possess different symmetries (permutations, translation-invariance, reflections, periodicity etc). These symmetries correspond to the Weyl group acting on the root space. By coding these symmetries to new coordinates we find 'premature' operators to these Hamiltonians,

Rational case—Weyl-invariant variables:

$$t_a^{(\Omega)}(x) = \sum_{\alpha\in\Omega} (\alpha, x)^a,$$

where a's are the *degrees* of the Weyl group W and α is an orbit. They are defined ambiguously depending on the orbit taken, but they *always lead to algebraic forms*,

Trigonometric case—trigonometric (periodic) Weyl-invariant variables:

$$\tau_a^{(\Omega)}(x) = \sum_{\alpha\in\Omega} \sin^a(\alpha, x),$$

Algebraic forms do not always appear in this case, only for very special combination of $\tau_a^{(\Omega)}$.

3.1. History.

Rational case.

• Flat-space metric $g^{\mu\nu}$ written in $t_a^{(\Omega)}$-coordinates has polynomial matrix elements [**1**].

• Laplacian and A_n-extended Laplacian (this means the A_n-rational model, or the Calogero model) in $t_a^{(\Omega)}$—invariants of A_n root space—have an algebraic form and are written in generators of the gl_n-algebra of 1st-order differential operators (1.3) [**12**].

• Laplacian and B_n-, C_n-, BC_n-extended Laplacian (in other words, B_n-, C_n-, BC_n-rational model) in $t_a^{(\Omega)}$—invariants of B_n-, C_n, BC_n root space—have an algebraic form and are written in generators of the gl_n-algebra of 1st-order differential operators (1.3) [**4**].

• Similarly for G_2 and F_4 rational models but with hidden algebras $g^{(2)}$ and $f^{(4)}$, resp. [**2, 8, 11**], as well as E_6, E_7, E_8 [**3**].

Trigonometric case.

• Flat-space metric $g^{\mu\nu}$ being written in $\tau_a^{(\Omega)}$-coordinates has polynomial matrix elements [**4, 12**].

• Laplacian and A_n-extended Laplacian (A_n-trigonometric model or Sutherland model) in $\tau_a^{(\Omega)}$—trigonometric invariants of A_n root space—have an algebraic form and are written in generators of the gl_n-algebra of 1st-order differential operators (1.3) [**12**].

• Laplacian and B_n-, C_n-, BC_n-extended Laplacian (B_n-, C_n-, BC_n-trigonometric model) in $\tau_a^{(\Omega)}$—trigonometric invariants of B_n, C_n, BC_n root space—have an algebraic form and are written in generators of the gl_n-algebra of 1st-order differential operators (1.3) [**4**].

• Similarly for G_2 and F_4 trigonometric models but with hidden algebras $g^{(2)}$ and $f^{(4)}$, resp. [**2, 8, 11**]

3.2. Algebraic forms. The algebraic forms of the Olshanetsky–Perelomov Hamiltonians are given by the following differential operator

$$h = \mathcal{A}_{ij}(\tau)\frac{\partial^2}{\partial\tau_i\partial\tau_j} + \mathcal{B}_i(\tau)\frac{\partial}{\partial\tau_i} \tag{3.1}$$

with polynomial coefficient functions:

(1) A_n rational case (Calogero model)

$$
\begin{aligned}
\mathcal{A}_{ij} &= \frac{(n-i+1)j}{n+1}\,\tau_i\tau_j + \sum_{l\geq\max(1,j-i)} (j-i-2l)\,\tau_{i+l}\,\tau_{j-l},\\
\mathcal{B}_i &= -\frac{1}{n+1}(1+\nu+\nu n)(n-i+2)(n-i+1)\,\tau_{i-1} + 2\omega(i+1)\tau_i,
\end{aligned}
\tag{3.2}
$$

where $i, j = 1, 2, \ldots, n$.

(2) A_n trigonometric case (Sutherland model)

$$
\begin{aligned}
\mathcal{A}_{ij} &= \frac{(n+1-i)j}{n+1}\tau_i\tau_j + \sum_{l\geq\max(1,j-i)} (j-i-2l)\tau_{i+l}\tau_{j-l},\\
\mathcal{B}_i &= \frac{1}{n+1}(1+\nu+\nu n)i(n+1-i)\tau_i,
\end{aligned}
\tag{3.3}
$$

where $i, j = 1, 2, \ldots, n$.

(3) BC_n rational model

$$
\begin{aligned}
\mathcal{A}_{ij} &= 4\sum_{l\geq 0}(2l+1+j-i)\tau_{i-l-1}\tau_{j+l},\\
\mathcal{B}_i &= 2[1+\nu_2+2\nu(n-i)](n-i+1)\tau_{i-1} - 4\omega i\tau_i,
\end{aligned}
\tag{3.4}
$$

where $i, j = 1, 2, \ldots, n$.

(4) BC_n trigonometric model

$$
\begin{aligned}
\mathcal{A}_{ij} &= n\tau_{i-1}\tau_{j-1} - \sum_{l\geq 0}[(i-l)\tau_{i-l}\tau_{j+l} + (l+j-1)\tau_{i-l-1}\tau_{j+l-1}\\
&\qquad - (i-2-l)\tau_{i-2-l}\tau_{j+l} - (l+j+1)\tau_{i-l-1}\tau_{j+l+1}],\\
\mathcal{B}_i &= \frac{\nu_3}{2}(i-n-1)\tau_{i-1} - \left[1+\nu_2+\frac{\nu_3}{2}+\nu(2n-i-1)\right]i\tau_i\\
&\qquad - \nu(n-i+1)(n-i+2)\tau_{i-2},
\end{aligned}
\tag{3.5}
$$

where $i, j = 1, 2, \ldots n$.

(5) G_2 rational model

$$
\begin{aligned}
&\mathcal{A}_{11} = 2\tau_1, \qquad \mathcal{A}_{12} = 12\tau_2, \qquad \mathcal{A}_{22} = -\tfrac{8}{3}\tau_1^2\tau_2,\\
&\mathcal{B}_1 = \tfrac{4}{3}\omega\tau_1 + 2(1+3\mu+3\nu), \qquad \mathcal{B}_2 = 4\omega\tau_2 - \tfrac{4}{3}(1+2\nu)\tau_1^2.
\end{aligned}
\tag{3.6}
$$

(6) G_2 trigonometric model

$$
\begin{aligned}
\mathcal{A}_{11} &= \left(2\tau_1 + \frac{\alpha^2}{2}\tau_1^2 - \frac{\alpha^4}{24}\tau_2\right), \qquad \mathcal{A}_{12} = \left(12 + \frac{8\alpha^2}{3}\tau_1\right)\tau_2,\\
\mathcal{A}_{22} &= -(\tfrac{8}{3}\tau_1^2\tau_2 - 2\alpha^2\tau_2^2),\\
\mathcal{B}_1 &= 2(1+3\mu+3\nu) + \tfrac{2}{3}(1+3\mu+4\nu)\alpha^2\tau_1,\\
\mathcal{B}_2 &= -\tfrac{4}{3}(1+2\nu)\tau_1^2 + \left[\tfrac{7}{3} + 4(\mu+\nu)\right]\alpha^2\tau_2.
\end{aligned}
\tag{3.7}
$$

Algebraic forms for the F_4 rational and trigonometric models as well as E_6, E_7, E_8 rational models can be found in [**2**] and [**3**], respectively.

All rational models possess a remarkable property—each of them is characterized by appearance of an infinite family of eigenstates depending on single variable, which is the second Weyl invariant t_2 (see definition above).[1] This family always

---

[1]The invariant t_2 has a unique feature—it is the same for all Weyl groups.

includes the ground state. Finding these states is reduced to solving a simple eigenvalue problem for one-dimensional differential operator, which can be written in the form

$$h_r = -2t_2\partial^2_{t_2t_2} + (4\omega t_2 - 1 - 2a_r)\partial_{t_2},$$

where the parameter a_r depends on the model studied. It can be easily demonstrated that the operator h_r is reduced to the Hamiltonian of the A_1 rational model

$$\mathcal{H}_{\mathrm{A}_1} = -\frac{1}{2}\sum_{i=1}^{2}\frac{\partial^2}{\partial {x_i}^2} + \frac{\omega^2}{2}(x_1 - x_2)^2 + \frac{g}{(x_1 - x_2)^2},$$

where the coupling constant $g = a_r(1 - a_r)$. Somehow it can be said that any rational model contains the A_1 rational model.

4. Perturbation theory (examples)

In this section we present several examples of perturbative calculations for the ground state for different perturbations. Other examples can be found in [**17**].

(1) G_2 rational model (see (3.1) with coefficients (3.6)).

Let us take τ_2 as a perturbation which in Cartesian coordinates looks like

$$v_1^{(r)} = \tau_2 = (x_1 - x_2)^2(x_2 - x_3)^2(x_1 - x_3)^2$$

The ground state is

$$\phi_0 = 1, \quad \epsilon_0 = 0.$$

After simple calculations we get

$$\begin{aligned} \phi_1 &= -\frac{(1+2\nu)}{8\omega^2}\tau_1^2 - \frac{1}{4\omega}\tau_2 + \frac{3}{8\omega^3}(1+2\nu)(2+3\mu+3\nu)\tau_1, \\ \epsilon_1 &= \frac{3}{4\omega^3}(1+2\nu)(1+3\mu+3\nu)(2+3\mu+3\nu). \end{aligned} \tag{4.1}$$

It is not very complicated to calculate several higher corrections.

(2) G_2 trigonometric model (see (3.1) with coefficients (3.7)).

(i) Take τ_1 as a perturbation, which in Cartesian coordinates is

$$v_1^{(t)} = \tau_1 = \frac{-2}{\alpha^2}\left[\sin^2\frac{\alpha}{2}(x_1 - x_2) + \sin^2\frac{\alpha}{2}(x_2 - x_3) + \sin^2\frac{\alpha}{2}(x_3 - x_1)\right]$$

The first correction to the ground state is given by

$$\phi_1 = -\frac{3}{2(1+3\mu+4\nu)\alpha^2}\tau_1,$$

and

$$\epsilon_1 = -\frac{3(1+3\mu+3\nu)}{(1+3\mu+4\nu)\alpha^2}. \tag{4.2}$$

It is worth noting that ϕ_1 depends on the single variable τ_1. However, this will not be the case for higher corrections they begin to depend on both τ_1, τ_2. The expression (4.2) has quite amusing property. If $\nu = 0$, which corresponds to the three-body Sutherland model, ϵ_1 does not depend on μ. Of course, this does not hold for higher corrections, but leads to a rather nontrivial property of some correlation function (see below).

(ii) Take τ_2 as a perturbation, which in Cartesian coordinates has the form

$$v_1^{(t)} = \tau_2 = \frac{16}{\alpha^6}[\sin\alpha(x_1 - x_2) + \sin\alpha(x_2 - x_3) + \sin\alpha(x_3 - x_1)]^2.$$

In the limit $\alpha \to 0$, the perturbation $v_1^{(t)}$ becomes $v_1^{(r)}$. The first correction to the ground state is given by

$$\phi_1 = -\frac{1}{N}\left[\frac{12(1+2\nu)}{\alpha^2}\tau_1^2 + 3(7+12\mu+16\nu)\tau_2 - 72\frac{(2+3\mu+3\nu)(1+2\nu)}{(1+3\mu+4\nu)\alpha^4}\tau_1\right]$$

and

$$\epsilon_1 = \frac{144(1+2\nu)(1+3\mu+3\nu)(2+3\mu+3\nu)}{(1+3\mu+4\nu)N\alpha^4}, \tag{4.3}$$

where $N = [7(7+12\mu+16\nu) - (1 - 84\mu - 82\nu - 144\mu^2 - 336\mu\nu - 192\nu^2)\alpha^2]$. It is not very complicated to calculate next several corrections.

(3) A_n rational model (in other words, the $(n+1)$-body Calogero problem, (see (3.1) with coefficients (3.2)). Let us take τ_3 as a perturbation. In Cartesian coordinates it looks like

$$v_1^{(r)} = \tau_3 = \sigma_4(y) = \sum_{i_1,i_2,i_3,i_4} y_{i_1}y_{i_2}y_{i_3}y_{i_4},$$

where $\sigma_4(y)$ is the fourth order elementary symmetric polynomial of the Perelomov coordinates as arguments,

$$Y = \sum x_i, \quad y_i = x_i - \frac{1}{n+1}Y, \quad i = 1, \ldots, (n+1).$$

An easy calculation gives a quite simple answer for the first correction

$$\begin{aligned} -\phi_1 &= \frac{1}{8\omega}\tau_3 + \frac{[1+\nu(n+1)](n-1)(n-2)}{32(n+1)\omega^2}\tau_1, \\ \epsilon_1 &= \frac{[1+\nu(n+1)]^2 n(n-1)(n-2)}{32(n+1)\omega^2}. \end{aligned} \tag{4.4}$$

It can be shown that for the nonphysical value $n = 1$ ('one-body' problem) the energy correction ϵ_1 as well as all other corrections vanish. The meaning of this result is not clear so far.

5. Generalized Dyson–Mehta integrals

Let us take one of the crystallographic root spaces R (A_n, B_n, C_n, D_n, BC_n, G_2, F_4, E_6, E_7, E_8). Suppose $P_c(t)$ is a Weyl-invariant polynomial of finite order in R and D_c = Weyl chamber. Then we define the integral

$$E_\nu(P_c) = \int_{D_c} P_c(t)(\text{Weyl det})^\nu \exp(-\omega t_2)\, dt,$$

where t_2 is 2nd invariant. Similarly one can introduce $Pt(\tau)$ as the Weyl-invariant trigonometric polynomial of finite order in R and D_t = Weyl alcove, and define the integral

$$T_\nu(P_t) = \int_{D_t} P_t(\tau)(\text{trig. Weyl det})^\nu\, d\tau.$$

E_ν, T_ν are related to the Selberg integrals and we call them *generalized Dyson–Mehta integrals*. If $P_c(t)$ $\big(P_t(\tau)\big) = 1$, they become the Dyson–Mehta integrals (for discussion see [**10**]).

THEOREM 3. (i) *the ratio* $E_\nu(P_c)/E_\nu(P_c=1)$ *is a rational function of* ν, ω *with rational coefficients,*
(ii) *the ratio* $T_\nu(P_t)/T_\nu(P_t=1)$ *is a rational function of* ν *with rational coefficients.*

PROOF. • The Weyl determinant defines the ground-state eigenfunction of the R-inspired rational (trigonometric) model,

• $\exists$ a set of Weyl invariants which as variables lead to an algebraic form of the Hamiltonian. They are generators of the algebra of Weyl-invariant polynomials,

• Following Theorem 2, the perturbation theory for any $P_c(t)$ $\big(P_t(\tau)\big)$ is algebraic. Hence the first energy correction to the ground state is found by pure algebraic means. Since the parameters enter into the algebraic forms of the unperturbed operators linearly or at most quadratically, it is evident that the answer can contain not more than rational function of parameters. □

It is worth recalling that for A_N,

$$(\text{Weyl det}) \doteq \prod_{i<j}^{N}(x_i - x_j)$$

$$(\text{trig. Weyl det}) \doteq \prod_{i<j}^{N}\sin(x_i - x_j)$$

are antisymmetric invariants.

A constructive way to realize Theorem 3 is to perturb one of the Olshanetsky–Perelomov Hamiltonians with a perturbation, which meets the conditions of the theorem. Then make two calculations of the first energy correction: (i) one is based on the algebraic perturbation theory and (ii) another is based on the Rayleigh–Schrödinger Perturbation Theory (RSPT). Now let us consider the examples presented in the previous section from this point of view.

Examples.
Case 2(i).

$$\frac{\int_{D_t(G_2)}[\sin^2\frac{1}{2}\alpha(x_1-x_2)+\sin^2\frac{1}{2}\alpha(x_2-x_3)+\sin^2\frac{1}{2}\alpha(x_3-x_1)]W^2(x_1,x_2,x_3)\,d^3x}{\int_{D_t(G_2)}W^2(x_1,x_2,x_3)\,d^3x} = \frac{3(1+3\mu+3\nu)}{2(1+3\mu+4\nu)},$$

where $D_t(G_2)$ is the G_2 Weyl alcove and

$$W(x_1,x_2,x_3) = \prod_{i<j}^{3}\sin^\nu\tfrac{1}{2}\alpha(x_i-x_j)\prod_{i<j}^{3}\sin^\mu\tfrac{1}{2}\alpha(x_i-x_j-2x_k).$$

Case 2(ii).

$$\frac{\int_{D_t(G_2)}[\sin\alpha(x_1-x_2)+\sin\alpha(x_2-x_3)+\sin\alpha(x_3-x_1)]^2W^2(x_1,x_2,x_3)\,d^3x}{\int_{D_t(G_2)}W^2(x_1,x_2,x_3)\,d^3x} = \frac{9\alpha^2(1+2\nu)(1+3\mu+3\nu)(2+3\mu+3\nu)}{(1+3\mu+4\nu)N},$$

Case 3.

$$\frac{\int_{D_c(\mathrm{A}_{n-1})} (\sum_{\{i_1,i_2,i_3,i_4\}} y_{i_1} y_{i_2} y_{i_3} y_{i_4}) \prod_{i<j} |y_i - y_j|^{2\nu} e^{-\omega \sum y_i^2} \, d^{n-1}y}{\int_{D_c(\mathrm{A}_{n-1})} \prod_{i<j} |y_i - y_j|^{2\nu} e^{-\omega \sum y_i^2} \, d^{n-1}y}$$
$$= \frac{1}{32\omega^2} \frac{(1+n\nu)^2 (n-1)(n-2)(n-3)}{n},$$

where $D_c(\mathrm{A}_{n-1})$ is the A_{n-1} Weyl chamber.

References

1. V. I. Arnol′d, *Wave front evolution and equivariant Morse lemma*, Comm. Pure Appl. Math. **29** (1976), no. 6, 557–582.
2. K. G. Boreskov, J. C. Lopez Vieyra, and A. V. Turbiner, *Solvability of* F_4 *integrable system*, Internat. J. Modern Phys. A **16** (2001), no. 29, 4769–4801.
3. K. G. Boreskov, A. V. Turbiner, and J. C. Lopez Vieyra, *Solvability of the Hamiltonians related to exceptional root spaces: rational case* (in progress)
4. L. Brink, A. Turbiner, and N. Wyllard, *Hidden algebras of the (super) Calogero and Sutherland models*, J. Math. Phys. **39** (1998), no. 3, 1285–1315.
5. F. Calogero, *Solution of a three-body problem in one dimension*, J. Math. Phys. **10** (1969), 2191–2196.
6. ———, *Ground state of a one-dimensional N-body problem*, J. Math. Phys. **10** (1969), 2197–2200.
7. ———, *Solution of the one-dimensional N-body problem with quadratic and/or inversely quadratic pair potentials*, J. Math. Phys. **12** (1971), 419–436
8. O. Haschke and W. Rühl, *Is it possible to construct exactly solvable models?*, Theoretical Physics. Fin de siècle (Wrocław, 1998), Lecture Notes in Phys., vol. 539, Springer, Berlin, 2000, pp. 118–140.
9. M. A. Olshanetsky and A. M. Perelomov, *Quantum completely integrable systems connected with semi-simple Lie algebras*, Lett. Math. Phys. **2** (1977/1978), no. 1, 7–13
10. ———, *Quantum integrable systems related to Lie algebras*, Phys. Rep. **94** (1983), no. 6, 313–404.
11. M. Rosenbaum, A. Turbiner, and A. Capella, *Solvability of the* G_2 *integrable system*, Internat. J. Modern Phys. A **13**, (1998), no. 22, 3885–3903.
12. W. Rühl and A. V. Turbiner, *Exact solvability of the Calogero and Sutherland models*, Modern Phys. Lett. A **10** (1995), no. 29, 2213–2222.
13. B. Sutherland, *Exact results for a quantum many-body problem in one dimension*, Phys. Rev. A **4** (1971), 2019–2021; II, Phys. Rev. **5** (1972), 1372–1376.
14. A. V. Turbiner, *Lie algebras and linear operators with invariant subspace*, Lie Algebras, Cohomology, and New Applications to Quantum Mechanics (Springfield, MO, 1992), Contemp. Math., vol. 160, Amer. Math. Soc., Providence, RI, 1994 pp. 263–310.
15. ———, *Lie-algebras and quasi-exactly-solvable differential equations*, CRC Handbook of Lie Group Analysis of Differential Equations. Vol. 3. New Trends in Theoretical Developments and Computational Methods, CRC Press, Boca Raton, FL, 1995, pp. 331–366.
16. ———, *Hidden algebra of three-body integrable systems*, Modern Phys. Lett. A **13** (1998), no. 18, 1473–1483.
17. ———, *Quantum many-body problems and perturbation theory*, Phys. Atomic Nuclei **65** (2002), no. 6, 1135–1143.

Institute for Theoretical and Experimental Physics, Moscow 117259, Russia
Current address: Instituto de Ciencias Nucleares, UNAM, A.P. 70-543, 04510 México, México
E-mail address: turbiner@nuclecu.unam.mx

Centre de Recherches Mathématiques
CRM Proceedings and Lecture Notes
Volume **37**, 2004

Separability and the Birkhoff–Gustavson Normalization of the Perturbed Harmonic Oscillators with Homogeneous Polynomial Potentials

Yoshio Uwano

ABSTRACT. In this paper, separability of the perturbed 2-dimensional isotropic harmonic oscillators with homogeneous polynomial potentials is characterized from their Birkhoff–Gustavson (BG) normalization, one of the conventional methods for *nonintegrable* Hamiltonian systems.

1. Introduction

The Bertrand-Darboux (BD) theorem is a very well-known theorem established more than a century ago [**4, 6, 10**], which characterizes separability and existence of constants of motion quadratic in momenta of simple dynamical systems on the Euclidean plane. As expected, the BD theorem has been playing a key role in various studies of *integrable systems* (see [**3, 4, 6, 10**] and references therein).

On turning to *nonintegrable systems*, the Birkhoff–Gustavson (BG) normalization is known as one of the conventional methods to treat them [**5**]: For a given system that can be normalized, the BG normalization provides a good account for the phase portrait in the regular regime.

Although directed to different characteristics of dynamical systems, those well-known methods have been encountered in the inverse problem of the BG normalization, posed by the author as follows ([**7, 9**]): *For a given polynomial* (*or power series*)[1] *Hamiltonian in the BG normal form* (*BGNF*), *identify all the possible Hamiltonians in polynomial or in power series which share the given BGNF.* In the inverse problem of the BG normalization of the perturbed isotropic harmonic oscillators (PHOs) with homogeneous polynomial potentials of *degree*-3,[2] the condition revealed in the BD theorem (BDC) has come out as follows:

THEOREM 1.1 ([**7**]). *A PHO with homogeneous polynomial potential of degree-3 shares its BGNF up to degree-4 with a PHO of degree-4 if and only if the PHO of*

2000 *Mathematics Subject Classification.* Primary 70H06, 70K45; Secondary 37J35, 37J46.
The author was supported in part by Grant-in-Aid Scientific Research No. 13660065, JSPS..
This is the final form of the paper.

[1]In discussing the BG normalization, we usually think of power series in the formal sense [**5**].

[2]Throughout this paper, we deal with 2-degrees of freedom systems only.

degree-3 satisfies the 'generic' BDC.[3] *The PHO of degree-4 corresponding to that PHO of degree-3 also satisfies the generic BDC.*

From now on, the perturbed 2-dimensional isotropic harmonic oscillator with a homogeneous polynomial potential of degree-δ will be abbreviated to a 'δ-PHO.' The aim of this paper is to report briefly that the following extension of Theorem 1.1 holds:[4]

THEOREM 1.2 (main theorem). *For any odd δ greater than or equal to 3, a δ-PHO shares its BGNF up to degree-$(2\delta-2)$ with a $(2\delta-2)$-PHO if and only if the δ-PHO is separable within a rotation of Cartesian coordinates. The $(2\delta-2)$-PHO sharing the BGNF with that δ-PHO is also separable within the same rotation of Cartesian coordinates.*

The organization of this paper is outlined as follows. In Section 2, the separability of the δ-PHOs is studied by applying the BD theorem to them. On associating the $2\times(\delta-1)$ matrix of the form

$$\mathcal{M}(K^{(\delta)}) = \begin{pmatrix} v_0^{(\delta)} - v_2^{(\delta)} & v_1^{(\delta)} - v_3^{(\delta)} & \cdots & v_{\delta-2}^{(\delta)} - v_\delta^{(\delta)} \\ 2v_1^{(\delta)} & 2v_2^{(\delta)} & \cdots & 2v_{\delta-1}^{(\delta)} \end{pmatrix} \tag{1.1}$$

with the δ-PHO Hamiltonian defined by

$$K^{(\delta)}(q,p) = \frac{1}{2}\sum_{j=1}^{2}(p_j^2+q_j^2) + V^{(\delta)}(q) \quad \text{and} \quad V^{(\delta)}(q) = \sum_{h=0}^{\delta} v_h^{(\delta)}\binom{\delta}{h} q_1^h q_2^{\delta-h}, \tag{1.2}$$

the separability of the δ-PHOs within rotations of Cartesian coordinates is shown to be equivalent to

$$\operatorname{rank}\mathcal{M}(K^{(\delta)}) = 1. \tag{1.3}$$

Since (1.3) with $\delta = 3, 4$ provides the 'generic' BDC for the 3- and 4-PHOs , the separability is taken as the extension of the 'generic' BDC. In Section 3, the BG normalization of the δ-PHOs is studied, which provides a plausible reason to extend the relation between the degrees, 3 and 4 of the PHOs in Theorem 1.1 to δ and $2\delta-2$ of those in Theorem 1.2. Section 4 is devoted to the proof of Theorem 1.2: The separability is shown to be sufficient for any δ-PHO to share its BGNF up to degree-$(2\delta-2)$ with a $(2\delta-2)$-PHO in Section 4.1, and is shown to be necessary in Section 4.2.

2. The separability of the δ-PHOs

2.1. The BD theorem for the δ-PHOs. To extend Theorem 1.1 to the PHOs of general degree, we wish to understand more the meaning of the 'generic' BDC for the 3- and 4-PHOs. We start with applying the BD theorem to the δ-PHOs.

THEOREM 2.1. *A δ-PHO admits a first integral quadratic in momenta if and only if it satisfies one of the followings:*

---

[3] See [**7**, Eqs. (54a) and (55b)].

[4] More detailed discussion will be made in a pair of subsequent papers, [**8**].

(I) *For odd* $\delta \geq 5$,

$$\operatorname{rank} \mathcal{M}(K^{(\delta)}) = 1, \tag{2.1}$$

where $\mathcal{M}(K^{(\delta)})$ *is the matrix associated with the* δ*-PHO by* (1.1) *and* (1.2).

(II) *For even* $\delta \geq 4$, *one of the following* (2.2) *and* (2.3);

$$\operatorname{rank} \mathcal{M}(K^{(\delta)}) = 1, \tag{2.2}$$

$$\begin{cases} v_{2h}^{(\delta)} = \{\binom{\delta/2}{h}/\binom{\delta}{2h}\} v_0^{(\delta)} \text{ with } v_0^{(\delta)} \neq 0 & (h = 1, \dots, \frac{\delta}{2}) \\ v_{2h'-1}^{(\delta)} = 0 & (h' = 1, \dots, \frac{\delta}{2}). \end{cases} \tag{2.3}$$

(III) *For* $\delta = 3$, *one of* (2.4) *and* (2.5);

$$\operatorname{rank} \mathcal{M}(K^{(3)}) = 1, \tag{2.4}$$

$$\operatorname{rank} \begin{pmatrix} 7v_1 & -v_0 + 6v_2 & -2v_1 + 5v_3 \\ -5v_0 + 2v_2 & -6v_1 + v_3 & -7v_2 \end{pmatrix} = 1. \tag{2.5}$$

PROOF. The proof is made straightforward by writing down explicitly the BDC [**6, 7**] in terms of $v_h^{(\delta)}$s.

The BDC: *There exist real-valued constants,* $(\alpha, \beta, \beta', \gamma, \gamma') \neq (0,0,0,0,0)$, *for which the potential function* $V(q)$ *of a given natural dynamical system satisfies*

$$\begin{aligned} \left(\frac{\partial^2 V}{\partial q_2^2} - \frac{\partial^2 V}{\partial q_1^2}\right)&(-2\alpha q_1 q_2 - \beta' q_2 - \beta q_1 + \gamma) \\ &+ 2\frac{\partial^2 V}{\partial q_1 \partial q_2}(\alpha q_2^2 - \alpha q_1^2 + \beta q_2 - \beta' q_1 + \gamma') \\ &\qquad + \frac{\partial V}{\partial q_1}(6\alpha q_2 + 3\beta) - \frac{\partial V}{\partial q_2}(6\alpha q_1 + 3\beta') = 0. \end{aligned} \tag{2.6}$$

(i) $\delta > 3$: On substituting

$$V(q) = \frac{1}{2}\sum_{j=1}^{2}(p_j^2 + q_j^2) + V^{(\delta)}(q) \tag{2.7}$$

with (1.2) into (2.6), the left-hand side, denoted by $\mathcal{L}^{(\delta)}$, of (2.6) is calculated to be

$$\mathcal{L}^{(\delta)} = \mathcal{L}_\delta^{(\delta)} + \mathcal{L}_{\delta-1}^{(\delta)} + \mathcal{L}_{\delta-2}^{(\delta)} + \mathcal{L}_1^{(\delta)} \tag{2.8}$$

with the homogeneous polynomial parts,

$$\mathcal{L}_1^{(\delta)} = 3\beta q_1 - 3\beta' q_2, \tag{2.9}$$

$$\begin{aligned} \mathcal{L}_{\delta-2}^{(\delta)} &= \delta(\delta-1)\sum_{h=0}^{\delta-2}\binom{\delta-2}{h}\{(v_h^{(\delta)} - v_{h+2}^{(\delta)})\gamma + 2v_{h+1}^{(\delta)}\gamma'\} q_1^h q_2^{\delta-2-h} \\ &= \delta(\delta-1)(\gamma, \gamma')\mathcal{M}(K^{(\delta)})\left(q_2^{\delta-2}, \dots, \binom{\delta-2}{h} q_1^h q_2^{\delta-2-h}, \dots, q_1^{\delta-2}\right)^T, \end{aligned} \tag{2.10}$$

$$(2.11)\quad \mathcal{L}^{(\delta)}_{\delta-1} = \beta\delta\Bigg[(2\delta+1)v_1^{(\delta)}q_2^{\delta-1} + \{(\delta+2)v_\delta^{(\delta)} - (\delta-1)v_{\delta-2}^{(\delta)}\}q_1^{\delta-1} + \sum_{h=1}^{\delta-2}\Bigg\{\left(3\delta\binom{\delta-1}{h} - (\delta-1)\binom{\delta-2}{h}\right)v_{h+1}^{(\delta)} - 3(\delta-1)\binom{\delta-2}{h-1}v_{h-1}^{(\delta)}\Bigg\}q_1^h q_2^{\delta-1-h}\Bigg]$$
$$- \beta'\delta\Bigg[\{(\delta+2)v_0^{(\delta)} - (\delta-1)v_2^{(\delta)}\}q_2^{\delta-1} + (2\delta+1)v_1^{(\delta)}q_1^{\delta-1} - \sum_{h=1}^{\delta-2}\Bigg\{\left(3\delta\binom{\delta-1}{h} - (\delta-1)\binom{\delta-2}{h-1}\right)v_h^{(\delta)} - 3(\delta-1)\binom{\delta-2}{h}v_{h+2}^{(\delta)}\Bigg\}q_1^h q_2^{\delta-1-h}\Bigg],$$

and

$$(2.12)\quad \mathcal{L}^{(\delta)}_{\delta} = \alpha\delta(\delta+2)\Bigg[v_1^{(\delta)}q_2^\delta - v_{\delta-1}^{(\delta)}q_1^\delta + \sum_{h=1}^{\delta-1}\Bigg\{\binom{\delta-1}{h}v_{h+1}^{(\delta)} - \binom{\delta-1}{h-1}v_{h-1}^{(\delta)}\Bigg\}q_1^h q_2^{\delta-h}\Bigg],$$

of degree-1, -$(\delta-2)$, -$(\delta-1)$ and -δ, respectively.[5]

(ii) $\delta = 3$: Substituting (2.7) with $\delta = 3$ into (2.6), we obtain

$$(2.13)\qquad \mathcal{L}^{(3)} = \mathcal{L}^{(3)}_3 + \mathcal{L}^{(3)}_2 + \mathcal{L}^{(3)}_1$$

with

$$(2.14)\qquad \mathcal{L}^{(3)}_1 = 3\{(\gamma,\gamma')\mathcal{M}(K^{(3)}) + (-\beta',\beta)\}\begin{pmatrix} q_2 \\ q_1 \end{pmatrix}$$

and

$$(2.15)\quad \mathcal{L}^{(3)}_2 = 3(\beta,\beta')\begin{pmatrix} 7v_1 & -v_0+6v_2 & -2v_1+5v_3 \\ -5v_0+2v_2 & -6v_1+v_3 & -7v_2 \end{pmatrix}(q_2^2, 2q_1q_2, q_1^2)^T,$$

where $\mathcal{L}^{(3)}_3$ is given by (2.12) with $\delta = 3$.

From the explicit expression of $\mathcal{L}^{(\delta)}$ thus obtained, we have Table 1, which classifies the possible choice of $(\alpha,\beta,\beta',\gamma,\gamma') \neq 0$.

The derivation of Table 1 will be given in more detail in [**8**]. □

TABLE 1. The BDC for the δ-PHOs.

	δ odd (≥ 5)	δ even (≥ 4)	$\delta = 3$
$\alpha = 0$, $(\beta,\beta') = (0,0)$, $(\gamma,\gamma') \neq (0,0)$	(2.1)	(2.2)	(2.4)
$\alpha = 0$, $(\beta,\beta') \neq (0,0)$, $(\gamma,\gamma') \neq (0,0)$	—	—	(2.5)
$\alpha \neq 0$, $(\beta,\beta') = (0,0)$, $(\gamma,\gamma') = (0,0)$	—	(2.3)	—

[5]The superscript ^{T} stands for the transpose throughout this paper.

2.2. The separability. As expected from the BD theorem, the classification, (2.1)–(2.5), of the BDC can be characterized from the separability viewpoint.

THEOREM 2.2. *A δ-PHO is separable*[6] *within a rotation of Cartesian coordinates if and only if* (1.3) *holds true.*

PROOF. From the Hamilton–Jacobi equation

$$\frac{1}{2}\sum_{j=1}^{2}\left(\frac{\partial S}{\partial q_j}\right)^2 + V^{(\delta)}(q) = E \quad (E: \text{energy value}) \tag{2.16}$$

for the δ-PHOs ($S(q)$: the generating function), it is easy to see that the separation of (2.16) with in rotations of Cartesian coordinates amounts to that of $V^{(\delta)}(q)$. Accordingly, let us assume that $V^{(\delta)}(q)$ is separable within a rotation

$$\kappa_\psi\colon (q,p) \to (\tilde{q},\tilde{p}) = (\sigma(\psi)q, \sigma(\psi)p) \tag{2.17}$$

with

$$\sigma(\psi) = \begin{pmatrix} \cos\psi & -\sin\psi \\ \sin\psi & \cos\psi \end{pmatrix} \quad (0 \le \psi < 2\pi). \tag{2.18}$$

Namely,

$$\widetilde{V}^{(\delta)}(\tilde{q}) = V^{(\delta)}(\sigma(\psi)^{-1}\tilde{q}) = \tilde{v}_0^{(\delta)}\tilde{q}_2^{\delta} + \tilde{v}_\delta^{(\delta)}\tilde{q}_1^{\delta}, \tag{2.19}$$

where $(\tilde{v}_0^{(\delta)}, \tilde{v}_\delta^{(\delta)}) \neq (0,0)$. Equations (2.17)–(2.19) are put together to yield the relation,

$$(4\sin 2\psi)(v_h^{(\delta)} - v_{h+2}^{(\delta)}) - (\cos 2\psi)(2v_{h+1}^{(\delta)}) = 0 \quad (h = 0,\ldots,\delta-2) \tag{2.20}$$

among $v_h^{(\delta)}$s, which immediately implies (1.3). The converse is easily shown by tracing back the discussion above. □

REMARK 2.3. It is also possible to characterize the other classes of δ-PHOs subject to BDC listed in Theorem 2.1 from the separation of variables viewpoint: The condition (2.3) is shown to be equivalent to the separability in the polar coordinates, and (2.5) in a off-centered parabolic coordinates [**8**].

Since the 'generic' BDC for the 3-PHOs and the 4-PHOs are equivalent to (1.3) with δ = 3, 4, we can thereby look the separability within rotations upon as the 'generic' BDC for δ-PHOs owing to Theorem 2.2.

3. The BG normalization of the δ-PHO

In this section, we proceed the BG normalization of the δ-PHOs, which provides us with a key other than Theorem 2.2 to extend Theorem 1.1 to Theorem 1.2.

We start by describing the way how the δ-PHO Hamiltonian $K^{(\delta)}(q,p)$ is brought into the BGNF. Let $G^{(\delta)}(\xi,\eta)$ be the BGNF of $K^{(\delta)}(q,p)$ and $W^{(\delta)}(q,\eta)$ be the generating function[7] used for the BG normalization, both of which are expressed

[6]As known well, a Hamiltonian system is said to be separable iff the Hamilton-Jacobi equation for that system is separable.

[7]$W^{(\delta)}(q,\eta)$ is said to be of the second-type since it is a function of the 'old' position variables q and the 'new' momentum ones η [**2**].

in power-series form

$$G^{(\delta)}(\xi,\eta)=\frac{1}{2}\sum_{j=1}^{2}(\eta_j^2+\xi_j^2)+\sum_{k=3}^{\infty}G_k^{(\delta)}(\xi,\eta) \tag{3.1}$$

and

$$W^{(\delta)}(q,\eta)=\sum_{j=1}^{2}q_j\eta_j+\sum_{k=3}^{\infty}W_k^{(\delta)}(q,\eta), \tag{3.2}$$

where $G_k^{(\delta)}(\xi,\eta)$ and $W_k^{(\delta)}(q,\eta)$ denote the homogeneous polynomial parts of degree-k of $G^{(\delta)}(\xi,\eta)$ and $W^{(\delta)}(q,\eta)$, respectively. We normalize $K^{(\delta)}(q,p)$ by applying the canonical transformation,

$$\tau\colon (q,p)\to(\xi,\eta)\quad \text{with } p=\frac{\partial W^{(\delta)}}{\partial q} \text{ and } \xi=\frac{\partial W^{(\delta)}}{\partial \eta} \tag{3.3}$$

associated with $W^{(\delta)}(q,\eta)$. Namely, $G^{(\delta)}(\xi,\eta)$ is determined by

$$G^{(\delta)}\left(\frac{\partial W^{(\delta)}}{\partial \eta},\eta\right)=K^{(\delta)}\left(q,\frac{\partial W^{(\delta)}}{\partial q}\right). \tag{3.4}$$

DEFINITION 3.1 (The BGNF). Let $G^{(\delta)}(\xi,\eta)$ be the power series (3.1), where each $G_k^{(\delta)}(\xi,\eta)$ is a homogeneous polynomial of degree-k ($k=3,4,\dots$) in (ξ,η).[8] Then $G^{(\delta)}(\xi,\eta)$ is said to be in the BGNF up to degree-ρ if and only if it satisfies

$$G_k^{(\delta)}(\xi,\eta)\in\ker D_{\xi,\eta}^{(k)}\quad (k=3,\dots,\rho), \tag{3.5}$$

where $D_{\xi,\eta}^{(k)}$ is the restriction of the linear differential operator

$$D_{\xi,\eta}=\sum_{j=1}^{2}\left(\xi_j\frac{\partial}{\partial\eta_j}-\eta_j\frac{\partial}{\partial\xi_j}\right) \tag{3.6}$$

on the vector space of homogeneous polynomials of degree-k in (ξ,η).

REMARK 3.2. The $D_{\xi,\eta}$ is understood as the Poisson derivation [**1**], $D_{\xi,\eta}=\{\frac{1}{2}\sum_{j=1}^{2}(\eta_j^2+\xi_j^2),\,\cdot\,\}$, associated with the isotropic harmonic oscillator.

The ordinary problem of the BG normalization of the δ-PHOs is posed as follows:[9]

DEFINITION 3.3 (The ordinary problem of degree-ρ, [**7, 9**]). Bring a given δ-PHO Hamiltonian $K^{(\delta)}(q,p)$ of the form (1.2) into the power series, $G^{(\delta)}(\xi,\eta)$, in the BGNF up to degree-ρ through (3.4), where generating function $W^{(\delta)}(q,\eta)$ of the second type in the form (3.2) is chosen to satisfy (3.4) and

$$W_k^{(\delta)}(q,\eta)\in\operatorname{image}D_{q,\eta}^{(\delta)}\quad (k=3,4,\dots,\rho). \tag{3.7}$$

The $D_{q,\eta}^{(k)}$ is defined by (3.6) with q in place of ξ.

REMARK 3.4. The condition (3.7) is very crucial to ensure the uniqueness of the outcome, $G^{(\delta)}(\xi,\eta)$, from $K^{(\delta)}(q,p)$ [**7, 9**].

---

[8] The homogeneous part of degree-2 in $G^{(\delta)}(\xi,\eta)$ is always in the isotropic harmonic oscillator form, due to (3.2).

[9] For the inverse problem of the BG normalization, see [**9**] and [**8**].

We are now in a position to present an explicit expression of the BGNF $G^{(\delta)}(\xi,\eta)$ of the δ-PHO Hamiltonian. A straightforward calculation of (3.4) shows the following:

LEMMA 3.5. *The BGNF, $G^{(\delta)}(\xi,\eta)$, of the δ-PHO Hamiltonian $K^{(\delta)}(q,p)$ in the form (1.2) takes the form*

$$G^{(\delta)}(\xi,\eta) = \frac{1}{2}\sum_{j=1}^{2}(\eta_j^2+\xi_j^2) + G_\delta^{(\delta)}(\xi,\eta) + G_{2\delta-2}^{(\delta)}(\xi,\eta) + o_{2\delta-1}(\xi,\eta), \tag{3.8}$$

where $o_{2\delta-1}(\xi,\eta)$ denotes a power series in (ξ,η) starting from degree-$(2\delta-1)$. The homogeneous polynomial part, $G_\delta^{(\delta)}(\xi,\eta)$, of degree-$\delta$ is given by

$$\begin{aligned} G_\delta^{(\delta)}(\xi,\eta) &= V^{(\delta)\ker}(\xi,\eta) \\ &= \begin{cases} 0 & (\delta \text{ odd}) \\ 2^{-\delta}\displaystyle\sum_{h=0}^{\delta}\binom{\delta}{h}v_h^{(\delta)}\Big[\sum_{m=M^\flat}^{M^\sharp}\sum_{n=N^\flat}^{N^\sharp}\binom{h}{m}\binom{\delta-h}{n}\zeta_1^m\zeta_2^n\bar\zeta_1^{h-m}\bar\zeta_2^{\delta-h-n}\Big] & (\delta \text{ even}), \end{cases} \end{aligned} \tag{3.9}$$

where $\zeta_j = \xi_j + i\eta_j$ $(j=1,\,2)$, and the range of the summation indices, m and n are determined by

$$\begin{cases} M^\flat = \max(0, h-\delta/2), & M^\sharp = \min(h,\delta/2), \\ N^\flat = \max(0,\delta/2-h), & N^\sharp = \min(\delta-h,\delta/2). \end{cases} \tag{3.10}$$

The superscript $^{\ker}$ *stands for taking the kernel component of $V^{(\delta)}(q)$ according to the action of $D_{q,\eta}^{(\delta)}$.*

The homogeneous polynomial part, $G_{2\delta-2}^{(\delta)}(\xi,\eta)$, of degree-$(2\delta-2)$ is calculated to be

$$G_{2\delta-2}^{(\delta)}(\xi,\eta) = \sum_{m=0}^{2\delta-2}\sum_{\ell=L^\flat}^{L^\sharp} c_{m,\ell}^{(\delta)}\zeta_1^\ell\zeta_2^{\delta-1-\ell}\bar\zeta_1^{m-\ell}\bar\zeta_2^{\delta-1-m+\ell} \tag{3.11}$$

with

$$\begin{aligned} c_{m,\ell}^{(\delta)} = \frac{2\delta^2}{4^\delta}\sum_{j=J^\flat}^{J^\sharp}\Big[&\binom{\delta-1}{j}\binom{\delta-1}{m-j}(v_j^{(\delta)}v_{m-j}^{(\delta)} + v_{j+1}^{(\delta)}v_{m-j+1}^{(\delta)}) \\ &\times\sum_{k=K^\flat}^{K^\sharp}\sum_{h=H^\flat}^{H^\sharp}\frac{\binom{j}{k}\binom{\delta-1-j}{h}\binom{m-j}{\ell-k}\binom{\delta-1-(m-j)}{\delta-1-\ell-h}}{\{(2(k+h+1)-\delta\}\{2(k+h)-\delta\}}\Big]. \end{aligned} \tag{3.12}$$

The ranges of the indices, h, j, k and ℓ, in (3.11) and (3.12) are given by

$$\begin{cases} H^\flat = \max(0,(m-j)-\ell), & H^\sharp = \min(\delta-1-\ell,\delta-1-j), \\ J^\flat = \max(0,m-(\delta-1)), & J^\sharp = \min(m,\delta-1), \\ K^\flat = \max(0,\ell-(m-j)), & K^\sharp = \min(j,\ell), \\ L^\flat = \max(0,m-(\delta-1)), & L^\sharp = \min(m,\delta-1) \end{cases} \tag{3.13}$$

respectively.

The proof is outlined in Appendix A and will be given in more detail in [**8**]. From Lemma 3.5, we obtain the following:

THEOREM 3.6. *Let* δ_1 *and* δ_2 *be any integers subject to*

$$3 \leq \delta_1 < \delta_2. \tag{3.14}$$

If the BGNF, $G^{(\delta_1)}(\xi, \eta)$*, of a* δ_1*-PHO Hamiltonian* $K^{(\delta_1)}(q, p)$ *coincides with the BGNF,* $G^{(\delta_2)}(\xi, \eta)$*, of a* δ_2*-PHO Hamiltonian* $K^{(\delta_2)}(q, p)$ *up to degree-*δ_2*, then* δ_1 *and* δ_2 *have to satisfy*

$$\delta_1 \textit{ is an odd integer} \quad \textit{and} \quad \delta_2 = 2\delta_1 - 2. \tag{3.15}$$

PROOF. Since the coincidence of the BGNFs is expressed as

$$G^{(\delta_1)}(\xi, \eta) - G^{(\delta_2)}(\xi, \eta) = o_{\delta_2+1}(\xi, \eta), \tag{3.16}$$

we obtain

$$G^{(\delta_1)}_{\delta_1}(\xi, \eta) = G^{(\delta_2)}_{\delta_1}(\xi, \eta) = 0 \tag{3.17}$$

from Lemma 3.5 as a necessary condition for (3.16). Equation (3.17) is put together with (3.9) to yield the first condition in (3.15). We derive the second one in turn. On recalling Lemma 3.5 again, the lowest nonvanishing homogeneous polynomial part of $G^{(\delta_1)}(\xi, \eta)$ turns out to be $G^{(\delta_1)}_{2\delta_1-2}(\xi, \eta)$ if δ_1 is odd, while $G^{(\delta_2)}_{\delta_2}(\xi, \eta)$ is the lowest one of $G^{(\delta_2)}(\xi, \eta)$. This shows the second equation of (3.15). □

Now that we have a pair of key Theorems 2.2 and 3.6, we are led to pose Theorem 1.2 as an extension of Theorem 1.1.

4. Proof of the main theorem

In this section, the proof of Theorem 1.2 is outlined. Throughout this section, we assume δ (≥ 3) to be odd.

4.1. The separability as a sufficiency. This subsection is devoted to show that the separability is a sufficient condition for a δ-PHO to share its BGNF with a $(2\delta - 2)$-PHO up to degree-$(2\delta - 2)$.

REMARK 4.1. Recalling the proof of Theorem 2.2, we see that the separability of the δ-PHOs[5] within rotations of Cartesian coordinates is equivalent to the separability of their Hamiltonians. We will use those equivalent expressions properly according to circumstances henceforce.

4.1.1. *The BG normalization of the* δ*-PHOs in separated form.* Let the δ-PHO be associated with the Hamiltonian in separated form,

$$K^{(\delta)}_{\text{sep}}(q, p) = \frac{1}{2}\sum_{j=1}^{2}(p_j^2 + q_j^2) + (v^{(\delta)}_{\text{sep},0} q_2^\delta + v^{(\delta)}_{\text{sep},\delta} q_1^\delta). \tag{4.1}$$

Then, applying (3.11) and (3.12) to $K_{\text{sep}}^{(\delta)}(q,p)$, we obtain the BGNF of $K_{\text{sep}}^{(\delta)}(q,p)$, denoted by $G_{\text{sep}}^{(\delta)}(\xi,\eta)$, to be in the following separated form,

$$(4.2)\quad G_{\text{sep}}^{(\delta)}(\xi,\eta) = \frac{1}{2}\sum_{j=1}^{2}(\eta_j^2+\xi_j^2) + \left\{\frac{2\delta^2}{4^\delta}\sum_{n=0}^{\delta-1}\frac{\binom{\delta-1}{n}^2}{\{2(n+1)-\delta\}(2n-\delta)}\right\} \times \{{v_{\text{sep},0}^{(\delta)}}^2(\zeta_2\bar{\zeta}_2)^{\delta-1} + {v_{\text{sep},\delta}^{(\delta)}}^2(\zeta_1\bar{\zeta}_1)^{\delta-1}\} + o_{2\delta-1}(\xi,\eta),$$

where $\zeta_j = \xi + \eta_j$ $(j = 1,\, 2)$. Recalling (3.8) and (3.9) with $2\delta-2$ in place of δ, we can find the unique $(2\delta-2)$-PHO Hamiltonian

$$(4.3)\quad K_{\text{sep}}^{(2\delta-2)}(q,p) = \frac{1}{2}\sum_{j=1}^{2}(p_j^2+q_j^2) + \left\{\frac{\delta^2}{2\binom{2\delta-2}{\delta-1}}\sum_{n=0}^{\delta-1}\frac{\binom{\delta-1}{n}^2}{\{2(n+1)-\delta\}(2n-\delta)}\right\} \times \{(v_{\text{sep},0}^{(\delta)})^2 q_2^{2\delta-2} + (v_{\text{sep},\delta}^{(\delta)})^2 q_1^{2\delta-2}\}$$

in separated form, whose BGNF coincides with $G_{\text{sep}}^{(\delta)}(\xi,\eta)$ up to degree-$(2\delta-2)$. To summarize, we have the following.

LEMMA 4.2. *For any δ-PHO Hamiltonian $K_{\text{sep}}^{(\delta)}(q,p)$ in separated form (4.1), there exists a unique $(2\delta-2)$-PHO Hamiltonian $K_{\text{sep}}^{(2\delta-2)}(q,p)$ in separated form (4.3) which shares the BGNF $G_{\text{sep}}^{(\delta)}(\xi,\eta)$ up to degree-$(2\delta-2)$ with $K_{\text{sep}}^{(\delta)}(q,p)$.*

4.1.2. *Proof of the sufficiency.* A key to prove the sufficiency is the commutativity of the BG normalization and the rotations of Cartesian coordinates:

LEMMA 4.3. *Let $G^{(\delta)}(\xi,\eta)$ be the BGNF of a δ-PHO Hamiltonian $K^{(\delta)}(q,p)$ up to degree-$(2\delta-2)$, which is associated with the generating function $W^{(\delta)}(q,\eta)$ (see (3.4)). Let $\tilde{G}^{(\delta)}(\tilde{\xi},\tilde{\eta})$, $\tilde{K}^{(\delta)}(\tilde{q},\tilde{p})$ and $\tilde{W}^{(\delta)}(\tilde{q},\tilde{\eta})$ be the power series defined by*

$$(4.4)\quad \begin{aligned} \tilde{K}^{(\delta)}(\tilde{q},\tilde{p}) &= K^{(\delta)}(\sigma^{-1}(\psi)\tilde{q},\sigma^{-1}(\psi)\tilde{p}) \\ \tilde{G}^{(\delta)}(\tilde{\xi},\tilde{\eta}) &= G^{(\delta)}(\sigma^{-1}(\psi)\tilde{\xi},\sigma^{-1}(\psi)\tilde{\eta}) \\ \tilde{W}^{(\delta)}(\tilde{q},\tilde{\eta}) &= W^{(\delta)}(\sigma^{-1}(\psi)\tilde{q},\sigma^{-1}(\psi)\tilde{\eta}), \end{aligned}$$

where $\sigma(\psi)$ is defined by (2.18). Then $\tilde{G}^{(\delta)}(\tilde{\xi},\tilde{\eta})$ is the BGNF of the δ-PHO Hamiltonian $\tilde{K}^{(\delta)}(\tilde{q},\tilde{p})$ up to degree-$(2\delta-2)$, which is brought through the canonical transformation, $(\tilde{q},\tilde{p}) \to (\tilde{\xi},\tilde{\eta})$, generated by $\tilde{W}^{(\delta)}(\tilde{q},\tilde{\eta})$.

PROOF. Due to the orthogonality of $\sigma(\psi)$, it is easily confirmed from (4.4) that $\tilde{K}^{(\delta)}(\tilde{q},\tilde{p})$, $\tilde{G}^{(\delta)}(\tilde{\xi},\tilde{\eta})$ and $\tilde{W}^{(\delta)}(\tilde{q},\tilde{\eta})$ are other δ-PHO Hamiltonian, BGNF up to degree-$(2\delta-2)$ and generating function of the second-type (cf (1.2), (3.1) and (3.2)), respectively. Further, the orthogonality of $\sigma(\psi)$ is put together with (4.4) to yield the equation,

$$(4.5)\quad \tilde{K}^{(\delta)}\left(\tilde{q},\frac{\partial\tilde{W}^{(\delta)}}{\partial\tilde{q}}\right) = \tilde{G}^{(\delta)}\left(\frac{\partial\tilde{W}^{(\delta)}}{\partial\tilde{\eta}},\tilde{\eta}\right),$$

from (3.4), so that $\tilde{G}(\tilde{\xi},\tilde{\eta})$ is the BGNF of $\tilde{K}(\tilde{q},\tilde{p})$. This completes the proof. □

We are at the final stage to prove the sufficiency of the separability in Theorem 1.2 now. Let us assume that $K^{(\delta)}(q,p)$ is separable within a rotation of

Cartesian coordinates: Namely, there exists the transformation κ_ψ with a suitable $\psi \in [0, 2\pi)$ (see (2.17) and (2.18)) which brings $K^{(\delta)}(q,p)$ to $K^{(\delta)}_{\rm sep}(\tilde{q},\tilde{p})$ through

$$K^{(\delta)}_{\rm sep}(\tilde{q},\tilde{p}) = K^{(\delta)}(\sigma(\psi)^{-1}\tilde{q}, \sigma(\psi)^{-1}\tilde{p}), \tag{4.6}$$

where $K^{(\delta)}_{\rm sep}(\tilde{q},\tilde{p})$ takes the separated form (4.1) with $(\tilde{q},\tilde{p})$ in place of (q,p). Then on applying Lemma 4.3 to the pair, $K^{(\delta)}_{\rm sep}(\tilde{q},\tilde{p})$ and $G^{(\delta)}_{\rm sep}(\tilde{\xi},\tilde{\eta})$, the BGNF $G^{(\delta)}(\xi,\eta)$ of $K^{(\delta)}(q,p)$ in Lemma 4.3 up to degree-$(2\delta-2)$ turns out to be

$$G^{(\delta)}(\xi,\eta) = G^{(\delta)}_{\rm sep}(\sigma(\psi)\xi, \sigma(\psi)\eta). \tag{4.7}$$

Further, according to Lemma 4.2, we can find a unique $(2\delta-2)$-PHO Hamiltonian in separated form, say $K^{(2\delta-2)}_{\rm sep}(\tilde{q},\tilde{p})$, sharing the BGNF $G^{(\delta)}_{\rm sep}(\tilde{\xi},\tilde{\eta})$ up to degree-$(2\delta-2)$ with $K^{(\delta)}_{\rm sep}(\tilde{q},\tilde{p})$. On defining the $(2\delta-2)$-PHO Hamiltonian $K^{(2\delta-2)}(q,p)$ and the BGNF $G^{(2\delta-2)}(\xi,\eta)$ by

$$\begin{aligned} K^{(2\delta-2)}(q,p) &= K^{(2\delta-2)}_{\rm sep}(\sigma(\psi)q, \sigma(\psi)p) \\ G^{(2\delta-2)}(\xi,\eta) &= G^{(\delta)}_{\rm sep}(\sigma(\psi)\xi, \sigma(\psi)\eta), \end{aligned} \tag{4.8}$$

Lemma 4.3 shows that $G^{(2\delta-2)}(\xi,\eta)$ is the BGNF of $K^{(2\delta-2)}(q,p)$ up to degree-$(2\delta-2)$. Equations (4.7) and (4.8) are put together to show the coincidence of $G^{(2\delta-2)}(\xi,\eta)$ with $G^{(\delta)}(\xi,\eta)$ up to degree-$(2\delta-2)$. To summarize, we have the following.

THEOREM 4.4. *Let δ be an odd integer greater than or equal to 3. If a δ-PHO Hamiltonian is separable (see Remark 4.1) within a rotation of Cartesian coordinates, there exists a unique $(2\delta-2)$-PHO Hamiltonian which shares the same BGNF up to degree-$(2\delta-2)$ with the δ-PHO. The $(2\delta-2)$-PHO is also separable within the same rotation of Cartesian coordinates.*

4.2. The separability as a necessity. This subsection is devoted to proving that separability is a necessary condition. Let us recall Lemma 3.5 and equate $G^{(\delta)}(\xi,\eta)$ with $G^{(2\delta-2)}(\xi,\eta)$ up to degree-$(2\delta-2)$. As a necessary and sufficient condition for the equation thus obtained, we have a number of equations

$$c^{(\delta)}_{m,\ell} = c^{(2\delta-2)}_{m,\ell} \quad (0 \le m \le 2\delta-2,\ L^\flat \le \ell \le L^\sharp). \tag{4.9}$$

Since the number of equations in (4.9) is large and since we have already shown separability as a sufficient condition, it would not be smart to study (4.9) for all the pairs of subscripts, (m,ℓ)s. Hence as a necessary condition for (4.9), we consider

$$c^{(\delta)}_{m,1}\tilde{c}^{(2\delta-2)}_{m,0} = c^{(\delta)}_{m,0}\tilde{c}^{(2\delta-2)}_{m,1} \quad (m = 2,3,\dots,\delta-1) \tag{4.10}$$

with

$$\tilde{c}^{(2\delta-2)}_{m,\ell} = 4^{1-\delta}\binom{2\delta-2}{m}\binom{m}{\ell}\binom{2\delta-2-m}{\delta-1-\ell} \quad (\ell = 0,1). \tag{4.11}$$

Note that we have $c^{(2\delta-2)}_{m,\ell} = v^{(2\delta-2)}_m \tilde{c}^{(2\delta-2)}_{m,\ell}$ for $\ell = 0,1$.

We wish to draw (1.3) as a necessary condition from (4.10). To do this, it is useful to prepare the notation,

$$\begin{aligned} U^{(\delta,m)}_j = {} & (v^{(\delta)}_j v^{(\delta)}_{m-j} + v^{(\delta)}_{j+1}v^{(\delta)}_{m-j+1}) \\ & - (v^{(\delta)}_{j+1}v^{(\delta)}_{m-(j+1)} + v^{(\delta)}_{(j+1)+1}v^{(\delta)}_{m-(j+1)+1}) \quad (j = 0,1,\dots,m-1). \end{aligned} \tag{4.12}$$

Using (3.9), (3.11) and (4.12), we can put (4.10) into the form

$$\left(\sum_{n=0}^{m} B_n^{(\delta,m)}\right)(v_m^{(\delta)}v_0^{(\delta)} + v_{m+1}^{(\delta)}v_1^{(\delta)}) + \sum_{j=0}^{m-1}\left(\sum_{n=0}^{j} B_n^{(\delta,m)}\right)U_j^{(\delta,m)} = 0 \qquad (m = 2,\dots,\delta-1), \tag{4.13}$$

where $B_n^{(\delta,m)}$s are defined to be

$$\begin{aligned}\left((\delta-1)^2\binom{\delta-2}{m-1}\right)^{-1} B_n^{(\delta,m)} &= \sum_{k=m-n-1}^{\delta-n-1}\frac{\binom{m-1}{n}\binom{\delta-m}{n+k+1-m}\binom{\delta-2}{k}}{\{2(k+1)-\delta\}(2k-\delta)} + \sum_{k=m-n-1}^{\delta-n-1}\frac{\binom{m-1}{n-1}\binom{\delta-m}{n+k+1-m}\binom{\delta-2}{k}}{\{2(k+2)-\delta\}\{2(k+1)-\delta\}} \\ &\quad - \sum_{k=m-n}^{\delta-h-1}\frac{\binom{m}{n}\binom{\delta-m-1}{n+k-m}\binom{\delta-1}{k}}{\{2(k+1)-\delta\}(2k-\delta)} \qquad (n \neq 0, m),\end{aligned} \tag{4.14}$$

$$\begin{aligned}\left((\delta-1)^2\binom{\delta-2}{m-1}\right)^{-1} B_0^{(\delta,m)} &= \sum_{k=m-1}^{\delta-2}\frac{\binom{\delta-m}{k+1-m}\binom{\delta-2}{k}}{\{2(k+1)-\delta\}(2k-\delta)} - \sum_{k=m}^{\delta-1}\frac{\binom{\delta-m-1}{k-m}\binom{\delta-1}{k}}{\{2(k+1)-\delta\}(2k-\delta)},\end{aligned} \tag{4.15}$$

$$\begin{aligned}\left((\delta-1)^2\binom{\delta-2}{m-1}\right)^{-1} B_m^{(\delta,m)} &= \sum_{k=0}^{\delta-m-1}\frac{\binom{\delta-m}{k+1}\binom{\delta-2}{k}}{\{2(k+2)-\delta\}\{2(k+1)-\delta\}} - \sum_{k=0}^{\delta-m-1}\frac{\binom{\delta-m-1}{k+1}\binom{\delta-1}{k}}{\{2(k+1)-\delta\}(2k-\delta)}.\end{aligned} \tag{4.16}$$

The definition (4.14)–(4.16) of $B_n^{(\delta,m)}$s and several well-known formulas for the binomial coefficients are put together to show

$$\sum_{n=0}^{m} B_n^{(\delta,m)} = 0 \quad (m = 2,\dots,\delta-1) \tag{4.17}$$

(see Appendix B), so that we can put (4.13) in the form

$$\sum_{j=0}^{m-1}\left(\sum_{n=0}^{j} B_n^{(\delta,m)}\right)U_j^{(\delta,m)} = 0 \quad (m = 2,\dots,\delta-1). \tag{4.18}$$

We verify (4.18) further by characterizing $U_j^{(\delta,m)}$ as the minors of $\mathcal{M}(K^{(\delta)})$ as follows. Denoting by $\Delta_{ab}^{(\delta)}$ the minor of $\mathcal{M}(K^{(\delta)})$ consisting of its a- and b-th columns $(1 \le a < b \le \delta - 1)$, we have

$$U_j^{(\delta,m)} = \begin{cases} \Delta_{j+1\,m-j}^{(\delta)} & (j = 0,\dots,[m/2]-1) \\ -\Delta_{m-j\,j+1}^{(\delta)} & (j = [m/2],\dots,m-1), \end{cases} \tag{4.19}$$

where $[m/2]$ stands for the integer part of $m/2$. Putting (4.19) and the symmetry,

$$\begin{aligned} U_j^{(\delta,m)} &= U_{m-j}^{(\delta,m)} \qquad (j = 0,1,\dots,m-1) \\ B_n^{(\delta,m)} &= B_{m-n}^{(\delta,m)} \qquad (n = 0,\dots,m), \end{aligned} \tag{4.20}$$

together, we can rewrite (4.18) as

$$\sum_{j=0}^{[m/2]-1}\left(\sum_{n=j+1}^{m-(j+1)} B_n^{(\delta,m)}\right)\Delta^{(\delta)}_{j+1\,m-j}=0 \quad (m=2,\dots,\delta-1). \tag{4.21}$$

From now on, we assume

$$(v_0^{(\delta)}-v_2^{(\delta)})^2+(v_1^{(\delta)})^2\neq 0 \tag{4.22}$$

on $\mathcal{M}(K^{(\delta)})$, without loss of generality (see Appendix C). We show the following lemma:

LEMMA 4.5. *Let $\mathcal{M}_k(K^{(\delta)})$ be the $2\times k$ submatrix*

$$\mathcal{M}_k(K^{(\delta)})=\begin{pmatrix} v_0^{(\delta)}-v_2^{(\delta)} & \cdots & v_{k-1}^{(\delta)}-v_{k+1}^{(\delta)} \\ v_1^{(\delta)} & \cdots & v_k^{(\delta)} \end{pmatrix} \quad (k=1,\dots,\delta-1) \tag{4.23}$$

of $\mathcal{M}(K^{(\delta)})$ subject to (4.22). If (4.21) with $m=k+1$ and $\operatorname{rank}\mathcal{M}_k(K^{(\delta)})=1$ holds true, then $\operatorname{rank}\mathcal{M}_{k+1}(K^{(\delta)})=1$ holds true.

PROOF. We write down one of the assumptions, (4.21) with $m=k+1$, more explicitly as

$$\begin{aligned}\sum_{j=0}^{[\frac{k+1}{2}]-1}&\left(\sum_{n=j+1}^{(k+1)-j-1} B_n^{(\delta,k+1)}\right)\Delta^{(\delta)}_{j+1\,k+1-j} \\ &=\left(\sum_{n=1}^{(k+1)+1} B_n^{(\delta,k+1)}\right)\Delta^{(\delta)}_{1\,k+1}+\sum_{j=1}^{[\frac{k+1}{2}]-1}\left(\sum_{n=j+1}^{(k+1)-j-1} B_n^{(\delta,k+1)}\right)\Delta^{(\delta)}_{j+1\,k+1-j}=0.\end{aligned} \tag{4.24}$$

Since the other assumption, $\operatorname{rank}\mathcal{M}_k(K^{(\delta)})=1$, implies

$$\Delta_{j+1\,k+1-j}=0 \quad (j=1,\dots,[(k+1)/2]-1), \tag{4.25}$$

we can bring (4.24) into

$$\left(\sum_{n=1}^{(k+1)+1} B_n^{(\delta,k+1)}\right)\Delta^{(\delta)}_{1\,k+1}=0. \tag{4.26}$$

so that we have $\Delta_{1\,k+1}=0$. The vanishing $\Delta_{1\,k+1}=0$ is put together with $\operatorname{rank}\mathcal{M}_k(K^{(\delta)})=1$ to show $\operatorname{rank}\mathcal{M}_{k+1}(K^{(\delta)})=1$. This completes the proof. □

We are at the final stage to draw (1.3) from (4.9): Let us consider (4.21) with $m=2$ for $K^{(\delta)}(q,p)$ subject to (4.22), which is written explicitly as $\Delta^{(\delta)}_{1\,2}=0$. This shows $\operatorname{rank}\mathcal{M}_2(K^{(\delta)})=1$ under (4.22). We can thereby start applying Lemma 4.5 to (4.21) recursively from $m=3$ to $m=\delta-1$, and finally reach to $\operatorname{rank}\mathcal{M}(K^{(\delta)})=\operatorname{rank}\mathcal{M}_{(\delta-1)+1}(K^{(\delta)})=1$ under (4.22). As for $K^{(\delta)}(q,p)$ not subject to (4.22), Appendix C shows that (1.3) is a necessary condition of (4.9). Recalling Theorem 2.2, we have the following.

THEOREM 4.6. *If a δ-PHO shares its BGNF up to degree-$(2\delta-2)$ with a $(2\delta-2)$-PHO, then the δ-PHO is separable within a rotation of Cartesian coordinates.*

Theorems 4.4 and 4.6 are put together to make our conclusion:

CONCLUSION. Our main theorem, Theorem 1.2, holds true.

Appendix A. Outline of the proof of Lemma 3.5

Since the BGNF of the 3-PHO Hamiltonians has been given explicitly[10] in [**7**], we focus our attention only on the case of $\delta \geq 4$ henceforth.

(i) Equating the homogeneous parts of degree-3 on the both sides of (3.4), we have $G_3^{(\delta)}(q,\eta) + (D_{q,\eta}^{(3)} W_3^{(\delta)})(q,\eta) = 0 \ (= H_3^{(\delta)}(q,\eta))$. On account of $\ker D_{q,\eta}^{(3)} = \{0\}$[11], the equation above for the degree-3 part is solved as $G_3^{(\delta)}(\xi,\eta) = 0$ and $W_3^{(\delta)}(q,\eta) = 0$. Then by induction, we can show

$$\text{(A.1)} \qquad G_\ell^{(\delta)}(\xi,\eta) = 0, \quad W_\ell^{(\delta)}(q,\eta) = 0 \qquad (\ell = 3,\ldots,\delta-1).$$

Under (A.1), we obtain $G_\delta^{(\delta)}(q,\eta) + (D_{q,\eta}^{(\delta)} W_\delta^{(\delta)})(q,\eta) = V^{(\delta)}(q) \ (= H_\delta^{(\delta)}(q,p))$ by calculating the degree-δ part of (3.4), which yields

$$\text{(A.2)} \qquad G_\delta^{(\delta)}(\xi,\eta) = V^{(\delta)\ker}(\xi,\eta), \qquad W_\delta^{(\delta)}(q,\eta) = (\tilde{D}_{q,\eta}^{(\delta)-1} V^{(\delta)\mathrm{image}})(q,\eta)$$

where $\tilde{D}_{q,\eta}^{(\delta)}$ denotes the restrict of $D_{q,\eta}^{(\delta)}$ on its image. This shows (3.9).

(ii) Under (A.1) and (A.2), we can show

$$\text{(A.3)} \qquad G_\ell^{(\delta)}(\xi,\eta) = 0, \quad W_\ell^{(\delta)}(q,\eta) = 0 \qquad (\ell = \delta+1,\ldots,2\delta-3)$$

by induction. On substituting (A.1)–(A.3) into (3.4), the degree-$(2\delta-2)$ part of (3.4) is calculated to be

$$\text{(A.4)} \qquad G_{2\delta-2}^{(\delta)}(q,\eta) + (D_{q,\eta}^{(\delta)} W_{2\delta-2}^{(\delta)})(q,\eta) = \frac{1}{2}\sum_{j=1}^{2}\left\{\left(\frac{\partial W_\delta^{(\delta)}}{\partial q_j}\right)^2 - \left(\frac{\partial W_\delta^{(\delta)}}{\partial \eta_j}\right)^2\right\}.$$

The final expression (3.11) with (3.12) is obtained by writing down explicitly the kernel component of the right-hand side of (A.4), which requires another very simple but long calculation. □

Appendix B. Proof of (4.17)

From (4.14)–(4.16), the right-hand side of (4.17) is put in a form

$$\text{(B.1)} \quad \left((\delta-1)^2\binom{\delta-2}{m-1}\right)^{-1}\sum_{n=0}^{m} B_n^{(\delta,m)}$$
$$= \sum_{k=0}^{\delta-2}\frac{\binom{\delta-2}{k}}{\{2(k+1)-\delta\}(2k-\delta)}\left(\sum_{n=N_1^\flat}^{N_1^\sharp}\binom{m-1}{n}\binom{\delta-m}{n+k+1-m}\right)$$
$$+ \sum_{k=0}^{\delta-2}\frac{\binom{\delta-2}{k}}{\{2(k+2)-\delta\}\{2(k+1)-\delta\}}\left(\sum_{n=N_2^\flat}^{N_2^\sharp}\binom{m-1}{n-1}\binom{\delta-m}{n+k+1-m}\right)$$
$$+ \sum_{k=0}^{\delta-1}\frac{\binom{\delta-1}{k}}{\{2(k+1)-\delta\}(2k-\delta)}\left(\sum_{n=N_3^\flat}^{N_3^\sharp}\binom{m}{n}\binom{\delta-m-1}{n+k-m}\right),$$

---

[10]The expression agrees with (3.8)–(3.11).

[11]There is no invariant homogeneous polynomials of odd-degree under the SO(2) action generated by $D_{q,\eta}$ viz. $\ker D_{q,\eta}^{(k)} = \{0\}$ for any odd k.

where

$$\text{(B.2)}\qquad \begin{aligned} N_1^\flat &= \max(0, m-1-k) & N_1^\sharp &= \min(m-1, \delta-1-k)\\ N_2^\flat &= \max(1, m-1-k) & N_2^\sharp &= \min(m, \delta-1-k)\\ N_3^\flat &= \max(0, m-k) & N_3^\sharp &= \min(m, \delta-1-k). \end{aligned}$$

For the sums with the summation index n on the right-hand side of (B.1), we have

$$\text{(B.3)}\qquad \begin{aligned} &\sum_{n=N_1^\flat}^{N_1^\sharp} \binom{m-1}{n}\binom{\delta-m}{n+k+1-m} = \sum_{n=N_3^\flat}^{N_3^\sharp} \binom{m}{n}\binom{\delta-m-1}{n+k-m} = \binom{\delta-1}{k}\\ &\sum_{n=N_2^\flat}^{N_2^\sharp} \binom{m-1}{n-1}\binom{\delta-m}{n+k+1-m} = \binom{\delta-1}{k+1}. \end{aligned}$$

Hence, (B.1)–(B.3) are put together to show

$$\text{(B.4)}\qquad \begin{aligned} &\frac{\sum_{n=0}^m B_n^{(\delta,m)}}{(\delta-1)^2\binom{\delta-2}{m-1}}\\ &= \Bigg[\sum_{k=0}^{\delta-2} \frac{\binom{\delta-1}{k}\binom{\delta-2}{k}}{\{2(k+1)-\delta\}(2k-\delta)} + \sum_{k=0}^{\delta-2} \frac{\binom{\delta-1}{k+1}\binom{\delta-2}{k}}{\{2(k+2)-\delta\}\{2(k+1)-\delta\}}\\ &\qquad\qquad + \sum_{k=0}^{\delta-1} \frac{\binom{\delta-1}{k}\binom{\delta-1}{k}}{\{2(k+1)-\delta\}(2k-\delta)}\Bigg]\\ &= \sum_{k=1}^{\delta-2} \frac{\binom{\delta-1}{k}\left\{\binom{\delta-2}{k} + \binom{\delta-2}{k-1} - \binom{\delta-1}{k}\right\}}{\{2(k+1)-\delta\}(2k-\delta)} = 0. \end{aligned}$$

Appendix C. SO(2) action to $\mathcal{M}(K^{(\delta)})$

To make the following discussion simple, we assume that δ is odd. Let $K'^{(\delta)}(q,p)$ be the δ-PHO Hamiltonian given by

$$\text{(C.1)}\qquad K'^{(\delta)}(q,p) = K^{(\delta)}(\sigma(\psi)^{-1}q, \sigma(\psi)^{-1}p),$$

where $K^{(\delta)}(q,p)$ is defined by (1.2), and $\sigma(\psi)$ by (2.18). It is then shown by a straightforward calculation that they are related as

$$\text{(C.2)}\qquad \mathcal{M}(K'^{(\delta)}) = \sigma(2\psi)\mathcal{M}(K^{(\delta)})R^{(\delta-2)}(\psi),$$

where $R^{(\delta-2)}(\psi)$ is the standard representation of SO(2) on the real vector space of homogeneous polynomials of degree-$(\delta-2)$.

Assume that $K^{(\delta)}(q,p)$ is not subject to (4.22). What we have to show is the existence of $\sigma(\psi)$ which brings $K^{(\delta)}(q,p)$ to $K'^{(\delta)}(q,p)$ with $\mathcal{M}(K'^{(\delta)})$ having nonvanishing first column. Since the nonexistence of such $\sigma(\psi)$ is equivalent to that the vector subspace, $\mathcal{N} = \operatorname{span}\{q_1^h q_2^{\delta-2-h}\}_{h=1,\dots,\delta-2}$, is an invariant subspace of the SO(2) action given by $R^{(\delta-2)}(\psi)$. However, this is not true: It is easily seen that any invariant subspace is given by a direct sum of the 2-dimensional subspaces each of which is spanned by $\operatorname{Re}\big((q_1+iq_2)^h(q_1-iq_2)^{\delta-2-h}\big)$ and $\operatorname{Im}\big((q_1+iq_2)^h(q_1-iq_2)^{\delta-2-h}\big)$ $(h = 0,\dots,(\delta-1)/2)$. Hence $\mathcal{N}$ is not SO(2)-invariant.

We are now at the final stage of explaining that the assumption (4.22) does not lose generality of our proof of necessity. Let consider the δ-PHO Hamiltonian $K^{(\delta)}(q,p)$ which is not subject to (4.22) and shares its BGNF with a $(2\delta-2)$-PHO $K^{(2\delta-2)}(q,p)$. As shown above, we can bring $K^{(\delta)}(q,p)$ to $K'^{(\delta)}(q,p)$ through (C.1) that satisfies (4.22) by adopting a suitable rotation with $\sigma(\psi)$. According to the commutativity of the BG normalization and the rotations shown in Section 3, $K'^{(\delta)}(q,p)$ shares its BGNF with a $(2\delta-2)$-PHO $K'^{(2\delta-2)}(q,p)$ other than $K^{(2\delta-2)}(q,p)$. Hence, the discussion in Section 4.2 is applied to $K'^{(\delta)}(q,p)$ to show the necessity of $\operatorname{rank}\mathcal{M}(K'^{(\delta)})=1$. Accordingly, the equation,

$$\operatorname{rank}\mathcal{M}(K'^{(\delta)}) = \operatorname{rank}\mathcal{M}(K^{(\delta)}), \tag{C.3}$$

following from (C.2) leads us to (1.3).

References

1. V. I. Arnol′d, *Mathematical methods of classical mechanics*, Grad. Texts in Math. vol. 60, Springer-Verlag, New York–Heidelberg, 1978.
2. H. Goldstein, *Classical mechanics*, Addison-Wesley, Cambridge, MA, 1951.
3. C. Grosche, G. S. Pogosyan and A. N. Sissakian, *Path integral discussion for Smorodinsky-Winternitz potentials. I. Two- and three-dimensional Euclidean space*, Fortschr. Phys. **43** (1995), no. 6, 453–521; II. *The two- and three-dimensional sphere*, 523–563.
4. J. Hietarinta, *Direct methods for the search of the second invariant*, Phys. Rep. **147** (1987), no. 2, 87–154.
5. J. K. Moser, *Lectures on Hamiltonian systems*, Mem. Amer. Math. Soc., vol. 81, Amer. Math. Soc., Providence, RI, 1968.
6. I. Marshall and S. Wojciechowski, *When is a Hamiltonian system separable*?, J. Math. Phys. **29** (1988), no. 6, 1338–1346.
7. Y. Uwano, *From the Birkhoff–Gustavson normalization to the Bertrand–Darboux integrability condition*, J. Phys. A **33** (2000), no. 37, 6635–6653.
8. ———, *Separability of the perturbed harmonic oscillators with homogeneous polynomial potentials from the Birkhoff–Gustavson normalization viewpoint* (in preparation).
9. Y. Uwano, N. Chekanov, V. Rostovtsev, and S. Vinitsky, *On normalization of a class of polynomial Hamiltonians: from ordinary and inverse points of view*, Computer Algebra in Scientific Computing—CACS'99 (Munich, 1999), Springer, Berlin, 1999, pp. 441–461.
10. E. T. Whittaker, *A treatise on the analytical dynamics of particles and rigid bodies*, Cambridge Univ. Press, Cambridge, 1937.

Department of Applied Mathematics and Physics, Kyoto University, Sakyo-ku, Kyoto 606-8501, Japan

E-mail address: uwano@amp.i.kyoto-u.ac.jp

Centre de Recherches Mathématiques
CRM Proceedings and Lecture Notes
Volume **37**, 2004

Integrability and Superintegrability without Separability

Josée Bérubé and Pavel Winternitz

Abstract. Integrable and superintegrable systems with second-order integrals of motion are investigated in the presence of a magnetic field. It is shown that such systems in classical and quantum mechanics differ. Moreover, superintegrability no longer implies multiseparability. A comparison is made with the case of purely scalar potentials, but higher-order integrals of motion.

1. Introduction

A large body of work exists on superintegrable systems in 2, 3 and n dimensions [**3,7–9,16–19,21,26,27,29,30,33**]. In most studies a Hamiltonian was postulated to have the form

$$H = \tfrac{1}{2}\vec{p}^{\,2} + V(\vec{r}). \tag{1.1}$$

The integrals were assumed to be second-order polynomials in the momenta

$$X_a = \sum_{i,j=1}^{n} A_a^{ik}(\vec{x})p_i p_k + g_a(\vec{x}), \qquad a = 1,\dots,n \tag{1.2}$$

$$Y_\mu = \sum_{i,j=1}^{n} a_\mu^{ik}(\vec{x})p_i p_k + k_\mu(\vec{x}), \qquad \mu = 1,\dots,k, \quad 1 \le k \le n-1 \tag{1.3}$$

satisfying

$$[H, X_a] = [H, Y_\mu] = 0, \qquad [X_a, X_b] = 0. \tag{1.4}$$

The above brackets are Poisson brackets in classical mechanics and Lie commutators in quantum mechanics. If n such integrals X_a exist (including the Hamiltonian), the system is integrable. If k further integral Y_μ exist, the system is superintegrable. If we have $k = n-1$, the system is maximally superintegrable. Notice that the integrals of motion X_a are in involution, whereas there is no such assumption for the integrals Y_μ. Moreover, there is no assumption on the commutators $[X_a, Y_\mu]$. It is always assumed that the integrals $\{X_1,\dots,X_n,Y_1,\dots,Y_k\}$ are functionally

2000 *Mathematics Subject Classification.* Primary 35Q40; Secondary 81R12.

The research of P. Winternitz was partly supported by research grants from NSERC of Canada and FQRNT du Québec.

This is the final form of the paper.

independent in classical mechanics. The corresponding operators in quantum mechanics are also supposed to be independent (in a sense that actually requires a more precise definition).

Some basic results valid for superintegrable systems of the above type are

(1) The potentials $V(\vec{x})$ allowing $n+k$ integrals of motion in classical and quantum mechanics coincide.
(2) In classical mechanics they allow additive separation of variables in the Hamilton–Jacobi equation in more than one system of coordinates. In quantum mechanics they allow multiplicative separation of variables in the Schrödinger equation in more than one system of coordinates. The separable coordinates in classical and quantum mechanics coincide. In other words, "quadratic superintegrability" implies "multiseparability."
(3) For maximally superintegrable systems in classical mechanics, all trajectories are periodic [**23**].
(4) In quantum mechanics, all known maximally superintegrable systems are exactly solvable [**27, 30**]. This means that their energy levels can be calculated algebraically, by solving algebraic, rather than transcendental equations.

The purpose of this presentation is to show that the first two properties do not hold if the Hamiltonian includes a vector potential, in addition to the scalar one, even if the integrals of motion are quadratic in the momenta.

Similarly, if we consider purely scalar potentials, as in (1.1), but allow for integrals of motion of order three, or higher, the first two properties of superintegrable systems do not hold.

Integrable systems with magnetic fields in classical mechanics have been studied elsewhere [**4, 22**]. A study of integrability and superintegrability with a magnetic field in quantum mechanics is much more recent [**2**]. For related studies of the separation of variables in the presence of vector potentials, see, e.g., [**1, 34, 35**].

2. First-order integrability and superintegrability with a magnetic field

The classical Hamiltonian that we will consider in this article is

$$H = \frac{1}{2}(p_1^2 + p_2^2) + \sum_{i=1}^{2} A_i(\vec{x})p_i + V(\vec{x}) \tag{2.1}$$

(We set $x_1 = x$, $x_2 = y$ whenever convenient). A first-order integral of motion will have the form

$$C = \sum_{i=1}^{2} f_i(\vec{x})\dot{x}_i + m(\vec{x}) \tag{2.2}$$

and must satisfy

$$\frac{dC}{dt} = \{H, C\} = 0 \tag{2.3}$$

on solutions of the equations of motion. In Newton form these equations are

$$\ddot{x} = -W_x + \Omega\dot{y}, \qquad \ddot{y} = -W_y - \Omega\dot{x} \tag{2.4}$$

with

$$W = V - \tfrac{1}{2}(A_1^2 + A_2^2), \qquad \Omega = A_{1,y} - A_{2,x}. \tag{2.5}$$

Notice that the magnetic field Ω and the effective potential W are gauge invariant quantities (contrary to the vector and scalar potentials $\vec{A}$ and V). Indeed a gauge transformation

$$V(x,y) \to V + (\vec{A}, \nabla\phi) + \tfrac{1}{2}(\nabla\phi)^2, \qquad \vec{A} \to \vec{A} + \nabla\phi \tag{2.6}$$

with $\phi(x,y)$ arbitrary does not change the equations of motion and does not influence any physics.

Equation (2.3) leads to six determining equations for the five unknown functions f_1, f_2, m, W and Ω.

In quantum mechanics we symmetrize the Hamiltonian (2.1) and the integral of motion (2.2), after replacing all velocities by momenta

$$\dot{x}_k = p_k + A_k, \qquad p_k = -i\hbar\frac{\partial}{\partial x_k}. \tag{2.7}$$

We have

$$H = -\frac{1}{2}\hbar^2\Delta - \frac{i\hbar}{2}\sum_k\left(A_k\frac{\partial}{\partial x_k} + \frac{\partial}{\partial x_k}A_k\right) + V(\vec{x}) \tag{2.8}$$

$$X = -\frac{i\hbar}{2}\sum_k(f_k\partial_{x_k} + \partial_{x_k}f_k) + \sum_k f_kA_k + m. \tag{2.9}$$

The commutativity condition $[X,H]=0$ provides six determining equations

$$f_{1,x} = 0, \qquad f_{2,y} = 0, \qquad f_{1y} + f_{2x} = 0 \tag{2.10}$$

$$-f_2\Omega + m_x = 0, \qquad f_1\Omega + m_y = 0, \qquad f_1W_x + f_2W_y = 0. \tag{2.11}$$

These equations do not contain the Planck constant $\hbar$ and hence coincide exactly with the classical determining equations, obtained from the requirement (2.3). Equations (2.10) imply

$$X = \alpha L_3 + \beta P_1 + \gamma P_2 + \alpha(x_2A_1 - x_1A_2) + \beta A_1 + \gamma A_2 + m, \tag{2.12}$$

where α, β and γ are constants and

$$P_1 = -i\hbar\partial_{x_1}, \qquad P_2 = -i\hbar\partial_{x_2}, \qquad L_3 = -i\hbar(x_2\partial_{x_1} - x_1\partial_{x_2}). \tag{2.13}$$

Thus, the leading part of the integral of motion X lies in the Euclidean Lie algebra e(2). Putting $f_1 = \alpha x_2 + \beta$, $f_2 = -\alpha x_1 + \gamma$ into (2.11) we obtain equations for Ω, W and m that are easy to solve. Two cases must be distinguished:

(1) $\alpha \neq 0$. We put $\alpha = 1$ and by translations set $\beta = \gamma = 0$. We find that the problem possesses a geometric symmetry, invariance under rotations:

$$\Omega = \Omega(r), \quad W = W(r), \quad r = \sqrt{x_1^2 + x_2^2}. \tag{2.14}$$

(2) $\alpha = 0$, $\beta^2 + \gamma^2 \neq 0$. By rotations and a normalization we set $\gamma = 1$, $\beta = 0$ and again obtain a geometric symmetry, invariance under translations in the y direction

$$\Omega = \Omega(x), \qquad W = W(x). \tag{2.15}$$

Thus, for first-order integrals there are no surprises. The magnetic field and effective potential in the classical and quantum cases coincide. First-order integrability implies a geometric symmetry: either translational, or rotational invariance.

Let us however consider the separation of variables in the Schrödinger equation in the case (2.15) (a translationally invariant magnetic field and effective potential). Using a gauge transformation to annul A_2, we obtain

$$A_1(x,y) = \Omega(x)y, \qquad A_2 = 0, \qquad V(x,y) = W(x) + \tfrac{1}{2}\Omega^2 y^2. \tag{2.16}$$

The Hamiltonian and first integral then are

$$\begin{aligned} H\psi &= \left[\frac{1}{2}(-i\hbar\partial_x + \Omega y)^2 - \frac{\hbar^2}{2}\partial_y^2 + W(x)\right]\psi = E\psi \\ \widehat{X}\psi &= \big(-i\hbar\partial_y + m(x)\big)\psi = k\psi \qquad \dot{m}(x) = \Omega(x). \end{aligned} \tag{2.17}$$

We see that the variables in the Schrödinger equation do not separate. They do however R-separate, i.e., we can write the solution as

$$\psi(x,y) = R(x,y)\psi_1(x,k,E)\psi_2(y,k,E), \tag{2.18}$$

where $R(x,y)$ is some universal factor (not depending on k and E). Substituting into Eq. (2.17), we find

$$\psi(x,y) = e^{-i(ym(x)/\hbar)}e^{i(ky/\hbar)}\psi_1(x) \tag{2.19}$$

where $\psi_1(x)$ satisfies

$$-\frac{\hbar^2}{2}\ddot{\psi}_1 + [\tfrac{1}{2}\big(-m(x)+k\big)^2 + W(x)]\psi_1 = E\psi_1. \tag{2.20}$$

We see that the function $R(x,y)$ in (2.19) is a phase factor, as long as $m(x)$ is real.

The system is first-order superintegrable if two independent first-order integrals exist. This is only possible if both Ω and W are constant

$$\Omega = \Omega_0, \qquad W = W_0. \tag{2.21}$$

Choosing a convenient gauge, we obtain the following Hamiltonian and three integrals of motion

$$H = \tfrac{1}{2}(p_1 + \Omega x_2)^2 + p_2^2 \tag{2.22}$$

$$X_1 = P_1, \qquad X_2 = p_2 + x_1\Omega, \qquad X_3 = L_3 - \tfrac{1}{2}(x_1^2 - x_2^2)\Omega. \tag{2.23}$$

The four integrals (2.22), (2.23) cannot be independent and indeed, both in classical and quantum mechanics they satisfy

$$X_1^2 + X_2^2 + 2\Omega_0 X_3 - 2H = 0. \tag{2.24}$$

Moreover, they satisfy the commutation relations of a central extension of the Euclidean Lie algebra:

$$[X_1, X_2] = -i\hbar\Omega, \quad [X_3, X_1] = -i\hbar X_2, \quad [X_3, X_2] = i\hbar X_1, \quad [X_i, \Omega] = 0. \tag{2.25}$$

The classical equations of motion (2.4) in this case are easily solved to give

$$x = x_0 + A\sin\Omega(t-t_0), \qquad y = y_0 + A\cos\Omega(t-t_0), \tag{2.26}$$

i.e, the particle rotates in a circle around a point (x_0, y_0) with constant angular velocity Ω and radius A. The Schrödinger equation separates in Cartesian coordinates

$$\psi(x,y) = e^{i(\lambda/h)x}f(y), \qquad f'' - \left\{\left(\frac{\Omega y + \lambda}{\hbar}\right)^2 - \frac{2E}{\hbar^2}\right\}f = 0 \tag{2.27}$$

and R-separates in polar coordinates

$$\psi(r,\phi) = e^{-(i/4)\Omega r^2 \sin^2\phi} J_m(kr) e^{im\phi}, \qquad k^2 = 2E + m\Omega \tag{2.28}$$

where $J_m(kr)$ is a Bessel function. The nonseparated function $R(x,y)$ is a phase factor.

3. Second-order integrability with a magnetic field

The Hamiltonian has the form (2.1), but we shall use the notations $x_1 \equiv x$, $x_2 \equiv y$. The second-order classical integral of motion will have the form

$$C = g_1(x,y)\dot{x}^2 + g_2(x,y)\dot{y}^2 + g_3(x,y)\dot{x}\dot{y} + k_1(x,y)\dot{x} + k_2(x,y)\dot{y} + m(x,y). \tag{3.1}$$

In order to treat the classical and quantum cases on the same footing, we replace the velocities by momenta, putting

$$\dot{x} = p_1 + A_1, \qquad \dot{y} = p_2 + A_2. \tag{3.2}$$

After symmetrizing, we obtain the operator corresponding to C in quantum mechanics as

$$\begin{aligned} X = &-\frac{1}{2}\hbar^2\{2g_1\partial_x^2 + 2g_2\partial_y^2 + 2g_3\partial_x\partial_y + (2g_{1,x} + g_{3,y})\partial_x \\ &\quad + (2g_{2,y} + g_{3,x})\partial_y + g_{1,xx} + g_{2,yy} + g_{3,xy}\} \\ &- \frac{i\hbar}{2}\{(4g_1A_1 + 2g_3A_2 + 2k_1)\partial_x + (4g_2A_2 + 2g_3A_1 + 2k_2)\partial_y \\ &\quad + 2g_1A_{1,x} + 2g_{1,x}A_1 + 2g_2A_{2,y} + 2g_{2,y}A_2 + g_3A_{1,y} + g_3A_{2,x} \\ &\quad + A_1g_{3,y} + A_2g_{3,x} + k_{1,x} + k_{2,y}\} \\ &+ g_1A_1^2 + g_2A_2^2 + g_3A_1A_2 + k_1A_1 + k_2A_2 + m. \end{aligned} \tag{3.3}$$

Imposing the commutation relation $[H,X] = 0$, we obtain ten determining equations for the operator X, namely

$$g_{1,x} = 0, \quad g_{2,y} = 0, \quad g_{1,y} + g_{3,x} = 0, \quad g_{2,x} + g_{3,y} = 0, \tag{3.4}$$

$$\begin{aligned} &k_{1x} - g_3\Omega = 0, \quad k_{2y} + g_3\Omega = 0, \quad 2\Omega(g_1 - g_2) + k_{1y} + k_{2x} = 0, \\ &2g_1W_x + g_3W_y + k_2\Omega - m_x = 0, \quad 2g_2W_y + g_3W_x - k_1\Omega - m_y = 0, \end{aligned} \tag{3.5}$$

$$k_1W_x + k_2W_y + \frac{\hbar^2}{4}(g_{2x}\Omega_y - g_{1y}\Omega_x) = 0. \tag{3.6}$$

In classical mechanics we impose the relation (2.3) on the integral (3.1). We also obtain ten equations, namely (3.4), (3.5) as they are and (3.6) with $\hbar = 0$. Thus, quadratic integrability in a magnetic field is not the same in classical and quantum mechanics! This is due to the last term in (3.6), proportional to the square of the Planck constant.

From (3.4) we obtain

$$g_1 = \alpha y^2 - \beta y + \delta, \quad g_2 = \alpha x^2 + \gamma x + \zeta, \quad g_3 = -2\alpha xy + \beta x - \gamma y + \xi, \tag{3.7}$$

and the operator (3.3) can be rewritten as

$$
\begin{aligned}
(3.8) \quad X = &\alpha[(L_3 + yA_1 - xA_2)^2 + \hbar^2] \\
&- \frac{1}{2}\beta[(L_3 + yA_1 - xA_2)(P_1 + A_1) + (P_1 + A)(L_3 + yA_1 - xA_2)] \\
&- \frac{1}{2}\gamma[(L_3 + yA_1 - xA_2)(P_2 + A_2) + (P_2 + A_2)(L_3 + yA_1 - xA_2)] \\
&+ \delta(P_1 + A_1)^2 + \zeta(P_2 + A_2)^2 + \xi(P_1 + A_1)(P_2 + A_2) \\
&- \frac{i\hbar}{2}(2k_1\partial_x + k_{1,x} + 2k_2\partial_y + k_{2,y}) + k_1A_1 + k_2A_2 + m.
\end{aligned}
$$

We see that the second-order terms in X lie in the enveloping algebra of the Euclidean Lie algebra (2.13). The quantities $\alpha, \ldots, \xi$ are constants, k_1, k_2 and m are functions of x and y.

Euclidean transformations of the plane will leave the form of H and C (or X) invariant, but will change the constants $\alpha, \ldots, \xi$ into new ones (they will also change the form of the as yet undertermined functions V, A_1, A_2, k_1, k_2 and m). Two quantities are invariant under Euclidean transformations and linear combinations with the Hamiltonian, namely **[24, 33]**

$$
(3.9) \qquad I_1 = \alpha, \qquad I_2 = \{[\alpha(\delta - \zeta) - \tfrac{1}{4}(\beta^2 - \gamma^2)]^2 + 2(2\alpha\xi - \beta\gamma)^2\}^{1/2}.
$$

The quantum integrals of motion X of Eq. (3.8) can be classified into four orbits under the action of the group $E(2)$, according to the values of I_1 and I_2. In the absence of a magnetic field each orbit corresponds to the separation of variables in the Schrödinger equation, or the Hamilton–Jacobi equation, in a specific coordinate system. We shall keep this terminology, calling the corresponding integrals Cartesian, polar, parabolic, or elliptic, respectively. The four orbits can be represented by the following operators:

(1) Cartesian orbit, $I_1 = I_2 = 0$

$$
(3.10) \quad X_{\mathrm{C}} = (P_1 + A_1)^2 - \frac{i\hbar}{2}(2k_1\partial_x + k_{1,x} + 2k_2\partial_y + k_{2,y}) + k_1A_1 + k_2A_2 + m,
$$

(2) Polar orbit, $I_1 \neq 0$, $I_2 = 0$

$$
\begin{aligned}
(3.11) \quad X_{\mathrm{R}} = (L_3 + yA_1 - xA_2)^2 + \hbar^2 - \frac{i\hbar}{2}&(2k_1\partial_x + k_{1,x} + 2k_2\partial_y + k_{2,y}) \\
&+ k_1A_1 + k_2A_2 + m,
\end{aligned}
$$

(3) Parabolic orbit, $I_1 = 0$, $I_2 \neq 0$

$$
\begin{aligned}
(3.12) \quad X_{\mathrm{P}} = -\frac{1}{2}&\{(L_3 + yA_1 - xA_2)(P_1 + A_1) + (P_1 + A_1)(L_3 + yA_1 - xA_2)\} \\
&- \frac{i\hbar}{2}(2k_1\partial_x + k_{1,x} + 2k_2\partial_y + k_{2,y}) + k_1A_1 + k_2A_2 + m,
\end{aligned}
$$

(4) Elliptic orbit, $I_1 \neq 0$, $I_2 \neq 0$

$$
\begin{aligned}
(3.13) \quad X_{\mathrm{E}} = &(L_3 + yA_1 - xA_2)^2 + \hbar^2 + \sigma[(P_1 + A_1)^2 - (P_2 + A_2)^2] \\
&- \frac{i\hbar}{2}(2k_1\partial_x + k_{1,x} + 2k_2\partial_y + k_{2,y}) + k_1A_1 + k_2A_2 + m, \\
&\qquad \sigma > 0, \ \sigma = \frac{I_2}{I_1^2}.
\end{aligned}
$$

Substituting expressions (3.7) into Eq. (3.6) we obtain

$$k_1 W_x + k_2 W_y + \frac{\hbar^2}{4}[(2\alpha x + \gamma)\Omega_y - (2\alpha y - \beta)\Omega_x] = 0. \tag{3.14}$$

This is the only equation that differentiates between classical and quantum integrability. We see that for $\alpha = \beta = \gamma = 0$, i.e., the case (3.10) of "Cartesian integrability," the term proportional to $\hbar^2$ drops out. In this case, quantum and classical integrability coincide.

4. Cartesian integrability and superintegrability

Let us now consider the Cartesian case (3.10), i.e., put $\alpha = \beta = \gamma = \zeta = \xi = 0$, $\delta = 1$ in (2.20). Equations (3.4) are already solved and, (3.5) and (3.6) reduce to

$$\Omega + k_{1,y} + k_{2,x} = 0, \qquad k_{1,x} = 0, \qquad k_{2,y} = 0, \tag{4.1}$$

$$k_1\Omega + m_y = 0, \qquad -k_2\Omega + m_x - W_x = 0, \tag{4.2}$$

$$k_1 W_x + k_2 W_y = 0. \tag{4.3}$$

Equations (4.1) are easily solved and imply

$$k_1 = -g'(y), \qquad k_2 = -\dot{f}(x), \qquad \Omega = \ddot{f} + g''. \tag{4.4}$$

Compatibility of the two equations (4.2) implies

$$W = fg'' + \ddot{f}g + u(x) + v(y),$$

where $f(x)$, $g(y)$, $u(x)$ and $v(y)$ are functions of one variable. The dots denote x-derivatives, the primes y-derivatives. Equation (4.3) restricts the function f and g to satisfy:

$$\ddot{f} = \alpha f^2 + \beta f + \gamma, \qquad g'' = -\alpha g^2 + \delta g + \zeta, \tag{4.5}$$

where α, β, γ, δ and ζ are arbitrary constants (not related to those of (3.7)). Thus, f and g are elliptic functions, or elementary functions that are degenerate or limiting cases of elliptic ones.

Finally, the magnetic field (throughout assumed to satisfy $\Omega \neq 0$) and effective potential are expressed in terms of f and g as

$$\begin{aligned} \Omega &= \alpha(f^2 - g^2) + \beta f + \delta g + \gamma + \zeta \\ W &= \frac{\alpha}{3}(g-f)^3 - \frac{\beta+\delta}{2}(g-f)^2 + (\gamma + \kappa - \xi)(g-f). \end{aligned} \tag{4.6}$$

The functions k_1, k_2 and m figuring in the integral of motion satisfy

$$\begin{aligned} &k_1 = -g', \qquad k_1 = -\dot{f} \\ m &= -\frac{\alpha}{3}(g^3 + 2f^3 - 3gf^2) + \beta(fg - f^2) \\ &\qquad + \tfrac{1}{2}\delta(g^2 - f^2) + \gamma(g - 2f) + \zeta g - \kappa f. \end{aligned} \tag{4.7}$$

In general Eq. (4.6) is not compatible with the separation of variables in the Hamilton–Jacobi, or Schrödinger equation. For numerical studies of classical motion in such systems, see [**4**].

Let us now turn to the question of Cartesian superintegrability. By this we mean that in addition to the operator X_1 of the form (3.10) a further second-order operator X_2 of the general form (3.8) exists and is polynomially independent of X_1 and H. We write the determining equations for X_2 and investigate their compatibility with those for X_1. It turns out [**2**] that these equations are compatible only if the magnetic field Ω and effective potential W are constant. We have already shown in Section 2 that this corresponds to first-order superintegrability. In this case two functionally independent first-order operators commuting with H exist. The second-order ones are polynomials in the first-order ones.

5. Polar integrability and superintegrability

Let us now consider the case when one second-order integral exists and is of the type X_R of Eq. (3.11). We introduce polar coordinates $x = r\cos\phi$, $y = r\sin\phi$, put $P = k_1\cos\phi + k_2\sin\phi$, $Q = -k_1\sin\phi + k_2\cos\phi$ and rewrite the determining equations (3.5) and (3.6) as

$$P_r = 0, \qquad P + Q_\phi = 0, \tag{5.1}$$

$$2r^3\Omega - P_\phi - rQ_r + Q = 0, \tag{5.2}$$

$$m_\phi - 2r^2 W_\phi + rP\Omega = 0, \qquad m_r - Q\Omega = 0, \tag{5.3}$$

$$\frac{\hbar^2}{2}\Omega_\phi + PW_r + \frac{1}{r}QW_\phi = 0. \tag{5.4}$$

Equation (5.4) contains a term proportional to $\hbar^2\Omega_\phi$. Thus, if we have $\Omega_\phi \neq 0$ the quantum and classical determining equations differ and hence their solutions will also differ.

To solve the system (5.1)–(5.4) we first put

$$P = -f'(\phi), \qquad Q = f(\phi) + R(r). \tag{5.5}$$

The primes and dots will denote ϕ and r derivatives, respectively. We assume throughout that the magnetic field satisfies $\Omega \neq 0$. From (5.2) we have

$$\Omega = -\frac{1}{2r^3}(f'' + f + R - r\dot{R}). \tag{5.6}$$

Compatibility of the two equations (5.3) leads to an equation for the effective potential $W(r,\phi)$, namely

$$2r^2 W_{r\phi} + 4rW_\phi + \frac{3f'}{2r^3}(f'' + f + R - r\dot{R}) + \frac{f'\ddot{R}}{2r} + \frac{1}{2r^3}(f+R)(f''' + f') = 0. \tag{5.7}$$

Let us first consider two special cases

(1) $f(\phi) + R(r) = 0$. Equation (5.6) then implies $\Omega = 0$, a case we are not interested in.
(2) $f'(\phi) = 0$. Equations (5.1)–(5.4) then imply $\Omega = \Omega(r)$, $W = W(r)$. As we have shown in Section 2, this means that a first-order integral exists and this case was already treated.

From now on we assume $f + R \neq 0$, $f' \neq 0$. To solve (5.7) we take W_ϕ from (5.4) and then eliminate W_ϕ and $W_{r\phi}$ from (5.7). We obtain the following equation for W

$$W_{rr} + \frac{(3(f+R) - r\dot{R})}{r(f+R)}W_r - \frac{\hbar^2\dot{R}(f''' + f')}{4r^3 f'(f+R)} + \frac{3(f+R)}{4r^6}(f'' + f + R - r\dot{R}) + \frac{(f+R)\ddot{R}}{4r^4} + \frac{(f+R)^2}{4r^6 f'}(f''' + f') = 0. \tag{5.8}$$

This has the form of an ordinary differential equation with r as a variable and ϕ as a parameter. Eq. (5.8) can be integrated and we obtain

$$(5.9)\quad \begin{aligned} W = &\frac{\hbar^2(f'''+f')}{8r^2f'} - \frac{3f(f''+f)}{32r^4} - \frac{f'''}{8f'}r^6\dot{u}^2 - \frac{r^2}{8}(r^3\ddot{u}+3r^2\dot{u})^2 \\ &- \frac{f^2(f'''+f')}{32f'r^4} - \frac{Ff}{2r^2} + \left(\frac{3f''}{8} + \frac{ff'''}{8f'}\right)r\dot{u} + \frac{3f''}{4}u \\ &- \frac{f}{4r}(r^3\ddot{u}+3r^2\dot{u}) + Fr^3\dot{u} + W_0, \end{aligned}$$

where $F(\phi)$ and $W_0(\phi)$ are arbitrary functions and we have defined $S(r)$ and $u(r)$ by the relations

$$(5.10)\quad \dot{S}(r) = \frac{1}{r^3}R(r), \qquad \dot{u}(r) = \frac{1}{r^3}S(r).$$

We now return to the original equations (5.1)–(5.4). We already have P, Q, Ω and W expressed in terms of functions of one variable only, namely $f(\phi)$, $F(\phi)$, $W_0(\phi)$ and, ultimately, $u(r)$ (in view of (5.10)). From (5.3) and (5.4) we express $m(r,\phi)$ as

$$(5.11)\quad m = \frac{ff''}{4r^2} + \frac{f^2}{4r^2} + \frac{fR}{2r^2} - \frac{f''S}{2} + \frac{R^2}{4r^2} + m_0(\phi), \quad m_0'(\phi) = 2f'F.$$

The function $u(r)$ must satisfy

$$(5.12)\quad \begin{aligned} \dddot{u} + \frac{6}{r}\ddot{u} - \frac{r^3\dot{u}^2}{2}\left(\frac{f'''}{f'}\right)'\frac{1}{f'} + \left(\frac{6}{r^2} + \frac{C}{2r^2f'} + \frac{4F'}{f'}\right)\dot{u} \\ + \frac{3f'''}{r^3f'}u + \frac{4A}{r^7f'} + \frac{4B}{r^5f'} + \frac{4W_0'}{r^3f'} = 0. \end{aligned}$$

In (5.12) we have introduced the following functions of ϕ

$$(5.13)\quad \begin{aligned} A &= -\frac{9ff'''}{32} - \frac{f^2}{32}\left(\frac{f'''}{f'}\right)' - \frac{15f'f''}{32} - \frac{3ff'}{4}, \\ B &= \frac{\hbar^2}{8}\left(\frac{f'''}{f'}\right)' - \frac{fF'}{2} - \frac{3f'F}{2}, \\ C &= 6f''' + f\left(\frac{f'''}{f'}\right)'. \end{aligned}$$

We treat (5.12) as an ODE for $u(r)$ and impose consistency conditions to ensure that $u(r)$ is indeed independent of ϕ. Two cases must be considered separately.

The first one is

$$(5.14)\quad \left[\left(\frac{f'''}{f'}\right)'\frac{1}{f'}\right]' = 0.$$

The analysis is rather lengthy [**2**], but leads to a simple result for Ω and W, namely

$$(5.15)\quad \Omega = 6ar^2 - b,$$

$$(5.16)\quad W = -2ar(C_1\cos(\phi) + C_2\sin(\phi)) + \frac{ab}{2}r^4 - 3acr^2 - a^2r^6.$$

We see that $\Omega = \Omega(r)$ (i.e., the magnetic field does not depend on the angle ϕ), hence the classical and quantum cases coincide.

The second case is

$$\left[\left(\frac{f'''}{f'}\right)'\frac{1}{f'}\right]' \neq 0. \tag{5.17}$$

Again a detailed analysis [**2**] leads to a relatively simple result, namely

$$W = \frac{\hbar^2}{8r^2}\left(1 + \frac{2f''}{f+a} - \frac{(f')^2}{(f+a)^2}\right) - \frac{(f+a)^2}{32r^4}\left(\frac{f'''}{f'} + 4\right) - \frac{3f''(f+a)}{32r^4} - \frac{C_1}{2r^2(f+a)^2} + C_2, \tag{5.18}$$

$$\Omega = -\frac{f'' + f + a}{2r^3} \tag{5.19}$$

where the function $f(\phi)$ satisfies

$$(f+a)^2\left(\frac{f'''}{f'}\right)' + 24f'(f+a) + 9f'''(f+a) + 15f'f'' = 0. \tag{5.20}$$

Equation (5.20) can be integrated twice. Putting $y = f + a$, we obtain a second-order ODE for y:

$$y'' = \frac{2}{y}(y')^2 - 3y + \frac{4A}{y} + \frac{B^2 - A^2}{y^3}, \tag{5.21}$$

where A and B are integration constants. This equation has a first integral, namely

$$y^4(y')^2 = -y^6 + 2Ay^4 + (B^2 - A^2)y^2 + K. \tag{5.22}$$

Equation (5.22) can be integrated in terms of elementary functions whenever the polynomial on the right-hand side has multiple roots. For instance, for $K = 0$ the general solution is

$$y = \pm\sqrt{A + B\sin 2(\phi - \phi_0)}. \tag{5.23}$$

Thus, we have a periodic finite solution with period $T = \pi$.

Let us now investigate the possibility of polar superintegrability. By this we mean that one first integral of the type (3.11) should exist, together with one other one of any type. We shall exclude the case when the magnetic field Ω and the effective potential W are constant, since in this case three first-order integrals exist. The additional integral cannot be Cartesian, as shown in Section 4. It can also not be of the elliptic type, since the existence of a polar integral, together with an elliptic one, implies the existence of a Cartesian integral. Finally, the second integral would have to be of the parabolic type. An analysis of the compatibility conditions shows that this is impossible [**2**].

6. Comparison with the case of a scalar potential and third-order integrals of motion

Let us now consider the "natural Hamiltonian"

$$H = -\frac{\hbar^2}{2}\Delta + V(x, y) \tag{6.1}$$

in two-dimensional Euclidean space E_2 and request that a third-order integral of motion

$$X = \sum_{i+j=0}^{3} \left\{ f_{ij}(x,y), p_1^i p_2^j \right\} \tag{6.2}$$

with p_k as in (2.7) should exist [**12**]. The curly bracket is an anticommutator and $f_{ij}(x,y)$ are some sufficiently smooth real functions. The commutation relation $[X,H]=0$ implies several general results:

(1) Even and odd powers in (6.2) can be considered separately.

(2) The leading third-order term in the integral (6.2) lies in the enveloping algebra of the Euclidean Lie algebra. Hence the operator X actually has the form

$$X = \sum_{i+j+k=3} A_{ijk}\{L_3^i, p_1^j p_2^k\} + \sum_{k=1}^{2}\{g_k(x,y), p_k\}, \tag{6.3}$$

where A_{ijk} are real constants. The constants A_{ijk} and the functions g_1, g_2 and V satisfy the following determining equations

$$g_{1,x} = 3f_1(y)V_x + f_2(x,y)V_y \tag{6.4}$$

$$g_{2,y} = f_3(x,y)V_x + 3f_4(x)V_y \tag{6.5}$$

$$g_{1,y} + g_{2,x} = 2\left[f_2(x,y)V_x + f_3(x,y)V_y\right] \tag{6.6}$$

$$\begin{aligned} g_1V_x + g_2V_y - \frac{\hbar^2}{4}[&f_1V_{xxx} + f_2V_{xxy} + f_3V_{xyy} + f_4V_{yyy} \\ &+ 8A_{300}(xV_y - yV_x) + 2(A_{210}V_x + A_{201}V_y)] = 0 \end{aligned} \tag{6.7}$$

where we have put

$$\begin{aligned} f_1(y) &= -A_{300}y^3 + A_{210}y^2 - A_{120}y + A_{030} \\ f_2(x,y) &= 3A_{300}xy^2 - 2A_{210}xy + A_{201}y^2 + A_{120}x - A_{111}y + A_{021} \\ f_3(x,y) &= -3A_{300}x^2y + A_{210}x^2 - 2A_{201}xy + A_{111}x - A_{102}y + A_{012} \\ f_4(x) &= A_{300}x^3 + A_{201}x^2 + A_{102}x + A_{003}. \end{aligned} \tag{6.8}$$

We see that (6.7) contains a term proportional to $\hbar^2$. Hence, classical and quantums integrability with a third-order integral of motion differ! This was first noticed by J. Hietarinta [**13, 14**] in a different context.

Equations (6.4)–(6.6) have not yet been solved in their generality. The problem of cubic integrals of motion in classical mechanics was posed and partially solved by Drach in 1935 [**5**]. This was done in a two-dimensional complex space and the adaptation to a real Euclidean space was not considered.

What has been studied so far is superintegrability with one third-order integral and one first, or second-order one [**10–12**]. We refer to the original articles and to the contributions of S. Gravel [**11**] and M. B. Sheftel [**28**] to this volume for further results.

The point that we are making here is that there are striking similarities between quadratic integrability and superintegrability in a magnetic field and cubic integrability and superintegrability in a purely scalar potential. In both cases purely quantum effects appear and in both cases the connection between superintegrability and multiseparability is lost.

For other perspectives on third-order integrals, see, e.g., [**6, 20, 25, 31**].

7. Conclusions

Our main conclusion is that while quadratic integrability with scalar potentials only is quite well understood, the same cannot be said about the case of vector potentials. Indeed, it is similar to the case of scalar potentials and integrals of order $n \geq 3$. We see as the main question: what do you do with integrals of motion in quantum mechanics that are third or higher-order differential operators?

References

1. S. Benenti, C. Chann, and G. Rastelli, *Variable separation for natural Hamiltonians with scalar and vector potentials on Riemannian manifolds*, J. Math. Phys. **42** (2001), no. 5, 2065–2091.
2. J. Bérubé and P. Winternitz, *Integrable and superintegrable systems in a magnetic field*, J. Math. Phys. **45** (2004), no. 5, 1959–1973.
3. J. A. Calzada, J. Negro, M. A. del Olmo, and M. A. Rodriguez, *Contraction of superintegrable Hamiltonian systems*, J. Math. Phys. **41** (2000), no. 1, 317–336.
4. B. Dorizzi, B. Grammaticos, A. Ramani, and P. Winternitz, *Integrable Hamiltonian systems with velocity dependent potentials*, J. Math. Phys. **26** (1985), no. 12, 3070–3079.
5. J. Drach, *Sur l'intégration logique des équations de la dynamique à deux variables: forces conservatrices, intégrales cubiques, mouvements dans le plan*, C.R. Acad. Sci. III **200** (1935), 22–27.
6. H. R. Dullin, V. S. Matveev, and P. J. Topalov, *On integrals of third degree in momenta*, Regul. Chaotic Dyn. **4** (1999), no. 3, 35–44.
7. N. W. Evans, *Superintegrability in classical mechanics*, Phys. Rev. A (3) **41** (1990), no. 10, 5666–5676.
8. ______, *Group theory of the Smorodinsky–Winternitz system*, J. Math. Phys. **32** (1991), no. 12, 3369–3375.
9. I. Friš, V. Mandrosov, Ya. A. Smorodinsky, M. Uhlíř, and P. Winternitz, *On higher symmetries in quantum mechanics*, Phys. Lett. **16** (1965), 354–356.
10. S. Gravel, *Hamiltonians separable in Cartesian coordinates and third-order integrals of motion*, J. Math. Phys. **45** (2004), no. 3, 1003–1019.
11. ______, *Cubic integrals of motion and quantum superintegrability*, in this volume.
12. S. Gravel and P. Winternitz, *Superintegrability with third-order integrals in quantum and classical mechanics*, J. Math. Phys. **43** (2002), no. 12, 5902–5912.
13. J. Hietarinta, *Classical versus quantum integrability*, J. Math. Phys. **25** (1984), no. 6, 1833–1840.
14. J. Hietarinta and B. Grammaticos, *On the $\hbar^2$-correction in quantum integrability*, J. Phys. A **22** (1989), no. 9, 1315–1322.
15. E. G. Kalnins, J. M. Kress, W. Miller, Jr., and P. Winternitz, *Superintegrable systems in Darboux spaces*, J. Math. Phys. **44** (2003), no. 12, 5811–5848.
16. E. G. Kalnins, J. M. Kress, G. S. Pogosyan, and W. Miller, Jr., *Completeness of superintegrability in two-dimensional constant curvature spaces*, J. Phys. A **34** (2001), no. 22, 4705–4720.
17. E. G. Kalnins, W. Miller, Jr., and P. Winternitz, *The group* 0(4)*, separation of variables and the hydrogen atom*, SIAM J. Appl. Math. **30** (1976), no. 4, 630–664.
18. E. G. Kalnins, G. C. Williams, W. Miller, Jr., and G. S. Pogosyan, *Superintegrability in three-dimensional Euclidean space*, J. Math. Phys. **40** (1999), no. 2, 708–725.
19. ______, *On superintegrable symmetry-breaking potentials in N-dimensional Euclidean space*, J. Phys. A **35** (2002), no. 22, 4755–4773.
20. M. Karlovini and K. Rosquist, *A unified treatment of cubic invariants at fixed and arbitrary energy*, J. Math. Phys. **41** (2000), no. 1, 370–384.
21. A. A. Makarov, Ya. A. Smorodinsky, Kh. Valiev, and P. Winternitz, *A systematic search for nonrelativistic systems with dynamical symmetries*, Nuovo Cimento A **52** (1967), 1061–1084.
22. E. McSween and P. Winternitz, *Integrable and superintegrable Hamiltonian systems in magnetic fields*, J. Math. Phys. **41** (2000), no. 5, 2957–2967.

23. N. N. Nehorošev, *Action-angle variables and their generalizations*, Trudy Moskov. Mat. Obšč. **26** (1972), 181–198 (Russian); English transl., Trans. Moscow Math. Soc. **26** (1972), 180–198.
24. G. Pogosyan, A. Sissakian, and P. Winternitz, *Separation of variables and Lie algebra contractions. Application to special functions*, Phys. Particles Nuclei **33** (2002), no. 7, 235–276.
25. M. F. Rañada, *Superintegrable $n = 2$ systems, quadratic constants of motion, and potentials of Drach*, J. Math. Phys. **38** (1997), no. 8, 4165–4178.
26. M. F. Rañada and M. Santander, *On harmonic oscillators on the two-dimensional sphere S^2 and the hyperbolic planet H^2*, J. Math. Phys. **43** (2002), no. 1, 431–451.
27. M. A. Rodriguez and P. Winternitz, *Quantum superintegrability and exact solvability in n dimensions*, J. Math. Phys. **43** (2002), no. 3, 1309–1322.
28. M. B. Sheftel, *On the classification of third-order integrals of motion in two-dimensional quantum mechanics*, in this volume.
29. M. B. Sheftel, P. Tempesta, and P. Winternitz, *Superintegrable systems in quantum mechanics and classical Lie theory*, J. Math. Phys. **42** (2001), no. 2, 659–673.
30. P. Tempesta, A. V. Turbiner, and P. Winternitz, *Exact solvability of superintegrable systems*, J. Math. Phys. **42** (2001), no. 9, 4248–4257.
31. A. V. Tsiganov, *Lax representation for an integrable motion on the sphere with a cubic second invariant*, Regul. Chaotic Dyn. **4** (1999), no. 3, 21–29.
32. P. Winternitz and I. Friš, *Invariant expansions of relativistic amplitudes and the subgroups of the proper Lorentz group*, Jadernaja Fiz. **1** (1965), 889–901 (Russian); English transl., Soviet J. Nuclear Phys. **1** (1965), 636–643.
33. P. Winternitz, Ya. A. Smorodinsky, M. Uhlir̆, and I. Friš, *Symmetry groups in classical and quantum mechanics*, Jadernaja Fiz. **4** (1966), 625–635 (Russian); English transl., Soviet J. Nuclear Phys. **4** (1967), 444–450.
34. A. Zhalij, *On separable Pauli equations*, J. Math. Phys. **43** (2002), no. 3, 1365–1389.
35. R. Zhdanov and A. Zhalij, *On separable Schrödinger equations*, J. Math. Phys. **40** (1999), no. 12, 6319–6338.

Département de mathématiques et de statistique and Centre de recherche mathématiques, Université de Montréal, C.P. 6128, succ. Centre-Ville, Montréal, QC H3C 3J7, Canada

Current address: Collège Jean-de-Brébeuf, 3200 chemin de la Côte-Sainte-Catherine, Montréal, QC H3T 1C1, Canada

E-mail address: jberube@brebeuf.qc.ca

Département de mathématiques et de statistique and Centre de recherche mathématiques, Université de Montréal, C.P. 6128, succ. Centre-Ville, Montréal, QC H3C 3J7, Canada

E-mail address: wintern@crm.umontreal.ca

Centre de Recherches Mathématiques
CRM Proceedings and Lecture Notes
Volume **37**, 2004

Applications of Crack in the Classification of Integrable Systems

Thomas Wolf

Abstract. With progress in the computerized solution of large bilinear algebraic systems a number of classifications of integrable equations and systems becomes possible. In this contribution investigations of integrable polynomial vector evolution equations, quadratic Hamiltonians with an e(3) Lie–Poisson structure and nonlocal $2+1$-dimensional evolution equationsequations are described. An overview of the computer algebra program used is given in the appendix.

1. Overview

The talk given by the author at the CRM workshop on Superintegrability in September 2002 and this related paper report on work in two subjects. One is the classification of integrable equations and systems. This includes collaborations with V. Sokolov and T. Tsuchida in an effort to classify integrable polynomial vector evolution equations, work with O. V. Efimovskaya on quadratic Hamiltonians with a e(3) Lie-Poisson structure and a computation dealing with nonlocal $2+1$-dimensional together with A. Mikhailov, V. V. Sokolov, and R. Hernandez Heredero. The other subject is the computer algebra package Crack which did the main computations in solving large bilinear algebraic systems. Although originally designed to solve overdetermined systems of partial differential equations a number of extensions made Crack a powerful tool for solving systems of bilinear algebraic equations. Such systems turn up in many different classification problems some of which were investigated by other participants of this workshop.

In the talk on which this article is based, a method to reduce the length of equations was presented which proved to be useful in solving the bilinear algebraic systems. Due to numerous questions asked about the computer program, a more complete overview is given in the appendix.

2000 *Mathematics Subject Classification.* 37J35, 37K10, 37N15, 35Q53, 35Q55.
This is the final form of the paper.

2. The classification of integrable vector evolution equations

The method of using symmetries to classify nonlinear evolutionary 1 + 1-dimensional PDEs is a most productive one and has been known for some time (see [**5, 13, 14, 18**]).

An extension of the simplest approach [**4, 9**] to the case of so-called vector evolution equations is described in work with Vladimir Sokolov [**21**]. Examples of vector evolution equations are two different vector generalizations

$$U_t = U_{xxx} + \langle U, U\rangle U_x, \tag{2.1}$$

$$U_t = U_{xxx} + \langle U, U\rangle U_x + \langle U, U_x\rangle U \tag{2.2}$$

of the mKdV equation. Here $U(t,x)$ is an N-component vector and $\langle \cdot, \cdot \rangle$ stands for the standard scalar product.

In the performed symmetry classification we considered equations of the form

$$U_t = f_n U_n + f_{n-1} U_{n-1} + \cdots + f_1 U_1 + f_0 U, \quad U_i = \frac{\partial^i U}{\partial x^i}. \tag{2.3}$$

where $U = (U^1, U^2, \ldots, U^N)$ is an unknown vector of arbitrary dimension N and coefficients f_i are polynomials of scalar products $\langle U_i, U_j\rangle$, $0 \le i \le j \le n$.

For more details on possible scalar products, orthogonal symmetry transformations and generality of N see [**21**].

All such equations were determined that admitted a symmetry of the same form

$$U_\tau = g_m U_m + g_{m-1} U_{m-1} + \cdots + g_1 U_1 + g_0 U,$$

where the differential order n of the equation and the order m of the symmetry have selected values. By taking f_i, g_j to be homogeneous polynomials of the scalar products one achieves that the symmetry condition $U_{t\tau} = U_{\tau t}$ yields an overdetermined system of bilinear algebraic conditions for the undetermined coefficients of both polynomials f_i and g_j.

The number of coefficients of the equation and symmetry is reduced further by a homogeneity assumption that the differential equations and symmetries are invariant under the scaling group

$$(x, t, U) \to (a^{-1}x, a^{-\mu}t, a^{\lambda}U)$$

with one value λ for all components of the vector U.

Finally, we restrict λ based on results for the scalar case proved in [**16**] that a λ-homogeneous polynomial equation with $\lambda > 0$ may possess a homogeneous polynomial higher symmetry only if $\lambda \in \{2, 1, \frac{1}{2}\}$.

The differential orders of the equation and symmetry that have been investigated are also motivated by the scalar case where infinite commutative hierarchies have either a lowest order of the equation equal to 2 and a symmetry of order 3 (in the following called type $(2,3)$ like the Burgers equation) or an equation of order 3 with a symmetry of order 5 (type $(3,5)$ like the Korteweg–de Vries (KdV) equation) or an equation of order 5 with a symmetry of order 7 (type $(5,7)$ like the Kaup–Kupershmidt and Sawada–Kotera equations).

For the nine resulting cases shown in Table 1 different computer programs generated the equations and symmetries, computed the commutator and formulated and solved the bilinear system automatically (apart from the case $\lambda = \frac{1}{2}$, type $(5,7)$ where the solution was not fully automatic).

TABLE 1. A classification of single vector equations.

λ	2			1			$\frac{1}{2}$		
orders of (eq, sym)	(2, 3)	(3, 5)	(5, 7)	(2, 3)	(3, 5)	(5, 7)	(2, 3)	(3, 5)	(5, 7)
# of unknowns (eq, sym, tot)	–	2	3, 7, 10	2, 3, 5	3, 9, 12	9, 24, 33	4, 8, 12	8, 27, 35	27, 82, 109
# of equations	–	4	34	5	26	198	21	129	927
total # of terms	–	4	162	9	121	3125	80	1603	52677
average # of terms/equation	–	1	4.7	1.8	4.6	15.8	3.8	12.4	56.8
time to formulate	–	0.2s	1.7s	0s	1.3s	1m 1s	1s	17s	13h 12m
time to solve	–	0s	0.7s	0s	0.5s	1m 15s	0.4s	32s	2 days
solutions	–	–	–	–	(2.1), (2.2)	–	–	(2.4)	–

TABLE 2. A classification of NLS-type systems of vector equations.

λ	1	$\frac{1}{2}$
orders of (eq, sym)	(2, 3)	(2, 3)
# of unknowns (eq, sym, tot)	9, 15, 24	53, 155, 208
# of equations	78	1206
total # of terms	242	28768
average # of terms/equation	3.1	23.8
time to formulate	2.3s	1h
time to solve	2.9s	26m 15s
# of solutions	2	7

COMMENTS. Times are measured on a 1.7GHz Pentium 4 running the 2002 version of the program CRACK in a 120MByte Reduce session under Linux. Non-linear $\lambda = 2$ equations of order 2 do not exist. All found solutions were already known: equations (2.1), (2.2) are vector generalizations of the mKdV equation, and equation

$$U_t = U_{xxx} + 3\langle U,U\rangle U_{xx} + 6\langle U,U_x\rangle U_x + 3\langle U,U\rangle^2 U_x + 3\langle U_x,U_x\rangle U \tag{2.4}$$

is a vector analogue (reported in [**20**]) of the Ibragimov–Shabat–Calogero equation [**2, 9**]. Solutions of order (5, 7) are only symmetries of found lower order equations and have therefore not been listed in the table.

3. NLS systems with two vector unknowns

The investigation of integrable vector NLS-type systems of the form

$$\begin{cases} U_t = U_{xx} + p_1 U_x + p_2 V_x + p_3 U + p_4 V, \\ V_t = -V_{xx} + p_5 U_x + p_6 V_x + p_7 U + p_8 V, \end{cases} \tag{3.1}$$

where U and V are vectors and the coefficients p_i are λ-homogeneous polynomials depending on all possible scalar products of vectors U, V, U_x, V_x was more succesful. For $\lambda = 2$ second order systems can only be linear but for $\lambda = 1$ or $\frac{1}{2}$ results are shown in Table 2. Just as in the scalar case (see [**14**]) a symmetry of the form

$$\begin{cases} U_\tau = U_{xxx} + q_1 U_{xx} + q_2 V_{xx} + q_3 U_x + q_4 V_x + q_5 U + q_6 V, \\ V_\tau = V_{xxx} + q_7 U_{xx} + q_8 V_{xx} + q_9 U_x + q_{10} V_x + q_{11} U + q_{12} V, \end{cases} \tag{3.2}$$

is assumed where the coefficients q_i are λ-homogeneous polynomials of all possible scalar products of U, V, U_x, V_x, U_{xx}, V_{xx}.

RESULTS. The 2 solutions for $\lambda = 1$ are known [**10, 11**]. They are special cases (see [**21**]) of a generalization of the NLS system by Svinolupov using a Jordan triplet system (in [**22**]). For $\lambda = \frac{1}{2}$, after identifying solutions related by $U \leftrightarrow V$, $t \leftrightarrow -t$ six solutions remain. Two of these systems had been known [**23**]. For the remaining four systems, Takayuki Tsuchida showed C-integrability for two of them and S-integrability for the other two.

4. Systems with one scalar and one vector unknown

Computations to classify single vector equations that involved an arbitrary constant vector did not give new results, but the possibility to apply an orthogonal transformation to make the constant vector equal to $(1, 0, 0, \dots)$ provided a natural split of the single vector equation into one for a scalar function u (equal to the former component U^1) and a new vector function U (equal the remaining components $(U^2, U^3, \dots)$). General investigations of systems with one scalar and one vector brought a surprisingly rich set of integrable systems. Results in the case $\lambda = 2$ were analyzed by Vladimir Sokolov and Takayuki Tsuchida. What appears to be a new solution in this class is the vector generalization

$$\begin{cases} u_t = u_{xxx} + 6uu_x - 12\langle U, U_x\rangle, \\ U_t = -2U_{xxx} - 6uU_x. \end{cases} \tag{4.1}$$

of the two-component coupled KdV system proposed by Hirota and Satsuma [**8**]. A Lax pair representation has been found by Takayuki Tsuchida.

Especially the case of $\lambda = 1$ with 25 solutions analyzed by Takayuki Tsuchida posed a major challenge. Not all systems are of interest. Some become triangular by defining $\langle U, U\rangle$ as a new scalar variable. Others are just the result of splitting a single vector equation into scalar + vector equations for the scalar U^1 and the vector $(U^2, U^3, \ldots, U^N)$. Among the interesting cases just one will be shown here:

$$\begin{cases} u_t = u_{xxx} - 6u^2 u_x + u_x\langle U, U\rangle + 2u\langle U, U_x\rangle + \langle U, U_{xx}\rangle + \langle U_x, U_x\rangle, \\ U_t = -2U_{xxx} - 6u_{xx}U - 6u_x U_x + 12uu_x U + 6u^2 U_x \\ \qquad\qquad + \langle U, U\rangle U_x - 2\langle U, U_x\rangle U. \end{cases} \tag{4.2}$$

Its integrability can be established (Tsuchida) through a change to new variables w and W

$$\begin{cases} w = u_x + u^2 + \frac{1}{6}\langle U, U\rangle, \\ W = U_x + 2uU, \end{cases} \tag{4.3}$$

which satisfy the following system:

$$\begin{cases} w_t = w_{xxx} - 6ww_x + 2\langle W, W_x\rangle, \\ W_t = -2W_{xxx} + 6wW_x. \end{cases} \tag{4.4}$$

This system coincides up to a scaling of variables with the coupled Hirota–Satsuma system (4.1) found in the $\lambda = 2$ investigation above.

The Miura-type transformation (4.3) is a generalization of the one for scalar U in [**3, 25**] and the one for two-component vector U in [**30**].

In the case when U is a scalar variable, one can set

$$u = -\frac{1}{2}(q + r), \quad U = \frac{\sqrt{6}}{2}\mathrm{i}(q - r)$$

and rewrite (4.2) as a system of symmetrically coupled mKdV equations:

$$\begin{cases} q_t = [-\frac{1}{2}q_{xx} + \frac{3}{2}r_{xx} + 3(q - r)q_x - 2r^3]_x, \\ r_t = [\frac{3}{2}q_{xx} - \frac{1}{2}r_{xx} - 3(q - r)r_x - 2q^3]_x. \end{cases} \tag{4.5}$$

This system is identical to [**6**, (63)] or [**7**, (3.22)]. It was found in connection with the Kac–Moody Lie algebras and written in the Hamiltonian form about twenty years ago (cf. the $\mathrm{C}_2^{(1)}$ case in [**25**] or the $\mathrm{B}_2^{(1)}$ case in [**3**]).

Table 3 gives an overview of systems for one scalar and one vector found by CRACK to have higher order symmetries. Details will be discussed in a future contribution [**24**].

5. Classification of Integrable Hamiltonians

In work done together with Olga V. Efimovskaya quadratic Hamiltonians H have been investigated that have the form

$$H = \langle U, AU\rangle + \langle U, BV\rangle + \langle V, CV\rangle + \langle N, U\rangle + \langle M, V\rangle, \tag{5.1}$$

where $U = (U_1, U_2, U_3)$ and $V = (V_1, V_2, V_3)$ are three-dimensional vectors, A, C are symmetric matrices, B is an arbitrary matrix and N, M are constant vectors. Such Hamiltonians are relevant in the dynamics of rigid bodies.

The equations of motion in the rigid body dynamics are defined by a linear Poisson bracket of the form

$$\{Y_i, Y_j\} = c_{ij}^k Y_k, \quad i, j, k = 1, \ldots, N \tag{5.2}$$

TABLE 3. A classification of systems of one scalar + one vector equations.

λ	2			1			$\frac{1}{2}$		
orders of (eq, sym)	(2, 3)	(2, 4)	(3, 5)	(2, 3)	(2, 4)	(3, 5)	(2, 3)	(2, 4)	(3, 5)
# of unknowns (eq, sym, tot)	5, 6, 11	5, 12, 17	6, 17, 23	10, 21, 31	10, 39, 49	21, 74, 95	15, 36, 51	15, 79, 94	36, 164, 200
# of equations	13	26	50	66	123	386	149	313	1154
total # of terms	34	77	218	341	770	5000	1093	3096	27695
average # of terms/equation	2.6	2.9	4.3	5.1	6.3	13	7.3	9.9	24
time to formulate	0.5s	1s	5s	1.8s	5s	2m 52s	8s	48s	2h 7m
time to solve	0.5s	0.4s	6.5s	29s	1m 58s	5h 47m	29s	3m 44s	1 day
# of solutions	0	0	4	3	3	25	0	0	2

λ	scalar: $\frac{1}{3}$, vec: $\frac{2}{3}$			scalar: $\frac{2}{3}$, vec: $\frac{1}{3}$		
orders of (eq, sym)	(2, 3)	(2, 4)	(3, 5)	(2, 3)	(2, 4)	(3, 5)
# of unknowns (eq,sym,tot)	10, 24, 34	10, 54, 64	24, 115, 139	13, 22, 35	13, 66, 79	22, 126, 148
# of equations	102	215	798	114	276	955
total # of terms	529	1462	12694	694	2435	17385
average # of terms/equation	5.2	6.8	16	6.1	8.8	18
time to formulate	3.2s	13s	23m 45s	6.3s	48s	41m 18s
time to solve	45s	1m 23s	1h 20m	22s	3m 40s	1h 7m
# of solutions	0	0	0	1	2	2

where c_{ij}^k are some constants. The evolution of dynamic variables $Y_1, \dots, Y_N$ is defined by the formula

$$\frac{d}{dt}Y_i = \{Y_i, H\},$$

where H is the Hamiltonian.

The skew-symmetricity and the Jacobi identity for the linear Poisson bracket is equivalent to the fact that c_{ij}^k are the structure constants of some Lie algebra. It is known that the Hamiltonian structure of most cases of rigid body dynamics can be defined by the linear Poisson brackets

$$\{U_i, U_j\} = \varepsilon_{ijk}U_k, \qquad \{U_i, V_j\} = \varepsilon_{ijk}V_k, \qquad \{V_i, V_j\} = 0, \tag{5.3}$$

corresponding to the Lie algebra e(3). For example, two classical problems with a Hamiltonian of form (5.1) and the Poisson structure (5.3) are

(1) The Kirchhoff problem (where $N = 0$, $M = 0$ in H),
(2) The problem of motion of a massive rigid body around a fixed point. In this case $B = C = N = 0$.

The bracket (5.3) possesses two Casimir functions:

$$J_1 = V_1^2 + V_2^2 + V_3^2, \qquad J_2 = U_1V_1 + U_2V_2 + U_3V_3. \tag{5.4}$$

Therefore to integrate a system on e(3) we need one additional first integral I, functionally independent of H, J_1, J_2. All Hamiltonians (5.1) admitting an additional polynomial first integral of first or second degree are well known. The main goal of our work is to find all Hamiltonians (5.1) that admit an additional polynomial integral of degree 3 or 4.

There exist two kinds of linear transformations of U and V which preserve the Poisson structure (5.3) and the form of H. The first kind is defined by

$$\overline{U} = TU, \qquad \overline{V} = TV, \tag{5.5}$$

where T is an arbitrary constant orthogonal matrix. The second kind is defined by

$$\overline{U} = U + SV, \tag{5.6}$$

where S is an arbitrary antisymmetric matrix. With the transformations (5.5) we reduce the matrix A to diagonal form: $A = \mathrm{diag}(a_1, a_2, a_3)$. Transformations (5.6) are usually used for a simplification of matrix B.

The generic case $a_1 \neq a_2 \neq a_3 \neq a_1$ has been investigated in detail and all cases when there exists an additional polynomial integral are known. In calculations mentioned below we study the case $a_1 = a_2 \neq a_3$. Using the transformation (5.6) we can set $b_{12} = b_{13} = b_{23} = 0$. Subtracting multiples of both Casimirs enables $b_{11} = c_{11} = 0$ and gives a Hamiltonian of the form

$$\begin{aligned} H = {} & U_1^2 + U_2^2 + a_3U_3^2 \\ & + 2b_{21}U_2V_1 + 2b_{31}U_3V_1 + 2b_{32}U_3V_2 + 2b_{22}U_2V_2 + 2b_{33}U_3V_3 \\ & + 2c_{12}V_1V_2 + 2c_{13}V_1V_3 + c_{22}V_2^2 + 2c_{23}V_2V_3 + c_{33}V_3^2 \\ & + p_1U_1 + p_2U_2 + p_3U_3 + q_1V_1 + q_2V_2 + q_3V_3. \end{aligned} \tag{5.7}$$

We consider Hamiltonians (5.7) that have an additional cubic integral. The Ansatz for a general first integral I of third degree involves 80 terms. Together with the 16 unknown constants in (5.7) the bilinear algebraic system which results from $\{H, I\} = 0$ involves 96 unknowns to be determined. The result of the computation is summarized in the following theorem.

THEOREM 1. • *The Hamiltonian* (5.7) *admits a polynomial integral of third degree iff it has the form*

$$H = U_1^2 + U_2^2 + s_1 U_3^2 + s_2 V_3 U_3 + s_3 V_3^2 + s_4 U_3 + s_5 V_3, \tag{5.8}$$

where s_i *are arbitrary parameters;*

• *The Hamiltonian* (5.7) *admits a polynomial integral of third degree on a special level* $J_2 = 0$ *of the Casimir function* (5.4) *iff either it has the form* (5.8) *or the form*

$$H = U_1^2 + U_2^2 + 4U_3^2 + 4(s_1 V_1 + s_2 V_2)U_3 - (s_1^2 + s_2^2)V_3^2 + s_3 U_3 + s_4 V_1 + s_5 V_2, \tag{5.9}$$

where s_i *are arbitrary parameters.*

Hamiltonian (5.8) is a generalization of the Hamiltonian for the Lagrange and Kirchhoff classical integrable cases admitting not only a third degree but also an additional first degree integral $I = U_3$. If $s_1 = s_2 = 0$ then Hamiltonian (5.9) describes the so-called Goryachev–Chaplygin case in the problem of the motion of a rigid body around a fixed point. The integrability of the general Hamiltonian (5.9) has been recently established in [**19**].[1]

6. Nonlocal 2 + 1-dimensional equations

The computation described in this section solves only a first special case of a wider problem. We still show it as it gives an example of how even nonlocal 2 + 1-dimensional classification problems can be reduced to the solution of bilinear algebraic systems for which CRACK can be useful.

The Kadomtsev–Petviashvili equation

$$u_{tx} = -(6uu_x + u_{xxx})_x + u_{yy}$$

can also be written as

$$u_t = -(6uu_x + u_{xxx}) + \Delta^2 u_x$$

with $\Delta = D_x^{-1} D_y$. A. Mikhailov and R. I. Yamilov observed in [**15**] that all known integrable 2 + 1-dimensional equations can be written as

$$u_t = \text{expression in } \Delta,\ D_x, \text{ and } u.$$

Based on this idea A. Mikhailov, V. V. Sokolov and R. Hernandez Heredero did work on classifying KdV-type integro-differential equations of the form

$$u_t = a u_{xxx} + b\Delta(u_{xxx}) + c\Delta^2(u_{xxx}) + \cdots + e\Delta^3(u_{xxx}) + \text{terms of lower } x\text{-order}$$

which have symmetries of the same form.

The special Ansatz for equation plus symmetry that has been investigated with CRACK is

$$\begin{aligned} u_t &= u_{xxx} + b\Delta(u_{xxx}) + c\Delta^2(u_{xxx}) + 0 \cdot \Delta^3(u_{xxx}) + e\Delta^{-1}(u_{xxx}) \\ &\qquad + \text{lower order} \\ u_\tau &= 0 \cdot u_{xxx} + b'\Delta(u_{xxx}) + c'\Delta^2(u_{xxx}) + \Delta^3(u_{xxx}) + e'\Delta^{-1}(u_{xxx}) \\ &\qquad + \text{lower order} \end{aligned} \tag{6.1}$$

[1]At the time of proof reading a paper discussing fourth degree first integrals had appeared [**29**].

with two coefficients equal to one and two zero coefficients, due to suitable linear combinations of both equations. The lower order terms can be at most of order 2 in x, linear or quadratic in u, and with Δ restricted as above.

For this Ansatz, which is the simplest in this class, the condition $u_{t\tau} - u_{\tau t} \equiv 0$ provides already 2865 bilinear algebraic conditions for the 70 unknowns $b, c, \ldots,$ and 70 unknowns $b', c', \ldots$.

According to CRACK only 2 solutions exist. One is the Boiti, Leon, Manna, Pempinelli equation (equation (2.12) in [**1**] with $a_0 = -1$, $a_2 = b_j = 0$)

$$u_t = u_{xxx} + \alpha D_x\big(u\Delta^{-1}(u)\big)$$

with symmetry

$$u_\tau = \Delta^3(u_{xxx}) + D_x\left(\Delta\big(u\Delta(u)\big)\right).$$

The second solution is the same for $u \rightarrow \Delta u$.

Other known 2-dimensional integrable equations, like KdV in the form $u_t = \Delta u_{xxx} + 4u\Delta u_x + 2u_x\Delta u$ which is a symmetry of the usual KdV equation would not have terms u_{xxx} and $\Delta^3(u_{xxx})$ as it was required in the special Ansatz (6.1).

Appendix A. A short description of CRACK

The following overview describes the November 2003 version of the program. Any identifiers or numbers in curled brackets { } provided at the end or within the following paragraphs refer to keywords, file names, module numbers or flags which can be looked up in the CRACK manual `crack.tex` (see below under 'Availability') or even be searched in the source code if needed.

Philosophy. The program CRACK is a computer algebra package written in REDUCE for the solution of overdetermined systems of algebraic, ordinary or partial differential equations with at most polynomial nonlinearity. It was originally developed to run automatically and effort was taken for the program to decide which computational steps are to be done next with a choice between integrations, separations, substitutions and investigation of integrability conditions. It is known from hand computations that the right sequence of operations with exactly the right equations at the right time is often crucial to avoid an explosion of the length of expressions. This statement remains true for the computerized solution of systems of equations as they become more complex. As a consequence more and more interactive access has been provided to inspect data, to specify how to proceed with the computation and how to control it. This allows the human intervention in critical stages of the computations. {off batch_mode}

General structure. A problem consists of a system of equations and a set of inequalities. With each equation are associated a short name and numerous data, like size, which functions, derivatives and variables occur but also which investigations have already been done with this equation and which not, in order to avoid unnecessary duplication of work. These data are constantly updated if the equation is modified in any way. An inequality can either be a single expression that must not vanish or a list of expressions of which at least one must not vanish.

A set of about 30 modules is available to integrate, substitute, decouple,... equations. A complete list can be inspected in interactive mode with the command p2, each operation is listed with the number it is called. All modules can be called interactively or automatically. Automatic computation is organized by a priority list of modules (each represented by a number) where modules are invoked in the order they appear in the priority list, each module trying to find equations in the system it can be applied to. If a module is not successful then the next module in the list is tried, if any one is successful then execution starts again at the beginning of the priority list. {prog_list_, default_proc_list_, full_proc_list_}

Because each module has access to all the data, it is enough to call a module by its number. For example, the input of the number 2 in interactive mode will start the direct separation module (see below) to look for a directly separable equation and will split it.

Modules. The following modules are represented by numbers in the priority list. Each module can appear with modifications under different numbers. For example, integration is available under 7, 24 and 25. Here 7 encodes an integration of short equations $0 = \partial^n f/\partial x^n$. 7 has highest priority of the three integrations. 24 encodes the integration of an equation that leads to the substitution of a function and 25 refers to any integration and has lowest priority.

Integration and separation. An early feature in the development of the package CRACK was the ability to integrate exact differential equations and some generalizations of them (see [**26**]). As a consequence of integrations an increasing number of functions of fewer variables is introduced which sooner or later produces equations with some independent variables occurring only explicitly and not as variables in functions. Such equations are split by the separation module.

Substitutions. Substitutions can have a dramatic effect on the size and complexity of systems. Therefore it is possible to have them not only done automatically but also controlled tightly, either by specifying exactly which unknown should be substituted where using which equation, or by picking a substitution out of a list of substitutions offered by the program {cs}. Substitutions to be performed automatically can be controlled with a number of filters, for example, by

- limiting the size of the equation to be used for substitution, {subst_}
- limiting the size of equations in which the substitution is to be done, {target_limit}
- allowing only linear equations to be used for substitutions, {lin_subst}
- allowing equations to increase in size only up to some factor in order for a substitution to be performed in that equation, {cost_limit}
- allowing a substitution for a function through an expression only if that expression involves exclusively functions of fewer variables, {less_vars}
- allowing substitutions only that do not lead to a case distinction coefficient $= 0$ or not,
- specifying whether extra effort should be spent to identify the substitution with the lowest bound on growth of the full system. {min_growth}

Substitution types are represented by different numbers depending on the subset of the above filters to be used. If a substitution type is to be done automatically, then from all possible substitutions passing all filters of this type that substitution is selected that leads to no subcases (if available) and that uses the shortest equation.

Factorization. It is very common that big algebraic systems contain equations that can be factorized. Factorizing an equation and setting the factors individually to zero simplifies the whole task, because factors are simpler expressions than the whole equation and set to zero they may lead to substitutions and thereby further simplifications. The downside is that if problems with, say 100 unknowns, need 40 case distinctions in order to be able to solve automatically for the remaining 60 unknowns, then this would require 2^{40} cases to be investigated which is impractical. The problem is to find the right balance, between delaying case distinctions in order not to generate too many cases and, on the other hand, introducing case distinctions as early as necessary in order to simplify the system. This simplification may be necessary to solve the system, but in any case it will speed up its solution (although at the price of having to solve a simplified system at least twice, depending on the number of factors).

For large systems with many factorisable equations the careful selection of the next equation to be factorized is important to gain the most from each factorization and to succeed with as few as possible factorizations. Criteria which give factors and therefore equations a higher priority are

- the number of equations in which this factor occurs,
- if the factor is a single unknown function or constant, then the number of times this unknown turns up in the whole system,
- the total degree of the factor,
- the number of factors of an equation,
- and others.

It also matters in which order the factors are set to zero. For example, the equation $0 = ab$ can be used to split into the 2 cases: (1) $a = 0$; (2) $a \neq 0$, $b = 0$ or to split into the 2 cases (1) $b = 0$; (2) $b \neq 0$, $a = 0$. If one of the 2 factors, say b, involves functions which occur only linearly then this property is to be preserved and these functions should be substituted as such substitutions preserve their linearity. But to have many such substitutions available, it is useful to know of many nonlinearly occurring functions to be nonzero as they occur as coefficients of the linearly occurring functions. In the above situation it is therefore better to do the first splitting (1) $a = 0$; (2) $a \neq 0$, $b = 0$ because $a \neq 0$ will be more useful for substitutions of linear functions than $b \neq 0$ would be.

An exception of this plausible rule occurs toward the end of all the substitutions of all the linearly occurring b_i, when some b_i are an overall factor to many equations. If one would then set, say, $b_{22} = 0$ as the second case in a factorization, the first case would generate as subcases factorizations of other equations where $b_{22} = 0$ would be the second case again and so on. To avoid this one should investigate $b_{22} = 0$ as the first case in the first factorization.

The only purpose of that little thought experiment was to show that simple questions, like which factored equation should be used first for case-distinctions and in which order to set factors to zero can already be difficult to answer in general.

Elimination (Gröbner basis) steps. To increase safety and avoid excessive expression swell one can, apart from the normal call {30}, request to do Gröbner basis computation steps only if they are simplification steps replacing an equation by a shorter equation. {27}

In a different version only steps are performed in which equations are included which do not contain more than 3 unknowns. This helps to focus on steps which are more likely to solve small sub-systems with readily available simple results. {57}

Often the computationally cheapest way to obtain a consistent (involutive) system of equations implies to change the ordering during the computation. This is the case when substitutions of functions are performed which are not ranked highest in a lexicographical ordering of functions. But CRACK also offers an interactive way to

- change the lexicographical ordering of variables, {ov}
- change the lexicographical ordering of functions, {of}
- give the differential order of derivatives a higher or lower priority in the total ordering than the lexicographical ordering of functions, {og}
- give either the total differential order of a derivative of a function a higher priority than the lexicographical ordering of the derivative of that function or to take the lexicographical ordering of derivatives as the only criterion. {of}

Solution of an underdetermined differential equation. When solving an overdetermined system of linear differential equations where the general solution involves free functions, then in the last computational step often a single equation for more than one function remains to be solved. Examples are the computation of symmetries and conservation laws of nonlinear differential equations which are linearizable. In CRACK two procedures are available, one for underdetermined linear ODEs {22} and one for linear PDEs, {23} both with nonconstant coefficients.

Indirect separation. Due to integrations new functions of fewer variables are introduced. Substituting functions may lead to equations where no function depends on all variables but all variables appear as variables to unknown functions, e.g., $0 = f(x) + g(y)$ although usually much more complicated with 10 or 20 independent variables and many functions depending on different combinations of these variables. Because no variable occurs only explicitly, direct separations mentioned above are not possible. Two different algorithms, one for linear indirectly separable equations {10, 26} and one for nonlinear directly separable equations {48} provide systematic ways of dealing with such equations.

Indirectly separable equations always result when an equation is integrated with respect to different variables, like $0 = f_{xy}$ to $f = g(x) + h(y)$ and a function, here $f(x, y)$, is substituted.

Function and variable transformations. In the interactive mode one can specify a transformation of the whole problem in which old functions and variables are expressed in a mix of new functions and variables.

Solution of first order linear PDE. If a system contains a single linear first order PDE for just one function then in an automatic step characteristic ODE-systems are generated, integrated if possible, a variable transformation for the whole system of equations is performed to have in the first order PDE only one single derivative and to make this PDE integrable for the integration modules. {61}

We conclude the listing of modules and continue with other aspects of CRACK.

Exploiting bilinearity. In bilinear algebraic problems we have 2 sets of variables $a_1, \ldots, a_m$ and $b_1, \ldots, b_n$ such that all equations have the form $\sum_{k=1}^{l} \gamma_k a_{i_k} b_{j_k} = 0$, $\gamma_k \in G$. Although the problem is linear in the a_i and linear in the b_j it still is a nonlinear problem. A guideline which helps keeping the structure of the system during computation relatively simple is to preserve the linearity of either the a_i or the b_j as long as possible. In classification problems of integrable systems the Ansatz for the symmetry/first integral involves usually more terms and therefore more constants (called b_j in applications of CRACK) than the Ansatz for the integrable system (with constants a_i). A good strategy therefore is to keep the system linear in the b_j during the computation, i.e., to

- substitute only a b_j in terms of a_i, b_k, or an a_i in terms of an a_k but not an a_i in terms of any b_k,
- do elimination steps for any b_j or for an a_i if the involved equations do not contain any b_k,

The proposed measures are effective not only for algebraic problems but for ODEs/ PDEs too (i.e., to preserve linearity of a subset of functions as long as possible). {flin_}

Flexible process control. Different types of overdetermined systems are more or less suited for an automatic solution. With the current version (2002) it is relatively safe to try solving large bilinear algebraic problems automatically. Another well suited area concerns overdetermined systems of linear PDEs. In contrast, nonlinear systems of PDEs most likely require a more tight interactive control. Different modes of operation are possible. One can

- perform one {a} or more computational steps {g} automatically, each step trying modules in the order defined by the current priority list {p1} until one module succeeds in its purpose,
- perform one module a specific number of times or as long as it is successful, {l}
- set a time limit until which the program should run automatically, {time_limit, limit_time}
- interrupt an on-going automatic computation and continue the computation interactively, {_stop_}
- arrange that the priority list of modules changes at a certain point in the computation when the system of equations has changed its character,
- induce a case distinction whether a user-given expression is zero or not. {44}
- have a module that changes the priority list proc_list_ dynamically, depending essentially on the size and difficulty of the system but also on the success rate of previous steps. {61, 62, 63}

Apart from flexible control over what kind of steps are to be done, the steps themselves can be controlled more or less too, e.g., whether equations are selected by the module or by the user.

Highest priority in the priority list have so-called to-do steps. The list of to-do steps is usually empty but can be filled by any successful step if it requires another specific step to follow instantly. For example, if a very simple equation $0 = f_x$ is integrated then the substitution of f should follow straight away, even if substitutions would have a low priority according to the current priority list.

Total data control. To make wise decisions of how to continue the computation in an interactive session one needs tools to inspect large systems of equations. Helpful commands in CRACK print

- equations, inequalities, functions and variables, {e, pi, f}
- the occurrence of all derivatives of selected functions in any equation, {v}
- a statistics summary of the equations of the system, {s}
- a matrix display of occurrences of unknowns in all equations, {pd}

- a count of the total number of appearances of all unknowns, {pf}
- the value of any LISP variable, {pv}
- the value of algebraic expressions that can be specified using equation names (e.g., `coeffn(e_5,df(f,x,y),2))`, {pe}
- not under-determined subsystems. {ss}

Inspecting a computation which already goes on for hours or a day and that has performed many thousand steps is time consuming. The task is made easier with the possibility to plot graphically as a function of time: the type of steps performed, the number of unknowns, the number of remaining equations, the number of terms in these equations and the memory usage. {ps}

When nonlinear systems are considered and many case distinctions and sub(sub...)case distinctions are made in a long computation then one easily loses track. With one command one can list all cases that have been considered so far with their assumption, the number of steps made until they are solved or until the next subcase distinction was made and the number of solutions contained in each completed case. {ls}

Safety. When working on large problems it may come to a stage where computational steps are necessary, like substitution, which are risky in the sense that they may simplify the problem but also complicate it by increasing its size. To avoid this risk a few safety features have been implemented.

- At any time during the computation one can save a backup of the complete current situation in a file and also load a backup interactively. {sb, rb} or save it automatically {66}
- All key strokes are automatically recorded in a list and are available after each interactive step, or when the computation has finished. This list can be fed into CRACK at the beginning of a new computation so that the same operations are performed automatically that were performed interactively before. The purpose is to be able to do an interactive exploration first and to repeat it afterward automatically without having to note with pen or pencil all steps that had been done. {history_, old_history}
- During an automatic computation the program might start a computational step which turns out to take far too long. It would be better to stop this computation and try something else instead. But in computer algebra with lots of global variables involved it is not straight forward to stop a computation in the middle of it. If one would use time as a criterion then it could happen that time is up during a garbage collection which to stop would be deadly for the session. CRACK allows to set a limit of garbage collections for any one of those computations that have the potential to last forever, like algebraic factorizations of large expressions. With such an arrangement an automatic computation can not get stuck anymore due to lengthy factorizations, searches for length reductions or elimination steps. {max_gc_elimin, max_gc_fac, max_gc_red_len, max_gc_short,max_gc_ss}
- Due to a recent (April 2002) initiative of Winfried Neun the parallel version of the computer algebra system REDUCE has been re-activated and is running on the Beowulf cluster at Brock University [**12**]. This allows conveniently (with a 2-letter command) to duplicate the current status of a CRACK computation to another computer, to try out there different operations (e.g., risky ones) until a viable way to continue the computation is found without endangering the original session. {pp}

Managing solutions. Nonlinear problems can have many solutions. The number of solutions found by CRACK can even be higher because to make progress CRACK may have factorized an equation and considered the two cases $a = 0$ and $a \neq 0$ whereas solutions in both cases could be merged to only one solution without any restriction for a. This merging of solutions can be accomplished with a separate program `merge_sol()` after the computation.

Another form of post-processing is the production of a web page for each solution, like `http://lie.math.brocku.ca/twolf/bl/v/v1l05o35-s1.html`.

If in the solution of overdetermined differential equations the program performs integrations of equations before the differential Gröbner basis was computed then in the final solution there may be redundant constants or functions of integration. Redundant constants or functions in a solution are not an error but they make solutions appear unnecessarily complicated. In a postprocessing step these functions and constants can be eliminated. {adjust_fnc, drop_const(), dropredundant()}

Parallelization. The availability of a parallel version of CRACK was mentioned above allowing to try out different ways to continue an ongoing computation. A different possibility to make use of a cluster of computers with CRACK is, to export automatically the investigation of sub-cases and sub-sub-cases to different computers to be solved in parallel.

It was explained above how factorizations may be necessary to make any progress but also their potential of exploding the time requirements. By running the computation on a cluster and being able to solve many more cases one can give factorizations a higher priority and capitalize on the benefit of factorizations, i.e. the simplification of the problem.

Relationship to Gröbner basis algorithms. For systems of equations in which the unknown constants or functions turn up only polynomially a well known method is able to check the consistency of the system. For algebraic systems this is the Gröbner Basis Method and for systems of differential equations this is the differential Gröbner Basis method. To guarantee the method to terminate a total ordering of unknowns and their derivatives has to be introduced. This ordering determines which highest powers of unknowns are to be eliminated next or which highest order derivatives have to be eliminated next using integrability conditions. Often such eliminations lead to exponential growth of the generated equations. In the package CRACK these computations are executed with only a low priority. A higher priority have operations which reduce the length of equations, irrespective of any orderings. Violating any ordering a finite number of times still guarantees a finite algorithm. The potential gain is large as described next.

Length reduction of equations. An algorithm designed originally to length-reduce differential equations proved to be essential in length reducing systems of bilinear algebraic equations or homogeneous equations which resulted from bilinear equations during the solution process.

The aim of the method is to find out whether one equation $0 = E_1$ can length reduce another one $0 = E_2$ by replacing E_2 through an appropriate linear combination $\alpha E_1 - \beta E_2$, $\beta \neq 0$. To find α, β one can divide each term of E_2 through each term of E_1 and count how often each quotient occurs. If a quotient α/β occurs m times then $\alpha E_1 - \beta E_2$ will have $\leq n_1 + n_2 - 2m$ terms because $2m$ terms will cancel each other. A length reduction is found if $n_1 + n_2 - 2m \leq \max(n_1, n_2)$.

The method becomes efficient after a few algorithmic refinements discussed in [**27**]. Length reduced equations

- are more likely to length reduce other equations,
- are much more likely to be factorisable,
- are more suited for substitutions as the substitution induces less growth of the whole system and introduces fewer new occurrences of functions in equations,
- are more likely to be integrable by being a total derivative or being an ODE if the system consists of differential equations,
- involve on average fewer unknowns and make the whole system more sparse. This sparseness can be used to plan better a sequence of eliminations.

Customization. The addition of new modules to perform new specialized computations is easy. The new module only has to accept as input the system of equations, list of inequalities, list of unknowns to be computed and provide output in a similar form. The module name has to be added to a list of all modules and a one line description has to be added to a list of descriptions. This makes it easy for users to add special techniques for the solution of systems with extra structure. A dummy template module {58} is already added and has only to be filled with content.

Debugging. A feature useful mainly for debugging is that in the middle of an ongoing interactive computation the program can be changed by loading a different version of CRACK procedures. Thus one could advance quickly close to the point in the execution where an error occurs, load a version of the faulty procedure that gives extensive output and watch how the fault happens before fixing it.

Two modules {60, 64} can check whether a solution defined in a external procedure is a solution of the systemthat is currently under investigation. By performing this test after each computational step it is possible to pinpoint any bug that introduces extra false solutions or that excludes proper solutions.

The possibility to interrupt REDUCE itself temporarily and to inspect the underlying LISP environment {br} or to execute LISP commands and to continue with the CRACK session afterward {pc} led to a few improvements and fixes in REDUCE itself.

Availability. The package CRACK including manual is freely avalaible from `http://lie.math.brocku.ca/twolf/crack/`. The author would be grateful if the CRACK manual or this paper is cited if CRACK has been used for any computations done for a publication.

References

1. M. Boiti, J. Leon, M. Manna, and F. Pempinelli, *On a spectral transform of a KDV-like equation related to the Schrödinger operator in the plane*, Inverse Problems **3** (1987), no. 1, 25–36.
2. F. Calogero, *The evolution partial differential equation* $u_t = u_{xxx} + 3(u_{xx}u^2 + 3u_x^2 u) + 3u_x u^4$, J. Math. Phys. **28** (1987), no. 3, 538–555.
3. V. G. Drinfel′d and V. V. Sokolov, *Lie algebras and equations of Korteweg–de Vries type*, Current Problems in Mathematics, vol. 24, Itogi Nauki i Tekhniki, Akad. Nauk SSSR, Vsesoyuz. Inst. Nauchn. i Tekhn. Inform., Moscow, 1984, pp. 81–180 (Russian); English transl., J. Soviet Math. **30** (1985), 1975–2036.
4. A. S. Fokas, *A symmetry approach to exactly solvable evolution equations*, J. Math. Phys. **21** (1980), no. 6, 1318–1325.
5. ———, *Symmetries and integrability*, Stud. Appl. Math. **77** (1987), no. 3, 253–299.

6. M. V. Foursov, *Classification of certain integrable coupled potential KdV and modified KdV-type equations*, J. Math. Phys. **41**, (2000), no. 9, 6173–6185.
7. M. V. Foursov and P. J. Olver, *On the classification of symmetrically-coupled integrable evolution equations*, Symmetries and Differential Equations (K. V. Andreev and Yu. V Shanko, eds.) Institute of Computational Modelling, Krasnoyarsk, 2000, pp. 244–248.
8. R. Hirota and J. Satsuma, *Soliton solutions of a coupled Korteweg–de Vries equation*, Phys. Lett. A **85** (1981), no. 8-9, 407–408
9. N. H. Ibragimov and A. B. Shabat, *Evolution equations with a nontrivial Lie–Bäcklund algebra*, Funktsional. Anal. i Prilozhen. **14** (1980), no. 1, 25–36 (Russian); English transl., Functional Anal. Appl. **14** (1980), no. 1, 19–28.
10. P. P. Kulish and E. K. Sklyanin, O(N)*-invariant nonlinear Schrödinger equation—a new completely integrable system*, Phys. Lett. A **84** (1981), no. 7, 349–352.
11. S. V. Manakov, *On theory of two-dimensional self-focusing of electromagnetic waves*, Ž. Èksper. Teoret. Fiz. **65** (1973), 505–516.
12. H. Melenk and W. Neun, *RR: Parallel symbolic algorithm support for PSL based REDUCE*, ZIB preprint, 2002, `www.zib.de/Optimization/Software/Reduce/moredocs/parallel_reduce.ps`.
13. A. V. Mikhailov, A. B. Shabat, and V. V. Sokolov, *The symmetry approach to classification of integrable equations*, What Is Integrability?, Springer Ser. Nonlinear Dynam., Springer, Berlin, 1991, pp. 115–184.
14. A. V. Mikhailov, A. B. Shabat, and R. I. Yamilov, *A symmetric approach to the classification of nonlinear equations. Complete lists of integrable systems*, Russian Math. Surveys **42** (1987), no. 4, 1–63.
15. A. V. Mikhailov and R. I. Yamilov, *Towards classification of* $(2+1)$*-dimensional integrable equations. Integrability conditions.* I, J. Phys. A **31** (1998), no. 31, 6707–6715.
16. J. A. Sanders and J. P. Wang, *On the integrability of homogeneous scalar evolution equations*, J. Differential Equations **147** (1998), no. 2, 410–434.
17. V. V. Sokolov, *A new integrable case for the Kirchhoff equation*, Teoret. Mat. Fiz. **129** (2001), no. 1, 31–37 (Russian); English transl., Theoret. and Math.Phys. **129** (2001), no. 1, 1335–1340.
18. V. V. Sokolov and A. B. Shabat, *Classification of integrable evolution equations*, Mathematical Physics Reviews. Vol. 4, Soviet Sci. Rev. Sect. C Math. Phys. Rev., vol. 4, Harwood Academic Publ., Chur, 1984, pp. 221–280.
19. V. V. .Sokolov and A. V. Tsyganov, *On Lax pairs for the generalized Kovalevskaya and Goryachev–Chaplygin tops*, Teoret. Mat. Fiz. **131** (2002), no. 1, 118–125 (Russian); English transl., Theoret. and Math. Phys. **131** (2002), no. 1, 543–549.
20. V. V. Sokolov and T. Wolf, *A symmetry test for quasilinear coupled systems*, Inverse Problems **15** (1999), no. 2, L5–L11.
21. ———, *Classification of integrable polynomial vector evolution equations*, J. Phys. A **34** (2001), no. 49, 11139–11148.
22. S. I. Svinolupov, *Generalized Schrödinger equations and Jordan pairs*, Comm. Math. Phys. **143** (1992), no. 3, 559–575.
23. T. Tsuchida and M. Wadati, *New integrable systems of derivative nonlinear Schrödinger equations with multiple components*, Phys. Lett. A **257** (1999), 53–64.
24. T. Tsuchida and T. Wolf, *Classification of integrable that combine scalar variables and vector variables*, Sūrikaisekikenkyūsho Kōkyūroku 1302 (2003), 68–90 (Japanese).
25. G. Wilson, *The affine Lie algebra* $C_2^{(1)}$ *and an equation of Hirota and Satsuma*, Phys. Lett. A **89** (1981), no. 7, 332–334.
26. T. Wolf, *The symbolic integration of exact PDEs*, J. Symbolic Comput. **30** (2000), no. 5, 619–629.
27. ———, *Size reduction and partial decoupling of systems of equations*, J. Symbolic Comput. **33** (2002), no. 3, 367–383.
28. ———, *The integration of systems of linear PDEs using conservation laws of syzygies*, J. Symbolic Comput. **35** (2003), no. 5, 499–526.
29. T. Wolf and O. V. Efimovskaya, *Classification of integrable quadratic Hamiltonians on* e(3), Regul. Chaotic Dyn. **8** (2003), no. 2, 155-162.
30. Y. Wu, X. Geng, X. Hu, and S. Zhu, *A generalized Hirota–Satsuma coupled Korteweg–de Vries equation and Miura transformations*, Phys. Lett. A **255** (1999), no. 4-6, 259–264.

Department of Mathematics, Brock University, 500 Glenridge Ave., St. Catharines, ON L2S 3A1, Canada
E-mail address: twolf@brocku.ca

Centre de Recherches Mathématiques
CRM Proceedings and Lecture Notes
Volume **37**, 2004

The Prolate Spheroidal Phenomenon as a Consequence of Bispectrality

F. Alberto Grünbaum and Milen Yakimov

Abstract. In this paper we announce that a very large class of integral operators derived from bispectral algebras of rank 1 and 2 (parametrized by Lagrangian Grassmannians of infinitely large size) posses commuting differential operators. The examples of Landau, Pollak, Slepian, and Tracy, Widom, used in time-band limiting and random matrix theory arise as special cases of this result.

1. Introduction

It was discovered by Landau, Pollak, Slepian, and Tracy, Widom, that certain integral operators associated to the Airy and Bessel special functions possess commuting differential operators. They found important applications of this to time-band limiting, and to the study of asymptotics of Fredholm determinants, relevant to scaling limits of random matrix models. We call this phenomenon the prolate spheroidal phenomenon.

On the other hand, the problem of bispectrality was posed [**7**] about 20 years ago by one of us (A. G.) and J. J. Duistermaat as a tool to understand this prolate spheroidal property of integral operators. The aim was to extend it to larger classes and search for possible applications. Despite the dramatic recent developments in the areas of random matrices and bispectrality the two problems remained isolated except for several common examples, see [**9–11**]. In addition only a few integral operators possessing a commuting differential operator were found.

Here we announce that any self-adjoint bispectral algebra of ordinary differential operators (see Definition 3.3 and Definition 4.3 and Section 2 for general definitions) of rank 1 and 2 induces an integral operator possessing the prolate spheroidal property. The kernel of such an operator is of the form

$$K(x,z) = \int_{\Gamma_2} \Psi(x,z)\Psi(y,z)\,\mathrm{d}z$$

where $\Psi(x,z)$ is the corresponding bispectral (eigen)function and Γ_2 is a contour in the complex plane with 1 or 2 end points. It acts on the space $L^2(\Gamma_1)$ again for

2000 *Mathematics Subject Classification.* Primary 47G10; Secondary 15A52, 37K10.
This is the final form of the paper.

a contour with the same property. The main results are stated in Theorem 3.8 and Theorem 4.8.

The integral operators of Landau, Pollak, Slepian, and Tracy, Widom correspond to the cases $\Psi(x,z) = \sqrt{xz}J_{\nu+1/2}(ixz)$ and $\Psi(x,z) = A(x+z)$ (in terms of the Bessel functions of first kind and the Airy function) which are known to be "basic" bispectral functions, in the sense that the other rank 1 and 2 bispectral functions are obtained from them by certain types of Darboux transformations (see [**3,4,7,14,23**]). Although in these cases [**15,16,19–21**] the commuting differential operator is of order 2, in general it is of arbitrarily large order.

In the rest of the introduction we describe our strategy for proving Theorem 3.8 and Theorem 4.8 which relies on a very interesting property of the "size" of bispectral algebras of rank 1 and 2.

Consider, more generally, a holomorphic function $\Psi(x,z)$ in some domain of $\mathbb{C} \times \mathbb{C}$ which is not an eigenfunction of any differential operator in x or z. Denote by $\mathcal{B}_\Psi$ the algebra of differential operators $R(x,\partial_x)$ with rational coefficients for which there exists an operator $S(z,\partial_z)$ with rational coefficients such that

$$R(x,\partial_x)\Psi(x,z) = S(z,\partial_z)\Psi(x,z). \tag{1.1}$$

The algebra of all differential operators $S(z,\partial_z)$ obtained in this way will be denoted by $\mathcal{C}_\Psi$. The equality

$$b_\Psi\big(R(x,\partial_x)\big) := S(z,\partial_z)$$

correctly defines an anti-isomorphism from $\mathcal{B}_\Psi$ to C_Ψ. Recall that such a function $\Psi(x,z)$ is called bispectral if both algebras $\mathcal{B}_\Psi$ and $\mathcal{C}_\Psi$ contain rational functions. The subalgebra of $\mathcal{B}_\Psi$ and $\mathcal{C}_\Psi$, consisting of differential operators in x and z for which $\Psi(x,z)$ is an eigenfunction, are commutative. Algebras obtained in this way are called bispectral algebras. Their rank (which is equal and is also called rank of the bispectral function $\Psi(x,z)$) is the greatest common divisor of the orders of the operators of these algebras.

We derive our main result from the following remarkable property:

PROPERTY. *Consider the $\mathbb{Z}_+ \times \mathbb{Z}_+$ filtration of the algebra $\mathcal{B}_\Psi$ given by*

$$\mathcal{B}_\Psi^{l_1,l_2} = \{R(x,\partial_x) \in \mathcal{B}_\Psi \mid \operatorname{ord} R(x,\partial_x) \leq 2l_1,\ \operatorname{ord}(b_\Psi R)(x,\partial_x) \leq 2l_2\}.$$

If $\Psi(x,z)$ is a bispectral function of rank $r = 1$ or 2 then the the size of the spaces $\mathcal{B}_\Psi^{l_1,l_2}$ in this filtration is large in the sense

$$\dim \mathcal{B}_\Psi^{l_1,l_2} \geq \frac{2}{r}(2l_1l_2 + l_1 + l_2) - \text{const}$$

where the constant is independent of l_1 and l_2.

For the basic bispectral functions $\exp(xz)$, $A(x+z)$, and $\sqrt{xz}J_{\nu+1/2}(ixz)$, $\nu \in \mathbb{C}\backslash\mathbb{Z}$ of rank 1 and 2, the dimension of these spaces is exactly equal to the right-hand side with const $= -1$. This is remarkable since it shows that for all bispectral functions $\Psi(x,z)$ of rank 1 and 2 the spaces of the above natural filtration of the algebra $\mathcal{B}_\Psi$ are almost of the same dimension as the spaces corresponding to the basic bispectral functions.

It is very interesting to understand the relation between our results and the approach of Adler, Shiota, and van Moerbeke [**1, 2**] to the Tracy–Widom system of differential equations via Virasoro constraints. One possible way would be to incorporate the representation theoretic meaning of the Bessel tau functions from [**5**] in terms of representations of the $W_{1+\infty}$ algebra, see [**8**].

Another important problem is to understand the relation of this work to the isomonodromic deformations approach to random matrices from the works of Palmer [**18**], Harnad, Tracy, and Widom [**13,22**], Its and Harnad [**12**], and Borodin and Deift [**6**].

The proofs of the results announced here will appear in a forthcoming publication.

Acknowledgments. M. Y. would like to thank the organizers of the Workshop on Superintegrability, September 2002, CRM, Montreal and especially John Harnad and Pavel Winternitz for the opportunity to participate at the conference and for their interest in this work.

2. Bispectrality and commutativity

Denote by $\mathcal{W}_{\text{rat}}$ the algebra of differential operators in one variable with rational coefficients. Denote by $a(\cdot)$ the formal adjoint of operators in $\mathcal{W}_{\text{rat}}$:

$$a\Big(\sum_{k=0}^{n} b_k(x)\partial_x^n\Big) = \sum_{k=0}^{n}(-\partial_x)^n b_k(x). \tag{2.1}$$

For any $\xi \in \mathbb{C}$ consider the map from the subalgebra of $\mathcal{W}_{\text{rat}}$ consisting of operators regular at ξ to the set of bidifferential operators at ξ

$$\phi_\xi\Big(\sum_{k=0}^{n} b_k(x)\partial_x^n\Big) = \sum_{k=0}^{n}\sum_{i=0}^{k-1}(-1)^i\partial_x^{k-i-1}\otimes\partial_x^i b_k(x)\Big|_{x=\xi}. \tag{2.2}$$

For an oriented contour Γ in $\mathbb{C}$ we will denote its endpoints by $e(\Gamma)$. For any $\xi \in e(\Gamma)$ set $\pi(\xi) = 1$ or 0 depending on whether ξ is a left or right endpoint of Γ.

Assume that

$$D(x,\partial_x) = \sum_{k=0}^{n} b_k(x)\partial_x^n \in \mathcal{W}_{\text{rat}}$$

is a differential operator which is regular along Γ. Let $f(x)$ and $g(x)$ be smooth functions on Γ that decrease rapidly when $x \to \infty$ (i.e., $\lim_{x\to\infty} p(x)f^{(k)}(x) = 0$ for any polynomial $p(x)$ and any integer k, similarly for $g(x)$).

By standard integration by parts one gets that

$$\int_\Gamma \big(D(x,\partial_x)f(x)\big)g(x)\,\mathrm{d}x = \sum_{\xi\in e(\Gamma)}(-1)^{\pi(\xi)}\phi_\xi\big(D(x,\partial_x)\big)\big(f(x)\otimes g(x)\big) + \int_\Gamma f(x)\big(aD(x,\partial_x)g(x)\big)\,\mathrm{d}x. \tag{2.3}$$

Let $\Psi(x,z)$ be a holomorphic function in some domain of $\mathbb{C}\times\mathbb{C}$ which is not an eigenfunction of a differential operator in x or z, as in the introduction. Recall the definition (1.1) of the algebras of differential operators $\mathcal{B}_\Psi$ and $\mathcal{C}_\Psi$. Assume that $\widetilde{\mathcal{B}}_\Psi$ and $\widetilde{\mathcal{C}}_\Psi$ are two subalgebras of $\mathcal{B}_\Psi$ and $\mathcal{C}_\Psi$, respectively, that are stable under the formal adjoint map (2.1) and such that

$$b_\Psi(\widetilde{\mathcal{B}}_\Psi) = \widetilde{\mathcal{C}}_\Psi.$$

PROPOSITION 2.1. *Let Γ_1 and Γ_2 be two contours in $\mathbb{C}$ such that $\Gamma_1 \times \Gamma_2$ is in the domain of $\Psi(x,z)$ and $\Psi(x,z)$ decreases rapidly when x or z go to ∞ along Γ_1 and Γ_2. If $D(x,\partial_x) \in \widetilde{\mathcal{B}}_\Psi$ is such that*

$$ab_\Psi\big(D(x,\partial_x)\big) = b_\Psi a\big(D(x,\partial_x)\big) \tag{2.4}$$

and

$$\phi_\xi(D(x,\partial_x)) = 0,\ \forall \xi \in e(\Gamma_1); \qquad \phi_\xi((bD)(z,\partial_z)) = 0,\ \forall \xi \in e(\Gamma_2) \tag{2.5}$$

then the integral operator with kernel

$$K(x,y) = \int_{\Gamma_2} \Psi(x,z)\Psi(y,z)\,\mathrm{d}z \tag{2.6}$$

on $L^2(\Gamma_1)$ commutes with the differential operator $D(x,\partial_x)$ with domain all smooth functions on Γ_1 that decrease rapidly as $x \to \infty$.

Let us call a differential operator $D(x,\partial_x) \in \mathcal{W}_{\mathrm{rat}}$ *formally symmetric* if

$$(aD)(x,\partial_x) = D(x,\partial_x)$$

and *formally skewsymmetric* if

$$(aD)(x,\partial_x) = -D(x,\partial_x).$$

LEMMA 2.2. (i) *A differential operator $D(x,\partial_x) \in \mathcal{W}_{\mathrm{rat}}$ is formally symmetric if and only if it has the form*

$$D(x,\partial_x) = \sum_{i=0}^{n} \partial_x^n c_i(x) \partial_x^n \tag{2.7}$$

for some integer n and some rational functions $c_i(x)$.

(ii) *The operator $D(x,\partial_x)$ given by (2.7) satisfies*

$$\phi_\xi\big(D(x,\partial_x)\big) = 0$$

for some fixed $\xi \in \mathbb{C}$ if and only if

$$(\partial_x^i c_k)(\xi) = 0 \quad \text{for } k = 1,\dots,n,\ i = 0,\dots,k-1.$$

For a given function $\Psi(x,z)$ as before let $\mathcal{B}_{\Psi,\mathrm{sym}}$ be the subalgebra of $\mathcal{B}_\Psi$ consisting of differential operators $R(x,\partial_x)$ for which both $R(x,\partial_x)$ and $(b_\Psi R)(z,\partial_z)$ are formally symmetric. Set also $\mathcal{C}_{\Psi,\mathrm{sym}} := b_\Psi\big(\mathcal{B}_{\Psi,\mathrm{sym}}\big)$. Define the vector spaces

$$\mathcal{B}^{l_1,l_2}_{\Psi,\mathrm{sym}} = \{R(x,\partial_x) \in \mathcal{B}_{\Psi,\mathrm{sym}} \mid \operatorname{ord} R(x,\partial_x) \le 2l_1 \text{ and } \operatorname{ord}(b_\Psi R)(z,\partial_z) \le 2l_2\}, \tag{2.8}$$

$$\mathcal{C}^{l_1,l_2}_{\Psi,\mathrm{sym}} = \{S(z,\partial_z) \in \mathcal{C}_{\Psi,\mathrm{sym}} \mid \operatorname{ord}(b_\Psi^{-1} S)(x,\partial_x) \le 2l_1 \text{ and } \operatorname{ord} S(z,\partial_z) \le 2l_2\}. \tag{2.9}$$

Clearly $\mathcal{C}^{l_1,l_2}_{\Psi,\mathrm{sym}} = b_\Psi(\mathcal{B}^{l_1,l_2}_{\Psi,\mathrm{sym}})$.

¿From Proposition 2.1 and Lemma 2.2 we obtain:

THEOREM 2.3. *Assume the conditions from Proposition 2.1 for the the function $\Psi(x,z)$ and the contours Γ_1 and Γ_2. If either of the following two conditions is satisfied then the integral operator with kernel (2.6) possesses a formally commuting symmetric differential operator of order less than or equal to $2l_1$ and domain—the space of rapidly decreasing smooth functions on Γ_1*

Condition (i)

$$\mathcal{B}^{l_1,l_2}_{\Psi,\mathrm{sym}} > l_1(l_1+1)e(\Gamma_1)/2 + l_2(l_2+1)e(\Gamma_2)/2 + 1,$$

Condition (ii) $-e(\Gamma_1) = e(\Gamma_1)$, $-e(\Gamma_2) = e(\Gamma_2)$, *all operators in* $\mathcal{B}_{\Psi,\mathrm{sym}}$ *are invariant under the transformation* $x \mapsto -x$, *and*

$$\mathcal{B}^{l_1,l_2}_{\Psi,\mathrm{sym}} > l_1(l_1+1)e(\Gamma_1)/4 + l_2(l_2+1)e(\Gamma_2)/4 + 1.$$

3. Integral operators associated to self-adjoint Darboux transformations of Airy functions

3.1. The Airy bispectral function. Denote by $A(x)$ the Airy function and set

$$\Psi_A(x,z) = A(x+z). \tag{3.1}$$

Recall that $A(x)$ decreases rapidly when $x \to \infty$ in the sector $-\pi/3 < \arg x < \pi/3$.

If $L_A(x,\partial_x)$ denotes the Airy differential operator

$$L_A(x,\partial_x) = \partial_x^2 - x$$

then $\Psi_A(x,z)$ satisfies

$$L_A(x,\partial_x)\Psi_A(x,z) = z\Psi_A(x,z),$$
$$\partial_x\Psi_A(x,z) = \partial_z\Psi_A(x,z),$$
$$x\Psi_A(x,z) = L_A(z,\partial_z)\Psi_A(x,z).$$

For shortness denote the algebras $\mathcal{B}_{\Psi_A}$ and $\mathcal{C}_{\Psi_A}$ of differential operators with rational coefficients associated to the Airy function $\Psi_A(x,z)$, recall (1.1), by $\mathcal{B}_A$ and $\mathcal{C}_A$. It is straightforward to deduce:

LEMMA 3.1. *The algebras* $\mathcal{B}_A$ *and* $\mathcal{C}_A$ *coincide with the Weyl algebra* $\mathcal{W}_{\mathrm{poly}}$ *of differential operators in one variable with polynomial coefficients. Moreover the anti-isomorphism* b_A *associated to the Airy function* $\Psi_A(x,z)$, *recall* (3.1), *is uniquely defined from the relations*

$$b_A(x) = \big(L_A(z,\partial_z)\big), \qquad b_A(\partial_x) = \partial_z, \qquad b_A\big(L_A(x,\partial_x)\big) = z.$$

3.2. Self-adjoint Darboux transformations from the Airy function. Note that

$$\mathbb{C}[x] = \mathcal{B}_A \cap \mathbb{C}(x)$$

and

$$\mathbb{C}[L_A(x,\partial_x)] = b_A^{-1}(\mathcal{C}_A \cap \mathbb{C}(z)).$$

The set of rational Darboux transformations $\mathcal{D}_A$ from the Airy function was defined in [**4**] as the set of functions $\Psi(x,z)$ for which there exist differential operators

$$P(x,\partial_x), Q(x,\partial_x) \in (\mathbb{C}[x]\backslash 0)^{-1}\mathcal{B}_A = \mathcal{W}_{\mathrm{rat}} \tag{3.2}$$

such that

$$f(L_A(x,\partial_x)) = Q(x,\partial_x)P(x,\partial_x), \tag{3.3}$$

$$\Psi(x,z) = \frac{1}{p(z)}P(x,\partial_x)\Psi_A(x,z), \tag{3.4}$$

for some polynomials $f(t)$ and $p(z)$. (The polynomial $p(z)$ is included for normalization purposes only.) The quotient ring of $\mathcal{B}_A$ by $\mathbb{C}[x]\backslash\{0\}$ in (3.2) is well defined since $\mathbb{C}[x]\backslash\{0\}$ is an Ore subset of $\mathcal{B}_A$, see [**17**].

It was also shown in [**4, 14**] and more conceptually proved in [**3**] that:

THEOREM 3.2. *All rational Darboux transformations from the Airy function $\Psi(x,z)$ are bispectral functions of rank* 2.

DEFINITION 3.3. Define the set $\mathcal{SD}_A$ of self-adjoint Darboux transformations from the Airy function $\Psi(x,z)$ to consist of those functions $\Psi(x,z)$ for which there exists a differential operator $P(x,\partial_x) \in \mathcal{W}_{\text{rat}}$ such that

$$g(L_A(x,\partial_x))^2 = (aP)(x,\partial_x)P(x,\partial_x), \tag{3.5}$$

$$\Psi(x,z) = \frac{1}{g(z)}P(x,\partial_x)\Psi_A(x,z), \tag{3.6}$$

for some polynomial $g(t)$.

In fact $\mathcal{SD}_A$ consists exactly of those $\Psi(x,z) \in \mathcal{D}_A$ for which $Q(x,\partial_x) = (aP)(x,\partial_x)$ in (3.3)–(3.4) with an appropriate normalization of the polynomial $p(z)$. One can show that as a consequence $f(t)$ is the square of some polynomial $g(t)$, compare to (3.5)–(3.6).

3.3. Size of the algebra $\mathcal{B}_A$ relative to the anti-isomorphism b_A. Consider the $\mathbb{Z}_+ \times \mathbb{Z}_+$ filtration of the algebra $\mathcal{B}_A$ associated to the Airy function $\Psi_A(x,z)$, defined analogously to (2.8) by

$$\mathcal{B}_A^{l_1,l_2} = \{R(x,\partial_x) \in \mathcal{B}_A \mid \operatorname{ord} R(x,\partial_x) \le 2l_1, \operatorname{ord}(b_AR)(z,\partial_z) \le 2l_2\}. \tag{3.7}$$

LEMMA 3.4. *The vector space $\mathcal{B}_A^{l_1,l_2}$ has a basis that consists of the differential operators*

$$\{x^m(L_A(x,\partial_x))^n \mid n \le l_1, m \le l_2\} \cup \{x^m\partial_x(L_A(x,\partial_x))^n \mid n < l_1, m < l_2\}.$$

Note that the formal adjoint anti-involution a of $\mathcal{W}_{\text{rat}}$ preserves the spaces $\mathcal{B}_A^{l_1,l_2}$. Since $a^2 = \text{id}$ the space $\mathcal{B}_A^{l_1,l_2}$ is the direct sum of the eigenspaces of a with eigenvalues ±1. Denote the eigenvalue 1 subspace of $\mathcal{B}_A^{l_1,l_2}$ by $\mathcal{B}_{A,\text{sym}}^{l_1,l_2}$. For the Airy function $\Psi_A(x,z)$ one has $ab_A = b_Aa$ and thus:

$$\begin{aligned}\mathcal{B}_{A,\text{sym}}^{l_1,l_2} = \{R(x,\partial_x) \in \mathcal{B}_A \mid \operatorname{ord} R(x,\partial_x) \le 2l_1,\ \operatorname{ord}(b_AR)(z,\partial_z) \le 2l_2,\\ aR(x,\partial_x) = R(x,\partial_x),\ a(b_AR(z,\partial_z)) = b_AR(z,\partial_z)\}.\end{aligned}$$

LEMMA 3.5. *The set of operators*

$$\{x^m(L_A(x,\partial_x))^n + (L_A(x,\partial_x))^n x^m \mid n \le l_1, m \le l_2\} \tag{3.8}$$

is a basis for the space $\mathcal{B}_{A,\text{sym}}^{l_1,l_2}$. In particular

$$\dim \mathcal{B}_{A,\text{sym}}^{l_1,l_2} = (l_1+1)(l_2+1).$$

3.4. Size of the algebra $\mathcal{B}_{\Psi,\mathrm{sym}}$ for a self-adjoint Darboux transformation from the Airy function, relative to the involution b_Ψ. Fix an arbitrary self-adjoint Darboux transformation from the Airy function $\Psi \in \mathcal{SD}_A$ given by (3.5)–(3.6) for some $P(x,\partial_x) \in \mathcal{W}_{\mathrm{rat}}$ and $g(t) \in \mathbb{C}[t]$. Let

$$P(x,\partial_x) = \frac{1}{v(x)} R(x,\partial_x) \tag{3.9}$$

for some $R(x,\partial_x) \in \mathcal{W}_{\mathrm{poly}} = \mathcal{B}_A$ and $v(x) \in \mathbb{C}[x]$. Set

$$\operatorname{ord} R(x,\partial_x) = \rho_1 \qquad \text{and} \qquad \operatorname{ord}(b_A R)(x,\partial_x) = \rho_2. \tag{3.10}$$

Denote

$$\mathcal{S}_{\Psi,1} = \operatorname{Span}\left\{ \frac{1}{v(x)} R(x,\partial_x) M(x,\partial_x)(aR)(x,\partial_x)\frac{1}{v(x)} \;\middle|\; M(x,\partial_x) \in \mathcal{B}_{A,\mathrm{sym}}^{l_1-\rho_1,l_2} \right\},$$
$$\mathcal{S}_{\Psi,2} = \operatorname{Span}\{ v(x) M(x,\partial_x) v(x) \mid M(x,\partial_x) \in \mathcal{B}_{A,\mathrm{sym}}^{l_1,l_2-\rho_2} \}.$$

PROPOSITION 3.6. *In the above setting:*

(i) *The spaces of differential operators $\mathcal{S}_{\Psi,1}$ and $\mathcal{S}_{\Psi,2}$ are subspaces of $\mathcal{B}_{\Psi,\mathrm{sym}}^{l_1,l_2}$.*
(ii) *The dimension of the intersection $\mathcal{S}_{\Psi,1} \cap \mathcal{S}_{\Psi,2}$ is less than or equal to*

$$(l_1 - \rho_1 + 1)(l_2 - \rho_2 + 1).$$

Theorem 4.2 of [**3**] shows that (3.5)–(3.6) is equivalent to

$$v(L_A(z,\partial_z))^2 = (ab_A R)(z,\partial_z)\frac{1}{g(z)^2}(b_A R)(z,\partial_z),$$
$$\Psi(x,z) = \frac{1}{v(x)g(z)}(b_A R)(z,\partial_z)\Psi_A(x,z).$$

(The hard step is to prove the first equality.) One can show that

$$b_\Psi\left(\frac{1}{v(x)}R(x,\partial_x)M(x,\partial_x)(aR)(x,\partial_x)\frac{1}{v(x)}\right) = g(z)(b_A M)(z,\partial_z)g(z)$$

for all operators $M(x,\partial_x) \in \mathcal{B}_A$. The fact that $\mathcal{S}_{\Psi,2}$ is a subspace of $\mathcal{B}_{\Psi,\mathrm{sym}}^{l_1,l_2}$ follows from this and the fact that $\mathcal{S}_{\Psi,1} \subset \mathcal{B}_{\Psi,\mathrm{sym}}^{l_1,l_2}$ by exchanging the roles of x and z.

THEOREM 3.7. *For any self-adjoint Darboux transformation from the Airy function $\Psi(x,z) \in \mathcal{SD}_A$ the dimension of the space of differential operators $\mathcal{S}_{\Psi,1} + \mathcal{S}_{\Psi,2}$ is greater than or equal to $(l_1+1)(l_2+1) - \rho_1\rho_2$. In particular*

$$\dim \mathcal{B}_{\Psi,\mathrm{sym}}^{l_1,l_2} \geq (l_1+1)(l_2+1) - \rho_1\rho_2. \tag{3.11}$$

Theorem 3.7 and Theorem 2.3 imply our final result for integral operators derived from Darboux transformations from the Airy function.

THEOREM 3.8. *Let $\Psi(x,z) \in \mathcal{SD}_A$ be a self-adjoint Darboux transformation from the Airy function, given by (3.5), (3.6), (3.9). Let Γ_1, Γ_2 be two connected contours in $\mathbb{C}$ that do not contain the roots of the polynomials $v(t)$ and $g(t)$ respectively and that begin at some finite points and go to infinity in the sector $-\pi/3 < \arg x < \pi/3$. Then the integral operator*

$$K(x,y) = \int_{\Gamma_2} \Psi(x,z)\Psi(y,z)\mathrm{d}z$$

on $L^2(\Gamma_1)$ commutes with a formally symmetric differential operator with rational coefficients of order less than or equal to $2(\rho_1\rho_2+1)$ and domain all smooth functions on Γ_1 that decrease rapidly as $x \to \infty$.

4. Integral operators associated to self-adjoint Darboux transformations of Bessel functions

4.1. The Bessel bispectral function. Denote by J_ν the standard Bessel functions of first kind. By abuse of notation the functions

$$\Psi_\nu(x,z) = (xz)^{1/2} J_{\nu+1/2}(ixz) \tag{4.1}$$

will be also called Bessel functions. Consider the Euler operator

$$D_x = x\partial_x$$

and the operators

$$L_\nu(x,\partial_x) = \partial_x^2 - \frac{\nu(\nu+1)}{x^2} = \frac{1}{x^2}(D_x+\nu)(D_x-\nu-1) \tag{4.2}$$

to be called Bessel operators. The Bessel functions satisfy the equations

$$L_\nu(x,\partial_x)\Psi_\nu(x,z) = z^2\Psi_\nu(x,z), \tag{4.3}$$

$$D_x\Psi_\nu(x,z) = D_z\Psi_\nu(x,z), \tag{4.4}$$

$$x^2\Psi_\nu(x,z) = L_\nu(z,\partial_z)\Psi_\nu(x,z). \tag{4.5}$$

For shortness the algebras $\mathcal{B}_{\Psi_\nu}$ and $\mathcal{C}_{\Psi_\nu}$, associated to the Bessel function $\Psi_\nu(x,z)$, recall (1.1), will be denoted by $\mathcal{B}_\nu$ and $\mathcal{C}_\nu$.

The Bessel functions corresponding to $\nu_1,\nu_2 \in \mathbb{C}$ that differ by an integer can be obtained by a Darboux transformation from each other:

$$\Psi_{\nu+1}(x,z) = \frac{1}{xz}(D_x-\nu-1)\Psi_\nu(x,z), \quad \Psi_\nu(x,z) = \frac{1}{xz}(D_x+\nu+1)\Psi_{\nu+1}(x,z),$$

which corresponds to the factorizations

$$L_\nu = x^{-1}(D_x+\nu+1)x^{-1}(D_x-\nu-1), \quad L_{\nu+1} = x(D_x-\nu-1)x^{-1}(D_x+\nu+1).$$

According to (4.3)–(4.5) the algebras of differential operators $\mathcal{B}_\nu$ and $\mathcal{C}_\nu$ contain the operators $L_\nu(x,\partial_x)$, D_x, and x^2. Denote their subalgebras generated by those operators by $\widetilde{\mathcal{B}}_\nu$ and $\widetilde{\mathcal{C}}_\nu$, respectively. Clearly (4.3)–(4.5) imply

$$b_\nu(\widetilde{\mathcal{B}}_\nu) = \widetilde{\mathcal{C}}_\nu.$$

Similarly to Proposition 2.4 in [**4**] one shows:

LEMMA 4.1. (i) *For $\nu \in \mathbb{C}\backslash\mathbb{Z}$ the algebras $\mathcal{B}_\nu$ and $\mathcal{C}_\nu$ are generated by the operators $L_\nu(x,\partial_x)$, D_x, and x^2, i.e.,*

$$\mathcal{B}_\nu = \widetilde{\mathcal{B}}_\nu, \qquad \mathcal{C}_\nu = \widetilde{\mathcal{C}}_\nu.$$

(ii) *For $\nu \in \mathbb{Z}$ the subalgebras $\widetilde{\mathcal{B}}_\nu$ and $\widetilde{\mathcal{C}}_\nu$ of $\mathcal{B}_\nu$ and $\mathcal{C}_\nu$ consist of exactly those differential operators in $\mathcal{B}_\nu$ and $\mathcal{C}_\nu$ that are invariant under the transformation $x \mapsto -x$.*

Note that

$$ab_\nu\big(P(x,\partial_x)\big) = b_\nu a\big(P(x,\partial_x)\big), \quad \forall P(x,\partial_x) \in \widetilde{\mathcal{B}}_\nu. \tag{4.6}$$

4.2. Self-adjoint Darboux transformations from the Bessel functions. Similarly to the Airy case we have

$$\mathbb{C}[x^2] = \widetilde{\mathcal{B}}_\nu \cap \mathbb{C}(x), \qquad \mathbb{C}[L_\nu(x, \partial_x)] = b_\nu^{-1}(\widetilde{\mathcal{C}}_\nu \cap \mathbb{C}(z)).$$

The set of rational Darboux transformations $\mathcal{D}_\nu$ from the Bessel function $\Psi_\nu(x, z)$ is defined to be the set of functions $\Psi(x, z)$ for which there exist differential operators

$$P(x, \partial_x), Q(x, \partial_x) \in (\mathbb{C}[x^2]\backslash\{0\})^{-1}\mathcal{B}_A \tag{4.7}$$

such that

$$f(L_\nu(x, \partial_x)) = Q(x, \partial_x)P(x, \partial_x), \tag{4.8}$$

$$\Psi(x, z) = \frac{1}{p(z)} P(x, \partial_x)\Psi_\nu(x, z), \tag{4.9}$$

for some polynomials $f(t)$ and $p(z)$. Again $(\mathbb{C}[x^2]\backslash\{0\})$ is an Ore subset of $\mathcal{B}_A$ and quotient ring in (4.7) makes sense.

The following theorem was proved in [**23**] for $\nu = 1$ and in [**3, 4**] in the general case.

THEOREM 4.2. *All rational Darboux transformations from the Bessel functions are bispectral of rank* 2 *if* $\nu \in \mathbb{C}\backslash\mathbb{Z}$ *and of rank* 1 *if* $\nu \in \mathbb{Z}$.

DEFINITION 4.3. Define the set of self-adjoint ("even, self-adjoint" in the case $\nu \in \mathbb{Z}$) Darboux transformations $\mathcal{SD}_\nu$ from the Bessel functions $\Psi_\nu(x, z)$ to consist of all functions $\Psi(x, z)$ for which there exists a differential operator

$$P(x, \partial_x) \in (\mathbb{C}[x^2]\backslash\{0\})^{-1}(\widetilde{\mathcal{B}}_\nu)$$

such that

$$f(L_\nu(x, \partial_x)) = (-1)^m (aP)(x, \partial_x)P(x, \partial_x) \tag{4.10}$$

$$\Psi(x, z) = \frac{1}{z^m g(z^2)} P(x, \partial_x)\Psi_\nu(x, z), \tag{4.11}$$

for a polynomial $f(t)$ of the form

$$f(t) = t^{2m} g(t^2)^2, \quad g(0) \neq 0; g(t) \in \mathbb{C}[t]. \tag{4.12}$$

Here the term "even" reflects the fact that for $\nu \in \mathbb{Z}$ *the algebras* $\mathcal{B}_\nu$ *and* $\mathcal{C}_\nu$ *are bigger than* $\widetilde{\mathcal{B}}_\nu$ *and* $\widetilde{\mathcal{C}}_\nu$. The reason for this terminology is explained below.

As in the Airy case the set $\mathcal{SD}_\nu$ consists of those rational Darboux transformations $\Psi(x, z)$ from the Bessel function $\Psi_\nu(x, z)$ for which in the notation (4.8)–(4.9)

$$Q(x, \partial_x) = (-1)^{\operatorname{ord} P}(aP)(x, \partial_x) \tag{4.13}$$

with the additional property in the case $\nu \in \mathbb{Z}$

$$P(-x, -\partial_x) = P(x, \partial_x). \tag{4.14}$$

As a consequence it is obtained that the polynomial $f(t)$ in (4.8) has the form (4.12) and an appropriate normalization of $p(z)$ is made.

Note that Lemma 4.1 implies that in the rank 2 case $\nu \in \mathbb{C}\backslash\mathbb{Z}$ the condition (4.14) is a consequence of (4.7). In the rank 1 case $\nu \in \mathbb{Z}$ the term "even" reflects this extra condition. It is needed since in the case of Darboux transformations from the Bessel function the prolate spheroidal property will be deduced from the second condition in Theorem 2.3.

4.3. Size of the algebras $\widetilde{B}_\nu$ relative to the anti-isomorphisms b_ν. Consider the $\mathbb{Z}_+ \times \mathbb{Z}_+$ filtration of the algebras $\widetilde{B}_\nu$

$$\widetilde{\mathcal{B}}_\nu^{l_1,l_2} = \{R(x,\partial_x) \in \widetilde{\mathcal{B}}_\nu \mid \operatorname{ord} R(x,\partial_x) \le 2l_1, \operatorname{ord}(b_\nu R)(z,\partial_z) \le 2l_2\} \tag{4.15}$$

The formal adjoint involution a of $\mathcal{W}_{\text{rat}}$ preserves the spaces $\mathcal{B}_\nu^{l_1,l_2}$. Similarly to Lemma 3.4 one shows:

LEMMA 4.4. *The vector space $\mathcal{B}_\nu^{l_1,l_2}$ has a basis that consists of the differential operators*

$$\{x^{2m}(L_\nu(x,\partial_x))^n \mid n \le l_1, m \le l_2\} \cup \{x^{2m} D_x (L_\nu(x,\partial_x))^n \mid n < l_1, m < l_2\}.$$

Since $a^2 = \text{id}$ the space $\widetilde{\mathcal{B}}_\nu^{l_1,l_2}$ is the direct sum of the eigenspaces of a with eigenvalues ± 1. The eigenvalue 1 subspace of $\widetilde{\mathcal{B}}_\nu^{l_1,l_2}$ will be denoted by $\widetilde{\mathcal{B}}_{\nu,\text{sym}}^{l_1,l_2}$. The commutativity (4.6) of a and b_ν on $\widetilde{\mathcal{B}_\nu}^{l_1,l_2}$ implies:

$$\widetilde{\mathcal{B}}_{\nu,\text{sym}}^{l_1,l_2} = \{R(x,\partial_x) \in \widetilde{\mathcal{B}}_\nu \mid \operatorname{ord} R(x,\partial_x) \le 2l_1,\ \operatorname{ord}(b_\nu R)(z,\partial_z) \le 2l_2,$$
$$aR(x,\partial_x) = R(x,\partial_x),\ a(b_\nu R(x,\partial_x)) = b_\nu R(x,\partial_x)\}.$$

LEMMA 4.5. *The set of operators*

$$\{x^{2m}(L_\nu(x,\partial_x))^n + (L_\nu(x,\partial_x))^n x^{2m} \mid n \le l_1, m \le l_2\} \tag{4.16}$$

is a basis for the space $\widetilde{\mathcal{B}}_{\nu,\text{sym}}^{l_1,l_2}$. In particular

$$\dim \widetilde{\mathcal{B}}_{\nu,\text{sym}}^{l_1,l_2} = (l_1+1)(l_2+1).$$

4.4. Size of the algebra $\mathcal{B}_{\Psi,\text{sym}}$ for an (even) self-adjoint Darboux transformation from a Bessel function. Fix an arbitrary self-adjoint (and in addition even in the rank 1 case $\nu \in \mathbb{Z}$) Darboux transformation from a Bessel function $\Psi_\nu(x,z)$, $\Psi \in \mathcal{SD}_\nu$ given by (4.10)–(4.11) for some $P(x,\partial_x) \in (\mathbb{C}[x^2]\backslash\{0\})^{-1}\widetilde{\mathcal{B}}_\nu$ and $f(t) = t^{2m} g(t^2)^2$, $g(t) \in \mathbb{C}[t]$. Let

$$P(x,\partial_x) = \frac{1}{v(x^2)} R(x,\partial_x) \tag{4.17}$$

for some $R(x,\partial_x) \in \widetilde{\mathcal{B}}_\nu$ and $v(x) \in \mathbb{C}[x]$. Set

$$\operatorname{ord} R(x,\partial_x) = \rho_1 \quad\text{and}\quad \operatorname{ord}(b_\nu R)(x,\partial_x) = \rho_2. \tag{4.18}$$

Denote

$$\mathcal{S}_{\Psi,1} = \operatorname{Span}\left\{ \frac{1}{v(x^2)} R(x,\partial_x) M(x,\partial_x)(aR)(x,\partial_x) \frac{1}{v(x^2)} \,\middle|\, M(x,\partial_x) \in \widetilde{\mathcal{B}}_{\nu,\text{sym}}^{l_1-\rho_1,l_2} \right\},$$
$$\mathcal{S}_{\Psi,2} = \operatorname{Span}\{v(x^2) M(x,\partial_x) v(x^2) \mid M(x,\partial_x) \in \widetilde{\mathcal{B}}_{\nu,\text{sym}}^{l_1,l_2-\rho_2}\}.$$

PROPOSITION 4.6. (i) *The spaces of differential operators $\mathcal{S}_{\Psi,1}$ and $\mathcal{S}_{\Psi,2}$ are subspaces of $\mathcal{B}_{\Psi,\text{sym}}^{l_1,l_2}$ and are invariant under the transformation $x \mapsto -x$.*

(ii) *The dimension of the intersection $\mathcal{S}_{\Psi,1} \cap \mathcal{S}_{\Psi,2}$ is less than or equal to*

$$(l_1 - \rho_1 + 1)(l_2 - \rho_2 + 1).$$

Finally we obtain the following theorem.

THEOREM 4.7. *For any self-adjoint (and even, self-adjoint in the case $\nu \in \mathbb{Z}$) Darboux transformation from a Bessel function $\Psi_\nu(x,z)$, $\Psi(x,z) \in \mathcal{SD}_\nu$ the space $\mathcal{S}_{\Psi,1} + \mathcal{S}_{\Psi,2}$ consists of differential operators invariant under the transformation $x \mapsto -x$*

$$\dim(\mathcal{S}_{\Psi,1} + \mathcal{S}_{\Psi,2}) \geq (l_1+1)(l_2+1) - \rho_1\rho_2.$$

In particular the dimension of the subspace of $\mathcal{B}^{l_1,l_2}_{\Psi,\mathrm{sym}}$ of differential operators invariant under $x \mapsto -x$ is greater than or equal to $(l_1+1)(l_2+1) - \rho_1\rho_2$.

Theorems 4.7 and 2.3 imply our final result for Darboux transformations from the Bessel functions:

THEOREM 4.8. *Let $\Psi(x,z) \in \mathcal{SD}_\nu$ be a self-adjoint (and in addition even if $\nu \in \mathbb{Z}$) Darboux transformation from the Bessel function $\Psi_\nu(x,z)$, given by (4.10), (4.11), (4.12), (4.17). Let Γ_1, Γ_2 be two connected finite contours that do not contain the roots of the polynomials $v(t)$ and $g(t)$, respectively and such that $e(\Gamma_i) = -e(\Gamma_i)$.*

Then the integral operator with kernel

$$K(x,y) = \int_{\Gamma_2} \Psi(x,z)\Psi(y,z)\mathrm{d}z$$

on $L^2(\Gamma_1)$ commutes with a formally symmetric differential operator with rational coefficients of order less than or equal to $2(\rho_1\rho_2+1)$ and domain all smooth functions on Γ_1.

References

1. M. Adler, T. Shiota, and P. van Moerbeke, *Random matrices, Virasoro algebra, and noncommutative KP*, Duke Math. J. **94** (1998), no. 2, 379–431.
2. M. Adler and P. van Moerbeke, *The spectrum of coupled random matrices*, Ann. Math. **149** (1999), no. 3, 931–976.
3. B. Bakalov, E Horozov, and M. Yakimov, *General methods for constructing bispectral operators*, Phys. Lett A. **222** (1996), no. 1-2, 59–66.
4. ———, *Bispectral algebras of commuting ordinary differential operators*, Comm. Math. Phys. **190** (1997), no. 2, 331–373.
5. ———, *Highest weight modules over the $W_{1+\infty}$ algebra and the bispectral problem*, Duke Math. J. **93** (1998), no. 1, 41–72.
6. A. Borodin and P. Deift, *Fredholm determinants, Jimbo–Miwa–Ueno τ-functions, and representation theory*, Comm. Pure Appl. Math. **55** (2002), no. 9, 1160–1230.
7. J. J. Duistermaat and F. A. Grünbaum, *Differential equations in the spectral parameter*, Comm. Math. Phys. **103** (1986), no. 2, 177–240.
8. E. Frenkel, V. Kac, A. Radul, and W. Wang, *$W_{1+\infty}$ and $W(\mathrm{gl}_N)$ algebra with central charge N*, Comm. Math. Phys. **170** (1995), no. 2, 337–357.
9. F. A. Grünbaum, *Time-band limiting and the bispectral problem*, Comm. Pure Appl. Math. **47** (1994), no. 3, 307–328.
10. ———, *Band-time-band limiting integral operators and commuting differential operators*, Algebra i Analiz **8** (1996), no. 1, 122–126 (Russian); English transl., St. Petersburg Math. J. **8** (1997), no. 1, 93–96.
11. ———, *Some bispectral musing*, The Bispectral Problem (Montréal, 1997), CRM Proc. Lecture Notes, vol. 14, Amer. Math. Soc., Providence, RI, 1998, pp. 31–45.
12. J. Harnad and A. R. Its, *Integrable Fredholm operators and dual isomonodromic deformations*, Comm. Math. Phys. **226** (2002), no. 3, 497–530.
13. J. Harnad, C. A. Tracy, and H. Widom, *Hamiltonian structure of equations appearing in random matrices*, Low-Dimensional Topology and Quantum Field Theory (Cambridge, 1992), NATO Adv. Sci. Inst. Ser. B Phys., vol. 315, Plenum, New York, 1993, pp. 231–245.
14. A. Kasman and M. Rothstein, *Bispectral Darboux transformations: the generalized Airy case*, Phys. D **102** (1997), no. 3-4, 159–176.

15. H. J. Landau and H. O. Pollak, *Prolate spheroidal wave functions, Fourier analysis and uncertainty.* II, Bell System Tech. J. **40**, (1961), 65–84.
16. ———, *Prolate spheroidal wave functions, Fourier analysis and uncertainty.* III, Bell System Tech. J. **41**, (1961), 1295–1336.
17. J. C. McConnell and J. C. Robson, *Noncommutative Noetherian rings*, Pure Appl. Math., Wiley, Chichester, 1987.
18. J. Palmer, *Deformation analysis of matrix models*, Phys. D **78** (1995), no. 3-4, 166–185.
19. D. Slepian and H. O. Pollak, *Prolate spheroidal wave functions, Fourier analysis and uncertainty.* I, Bell System Tech. J. **40**, (1961), 43–63.
20. C. A. Tracy and H. Widom, *Level spacing distributions and the Airy kernel*, Comm. Math. Phys. **159** (1994), no. 1, 151–174.
21. ———, *Level spacing distributions and the Airy kernel*, Comm. Math. Phys. **161** (1994), no. 2, 289–309.
22. ———, *Fredholm determinants, differential equations and matrix models*, Comm. Math. Phys. **163** (1994), no. 1, 33–72.
23. G. Wilson, *Bispectral algebras of commuting ordinary differential operators*, J. Reine Angew. Math. **442** (1993), 177–204.

Department of Mathematics, University of California at Berkeley, Berkeley, CA 94720, USA

E-mail address: grunbaum@math.berkeley.edu

Department of Mathematics, University of California at Santa Barbara, Santa Barbara, CA 93106, USA

E-mail address: yakimov@math.ucsb.edu

Centre de Recherches Mathématiques
CRM Proceedings and Lecture Notes
Volume **37**, 2004

On a Trigonometric Analogue of Atiyah–Hitchin Bracket

Oksana Yermolayeva

ABSTRACT. The space of rational functions has a natural Poisson structure discovered by Atiyah and Hitchin. We consider its derivation by means of a quadratic r-matrix structure which naturally arises for the corresponding scattering problem and provide a trigonometric analogue.

1. Poisson brackets on space of rational maps

In the work by Atiyah and Hitchin [**1**] a symplectic structure on the space R_N of rational functions of the form

$$S(\lambda) = \sum_{i=0}^{N-1} \frac{\rho_i}{\lambda - \beta_i} \tag{1.1}$$

was introduced.

To describe this symplectic structure we represent function $S(\lambda)$ as a ratio of two polynomials

$$S(\lambda) = \frac{p(\lambda)}{q(\lambda)} \tag{1.2}$$

where $q(\lambda) = (\lambda - \beta_1) \cdots (\lambda - \beta_N)$. Then the Atiyah–Hitchin symplectic form looks as follows:

$$\omega = \sum_{i=1}^{N} \frac{dp(\beta_i) \wedge d\beta_i}{p(\beta_i)} \tag{1.3}$$

The symplectic structure implies the following Poisson brackets

$$\{p(\beta_m), p(\beta_n)\} = 0, \qquad \{\beta_m, \beta_n\} = 0 \tag{1.4}$$

$$\{p(\beta_m), \beta_n\} = p(\beta_m)\delta_m{}^n \tag{1.5}$$

These Poisson brackets in turn imply brackets between the polynomials $p(\lambda)$ and $q(\lambda)$:

$$\{q(\lambda), q(\mu)\} = 0, \qquad \{p(\lambda), p(\mu)\} = 0 \tag{1.6}$$

1991 *Mathematics Subject Classification.* Primary 37J35; Secondary 53D17, 70H06.

The author thanks Professors J. Harnad, D. Korotkin, and K. Takasaki for helpful discussions.
This is the final form of the paper.

$$\{p(\lambda), q(\mu)\} = \frac{p(\lambda)q(\mu) - q(\lambda)p(\mu)}{\lambda - \mu} \tag{1.7}$$

It was shown by Gekhtman and Faybusovich in [**3**] that the relations (1.6) and (1.7) are equivalent to the following brackets on the space of rational functions R_N (1.3):

$$\{S(\lambda), S(\mu)\} = \frac{(S(\lambda) - S(\mu))^2}{\lambda - \mu} \tag{1.8}$$

In fact, the bracket (1.8) can obviously be extended to the space of all rational functions, i.e., the ratio of two polynomials of arbitrary degree. However the requirement of distinct roots remains essential. As was pointed out by K. Takasaki [**4**], the Gekhtman–Faybusovich bracket (1.8) is naturally related to the rational quadratic Sklyanin bracket

$$\{\overset{1}{T}(\lambda), \overset{2}{T}(\mu)\} = [\overset{12}{r}(\lambda - \mu), \overset{1}{T}(\lambda)\overset{2}{T}(\mu)] \tag{1.9}$$

where

$$\overset{12}{r}(\lambda) = \frac{\Pi}{\lambda} = \frac{I \otimes I + \sigma_\alpha \otimes \sigma_\alpha}{2\lambda} \tag{1.10}$$

is the classical rational r-matrix and $\Pi(u \otimes v) = v \otimes u$ is a permutation matrix.

If we parametrize matrix $T(\lambda)$ to be:

$$T(\lambda) = \begin{pmatrix} p(\lambda) & \tilde{p}(\lambda) \\ q(\lambda) & \tilde{q}(\lambda) \end{pmatrix} \tag{1.11}$$

then the Sklyanin bracket (1.9) implies the following Poisson brackets between the matrix elements of $T(\lambda)$:

$$\{q(\lambda), q(\mu)\} = 0, \qquad \{p(\lambda), p(\mu)\} = 0 \tag{1.12}$$

$$\{p(\lambda), q(\mu)\} = \frac{p(\lambda)q(\mu) - q(\lambda)p(\mu)}{\lambda - \mu} \tag{1.13}$$

$$\{\tilde{q}(\lambda), \tilde{q}(\mu)\} = 0, \qquad \{\tilde{p}(\lambda), \tilde{p}(\mu)\} = 0 \tag{1.14}$$

$$\{\tilde{p}(\lambda), \tilde{q}(\mu)\} = \frac{\tilde{p}(\lambda)\tilde{q}(\mu) - \tilde{q}(\lambda)\tilde{p}(\mu)}{\lambda - \mu} \tag{1.15}$$

$$\{\tilde{q}(\lambda), q(\mu)\} = \frac{\tilde{q}(\lambda)q(\mu) - q(\lambda)\tilde{q}(\mu)}{\lambda - \mu}, \quad \{\tilde{p}(\lambda), p(\mu)\} = \frac{\tilde{p}(\lambda)p(\mu) - p(\lambda)\tilde{p}(\mu)}{\lambda - \mu} \tag{1.16}$$

$$\{p(\lambda), \tilde{q}(\mu)\} = \frac{q(\lambda)\tilde{p}(\mu) - \tilde{p}(\lambda)q(\mu)}{\lambda - \mu} \tag{1.17}$$

$$\{\tilde{p}(\lambda), q(\mu)\} = \frac{\tilde{q}(\lambda)p(\mu) - p(\lambda)\tilde{q}(\mu)}{\lambda - \mu} \tag{1.18}$$

Notice, that the brackets (1.12)) and (1.13) coincide with the Gekhtman–Faybusovich brackets (1.6), (1.7). Therefore, the Atiyah–Hitchin Poisson structure coincides with the Sklyanin bracket between elements T_{11} and T_{21} for polynomial dependence of matrix T on λ.

This coincidence leads to several natural questions. The Sklyanin bracket (1.9) arises in inverse scattering method as Poisson structure on scattering matrix $T(\lambda)$ implied by fundamental Poisson structure between physical field in models of non-linear Schrödinger type. On the other hand, the Atiyah-Hitchin bracket is also defined as a bracket on the space of scattering matrices related to solutions of the Bogomolny equations [**1**].

However, in Atiyah–Hitchin framework it remains unclear how this Poisson structure is related to fundamental Poisson bracket between physical fields A and ϕ; close relationship between Atiyah-Hitchin and Sklyanin brackets suggests that there should exist some kind of derivation of the Atiyah-Hitchin structure from the brackets on A and ϕ; however, so far we were unable to find it. Instead in these note we shall extend this observation to trigonometric case and show how to derive natural Poisson structure on the space of trigonometric rational functions starting from Sklyanin bracket with trigonometric r-matrix.

2. Trigonometric generalizations of Atiyah–Hitchin and Gekhtman–Faybusovich brackets

Now, consider a space of meromorphic functions $p(\lambda)/q(\lambda)$ being the ratio of trigonometric polynomials

$$p(\lambda) = \prod_{i=1}^{N-1} \sin(\lambda - \alpha_i) \tag{2.1}$$

$$q(\lambda) = \prod_{k=1}^{N} \sin(\lambda - \lambda_k) \tag{2.2}$$

The symplectic structure here is of the same form as before

$$\Omega = \sum\_i = 1^N \frac{dp(\lambda_i) \wedge d\lambda_i}{p(\lambda_i)} \tag{2.3}$$

Consider the following 2×2 matrix

$$T(\lambda) = \begin{pmatrix} p(\lambda) & \tilde{p}(\lambda) \\ q(\lambda) & \tilde{q}(\lambda) \end{pmatrix}$$

with polynomials $p(\lambda)$ and $q(\lambda)$ are of the form as above. Then with respect to trigonometric r-matrix defined as

$$r(\lambda) = \frac{1}{2\sin(\lambda)}[\sigma_1 \otimes \sigma_1 + \sigma_2 \otimes \sigma_2 + \cos(\lambda)\sigma_3 \otimes \sigma_3] \tag{2.4}$$

the quadratic Sklyanin Poisson bracket on matrices $T(\lambda)$ and $T(\mu)$ provides the following relations on the space of trigonometric polynomials

$$\{p(\lambda), p(\mu)\} = 0 \tag{2.5}$$

$$\{q(\lambda), q(\mu)\} = 0 \tag{2.6}$$

$$\{q(\mu), p(\lambda)\} = \frac{q(\lambda)p(\mu) - \cos(\lambda - \mu)q(\lambda)p(\mu)}{\sin(\lambda - \mu)} \tag{2.7}$$

which lead in their turn to the following generalization of the Gekhtman–Faybusovich bracket on the space of rational trigonometric functions:

$$\{S(\lambda), S(\mu)\} = \frac{\left(S(\lambda) - S(\mu)\right)^2}{\sin(\lambda - \mu)} + 2S(\lambda)S(\mu)(\mathrm{ctg}(\lambda - \mu) - 1) \tag{2.8}$$

It is easy to see that in the rational limit this bracket coincide with (1.8).

References

1. M. Atiyah and N. Hitchin, *The geometry and dynamics of magnetic monopoles*, M. B. Porter Lectures, Princeton Univ. Press, Princeton, NJ, 1988.
2. L. D. Faddeev and L. A. Takhtajan,*Hamiltonian methods in the theory of solitons*, Springer Ser. Soviet Math., Springer-Verlag, Berlin, 1987.
3. M. Gekhtman and L. Faybusovich, *Poisson brackets on rational functions and multi-Hamiltonian structures for integrable lattices*, Phys. Lett. A **272** (2000), no. 4, 236–244.
4. K. Takasaki (private discussion).
5. K. Takasaki and T. Takebe, *An integrable system on the moduli space of rational functions and its variants*, J. Geom. Phys. **47** (2003), no. 1, 1–20.

Centre de recherches mathématiques, Université de Montréal, C.P. 6128, succ. Centre-ville, Montréal, QC H3C 3J7, Canada and Department of Mathematics and Statistics, Concordia University, 7141 Sherbrooke West, Montréal, QC H4B 1R6, Canada

E-mail address: yermolae@crm.umontreal.ca

Centre de Recherches Mathématiques
CRM Proceedings and Lecture Notes
Volume **37**, 2004

Separation of Variables in Time-Dependent Schrödinger Equations

Alexander Zhalij and Renat Zhdanov

Abstract. We classify $(1+3)$-dimensional Schrödinger equations for a particle interacting with an electromagnetic field that are solvable by the method of separation of variables into second-order ordinary differential equations. It is established, in particular, that the necessary condition for the Schrödinger equation to be separable is that the magnetic field must be independent of the spatial variables. We describe vector-potentials that (a) provide the separability of the Schrödinger equation, (b) satisfy vacuum Maxwell equations without currents, and (c) describe a nonzero magnetic field. Furthermore, we apply the results obtained for separating variables to the Hamilton–Jacobi equation.

1. Introduction

The principal object of study in the present paper is a problem of the separation of variables in the Schrödinger equation (SE) for a particle interacting with an electromagnetic field,

$$\left(i\frac{\partial}{\partial t} - eA_0(t,\vec{x}) - \left(i\vec{\nabla} - e\vec{A}(t,\vec{x})\right)^2\right)\psi(t,\vec{x}) = 0. \tag{1.1}$$

where $A = (A_0, A_1, A_2, A_3)$ is the vector potential of the electromagnetic field, $e = \text{const}$.

As this equation has variable coefficients, a natural question arises: which equations of the form (1.1) are separable, namely, which potentials A_0, $\vec{A}$ allow for separability of the SE in some curvilinear coordinate system?

Winternitz et al. [**7, 11**] started a systematic study of potentials for which the stationary SE in two and three dimensions admits the separation of variables. This approach is based on the fact that a solution with separated variables is a common eigenfunction of first- or second-order differential operators, which commute with each other and with the operator of the equation under consideration.

2000 *Mathematics Subject Classification.* 35Q40, 81Q05.
The first author was supported in part by NSERC-NATO Fellowship.
This paper was finished while the first author was staying at the Centre de recherches mathématiques (Université de Montréal) as an NSERC-NATO Fellow. It is a pleasure for him to thank Professor Pavel Winternitz for hospitality and very useful comments.
This is the final form of the paper.

This approach to separation of variables was further developed by Kalnins and Miller [**5, 8**]. See [**5**] and references contained therein for applications of this approach to equations of form (1.1).

Independently, the symmetry approach to the separation of variables in the equations of quantum mechanics and quantum field theory was developed by Shapovalov [**9**] (who was the first to give a systematic treatment of the problem of variable separation in the Dirac equation using its non-Lie symmetry) and by Bagrov with collaborators [**1**]. Shapovalov and Sukhomlin [**10**] have obtained some separable SEs of the form (1.1), however, their results are not complete.

In this paper we propose an alternative approach to the problem of separation of variables in the equation (1.1). The results announced here were partially published in the papers [**12, 13**].

With all the variety of approaches to the separation of variables in PDEs one can notice three general principles, namely,

(a) Representation of a solution to be found in a separated (factorized) form via several functions of one variable.
(b) Requirement that the above mentioned functions of one variable should satisfy some ordinary differential equations.
(c) Dependence of the solution on several arbitrary (continuous or discrete) parameters, called spectral parameters, or separation constants.

By a proper formalization of the above features we will formulate an algorithm for variable separation in the SE with a vector-potential.

To have right to talk about the description of *all* potentials and *all* coordinate systems enabling us to separate the SE, one needs to provide a rigorous definition of the separation of variables. The definition we intend to use is based on ideas contained in a paper by Koornwinder [**6**].

Let us introduce a new coordinate system $\{t, \omega_a = \omega_a(t, \vec{x}), a = 1, 2, 3\}$, where ω_a are real-valued functions, functionally independent with respect to the spatial variables x_1, x_2, x_3:

$$\det\left\|\frac{\partial(\omega_1, \omega_2, \omega_3)}{\partial(x_1, x_2, x_3)}\right\| \neq 0. \tag{1.2}$$

For the solution to be found we adopt the following separation Ansatz:

$$\psi(t, \vec{x}) = Q(t, \vec{x})\varphi_0(t, \vec{\lambda})\varphi_1(\omega_1(t, \vec{x}), \vec{\lambda})\varphi_2(\omega_2(t, \vec{x}), \vec{\lambda})\varphi_3(\omega_3(t, \vec{x}), \vec{\lambda}), \tag{1.3}$$

where Q, φ_μ ($\mu = 0, 1, 2, 3$) are smooth functions of the indicated variables.

DEFINITION 1.1. We say that the SE (1.1) admits separation of variables in a coordinate system $\{t, \omega_a = \omega_a(t, \vec{x}), a = 1, 2, 3\}$, if there exists a function $Q(t, \vec{x})$ and four ordinary differential equations

$$\varphi_0' = U_0(t, \varphi_0; \vec{\lambda}), \quad \varphi_a'' = U_a(\omega_a, \varphi_a, \varphi_a'; \vec{\lambda}), \ a = 1, 2, 3. \tag{1.4}$$

jointly depending on three independent parameters λ_1, λ_2, λ_3 (separation constants), such that, for each triplet $(\lambda_1, \lambda_2, \lambda_3)$ and for each set of solutions $\varphi_0(t)$, $\varphi_1(\omega_1)$, $\varphi_2(\omega_2)$, $\varphi_3(\omega_3)$ of (1.4), the function (1.3) is a solution of (1.1).

In the above formulas $U_0, \dots, U_3$ are some smooth functions of the indicated variables.

DEFINITION 1.2. Three parameters λ_1, λ_2, λ_3 in (1.4) are called independent, if the equality

$$\operatorname{rank}\left\| \frac{\partial(U_0, U_1, U_2, U_3)}{\partial(\lambda_1, \lambda_2, \lambda_3)} \right\| = 3. \tag{1.5}$$

holds, whenever $\varphi_0(t)\varphi_1(\omega_1)\varphi_2(\omega_2)\varphi_3(\omega_3) \neq 0$.

Condition (1.5) secures the essential dependence of a solution with separated variables on the separation constants $\vec{\lambda}$.

Definition 1.1 is quite algorithmic in the sense that it contains a regular algorithm of variable separation in the SE (1.1). Formulas (1.3)–(1.5) form the input data of the method. The principal steps of the procedure of variable separation in the SE (1.1) are as follows.

(1) We insert the Ansatz (1.3) into the SE and express the derivatives φ_0', φ_1'', φ_2'', φ_3'' in terms of functions φ_0, φ_1, φ_2, φ_3, φ_1', φ_2', φ_3', using equations (1.4).

(2) We split the equality obtained with respect to the variables φ_0, φ_1, φ_2, φ_3, φ_1', φ_2', φ_3', λ_1, λ_2, λ_3, i.e., we demand that the obtained equality is transformed into identity with respect to these variables.

(3) After splitting we get an overdetermined system of nonlinear partial differential equations for the unknown functions Q, ω_1, ω_2, ω_3, A_0, A_1, A_2, A_3. Solving the system yields an exhaustive description of vector-potentials $A(t, \vec{x})$ providing separability of the SE and also the corresponding coordinate systems.

Note, that putting $Q = 1$, $\omega_a = x_a$, $a = 1, 2, 3$ in (1.3) yields the standard separation of variables in the Cartesian coordinate system. Next, choosing the spherical coordinates as ω_1, ω_2, ω_3 we arrive at the separation of variables in the spherical coordinate system and so on. The principal task is describing all possible forms of the functions Q, ω_a, $a = 1, 2, 3$, that provide separability of the SE in the sense of the definition given above. The solution of this problem, in turn, requires describing the functions $A_0, \ldots, A_3$ that enable the separation of variables in the SE in the corresponding coordinate system. More precisely, we will need to describe *all* cases of coefficients $A_0, \ldots, A_3$, for which the corresponding SE (1.1) is separable (in the sense of Definition 1.1) in at least one coordinate system.

Note, that formulas (1.3)–(1.5) form the input data of the method. We can change these conditions and thereby modify the definition of separation of variables. For instance, we can change the order of the reduced equations (1.4), or the number of essential parameters. So, our claim of obtaining the *complete description* of vector-potentials and coordinate systems providing separation of variables in (1.1) makes sense only within the framework of Definition 1.1. If one uses a more general definition, it might be possible to construct new coordinate systems and vector-potentials providing separability of equation (1.1). However, all solutions of the SE with separated variables known to us fit into the above suggested scheme.

Next, we introduce an equivalence relation on the set of all vector-potentials $A_0(t, \vec{x})$, $\vec{A}(t, \vec{x})$ providing separability of (1.1), on the sets of solutions with separated variables and the corresponding coordinate systems.

DEFINITION 1.3. We say that two vector-potentials $A(t, \vec{x})$ and $A'(t, \vec{x})$ are equivalent if they are transformed one into another by the gauge transformation

$$\vec{A} \to \vec{A}' = \vec{A} + \vec{\nabla} f, \qquad A_0 \to A_0' = A_0 - \frac{\partial f}{\partial t}, \tag{1.6}$$

where $f = f(t, \vec{x})$ is an arbitrary smooth function.

For the SEs to be invariant with respect to the above transformation, the wave function $\psi(t, \vec{x})$ is to be transformed according to the rule

$$\psi \to \psi' = \psi \exp(ief) \tag{1.7}$$

Indeed, if the transformations (1.6)–(1.7) in the SE (1.1) are performed, we obtain the initial equation, provided we replace the functions $\vec{A}$, A_0, ψ by $\vec{A}'$, A_0', ψ'.

Note that the system of PDEs (1.1) admits a wider equivalence group from the point of view of the standard theory of partial differential equations [**10**]. However, this group cannot be regarded as an equivalence group within the context of quantum mechanics, since allowed transformations of the wave function must preserve the probability density $\psi^*\psi$. It is straightforward to check that the wider Shapovalov and Sukhomlin equivalence group breaks this rule, because it, generally speaking, does not preserve $\psi^*\psi$. For this reason, we restrict our considerations to gauge transformations only.

DEFINITION 1.4. Two coordinate systems t, ω_1, ω_2, ω_3 and t', ω_1', ω_2', ω_3' are called equivalent if they give equivalent solutions with separated variables. In particular, two coordinate systems are equivalent if the corresponding Ansätze (1.3) are transformed one into another by reversible transformations of the form

$$t \to t' = f_0(t), \qquad \omega_a \to \omega_a' = f_a(\omega_a), \quad a = 1, 2, 3, \tag{1.8}$$

$$Q \to Q' = Ql_0(t)l_1(\omega_1)l_2(\omega_2)l_3(\omega_3), \tag{1.9}$$

where $f_0, \dots, f_3$ and $l_0, \dots, l_3$ are some smooth functions of the indicated variables.

Indeed, transformations (1.9) preserve the form of Ansätze (1.3). Hence, after completing the procedure of separation of variables in these coordinate systems, we obtain the same solutions with separated variables.

Within these equivalence relations we can always choose the reduced equations (1.4) to be

$$i\varphi_0' = (T_0(t) - T_i(t)\lambda_i)\varphi_0, \qquad \varphi_a'' = (F_{a0}(\omega_a) + F_{ai}(\omega_a)\lambda_i)\varphi_a, \tag{1.10}$$

where T_0, T_i, F_{a0}, F_{ai} are some smooth functions of the indicated variables, $a = 1$, 2, 3.

Having performed the first two steps of the above algorithm we obtain a system of nonlinear PDEs for the unknown functions Q, ω_1, ω_2, ω_3 in the form

$$\frac{\partial \omega_i}{\partial x_a}\frac{\partial \omega_j}{\partial x_a} = 0, \quad i \neq j, \; i, j = 1, 2, 3; \tag{1.11}$$

$$\sum_{i=1}^{3} F_{ia}(\omega_i)\frac{\partial \omega_i}{\partial x_j}\frac{\partial \omega_i}{\partial x_j} = T_a(t), \quad a = 1, 2, 3; \tag{1.12}$$

$$2\left(\frac{\partial Q}{\partial x_j} + ieQA_j\right)\frac{\partial \omega_a}{\partial x_j} + Q\left(i\frac{\partial \omega_a}{\partial t} + \Delta\omega_a\right) = 0, \quad a = 1, 2, 3; \tag{1.13}$$

$$Q\sum_{i=1}^{3} F_{i0}(\omega_i)\frac{\partial \omega_i}{\partial x_j}\frac{\partial \omega_i}{\partial x_j} + i\frac{\partial Q}{\partial t} + \Delta Q + 2ieA_a\frac{\partial Q}{\partial x_a} + Q\left(T_0(t) + ie\frac{\partial A_a}{\partial x_a} - eA_0 - e^2A_aA_a\right) = 0. \tag{1.14}$$

Hereafter the summation over the repeated Latin indices from 1 to 3 is implied.

Thus the problem of variable separation in the SE reduces to integrating a system of nonlinear PDEs for eight unknown functions A_0, A_1, A_2, A_3, Q, ω_1, ω_2, ω_3 of four variables t, $\vec{x}$. What is more, some coefficients are arbitrary functions which should be determined in the process of integrating of the system of PDEs (1.11)–(1.14). Roughly speaking, to solve a linear equation we have to solve a system of nonlinear equations.

2. Connection with the symmetry approach

Let us briefly analyze the connection between separability the SE (1.1) and its symmetry properties.

DEFINITION 2.1. The linear second order differential operator

$$L = k_{ab}(\vec{x},t)\frac{\partial^2}{\partial x_a \partial x_b} + m_a(\vec{x},t)\frac{\partial}{\partial x_a} + n(\vec{x},t),$$

where k_{ab}, m_a, n $(a, b = 1, 2, 3)$ are smooth functions of $\vec{x}$, t, is a *symmetry operator* for the SE (1.1), if

$$[L,S] = R(\vec{x},t)S, \tag{2.1}$$

where $S = p_0 - p_a p_a$ is the operator of equation (1.1), $[L,S] = LS - SL$ is the commutator of operators L and S, and $R(\vec{x},t)$ is some smooth function, which depends on the coefficients of the operator L.

THEOREM 2.2. *Let the SE* (1.1) *admit separation of variables in the sense of Definition* 1.1. *Then each solution with separated variables is a common eigenfunction of three mutually commuting linear second-order symmetry operators of* (1.1), *the separation constants* λ_1, λ_2, λ_3 *being their eigenvalues.*

PROOF. Let us make the following change of variables in (1.1)

$$\psi(t,\vec{x}) = Q(t,\vec{x})\Psi\big(t,\omega_1(t,\vec{x}),\omega_2(t,\vec{x}),\omega_3(t,\vec{x})\big), \tag{2.2}$$

where Q, ω_1, ω_2, ω_3 is an arbitrary solution of the system of PDE (1.11)–(1.14). Substituting the expression (2.2) into (1.1) and taking into account (1.11)–(1.14), we get

$$i\frac{\partial\Psi}{\partial t} + \sum_{a=1}^{3}\left(\frac{\partial^2\Psi}{\partial\omega_a^2} - F_{a0}(\omega_a)\Psi\right)\frac{\partial\omega_a}{\partial x_c}\frac{\partial\omega_a}{\partial x_c} - T_0(t)\Psi = 0.$$

Note, that condition (1.5) and equations (1.10), (1.12) give the condition

$$\det \|F_{ab}\|_{a,b=1}^{3} \neq 0. \tag{2.3}$$

Solving equation (1.12) with respect to

$$\frac{\partial\omega_1}{\partial x_c}\frac{\partial\omega_1}{\partial x_c}, \qquad \frac{\partial\omega_2}{\partial x_c}\frac{\partial\omega_2}{\partial x_c}, \qquad \frac{\partial\omega_3}{\partial x_c}\frac{\partial\omega_3}{\partial x_c},$$

we have

$$\frac{\partial\omega_a}{\partial x_c}\frac{\partial\omega_a}{\partial x_c} = G_{ba}T_b, \quad a = 1,2,3,$$

where G_{ba} are components of the matrix, inverse to matrix $\|F_{ab}\|$. Thus, in new coordinates t, ω_1, ω_2, ω_3, $\Psi(t,\omega_1,\omega_2,\omega_3)$ (1.1) takes the form

$$i\frac{\partial\Psi}{\partial t} + \sum_{a=1}^{3}\left(\frac{\partial^2\Psi}{\partial\omega_a^2} - F_{a0}(\omega_a)\Psi\right)G_{ba}T_b(t) - T_0(t)\Psi = 0. \tag{2.4}$$

Now we can construct the triplet of symmetry operators. Let us consider the system of equations (1.10) for the functions $\varphi_1(\omega_1)$, $\varphi_2(\omega_2)$, $\varphi_3(\omega_3)$. After multiplying the first one by $\varphi_0\varphi_2\varphi_3$, the second one by $\varphi_0\varphi_1\varphi_3$, and the third one by $\varphi_0\varphi_1\varphi_2$ we obtain

$$\frac{\partial^2\Psi}{\partial\omega_a^2} = (F_{a0}(\omega_a) + F_{ab}(\omega_a)\lambda_b)\Psi, \quad a = 1, 2, 3. \tag{2.5}$$

In virtue of condition (2.3) we can solve the system (2.5) with respect to $\lambda_b\Psi$, $b = 1$, 2, 3

$$\sum_{a=1}^{3} G_{ba}\left(\frac{\partial^2\Psi}{\partial\omega_a^2} - F_{a0}(\omega_a)\Psi\right) = \lambda_b\Psi, \quad b = 1, 2, 3,$$

The solution with separated variables

$$\Psi = \varphi_0(t)\varphi_1(\omega_1)\varphi_2(\omega_2)\varphi_3(\omega_3)$$

is a common eigenfunction of the three operators

$$P_a = \sum_{b=1}^{3} G_{ab}\left(\frac{\partial^2}{\partial\omega_b^2} - F_{b0}(\omega_b)\right), \quad a = 1, 2, 3, \tag{2.6}$$

(G_{ab} is matrix, inverse to $\|F_{ab}(\omega_a)\|$) and moreover the equality

$$P_a\Psi = \lambda_a\Psi. \tag{2.7}$$

holds.

By a direct (and very cumbersome) computation one can check, that the second order differential operators P_1, P_2, P_3 commute pairwise for arbitrary functions $F_{ab}(\omega_a)$, $F_{a0}(\omega_a)$, a, $b = 1$, 2, 3, i.e.,

$$[P_a, P_b] = P_aP_b - P_bP_a = 0, \quad a, b = 1, 2, 3. \tag{2.8}$$

After being rewritten in terms of the operators P_1, P_2, P_3 (2.4) reads

$$i\frac{\partial\Psi}{\partial t} + T_a(t)P_a\Psi - T_0(t)\Psi = 0.$$

Since the relations

$$\left[i\frac{\partial}{\partial t} + T_a(t)P_a - T_0(t), P_a\right] = 0, \quad a = 1, 2, 3, \tag{2.9}$$

hold, operators P_1, P_2, P_3 are mutually commuting symmetry operators for (2.4). If we designate by P_1', P_2', P_3' the operators P_1, P_2, P_3 written in the initial variables $t, \vec{x}, \psi$, then from (2.7)–(2.9) we get the following equalities

$$[p_0 - p_cp_c, P_a'] = 0, \quad P_a'\psi = \lambda_a\psi, \quad [P_a', P_b'] = 0, \qquad a, b = 1, 2, 3,$$

where $\psi = Q(t, \vec{x})\varphi_0(t)\varphi_1(\omega_1)\varphi_2(\omega_2)\varphi_3(\omega_3)$, and $p_0 - p_cp_c$ is operator of (1.1). □

Thus there are two possible approaches to variable separation in linear PDEs which are based on their symmetry properties. The first one is to start with a set of commuting symmetry operators of the equation under study and to finish by constructing the coordinate systems (Winternitz et al. [**7, 11**], Kalnins and Miller [**5, 8**], Shapovalov [**9, 10**]).

Another approach, that proposed in papers [**12, 13**], is closer to the original understanding of the separation of variables in PDE. A desired form (1.3) of a solution with separated variables is postulated and then it turns out that the solution obtained can be related to a set of mutually commuting symmetry operators of the equations under consideration.

Both approaches have their merits and drawbacks. We think that the utilization of the first approach is the only way to separate variables in multicomponent systems of PDEs. But to separate variables with one dependent variable it is preferable to apply the second approach, since a computation of symmetry operators is an extra step which is not, in fact, necessary for obtaining solution with separated variables.

3. Main results

We do not give the computations needed for integrating of the system of PDEs (1.11)–(1.14) in full detail, because they are very cumbersome. The details can be found in the paper [**13**].

Integration of the system (1.11)–(1.12) gives the most general form of separable coordinate systems. Its general form $\vec{\omega} = \vec{\omega}(t, \vec{x})$ is given implicitly within the equivalence relation (1.8) by the following formulas:

$$\vec{x} = \mathcal{O}(t)\mathcal{L}(t)\big(\vec{z}(\vec{\omega}) + \vec{v}(t)\big), \tag{3.1}$$

Here $\mathcal{O}(t)$ is a time-dependent 3×3 orthogonal matrix with Euler angles $\alpha(t)$, $\beta(t)$, $\gamma(t)$:

$$\mathcal{O}(t) = \begin{pmatrix} \cos\alpha\cos\beta - \sin\alpha\sin\beta\cos\gamma & -\cos\alpha\sin\beta - \sin\alpha\cos\beta\cos\gamma & \sin\alpha\sin\gamma \\ \sin\alpha\cos\beta + \cos\alpha\sin\beta\cos\gamma & -\sin\alpha\sin\beta + \cos\alpha\cos\beta\cos\gamma & -\cos\alpha\sin\gamma \\ \sin\beta\sin\gamma & \cos\beta\sin\gamma & \cos\gamma \end{pmatrix};$$

$\vec{v}(t)$ stands for the vector-column whose entries $v_1(t)$, $v_2(t)$, $v_3(t)$ are arbitrary smooth functions of t; $\vec{z} = \vec{z}(\vec{\omega})$ is given by one of the eleven formulas

(1) Cartesian coordinate system,

$$z_1 = \omega_1, \quad z_2 = \omega_2, \quad z_3 = \omega_3, \qquad \omega_1, \omega_2, \omega_3 \in \mathbb{R}.$$

(2) Cylindrical coordinate system,

$$z_1 = e^{\omega_1}\cos\omega_2, \quad z_2 = e^{\omega_1}\sin\omega_2, \quad z_3 = \omega_3, \qquad 0 \le \omega_2 < 2\pi,\ \omega_1, \omega_3 \in \mathbb{R}.$$

(3) Parabolic cylindrical coordinate system,

$$z_1 = (\omega_1^2 - \omega_2^2)/2, \quad z_2 = \omega_1\omega_2, \quad z_3 = \omega_3, \qquad \omega_1 > 0,\ \omega_2, \omega_3 \in \mathbb{R}.$$

(4) Elliptic cylindrical coordinate system,

$$z_1 = a\cosh\omega_1\cos\omega_2, \quad z_2 = a\sinh\omega_1\sin\omega_2, \quad z_3 = \omega_3,$$
$$\omega_1 > 0,\ -\pi < \omega_2 \le \pi,\ \omega_3 \in \mathbb{R},\ a > 0.$$

(5) Spherical coordinate system,

$$z_1 = \omega_1^{-1}\operatorname{sech}\omega_2\cos\omega_3, \quad z_2 = \omega_1^{-1}\operatorname{sech}\omega_2\sin\omega_3, \quad z_3 = \omega_1^{-1}\tanh\omega_2,$$
$$\omega_1 > 0,\ \omega_2 \in \mathbb{R},\ 0 \le \omega_3 < 2\pi.$$

(6) Prolate spheroidal coordinate system,

$$z_1 = a\operatorname{csch}\omega_1\operatorname{sech}\omega_2\cos\omega_3, \quad z_2 = a\operatorname{csch}\omega_1\operatorname{sech}\omega_2\sin\omega_3,$$
$$z_3 = a\coth\omega_1\tanh\omega_2,$$
$$\omega_1 > 0,\ \omega_2 \in \mathbb{R},\ 0 \le \omega_3 < 2\pi,\ a > 0.$$

(7) Oblate spheroidal coordinate system,

$$z_1 = a\csc\omega_1 \operatorname{sech}\omega_2 \cos\omega_3, \quad z_2 = a\csc\omega_1 \operatorname{sech}\omega_2 \sin\omega_3,$$
$$z_3 = a\cot\omega_1 \tanh\omega_2,$$
$$0 < \omega_1 < \pi/2,\ \omega_2 \in \mathbb{R},\ 0 \le \omega_3 < 2\pi,\ a > 0.$$

(8) Parabolic coordinate system,

$$z_1 = e^{\omega_1+\omega_2}\cos\omega_3, \quad z_2 = e^{\omega_1+\omega_2}\sin\omega_3, \quad z_3 = (e^{2\omega_1} - e^{2\omega_2})/2,$$
$$\omega_1, \omega_2 \in \mathbb{R},\ 0 \le \omega_3 \le 2\pi.$$

(9) Paraboloidal coordinate system,

$$z_1 = 2a\cosh\omega_1 \cos\omega_2 \sinh\omega_3, \quad z_2 = 2a\sinh\omega_1 \sin\omega_2 \cosh\omega_3,$$
$$z_3 = a(\cosh 2\omega_1 + \cos 2\omega_2 - \cosh 2\omega_3)/2,$$
$$\omega_1, \omega_3 \in \mathbb{R},\ 0 \le \omega_2 < \pi,\ a > 0.$$

(10) Ellipsoidal coordinate system,

$$z_1 = a\frac{1}{\operatorname{sn}(\omega_1, k)}\operatorname{dn}(\omega_2, k')\operatorname{sn}\omega_3, k), \quad z_2 = a\frac{\operatorname{dn}(\omega_1, k)}{\operatorname{sn}(\omega_1, k)}\operatorname{cn}(\omega_2, k')\operatorname{cn}(\omega_3, k),$$
$$z_3 = a\frac{\operatorname{cn}(\omega_1, k)}{\operatorname{sn}(\omega_1, k)}\operatorname{sn}(\omega_2, k')\operatorname{dn}(\omega_3, k),$$
$$0 < \omega_1 < K,\ -K' \le \omega_2 \le K',\ 0 \le \omega_3 \le 4K,\ a > 0.$$

(11) Conical coordinate system,

$$z_1 = \omega_1^{-1}\operatorname{dn}(\omega_2, k')\operatorname{sn}(\omega_3, k), \quad z_2 = \omega_1^{-1}\operatorname{cn}(\omega_2, k')\operatorname{cn}(\omega_3, k),$$
$$z_3 = \omega_1^{-1}\operatorname{sn}(\omega_2, k')\operatorname{dn}(\omega_3, k),$$
$$\omega_1 > 0,\ -K' \le \omega_2 \le K',\ 0 \le \omega_3 \le 4K.$$

and $\mathcal{L}(t)$ is a 3×3 diagonal matrix

$$\mathcal{L}(t) = \begin{pmatrix} l_1(t) & 0 & 0 \\ 0 & l_2(t) & 0 \\ 0 & 0 & l_3(t) \end{pmatrix}, \tag{3.2}$$

where $l_1(t)$, $l_2(t)$, $l_3(t)$ are arbitrary nonzero smooth functions that satisfy the following conditions

- $l_1(t) = l_2(t)$ for the partially split coordinate systems (cases 2–4 in the above list),
- $l_1(t) = l_2(t) = l_3(t)$ for nonsplit coordinate systems (cases 5–11 in in the above list).

Note that we have chosen the coordinate systems ω_1, ω_2, ω_3 by means of the equivalence relation (1.8) in such a way that the relations

$$\Delta\omega_a = 0, \quad a = 1, 2, 3 \tag{3.3}$$

hold for all eleven coordinate systems.

From a geometric point of view the right-hand side of formula (3.1) is a result of the application to vector $\vec{z}(\vec{\omega})$ of the following time-dependent transformations performed one after another:

(1) translations $\vec{z} \to \vec{z}' = \vec{z} + \vec{v}(t)$,

(2) dilatations $\vec{z} \to \vec{z}' = \mathcal{L}(t)\vec{z}$,
(3) three-dimensional rotations $\vec{z} \to \vec{z}' = \mathcal{O}(t)\vec{z}$ with Euler angles $\alpha(t)$, $\beta(t)$, $\gamma(t)$.

Together with the rotations the following vector $\overrightarrow{\Omega}(t) = (\Omega_1, \Omega_2, \Omega_3)$ is considered

$$\begin{aligned} \Omega_1(t) &= \dot{\gamma}(t)\cos\alpha(t) + \dot{\beta}(t)\sin\alpha(t)\sin\gamma(t), \\ \Omega_2(t) &= \dot{\gamma}(t)\sin\alpha(t) - \dot{\beta}(t)\cos\alpha(t)\sin\gamma(t), \\ \Omega_3(t) &= \dot{\alpha}(t) + \dot{\beta}(t)\cos\gamma(t), \end{aligned} \tag{3.4}$$

that is directed along the instantaneous axis of rotation and is called the *angular velocity vector*.

After integrating system (1.11)–(1.12) it is not difficult to integrate with the use of (3.3) the remaining equations (1.13)–(1.14) in the system under study. They can be regarded as algebraic equations for the functions $A_a(t, \vec{x})$ ($a = 1, 2, 3$) and $A_0(t, \vec{x})$, correspondingly.

THEOREM 3.1. *The SE* (1.1) *admits separation of variables* (*in the sense of Definition* 1.1) *if and only if it is gauge equivalent to a SE where*

- *the magnetic field* $\overrightarrow{H} = \operatorname{rot}\vec{A}$ *is independent of the spatial variables,*
- *the space-like components* A_1, A_2, A_3 *of the vector-potential of the electromagnetic field are given by*

$$\vec{A}(t, \vec{x}) = \frac{1}{2}\begin{pmatrix} 0 & -H_3(t) & H_2(t) \\ H_3(t) & 0 & -H_1(t) \\ -H_2(t) & H_1(t) & 0 \end{pmatrix}\vec{x} = \frac{1}{2}\overrightarrow{H}(t) \times \vec{x}, \tag{3.5}$$

where the symbol $\times$ *denotes the vector product,*
- *the time-like component* A_0 *is given by formula*

$$eA_0(t, \vec{x}) = \sum_{a=1}^{3} F_{a0}(\omega_a)\frac{\partial\omega_a}{\partial x_b}\frac{\partial\omega_a}{\partial x_b} - T_0(t) - e^2 A_b A_b - \frac{1}{4}P. \tag{3.6}$$

Here T_0, F_{a0} are some smooth functions of the indicated variables, and P is some second-order polynomial in the new variables $\vec{\omega}$ with coefficients that are functions of t (for more details see [**12**]).

Furthermore, we have proved the following theorem:

THEOREM 3.2. *Let the SE* (1.1) *admit separation of variables in some nonstationary coordinate system* t, $\omega_a = \omega_a(t, \vec{x})$, $a = 1, 2, 3$, *where the functions* $\omega_1(t, \vec{x})$, $\omega_2(t, \vec{x})$, $\omega_3(t, \vec{x})$ *are given implicitly by formulas* (3.1)–(3.2). *Then the angular velocity vector* (3.4) *of the rotation of this coordinate system equals* $-e\overrightarrow{H}$, *where* $\overrightarrow{H} = \operatorname{rot}\vec{A}$ *is magnetic field.*

It follows from the last theorem that a necessary condition for the SE (1.1) with nonzero magnetic field $\overrightarrow{H}$ to be separable (in the sense of Definition 1.1) is that the angular velocity vector (3.4) of rotation of the separation coordinate system (3.1)–(3.2) has to be nonzero.

We have proved, that the magnetic field $\overrightarrow{H} = \operatorname{rot}\vec{A}$ has to be independent of the spatial variables. Next, we have eleven classes of potentials $A_0(t, \vec{x})$, corresponding to eleven classes of coordinate systems t, $\omega_a = \omega_a(t, \vec{x})$, $a = 1, 2, 3$, where the functions $\omega_1(t, \vec{x})$, $\omega_2(t, \vec{x})$, $\omega_3(t, \vec{x})$ are given implicitly by formulas (3.1)–(3.2).

We will finish this section with the following remark. It follows from Theorem 3.1 that the choice of magnetic fields $\vec{H}$ allowing for variable separation in the corresponding SE is very restricted. Namely, the magnetic field should be independent of the spatial variables x_1, x_2, x_3 in order to provide the separability of the SE (1.1) into three second-order ordinary differential equations of the form (1.4). However, if we allow for the separation equations to be of a lower order, then additional possibilities for variable separation in the SE arise. As an example, we give the vector potential

$$A(t, \vec{x}) = \left(A_0\left(\sqrt{x_1^2 + x_2^2}\right), 0, 0, A_3\left(\sqrt{x_1^2 + x_2^2}\right)\right),$$

where A_0, A_3 are arbitrary smooth functions. The SE (1.1) with this vector-potential separates in the cylindrical coordinate system

$$t, \qquad \omega_1 = \ln\left(\sqrt{x_1^2 + x_2^2}\right), \qquad \omega_2 = \arctan(x_1/x_2), \qquad \omega_3 = x_3$$

into two first-order and one second-order ordinary differential equations. The corresponding magnetic field $\vec{H} = \operatorname{rot} \vec{A}$ is evidently x-dependent. In this respect, let us also mention the recent paper by Benenti at al. [**2**], where the problem of separation of variables in the stationary Hamilton–Jacobi equation with vector-potential has been studied. They have presented a number of vector-potentials, for which the Hamilton–Jacobi equation is separable, and the corresponding magnetic fields are inhomogeneous ones. These potentials allow for separation of variables in the stationary SEs with vector-potentials as well (in the next section we will point out the relationship between the separation of variables in the Schrödinger and Hamilton-Jacobi equations). These facts, as well as paper [**10**], imply that it is important to apply our approach to classify the non-stationary SEs of the form (1.1), which admit separation of variables into first- and second-order ordinary differential equations. Here we give the classification results for the case, when all the reduced equations are second-order ones. We intend to address this problem in a future publication.

4. Separation of variables in the Hamilton–Jacobi equation

It is well known [**3**] that there exists a deep connection between the separation of variables in the Schrödinger and Hamilton–Jacobi equations. The Hamilton–Jacobi equation,

$$u_t + eA_0 + (u_{x_a} + eA_a)(u_{x_a} + eA_a) = 0, \tag{4.1}$$

separates in any coordinate system providing separability of the Schrödinger equations (1.1) and, what is more, the inverse assertion is not true. We will make use of this connection in order to classify separable Hamilton–Jacobi equations.

First we fix the usual form of the separation Ansatz for the Hamilton–Jacobi equation,

$$u(t, \vec{x}) = S(t, \vec{x}) + \varphi_0(t) + \sum_{i=1}^{3} \varphi_i\big(\omega_i(t, \vec{x})\big), \tag{4.2}$$

and, furthermore, fix the form of the ordinary differential equations for φ_0, φ_1, φ_2, φ_3,

$$\varphi_0' = -T_0(t) - T_i(t)\lambda_i, \qquad \varphi_a' = (-F_{a0}(\omega_a) + F_{ai}(\omega_a)\lambda_i)^{1/2}. \tag{4.3}$$

Inserting the Ansatz (4.2) into equation (4.1), eliminating the first derivatives of the functions φ_0, φ_1, φ_2, φ_3 with the use of the above equations and splitting with respect to the variables φ_0, φ_1, φ_2, φ_3, λ_1, λ_2, λ_3, we arrive at the following system of nonlinear partial differential equations for the functions S, ω_1, ω_2, ω_3:

$$\begin{aligned}
&\frac{\partial\omega_i}{\partial x_a}\frac{\partial\omega_j}{\partial x_a} = 0, \quad i \neq j,\ i,j = 1,2,3;\\
&\sum_{i=1}^{3} F_{ia}(\omega_i)\frac{\partial\omega_i}{\partial x_j}\frac{\partial\omega_i}{\partial x_j} = T_a(t), \quad a = 1,2,3;\\
&2\left(\frac{\partial S}{\partial x_j} + eA_j\right)\frac{\partial\omega_a}{\partial x_j} + \frac{\partial\omega_a}{\partial t} = 0, \quad a = 1,2,3;\\
&-\sum_{i=1}^{3} F_{i0}(\omega_i)\frac{\partial\omega_i}{\partial x_j}\frac{\partial\omega_i}{\partial x_j} + \frac{\partial S}{\partial t} + 2eA_a\frac{\partial S}{\partial x_a} + \frac{\partial S}{\partial x_a}\frac{\partial S}{\partial x_a}\\
&\qquad - T_0(t) + eA_0 + e^2A_aA_a = 0.
\end{aligned} \tag{4.4}$$

It is not difficult to check that the change of variables,

$$Q(t,\vec{x}) = \exp\big(iS(t,\vec{x})\big), \tag{4.5}$$

in (1.11)–(1.14) yields a system that coincides with (4.4) with the exception of the last equation, where an additional term $-i(\Delta S + eA_{ax_a})$ appears. But this term is a function of t only and is absorbed by T_0. Consequently, all the results on variable separation for the SE apply to the case of the Hamilton–Jacobi equation (4.1) as well.

5. Separation of variables in the Schrödinger-Maxwell system

The expressions (3.5), (3.6) give the most general form of the vector-potential of the electromagnetic field, providing separability of the corresponding SEs. In view of the generality of the results, these expressions are cumbersome, and their physical interpretation is somewhat difficult. Therefore, it would be interesting to know the form of these potentials under certain physical restrictions. The most natural restriction is that the vector-potential satisfies the vacuum Maxwell equations without currents.

In this section we describe all explicit forms of the vector-potentials $A(t,\vec{x})$ that

(a) Allow separability of the SE,
(b) Satisfy vacuum Maxwell equations without currents,
(c) Describe the non-zero magnetic field.

Omitting the details of the calculations we present below the results. Note, when presenting lists of the vector-potentials $A(t,\vec{x})$ we use invariance of the system of the Schrödinger and Maxwell equations with respect to the groups of rotations by spatial variables x_1, x_2, x_3 and translations of the all variables t, x_1, x_2, x_3.

Case of nonstationary magnetic field.

$$\begin{aligned}
e\overrightarrow{H} &= (0, 0, At + B),\\
eA_0 &= -\frac{k}{2}(x_1^2 + x_2^2 - 2x_3^2) + a_1x_1 + a_2x_2 + a_3x_3,
\end{aligned}$$

where A, B, k, a_1, a_2, a_3 are arbitrary real constants.

The coordinate system is

$$\vec{x} = \mathcal{L}\mathcal{O}(\vec{z} + \vec{v}).$$

Here $\mathcal{O}$ is a time-dependent 3×3 orthogonal matrix $\mathcal{O}(\alpha, \beta, \gamma)$, where

$$\alpha = -\tfrac{1}{2}At^2 - Bt, \qquad \beta = 0, \qquad \gamma = 0;$$

$\vec{z}$ is the Cartesian, cylindrical or elliptic cylindrical coordinates; $\mathcal{L}$ is the 3×3 diagonal matrix

$$\mathcal{L} = \begin{pmatrix} l(t) & 0 & 0 \\ 0 & l(t) & 0 \\ 0 & 0 & l_3(t) \end{pmatrix},$$

and $\vec{v}(t)$ is vector-column $\vec{v}(t) = (v_1, v_2, v_3)^T$ where functions $l(t)$, $l_3(t)$, $v_1(t)$, $v_2(t)$, $v_3(t)$ are solutions of the following system of ordinary differential equations:

$$2\frac{c}{l^4} - \frac{1}{2}\frac{\ddot{l}}{l} + k = \frac{1}{2}(At + B)^2, \qquad \frac{c_3}{l_3^4} - \frac{1}{4}\frac{\ddot{l}_3}{l_3} = k,$$

$$l\ddot{v}_1 + 2\dot{l}\dot{v}_1 + 4c\frac{v_1}{l^3} - 2c_{11}\frac{1}{l} = -2(a_1 \cos\alpha + a_2 \sin\alpha),$$

$$l\ddot{v}_2 + 2\dot{l}\dot{v}_2 + 4c\frac{v_2}{l^3} - 2c_{12}\frac{1}{l} = -2(-a_1 \sin\alpha + a_2 \cos\alpha),$$

$$l_3\ddot{v}_3 + 2\dot{l}_3\dot{v}_3 + 4c_3\frac{v_3}{l_3^3} - 2c_{13}\frac{1}{l_3} = -2a_3.$$

Here c, c_3, c_{11}, c_{12}, c_{13} are arbitrary real constants.

Cases of stationary magnetic field.

CASE 1.

$$e\vec{H} = (0, 0, k), \quad k = \text{const} \neq 0;$$

$$eA_0 = -\frac{k^2}{12}(x_1^2 + x_2^2 - 2x_3^2) + a_1x_1 + a_2x_2 + a_3x_3,$$

where $\vec{a} = (a_1, a_2, a_3)$ is constant vector.

The coordinate system is

$$\vec{x} = l\mathcal{O}(\vec{z} + \vec{v}).$$

Here $\mathcal{O}$ is a time-dependent 3×3 orthogonal matrix $\mathcal{O}(\alpha, \beta, \gamma)$, where $\alpha = -kt$, $\beta = \text{const}$, $\gamma = \text{const}$; $\vec{z}$ is one of eleven coordinate systems, listed above; function $l(t)$ is solution of the equation

$$k^2 + \frac{3}{2}\frac{\ddot{l}}{l} = \frac{c}{l^4}$$

given by one of the formulas:

$$c = \mp 1, \qquad l^2 = \sqrt{C_1^2 \pm \frac{1}{k^2}\sin\Big(2\sqrt{\tfrac{2}{3}}kt\Big)} + C_1,$$

for all coordinate systems with the exception of parabolic cylindrical and parabolic ones and

$$c = 0, \qquad l = C_1 \sin\Big(\sqrt{\tfrac{2}{3}}kt\Big)$$

for all eleven coordinate systems. Here C_1 is an arbitrary real constant. Vector $\vec{v}$ is a solution of the following system of ordinary differential equations:

$$3l\ddot{\vec{v}} + 6\dot{l}\dot{\vec{v}} + \frac{2c}{l^3}\vec{v} = -6\mathcal{O}^{-1}\vec{a}.$$

CASE 2.

$$e\vec{H} = (0,0,k), \quad k = \text{const} \neq 0;$$

$$eA_0 = \frac{a}{\sqrt{x_1^2+x_2^2+x_3^2}} - \frac{k^2}{12}(x_1^2+x_2^2-2x_3^2), \quad a = \text{const} \neq 0.$$

The coordinate system is

$$\vec{x} = \mathcal{O}\vec{z}.$$

Here $\mathcal{O}$ is a time-dependent 3×3 orthogonal matrix $\mathcal{O}(\alpha,\beta,\gamma)$, where $\alpha = -kt$, $\beta = \text{const}$, $\gamma = \text{const}$ and $\vec{z}$ is one of the following coordinate systems: spherical, prolate spheroidal II (where one should replace z_3 by $z_3 = a(\coth\omega_1 \tanh\omega_2 \pm 1)$), and conical.

CASE 3.

$$e\vec{H} = (0,0,k), \quad k = \text{const} \neq 0;$$

$$eA_0 = -\frac{k^2}{12}(x_1^2+x_2^2-2x_3^2) + \frac{a_1}{r} + a_2\frac{x_3}{r^3} + \frac{a_3}{r^2}\left(\frac{x_3}{2r}\ln\frac{r+x_3}{r-x_3} - 1\right),$$

where $r = \sqrt{x_1^2+x_2^2+x_3^2}$ and a_1, a_2, a_3 are real constant numbers.

The coordinate system is

$$\vec{x} = l\mathcal{O}\vec{z}.$$

Here $\mathcal{O}$ is a time-dependent 3×3 orthogonal matrix $\mathcal{O}(\alpha,\beta,\gamma)$, where $\alpha = -kt$, $\beta = \gamma = 0$; $\vec{z}$ is the spherical coordinates and function $l(t)$ is given by

$$l^2 = \sqrt{C_1^2 \pm \frac{1}{k^2}}\sin\left(2\sqrt{\tfrac{2}{3}}kt\right) + C_1, \quad \text{or} \quad l = C_1\sin\left(\sqrt{\tfrac{2}{3}}kt\right)$$

under condition $a_1 = 0$ and $l = 1$ under condition $a_1 \neq 0$. Here C_1 is an arbitrary real constant.

CASE 4.

$$e\vec{H} = (0,0,k), \quad k = \text{const} \neq 0;$$

$$eA_0 = -\frac{k^2}{12}(x_1^2+x_2^2-2x_3^2) + \frac{a_1}{r^+} + \frac{a_2}{r^-} + a_3\left(\frac{1}{r^+}\operatorname{arc\,tanh}\frac{x_3^+}{r^+} - \frac{1}{r^-}\operatorname{arc\,tanh}\frac{x_3^-}{r^-}\right),$$

where $x_3^\pm = x_3 \pm a$ and $r^\pm = \sqrt{x_1^2+x_2^2+(x_3\pm a)^2}$, and a, a_1, a_2, a_3 are arbitrary real constants. The coordinate system is

$$\vec{x} = \mathcal{O}\vec{z}.$$

Here $\mathcal{O}$ is a time-dependent 3×3 orthogonal matrix $\mathcal{O}(\alpha,\beta,\gamma)$, where $\alpha = -kt$, $\beta = \gamma = 0$ and $\vec{z}$ is the prolate spheroidal coordinates.

CASE 5.

$$e\vec{H} = (0,0,k), \quad k = \text{const} \neq 0;$$

$$eA_0 = -\frac{k^2}{12}(x_1^2 + x_2^2 - 2x_3^2) + 2a_1 a\frac{f_1}{f} + 2a_2\frac{x_3}{ff_1} - 2a_3\left(a\frac{f_1}{f}\operatorname{arc\,cot} f_1 - \frac{x_3}{ff_1}\operatorname{arc\,tanh}\frac{x_3}{af_1}\right),$$

where

$$f = \sqrt{(a^2 - r^2)^2 + 4a^2x_3^2}, \qquad f_1 = \sqrt{\frac{-a^2 + r^2 + f}{2a^2}}, \qquad r = \sqrt{x_1^2 + x_2^2 + x_3^2},$$

and a, a_1, a_2, a_3 are arbitrary real constants. The coordinate system is

$$\vec{x} = \mathcal{O}\vec{z}.$$

Here $\mathcal{O}$ is a time-dependent 3×3 orthogonal matrix $\mathcal{O}(\alpha, \beta, \gamma)$, where $\alpha = -kt$, $\beta = \gamma = 0$ and $\vec{z}$ is the oblate spheroidal coordinates.

Note that expression for A_0 can be rewritten in the form

$$eA_0 = -\frac{k^2}{12}(x_1^2 + x_2^2 - 2x_3^2) + \frac{a_1 + ia_2}{\tilde{r}^+} + \frac{a_1 - ia_2}{\tilde{r}^-} + ia_3\left(\frac{1}{\tilde{r}^+}\operatorname{arc\,tanh}\frac{\tilde{x}_3^+}{\tilde{r}^+} - \frac{1}{\tilde{r}^-}\operatorname{arc\,tanh}\frac{\tilde{x}_3^-}{\tilde{r}^-}\right),$$

where $\tilde{x}_3^\pm = x_3 \pm ia$ and $\tilde{r}^\pm = \sqrt{x_1^2 + x_2^2 + (x_3 \pm ia)^2}$.

CASE 6.

$$e\overrightarrow{H} = (0, 0, k), \quad k = \text{const} \neq 0;$$
$$eA_0 = -\frac{k^2}{6}(x_1^2 + x_2^2 - 2x_3^2) + \frac{a_1}{r} + a_2x_3 + \frac{a_3}{r}\ln\frac{r + x_3}{r - x_3},$$

where $r = \sqrt{x_1^2 + x_2^2 + x_3^2}$ and a_1, a_2, a_3 are arbitrary real constants.

The coordinate system is

$$\vec{x} = \mathcal{O}\vec{z}.$$

Here $\mathcal{O}$ is a time-dependent 3×3 orthogonal matrix $\mathcal{O}(\alpha, \beta, \gamma)$, where $\alpha = -kt$, $\beta = \gamma = 0$ and $\vec{z}$ is the parabolic coordinates.

CASE 7.

$$e\overrightarrow{H} = (0, 0, k), \quad k = \text{const} \neq 0;$$
$$eA_0 = -\frac{q}{2}(x_1^2 + x_2^2 - 2x_3^2) + a\ln(x_1 + x_2) + a_3x_3,$$

where k, a, a_3 are arbitrary real constants.

The coordinate system is

$$x_1 = e^{\omega_1}\cos(\omega_1 - kt), \qquad x_2 = e^{\omega_1}\sin(\omega_1 - kt), \qquad x_3 = l_3\omega_3 + v_3,$$

where l_3, v_3 are solutions of the system of ordinary differential equations

$$\frac{c_3}{l_3^4} - \frac{1}{4}\frac{\ddot{l}_3}{l_3} = q, \qquad l_3\ddot{v}_3 + 2\dot{l}_3\dot{v}_3 + 4c_3\frac{v_3}{l_3^3} - 2c_{13}\frac{1}{l_3} = -2a_3,$$

where c_3 and c_{13} are arbitrary real constants.

Case of nonstationary magnetic field and cases 1–2 of stationary magnetic field allow for separability of the SE in more then one coordinate systems, and these cases are superintegrable ones. The corresponding sets of second-order symmetry operators can be constructed in the explicit form with the help of Theorem 2.2.

Note that some of the potentials obtained have a clear physical meaning. For instance, cases 2 and 3 under condition $k = a_2 = a_3 = 0$ give the standard Coulomb potential. Case 4 under condition $k = a_3 = 0$ gives the potential for a well-known two-center Kepler problem, i.e., the problem of finding wave functions of an electron moving in the field of two fixed Coulomb centers with charges a_1, a_2 and intercenter distance $2a$ (the model of a ionized hydrogen molecule). Coulson and Joseph [**4**] showed that the corresponding Schrödinger equation admits separation of variables in the prolate coordinate system only. We obtained this potential as a particular case of the more general potential.

References

1. V. G. Bagrov and D. M. Gitman, *Exact solutions of relativistic wave equations.* Math. Appl. (Soviet Ser.), vol. 39. Kluwer Academic Publishers, Dordrecht, 1990.
2. S. Benenti, C. Chanu, and G. Rastelli, *Variable separation for natural Hamiltonians with scalar and vector potentials on Riemannian manifolds*, J. Math. Phys. **42** (2001), no. 5, 2065–2091.
3. ———, *Remarks on the connection between the additive separation of the Hamilton-Jacobi equation and the multiplicative separation of the Schrodinger equation. I. The completeness and the Robertson conditions*, J. Math. Phys. **43** (2002), no. 10, 5183–5222; II. *First integrals and symmetry operators*, 5223–5253.
4. C. A. Coulson and A. Joseph, *Constant of the motion for the two-centre Kepler problem*, Internat. J. Quantum Chem. **1** (1967), 337–347.
5. E. G. Kalnins and W. Miller, Jr., *R-separation of variables for the time-dependent Hamilton-Jacobi and Schrödinger equations*, J. Math. Phys. **28** (1987), no. 5, 1005–1015.
6. T. H. Koornwinder, *A precise definition of separation of variables*, Geometrical Approaches to Differential Equations (Scheveningen, 1979), Lecture Notes in Math., vol. 810, Springer, Berlin, 1980, pp. 240–263.
7. A. A. Makarov, J. A. Smorodinsky, Kh. Valiev, and P. Winternitz, *Systematic search for nonrelativistic systems with dynamical symmetries*, Nuovo Cimento A **52** (1967), 1061–1084.
8. W. Miller, Jr., *Symmetry and separation of variables*, Encyclopedia Math. Appl., vol. 4, Addison-Wesley, Reading, MA–London–Amsterdam, 1977.
9. V. N. Shapovalov, *Separation of variables in second-order linear differential equations*, Differentsial′nye Uravneniya **16** (1980), no. 10, 1864—1874 (Russian); English transl., Differential Equations **16** (1980), no. 10, 1212–1220.
10. V. N. Shapovalov and N. B. Sukhomlin, *Separation of variables in the nonstationary Schrödinger equation*, Izv. Vyssh. Uchebn. Zaved., Fiz. (1974), no. 12, 100–105 (Russian); English transl., Soviet Phys. J. **17** (1976), 1718–??.
11. P. Winternitz, J. Smorodinsky, M. Uhlíř, and I. Friš, *Symmetry groups in classical and quantum mechanics*, Jadernaja Fiz. **4** (1966), 625–635 (Russian); English transl., Soviet J. Nuclear Phys. **4** (1967), 444–450.
12. A. Zhalij, *On separable Pauli equations*, J. Math. Phys. **43** (2002), no. 3, 1365–1389.
13. R. Zhdanov and A. Zhalij, *On separable Schrödinger equations*, J. Math. Phys. **40** (1999), no. 12, 6319–6338.

Inst. of Mathematics, Ukrainian National Academy of Sciences, 3 Tereshchenkivska St., 01601, Kyiv-4, Ukraine

Current address: Centre de recherches mathématiques, C.P. 6128, succ. Centre-ville, Montréal, QC H3C 3J7, Canada

E-mail address: zhaliy@imath.kiev.ua, zhalij@crm.umontreal.ca

Inst. of Mathematics, Ukrainian National Academy of Sciences, 3 Tereshchenkivska St., 01601, Kyiv-4, Ukraine

E-mail address: renat@imath.kiev.ua

Centre de Recherches Mathématiques
CRM Proceedings and Lecture Notes
Volume **37**, 2004

New Types of Solvability in PT Symmetric Quantum Theory

Miloslav Znojil

ABSTRACT. The characteristic anti-linear (parity/time reversal, PT) symmetry of non-Hermitian Hamiltonians with real energies is presented as a source of two new forms of solvability of Schrödinger's bound-state problems. In detail we describe (1) their very specific semi-exact solvability (SES) and (2) their innovated variational tractability. SES technicalities are discussed via charged oscillator example. In a broader context, speculations are added concerning possible relationship between PT symmetry, solvability and superintegrability.

PT symmetric quantum mechanics: A brief introduction

Prehistory: Isospectral operators. Origins of the popular Bender's and Boettcher's PT symmetric quantum mechanics [**3, 5**] lie in perturbation theory [**8, 13**]. For an elementary illustration one may recollect the pioneering paper by Buslaev and Grecchi [**7**] who proved the isospectrality of the Hermitian, spherically symmetric D-dimensional perturbed harmonic oscillator

$$H^{(\mathrm{PHO})}(g) = \frac{1}{2}\Big(-\Delta + \sum_{j=1}^{D} x_j^2\Big) + g^2\Big(\sum_{j=1}^{D} x_j^2\Big)^2$$

(in its mth partial-wave projection at any $m = 0, 1, \dots$) with its *non-Hermitian* "unstable" anharmonic partner(s) in one dimension,

$$H^{(\mathrm{UAO})}(g) = -\frac{d^2}{dz^2} + z^2(igz-1)^2 - (D+2m-2)(igz-\tfrac{1}{2}), \quad g > 0.$$

This means that the physics (and, in particular, the reality of energies) remains unchanged while the mathematics itself is significantly simplified when one weakens the Hermiticity of $H^{(\mathrm{PHO})}(g) = [H^{(\mathrm{PHO})}(g)]^\dagger$ to the mere PT symmetry of its spectrally equivalent partners [**10**]. The latter operators commute with the product PT of parity and time reversal. In the modern language of review [**17**], one should rather speak about the pseudo-Hermiticity defined by the relation

$$H^{(\mathrm{UAO})}(g) = P[H^{(\mathrm{UAO})}(g)]^\dagger P.$$

2000 *Mathematics Subject Classification.* Primary 34M15, 35Q40; Secondary 81Q05, 81R40.

The author was supported in part by AS CR Grant #A1048302.

This is the final form of the paper.

Recent progress: Exactly solvable examples. Paper [**7**] extends the proof to the isospectrality between $H^{(\mathrm{PHO})}(g)$ and $H^{(\mathrm{UAO})}(g)$ to the imaginary couplings $g = ih$. The spectrum of the resulting self-adjoint double-well $H^{(\mathrm{UAO})}(ih)$ in one dimension (famous for its perceivably hindered perturbative tractability) may be then deduced from the "unstable" modification $H^{(\mathrm{PHO})}(ih)$ of the central quartic oscillator in D dimensions. The latter, non-Hermitian PT symmetric model is sufficiently simple in its ordinary differential "radial" representation of [**7**],

$$H_\varepsilon^{(\mathrm{URO})}(h) = -\frac{d^2}{dr^2} + \frac{\ell(\ell+1)}{r^2} + r^2 - h^2 r^4,$$
$$r = r(x) = x - i\varepsilon,\ x \in (-\infty, \infty),\ h > 0,\ \varepsilon > 0.$$

In the context of the present proceedings this operator admits an immediate reinterpretation as a specific perturbed and regularized form of the most common superintegrable oscillator $H_0^{(\mathrm{URO})}(0)$ of Smorodinsky and Winternitz [**11, 14**] in any number of dimensions and after its separation in Cartesian coordinates.

The latter observation is inspiring since it is known that the concepts of (maximal) superintegrability and (exact) solvability are closely related at $\varepsilon = 0$ [**20**]. In such a setting the weakening $\varepsilon > 0$ of the Hermiticity extends the class of the solvable models [**29, 32**] as well as their interpretation in terms of Lie algebras [**16**] and evokes a number of new open questions. We have addressed the first few of them in [**27**] where we derived the exact spectrum and wave functions for the complexified SW oscillator $H_\varepsilon^{(\mathrm{URO})}(0)$. Our next move in this direction was devoted to the closely related complexified (though still separable) Calogero model of three particles [**36**] where an overlap of the integrability with PT symmetry acted as a generator of the new solvable self-adjoint Hamiltonians [**37**]. This experience makes the further study of PT symmetric models of (super)integrable type very promising. In the present paper, we are going to describe some new results achieved in this direction.

Running project: Partially solvable models. In place of the (not so easily feasible) analysis of the traditional quartic perturbations, we shall pay attention to the slightly simpler charged oscillators characterized, first of all, by their incompletely or quasi-exactly solvable[1] (QES) status. From the historical perspective, the acceptance of this concept was comparatively dramatic. In a prelude, various potentials have been shown partially solvable in terms of elementary functions. Thirty years ago, this period has been initiated by a two-page remark by André Hautot [**15**] who noticed that the charged oscillator possesses arbitrary finite multiplets of exact Sturmian solutions in two ($D = 2$) and three ($D = 3$) dimensions. The Hautot's choice of the elementary wave-function ansatz

$$\varphi(r) = e^{-r^2/2 - gr} \sum_{n=0}^{N} h_n r^{n+\kappa}$$

(with arbitrary N) preceded the discoveries of the QES sextic oscillator [**19**], of nonpolynomial QES anharmonicities [**23**] etc. During the "golden age" of the development of the subject, the existence of the common Lie-algebraic background of all the QES systems has been emphasized [**21**]. A summary of the "state of the art" up to the early nineties has been offered by Alex Ushveridze in his monograph [**22**]

[1] term coined by A. Turbiner, meaning an elementary solvability for finite multiplets of states

TABLE 1. Tentative classification of solvability.

class	quasi-exact	semi-exact	exact
solutions available	finite set	infinitely many	all
range of couplings	restricted	restricted	any
illustrative example	$H^{(\mathrm{BBO})}(a,b)$	see below	$H_\varepsilon^{(\mathrm{URO})}(0)$

where the quartic polynomial oscillator is mentioned as a typical system *without* quasi-exact solvability.

A change of the approach to QES models has been initiated by Bender and Boettcher [**4**] who revealed and described the QES solutions for the "unstable" quartic polynomial oscillators. Their construction has been extended to all the partial waves in [**30**]. In paper [**28**] we made the next move and turned attention to PT symmetrization of the Hautot's charged oscillator (cf. Appendix A). In the present continuation of this effort, we shall complete the picture by its extension, i.e., to all the partial waves (cf. Appendix B). In Section 1 we describe our main result, viz., the quasi-even solutions which remained unnoticed in [**28**]. Unexpectedly, these "facilitated QES" states are available in *infinite* multiplets[2], all the elements of which are, basically, nonnumerical. In a way emphasized in Section 2 these sets may easily serve as certain nonstandard bases in some "sufficiently large" subspaces of Hilbert space. For this reason, we suggest to call them semi-exactly solvable (cf. Table 1, abbreviation SES).

1. Quasi-exact solvability on complex contours

1.1. Three-term recurrences. Let us consider the Schrödinger equation

$$\left[-\frac{d^2}{dx^2}+\frac{\ell(\ell+1)}{r^2(x)}+i\frac{F}{r(x)}+2ibr(x)+r^2(x)\right]\psi_n(x)=E_n\psi_n(x) \tag{1.1}$$
$$\ell=\ell(L)=(L-1)/2,\ \ L=D-2,D,D+2,\ldots .$$

It contrast to the Hautot's QES problem of Appendix A it works with the purely imaginary charge and with a constant complex shift of coordinates $r=r(x)=x-i\varepsilon$ (cf. [**1**]). It also generalizes the $\ell=0$ problem of [**28**] so that we must employ the more powerful ansatz

$$\psi_n(r)=e^{-r^2/2-ibr}\sum_{n=0}^{N}(ir)^{n-\ell}p_n. \tag{1.2}$$

After its insertion, the differential form of our PT symmetric bound state problem (1.1) is replaced by the finite-dimensional matrix equation,

$$\begin{pmatrix} B_0 & C_0 & & & \\ A_1 & B_1 & C_1 & & \\ & \ddots & \ddots & \ddots & \\ & & A_{N-1} & B_{N-1} & C_{N-1} \\ & & & A_N & B_N \\ & & & & A_{N+1} \end{pmatrix}\begin{pmatrix} p_0\\ p_2\\ \vdots\\ p_N\end{pmatrix}=0,\qquad N<\infty \tag{1.3}$$

[2]such a feature is much more characteristic for the completely solvable models

with the *real* matrix elements,

$$A_n = b^2 + 2n - L - E, \quad B_n = -(2n+1-L)b - F, \quad C_n = (n+1)(n+1-L),$$
$$L = 2\ell + 1, \ n = 0, 1, \ldots .$$

The last line of (1.3) represents a separate condition $A_{N+1} = 0$ which gives our energies in closed form,

$$E = E_N = 2N + 2 - L + b^2, \quad N = 0, 1, \ldots . \tag{1.4}$$

The key consequence of this easy but important simplification lies in the emergence of a zero $C_{L-1} = 0$ in the upper diagonal of the square matrix of the simplified system (1.3). In Appendix B the presence of this zero enables us to construct the quasi-odd solutions at all ℓ.

1.2. The new, quasi-even solutions. All the coefficients in the polynomial wave functions (1.2) may be defined by closed formula (B.1) displayed in Appendix B. Let us now turn our attention to the quasi-even states. The superscript $^{(+)}$ will be introduced to mark their even quasi-parity which may be characterized, for our present purposes, by the simple negation of the odd quasi-parity condition (B.2),

$$|p_0^{(+)}| + |p_1^{(+)}| + \cdots + |p_{L-1}^{(+)}| > 0. \tag{1.5}$$

For quasi-even states, the vanishing matrix element $C_{L-1}^{(+)} = 0$ plays a different role. It separates again the two subsets of equations but the upper one ceases to be trivial. This means that the related subdeterminant must vanish,

$$S^{(+)} = \det \begin{pmatrix} B_0 & C_0 & & & \\ A_1 & B_1 & C_1 & & \\ & \ddots & \ddots & \ddots & \\ & & A_{L-2} & B_{L-2} & C_{L-2} \\ & & & A_{L-1} & B_{L-1} \end{pmatrix} = 0. \tag{1.6}$$

Up to the exceptional, degenerate cases where Eqs. (1.6) and (B.3) hold at the same time, we may drop the nonvanishing factor $S^{(-)} \neq 0$ and determine all the charges $F = F_{N,k}^{(+)}$ as roots of polynomial (1.6).

The $^{(+)}$-superscripted wave function coefficients are determined by Eq. (B.1), this time in the full range of indices $j = N, N-1, \ldots, 1$. As long as the dimension of these determinants grows with the difference $N - j$, we shall recommend a return to the recurrences

$$\begin{pmatrix} -2N & \beta_1 - F & 4 - 2L & & & \\ & 2 - 2N & \beta_2 - F & 9 - 3L & & \\ & & \ddots & \ddots & \ddots & \\ & & & -4 & \beta_{N-1} - F & N^2 - NL \\ & & & & -2 & \beta_N - F \end{pmatrix} \begin{pmatrix} p_0^{(+)} \\ p_1^{(+)} \\ \vdots \\ p_{N-1}^{(+)} \\ p_N^{(+)} \end{pmatrix} = 0$$

to be read from below upward[3]. The process must be initiated at $p_N^{(+)} \neq 0$. The vanishing $C_{L-1} = 0$ does not play any role in it.

[3]We abbreviated $\beta_n \equiv -(2n + 1 - L)b$ for the sake of brevity.

Roots of the SES secular determinants

1.3. Dimension-independence. In every partial wave, the size L of the matrix in Eq. (1.6) coincides with the degree of the resulting secular polynomial. This value does not change with the growth of the dimension $N+1$ of the vector $\vec{p}$. Thus, the degree N of our polynomial wave functions enters our tridiagonal secular equation as a mere parameter. Containing extremely elementary matrix elements, the fully explicit form of this equation reads

$$\det \begin{pmatrix} (L-1)b-F & 1-L & & & \\ -2N & (L-3)b-F & \ddots & & \\ & -2N+2 & \ddots & 4-2L & \\ & & \ddots & (3-L)b-F & 1-L \\ & & & -2(N+2-L) & (1-L)b-F \end{pmatrix} = 0$$

and specifies the family of the admissible charges $F = F^{(+)}_{N,k}$ at any integer $L > 0$ and index $k = 1, 2, \ldots, L$. The first few partial waves are exceptional since their eigencharges may in principle be defined by closed formulas at $L = 1$, $L = 2$, $L = 3$ and $L = 4$. At the simplest choice of $L = 1$ (which may mean both the s-wave in three dimensions and an even state at $D = 0$), one does not obtain anything new. Secular equation (1.6) provides the single root $F^{(+)}_{N,1}$ which is equal to zero at all N. In the limit $\varepsilon \to 0$, the quasi-even $L = 1$ solutions converge to the well known Hermite polynomials. One just rediscovers the exactly solvable harmonic oscillator basis at even parity. We must reemphasize that at any $L \geq 0$ the number of our quasi-even roots *does not change* with the growth of the dimension N. This is of paramount importance since the practical determination of eigencharges is performed for all the indices N at once. Each of these families numbered by k contains *infinitely many* elements numbered by the first subscript. This might facilitate their future applications (cf. Section 2 below).

1.4. The first nontrivial generalization of oscillator basis: $L = 2$. Let us move to the index $L = 2$ giving the p-wave in two dimensions or to the s-wave in four dimensions. The related $\ell = \frac{1}{2}$ wave functions may be chosen as compatible with the even-quasi-parity criterion (1.5),

$$p_0^{(+)}(N,k) = -\frac{1}{2N}[b + F_{N,k}]p_1^{(+)}(N,k), \quad k = 1, 2.$$

Secular equation (1.6) reads

$$\det \begin{pmatrix} b-F & -1 \\ -2N & -b-F \end{pmatrix} = F^2 - b^2 - 2N = 0$$

and has two roots

$$F^{(+)}_{N,1} = \sqrt{(b^2+2N)}, \qquad F^{(+)}_{N,2} = -\sqrt{(b^2+2N)}. \tag{1.7}$$

[Note the contrast with the necessity of searching for roots of the polynomial (B.3) of the $N-L+1$st degree!] We encounter the nonvanishing charges for the first time. They grow with the increasing size of the shift b and with the quantum number N (i.e., with the energy). One finds their set more similar to the "$F = 0$ line" of the harmonic oscillator than to the QES quasi-odd roots of Appendix B. Indeed, in the latter case the roots $F^{(-)}_{N,k}$ must be determined by the method which depends on N.

In this sense we may now speak about the most natural and unique non-Hermitian generalization of the harmonic oscillator even-parity basis of $L_2(0,\infty)$.

1.5. Cardano charges at $L = 3$. At $L = 3$ the comparatively compact form of our secular equation

$$\det \begin{pmatrix} 2b - F & -2 & 0 \\ -2N & -F & -2 \\ 0 & -2N+2 & -2b - F \end{pmatrix} = 0 \tag{1.8}$$

enables us to search for the triplet of charges $\{F_{N,1}, F_{N,2}, F_{N,3}\}$ via the closed (so called Cardano) formulas. One of alternative strategies may consist in a transition from the SES eigencharges $F = F_{N,k}(b)$ to the inverse functions $b = b_{N',k'}(F)$. Such a trick lowers (by one) the degree of the secular polynomial at the odd values of L and, hence, extends the solvable class up to $L = 5$. A serious shortcoming of such an approach may be seen in the parallel deformation of the energies (1.4) which would change with the shift b.

In the similar spirit, another simplification of formulas may be based on the formal elimination of N. The consequences may be illustrated by the secular $L = 3$ determinant (1.8) which represents the mere linear problem for N, with the unique solution

$$N = -\frac{1}{8F}(4Fb^2 + 8b - F^3 - 4F) = \frac{1}{2}\left[\left(\frac{1}{F} + \frac{F}{2}\right)^2 - \left(b + \frac{1}{F}\right)^2\right]. \tag{1.9}$$

This formula suggests a simultaneous tuning of both the shifts $b_{[1,2]}$ and the related charges $F_{[1,2]}$. This may be achieved, say, by their hyperbolic reparametrization with $F(t) = \sqrt{2}e^t$ and $\cosh t = \sqrt{N} \cosh \alpha(t)$ etc.. In terms of the parameters $t = t_{N,k}$ and/or $\alpha = \alpha_{N,k}$, the spectrum of energies becomes deformed by the induced parametric dependence of $b = b_{[1,2]}(t) = \pm\sqrt{2\cosh^2 t - 2N} - e^{-t}/\sqrt{2}$. Purely formally, the elimination of N might extend the use of closed formulas up to $L = 9$.

Higher partial waves

1.6. Numerical methods. A shortcoming of the present SES construction lies in the growth of its complexity at the large L. At $L \geq 5$ one already cannot generate closed formulas for eigencharges, and the purely numerical search for the roots F is necessary at all the very large $L \gg 1$. In an illustration using $L = 3$, and $b = 5$ one gets the three eigencharges

$$F = \{10.757, -10.400, -0.35755\}$$

at the smallest possible $N = 2$. They smoothly grow to the values

$$F = \{89.98, -89.975, -0.0049407\} \tag{1.10}$$

evaluated at very large $N = 1000$. This type of calculation is very quick and its results exhibit a smooth N-dependence sampled in Table 2.

1.7. Large N expansions. Any information concerning the eigenvalues $F = F(b)$ and/or $b = b(F)$ may shorten the necessary computations. At the intermediate

TABLE 2. N-dependence of eigencharges at $L = 4$ and $b = 5$

N	$F_{N,k}$			
3	−15.611	−5.9279	4.8887	16.651
30	−27.149	−9.2909	8.9294	27.511
300	−74.856	−24.984	24.936	74.904
3000	−232.82	−77.610	77.605	232.83
30000	−734.99	−245.00	245.00	734.99

$L = 5$ (which corresponds to the d-wave in three dimensions) we may eliminate, for example, the product $512\,FN^{\pm}$ which is equal to

$$\left[-768b - 256Fb^2 + 768F + 40F^3 \pm 24\sqrt{(1024b^2 + 192bF^3 + 512F^2 + F^6)}\right]$$

and use this formula as a constraint. Fortunately, closed formulas of this type may also open the door to the perturbative methods.

Empirically, many numerically computed roots $F_{N,k}$ are very smooth functions of N, especially in the asymptotic domain where they may be well approximated by their available large-N estimates. For example, the exact result (1.10) already lies very close to its leading-order analytic estimate

$$F \approx \{\sqrt{8N}, -\sqrt{8N}, -b/N\} \approx \{89.44, -89.44, -0.005\}.$$

At the intermediate values of the dimension N, precision may be still insufficient but one can easily evaluate the large$-N$ corrections. For illustration, let us return to Eq. (1.9) and rewrite it, in the case of its smallest root $F = -\hat{b}\widehat{F}/N$, as the strictly equivalent formula

$$\widehat{F} = 1 + \frac{\beta}{N^3}\widehat{F}^3, \qquad \beta = \frac{\hat{b}^3}{8b}, \qquad \hat{b} = \frac{b}{1 + (b^2-1)/2N}, \qquad L = 3.$$

It is suitable for iterations which represent our root as the following power series,

$$\widehat{F} = 1 + \frac{\beta}{N^3} + 3\frac{\beta^2}{N^6} + 12\frac{\beta^3}{N^9} + 28\frac{\beta^4}{N^{12}} + \cdots.$$

Also the other two $L = 3$ roots may be represented by the similar asymptotic series.

At the higher partial waves the same method works with a comparable efficiency. At $L = 4$, for example, the use of the variable $M = \sqrt{2N + b^2 - 2} > 0$ compactifies secular equation (1.6),

$$(F^2 - M^2)(F^2 - 9M^2) = 36 - 48bF.$$

In accord with Table 2, its asymptotically dominant $O(M^2)$ part determines the four distinct leading-order asymptotics of $F \sim \varrho M$ where $\varrho = \pm 1, \pm 3$. Once we renormalize $F = \varrho M\sqrt{1+R}$, the four exact charges obey the relation

$$R = \frac{48b\sqrt{1+R}}{8|\varrho| - 16 + \varrho^2 R}\frac{1}{\varrho M^3} + \frac{36}{8|\varrho| - 16 + \varrho^2 R}\frac{1}{\varrho^2 M^4}, \qquad L = 4.$$

Its iterations evaluate the correction term R with an astounding efficiency.

2. Quasi-variational solvability on model spaces

"Sufficiently large" finite subspaces in Hilbert space.

2.1. Left and right SES solutions. A full-fledged applicability of the standard sets of QES wave functions is weakened by their incompleteness. Moreover, all the Hermitian QES Hamiltonians are usually interpreted as possessing just a finite multiplet of elementary bound states (at a given, special QES coupling) or Sturmians (cf. their example in Appendix B). For this reason, even the study of their own small perturbations is not easy at all [**25**]. Still, the sets of QES solutions themselves become large enough when we take into consideration *both* their fixed-coupling and fixed-energy subsets. In practice, such a philosophy is proved useful for approximation purposes [**6**].

In this section, we are going to return to the latter idea in the present broader, PT symmetric SES context. For the sake of definiteness let us reconsider Hamiltonian of Eq. (1.1) as if it were defined at *any* charge F standing at the Coulomb-like force $W(r) = ir^{-1}$. Then we may denote $H = H(F) = H(0) + FW$ and, in the spirit of the Dirac's bra and ket notation, abbreviate $|\psi^{(-)}_{M,k}\rangle = |M, k\rangle$. This enables us to rewrite the special SES version of Eq. (1.1) in shorthand,

$$H(0)|M,k\rangle + W|M,k\rangle F_{M,k} = |M,k\rangle E_M, \quad k = 1,\ldots,L, \;\; M = 0,1,\ldots. \tag{2.1}$$

Our PT symmetric Hamiltonian is non-Hermitian so that its left and right eigenstates will differ in general. One has to complement Eq. (2.1) by its counterpart

$$\langle N,j|H(0) + F_{N,j}\langle N,j|W = E_N\langle N,j|, \quad j = 1,\ldots,L, \;\; N = 0,1,\ldots, \tag{2.2}$$

where the same operators act to the left. After a return to the differential form of this problem (1.1), a transition between Hermitian-conjugate equations (2.1) and (2.2) may be reinterpreted as a certain reflection R in the space of the parameters,

$$R : \{F \leftrightarrow -F, b \leftrightarrow -b, \varepsilon \leftrightarrow -\varepsilon\}. \tag{2.3}$$

In an illustration using $L = 2$, our SES states of Section 1 may be represented by a discrete set of points in the energy-charge plane. The values of their coordinates $(E_N, F_{N,k})$ are specified by the respective *closed formulas* (1.4) and (1.7). These points may be perceived as located on the two (viz. left and right) branches of a single hyperbolic curve. These two infinitely large multiplets are mutually related by the transformation R of Eq. (2.3) so that $R|N,1\rangle = |N,2\rangle$ and vice versa.

2.2. Generalized Sturmian SES states as a basis. There is no a priori reason for a bi-orthogonality between our multi-indexed bras $\langle N,k| \equiv \langle A|$ and ket vectors $|N',k'\rangle \equiv |b\rangle$ so that their overlaps

$$Q_{A,b} = \langle A|b\rangle$$

form a nondiagonal and asymmetric matrix in general. We have to assume that after a finite-dimensional truncation, this overlap matrix is invertible. Only in such a case we may define the inverse $R = Q^{-1}$ and introduce an identity projector in a "sufficiently large" subspace of Hilbert space,

$$I = \sum_{\substack{a \in J_{\text{ket}} \\ B \in J_{\text{bra}}}} |a\rangle R_{a,B}\langle B|. \tag{2.4}$$

In a constructive mood, let us return to equations (2.1) and (2.2) and imagine that they share all their eigenenergies and eigencharges. This enables us to write down the following two alternative matrix equations

$$\langle N,j|H(0)|M,k\rangle = \langle N,j|M,k\rangle E_M - \langle N,j|W|M,k\rangle F_{M,k} \tag{2.5}$$

$$\langle N,j|H(0)|M,k\rangle = E_N(\langle N,j|M,k\rangle - F_{N,j}\langle N,j|W|M,k\rangle \tag{2.6}$$

with $(N,j) \in J_{\text{bra}}$ and $(M,k) \in J_{\text{ket}}$. Their subtraction gives the constraint

$$(F_{M,k} - F_{N,j})\langle N,j|W|M,k\rangle = (E_M - E_N)Q_{(N,j),(M,k)}. \tag{2.7}$$

Due to the way of its derivation, this relation may be understood as an immediate generalization of bi-orthogonality. In particular, we see that within the subspace of a single Sturmian multiplet (i.e., for $M = N$), the left-hand side expression must be a diagonal matrix with respect to its second indices. For our present purposes we shall abbreviate $\langle N,j|W|N,j\rangle \equiv w_{N,j}$.

Non-QES bound-state problems

2.3. Matrix equations. The knowledge of a basis can facilitate the study of many perturbed Hamiltonians via textbook perturbation recipes [**12**]. The use of the present SES bases might lead to a new progress in this area. Let us recall the respective right and left forms of any Schrödinger non-QES bound state problem with the above-mentioned structure,

$$[H(0) + FW]|\Psi\rangle = E(F)|\Psi\rangle,$$
$$\langle\Psi|[H(0) + FW] = E(F)\langle\Psi|.$$

Assuming that $F \neq F^{(\text{QES})}$ and using Eq. (2.4) we may insert

$$|\Psi\rangle = \sum_{\substack{a\in J_{\text{ket}}\\ B\in J_{\text{bra}}}} |a\rangle R_{a,B}\langle B|\Psi\rangle = \sum_{a\in J_{\text{ket}}} |a\rangle h_a,$$

$$\langle\Psi| = \sum_{\substack{a\in J_{\text{ket}}\\ B\in J_{\text{bra}}}} \langle\Psi|a\rangle R_{a,B}\langle B| = \sum_{B\in J_{\text{bra}}} g_B\langle B|$$

and arrive at the double eigenproblem

$$\sum_{\substack{b\in J_{\text{ket}}\\ C\in J_{\text{bra}}}} \langle A|H(F)|b\rangle R_{b,C}\langle C|\Psi\rangle = E\langle A|\Psi\rangle, \qquad A \in J_{\text{bra}}$$

(plus its—omitted—conjugated companion), i.e., at the two conjugate linear algebraic systems of equations written in terms of the same matrix,

$$\begin{gathered} \mathbf{Z}(E,F)\vec{h} = 0, \qquad \vec{g}^{\dagger}\mathbf{Z}(E,F) = 0 \\ \mathbf{Z}_{A,b}(E,F) = \langle A|H(0)|b\rangle - E\langle A|b\rangle + F\langle A|W|b\rangle. \end{gathered} \tag{2.8}$$

Now, one could introduce a small parameter $\lambda = F - F^{(\text{QES})}$ and try to construct, say, $E = E(\lambda)$ in the form of a power series in λ.

2.4. Matrix elements. Long before any numerical determination of bound states, we must evaluate *all* the necessary matrix elements as an input. In practice, the latter step is usually the most time-consuming part of the algorithm. One has to optimize it in the present QES setting, therefore.

In the first step we recall Eq. (2.6) and eliminate all the matrix elements of $H(0)$. This means that in Eq. (2.8) we reduce the costly input to the mere evaluation of the matrix elements of the weakly singular Coulomb potential $W(r) = i/r$,

$$\mathbf{Z}_{A,b}(E,F) = (F - F_A)\langle A|W|b\rangle - (E - E_A)\langle A|b\rangle.$$

In the second step we keep $M \neq N$ (i.e., we stay out of the Sturmian subspaces or diagonal blocks in the matrix $\mathbf{Z}$) and postulate the absence of a random degeneracy of charges. This means $F_{M,k} \neq F_{N,j}$ so that we are permitted to rearrange the bi-orthogonality-like relation (2.7) into a further reduction of the necessary input information,

$$\langle N,j|W|M,k\rangle = \frac{E_M - E_N}{F_{M,k} - F_{N,j}} Q_{(N,j),(M,k)}, \qquad M \neq N.$$

In this way we arrive at the final form of our linear Schrödinger non-QES algebraic problem for the right eigenvectors,

$$w_{N,j}h_{N,j} + \sum_{\substack{K,p\\ K\neq N}} \frac{E_N - E_K}{F_{N,j} - F_{K,p}} Q_{(N,j),(K,p)}\, h_{K,p} = \frac{E - E_N}{F - F_{N,j}} \sum_{M,k} Q_{(N,j),(M,k)} h_{M,k},$$
$$j = 1,2,\dots,L, \;\; N = 0,1,\dots .$$

For left eigenvectors, the system of equations is very similar though not equivalent,

$$w_{N,j}g_{N,j} + \sum_{\substack{K,p\\ K\neq N}} \frac{E_N - E_K}{F_{N,j} - F_{K,p}} Q_{(K,p),(N,j)} g_{K,p} = \frac{E - E_N}{F - F_{N,j}} \sum_{M,k} Q_{(M,k),(N,j)}\, g_{M,k},$$
$$j = 1,2,\dots,L, \;\; N = 0,1,\dots .$$

To solve any of these systems, say, by a perturbation technique, we just need to know the overlaps Q and the vector of Coulombic elements $w_{N,j}$.

3. Concluding questions and remarks

3.1. Does the quantum quasi-exact solvability have a classical analogue? In some studies[4] a new relationship has been traced between classical and quantum mechanics. Its core may be seen in a correlation between the concepts of *integrability* and *solvability*. Needless to repeat, the former feature plays a key role in classical systems while its versions known as superintegrability and maximal superintegrability acquire more relevance in quantized world.

For many quantum Hamiltonians the latter qualities also seem related to a purely algebraic property of quasi-exact solvability [**18**]. In such a context, it is extremely exciting to ask questions about the robustness of the latter correlation *and* about the natural ways of the definition of the solvability itself. In a purely pragmatic manner, PT symmetrization could offer here another bridge toward the new solvable models. It is evident that after one weakens the traditional requirement of Hermiticity, many new QES models may be found, indeed [**9, 31**].

In the present paper we have seen that for the D dimensional central Coulomb plus shifted harmonic oscillator its quasi-exact solvability acquires a fairly unusual

[4]as reviewed, e.g., by Pavel Winternitz in this volume [**24**]

modified form. A significant difference appears between the quasi-even and quasi-odd states, where only the properties of the latter set remain standard. The facilitated construction and unexpectedly nonnumerical character of the former quasi-even family make it similar to the current complete oscillator basis on half-axis. This locates our new "semi-exact" solutions in a gap between their older exact and quasi-exact neighbors.

3.2. Do we need more semi-exactly solvable models? Our presentation of details started from the reduction of the four-term recurrences of our "paper I" [**28**] to their improved three-term form. This was rendered possible by our new ansatz which also proved able to reproduce all the available older results. We underlined that one of the most characteristic shortcomings of the standard QES equations lies in the necessity of solving the adjacent linear algebraic N-dimensional eigenvalue problem of growing size N which, in our particular example, selects the admissible charges.

In this context it is important that we succeeded in revealing the existence of a new, "facilitated" or "semi-exact" elementary solvability emerging when the quasi parity was assumed even. *Infinitely many* of these states acquire the exact polynomial form in each (= mth) partial wave and at any element $E = E(N)$ of an equidistant set of the energies. In fact, our states form the Sturmian L-plets (numbered by $N = 0, 1, \dots$) at charges $F_{N,k}$ numbered by $k = 1, 2, \dots, L$ where $L = D - 2 + 2m$ is N-independent.

The well-known even-parity basis on $L_2(0, \infty)$ (made of Hermite polynomials) is reobtained in the simplest $L = 1$ special case where the resulting (single) QES charge is zero, $F = 0$. In the first nontrivial $L = 2$ case we have got the two alternative eligible infinite series of states at charges $F = \pm\sqrt{b^2 + 2N}$ (where b is the shift). We have shown that and how the similar sets might serve as a source of the new matrix reformulations and approximation techniques within perturbation theory or variational considerations.

Our main attention has been paid to the generalized QES ansatz admitting solutions with both quasi-parities. Our main result, viz., a new version of the SES recurrent construction is "almost exact," first of all, due to its N-independence. In the first few lowest partial waves our infinite families of semi-exact states proved very transparent and close, by their applicability, to the basis of harmonic oscillators. We revealed that their construction is amenable to an efficient approximative treatment even at the higher partial waves via the Taylor-series expansions similar to the common large-N perturbation theory. This might not only initiate the study of the old models in the non-QES domains of couplings but also offer a new motivation for an intensified search for the new SES models.

3.3. Will our new SES bases find efficient practical applications? The matrices of overlaps $Q_{A,b}$ are, presumably, nondiagonal. In the other words, our SES basis states $|N, k\rangle$ and $|N, k\rangle\rangle$ are *not* mutually bi-orthogonal. Any deeper insight in their overlaps would be appreciated in applications, therefore. In the future, this could facilitate, say, a search for a non-Hermitian analogue of the QES + extrapolation trick as suggested by Burrows et al [**6**] for the Hermitian sextic anharmonicities.

Our formalism is not yet fully prepared for initiation of any practical numerical or perturbative calculations in the non-QES regime of course. At the same time, we already succeeded in a drastic reduction of the number of the necessary input matrix elements. This supports our belief that our present constructions represent a valuable key step toward deeper understanding of relations between the degree of integrability in classical mechanics and an extent of exact solvability in quantum mechanics. At this moment we have to reemphasize the definite progress in our understanding of properties of the "forgotten" quasi-even solutions, their quasi-exact solvability of which may be characterized as "significantly facilitated." They are distributed in the coupling-energy plane along curves which may be interpreted as a common generalization of the usual bound-state straight lines[5] and the straight lines of the so called Sturmian states[6]. In this sense, their families may be understood as certain generalized Sturmian families defined on some curves in the plane of charge and energy. These sets contain infinitely many elements and definitely resemble and generalize the current types of bases.

3.4. Could the complexification serve as a regularization recipe for SW models? In connection with the strongly singular behavior of many classically superintegrable models at $r = 0$, a new role of the present weakening of their Hermiticity may be also sought in its parallel regularization effect. In this role, the PT symmetric regularization proved very useful and really vital in the various PT symmetrized versions of the so called supersymmetric (SUSY) quantum mechanics [**2, 35**]. For example, in accord with [**33**] the SW Hamiltonians $H = p^2 + r^2 + G/r^2$ may form the *formal* SUSY partners $H_{(L,R)}$. Once their singularity at $r = 0$ is circumvented via the complex shift of the coordinate axis, their spectra remain real and discrete and exhibit still the usual SUSY-isospectrality pattern in the so called unbroken PT symmetry domain with $G > -\frac{1}{4}$.

In [**34**] we have shown that beyond the latter domain, the PT symmetry itself is completely broken but the spectra stay partially real on a SUSY-generated set of complex G. In addition, an alternative SUSY scheme with equal spikes, $G_{(L)} = G_{(R)}$ may be introduced giving an innovated super-Hamiltonian factorized in terms of certain creation and annihilation differential operators of the second order which conserve the quasi-parity and are mutually not adjoint. Together with the original Hamiltonian H the latter two operators commute in accord with the Lie algebra $\mathrm{sl}(2, \mathbb{R})$.

All these preliminary results and hints are encouraging and suggest that in the context of the Calogero and/or Smorodinsky–Winternitz superintegrable models with strong barriers the PT symmetry-induced tunneling effects did not say their last word yet.

Appendix A. Charged harmonic oscillator and the concept of quasi-parity

As we already mentioned, the shifted and charged harmonic oscillator of [**15**] is a characteristic illustrative example of QES model in quantum mechanics. Its

[5] characterized by the fixed couplings and variable energies and exemplified by the spectrum of the exactly solvable harmonic oscillator or by the Singh's sextic QES oscillators [**19**]

[6] characterized by the variable couplings and fixed energies and exemplified by the exactly solvable zero-order models in [**26**] or by the Flessas' nonpolynomial QES oscillators [**23**]

Hermitian Schrödinger equation

$$\left[-\frac{d^2}{dr^2}+\frac{\ell(\ell+1)}{r^2}+\frac{f}{r}+2gr+r^2\right]\varphi(r)=E\varphi(r),\quad r\in(0,\infty) \tag{A.1}$$

has been replaced by its PT-symmetrized version (1.1) in [**28**]. After the severe restriction of the potential to s-waves in three dimensions (i.e., to $D=3$ and $\ell=0$, motivated by technical reasons) we succeeded in confirming its quasi-exact solvability in non-Hermitian regime. In the present continuation of paper [**28**] this result is extended to all the partial waves $\ell=(D-3)/2,\ (D-3)/2+1,\dots,$ and dimensions $D=3,\ 4,\dots$.

The first hint indicating the open possibilities in connection with Eq. (A.1) appeared in [**27**] where the concept of quasi-parity has been introduced. One was able to conclude there that all the solutions of the differential Schrödinger equations of the radial type prove separated into quasi-even or quasi-odd states $\psi_n^{(\pm)}(x)$ in a way which complements the quasi-odd solutions of [**28**] by their present "forgotten" partners with quasi-even symmetry. The key notion of the quasi-parity itself may be defined either via asymptotics or, more easily, via the specific behavior of the wave functions near the singularity at $r(x)\sim 0$,

$$\psi_n^{(+)}(x)\sim(x-i\varepsilon)^{-\ell},\quad \psi_n^{(-)}(x)\sim(x-i\varepsilon)^{\ell+1},\qquad x\sim i\varepsilon.$$

The name suggests that the quasi-parity coincides with the ordinary parity after limiting transition $\varepsilon\to 0$ to a Hermitian Hamiltonian. In this limit, the quasi-even solutions themselves need not lose their normalizability (and may remain physical) whenever the strength of their singular repulsion is sufficiently small (cf. an explicit illustration in [**33**]).

Appendix B. QES states with odd quasi-parity

Formula (1.4) for the energies is independent of the charge and was already known to Hautot [**15**]. Thus, the energy is a constant for all the solutions of Eq. (1.3). The multiplets of bound states of such a type[7] find applications, say, in perturbation theory [**26**]. Up to a free normalization $p_N\neq 0$, their present closed definition

$$p_{j-1}^{(-)}=\frac{p_N^{(-)}}{(-A_j)(-A_{j+1})\cdots(-A_N)}\det\begin{pmatrix} B_j & C_j & & & \\ A_{j+1} & B_{j+1} & C_{j+1} & & \\ & \ddots & \ddots & \ddots & \\ & & A_{N-1} & B_{N-1} & C_{N-1}\\ & & & A_N & B_N\end{pmatrix} \tag{B.1}$$

is unique at any energy E, shift b and charge F.

Our new ansatz (1.2) is sufficiently flexible and reproduces all the older quasi-odd QES solutions of [**15, 28**]. Indeed, we may mark them by the superscript $^{(-)}$ and characterize their set by the following boundary condition,

$$p_0^{(-)}=p_1^{(-)}=\cdots=p_{L-1}^{(-)}=0,\quad p_L^{(-)}\neq 0 \tag{B.2}$$

which reduces the range of the indices in formula (B.1), $j=N,N-1,\dots,L$. We have to guarantee that the QES recurrent recipe terminates, i.e., that the value of

[7]defined at N different charges $F_{N,j}$ and called Sturmians

$p_{L-1}^{(-)}$ vanishes. This condition has the secular form

$$S^{(-)} = \det \begin{pmatrix} B_L & C_L & & & \\ A_{L+1} & B_{L+1} & C_{L+1} & & \\ & \ddots & \ddots & \ddots & \\ & & A_{N-1} & B_{N-1} & C_{N-1} \\ & & & A_N & B_N \end{pmatrix} = 0 \tag{B.3}$$

and its occurrence is characteristic for all QES systems [**22**]. At every admissible energy it specifies a multiplet of the admissible charges $F = F_{N,j}^{(-)}$ numbered by the second subscript $j = 1, 2, \ldots, N - L + 1$. At each N and L our secular polynomial is of degree $N - L + 1$. Of course, the practical determination of the quasi-odd QES eigencharges is a purely numerical procedure for $N > L + 3$. Fortunately, as we have seen in Section 1, this shortcoming may be suppressed in the quasi-even case.

References

1. Z. Ahmed, *Pseudo-Hermiticity of Hamiltonians under imaginary shift of the coordinate: real spectrum of complex potentials*, Phys. Lett. A **290** (2001), no. 1-2, 19–22.
2. B. Bagchi, S. Mallik, and C. Quesne, *Complexified PSUSY and SSUSY interpretations of some PT-symmetric Hamiltonians possessing two series of real energy eigenvalues*, Internat. J. Modern Phys. A **17** (2002), no. 1, 51–72.
3. C. M. Bender and S. Boettcher, *Real spectra in non-Hermitian Hamiltonians having* PT *symmetry*, Phys. Rev. Lett. **80** (1998), no. 24, 5243–5246.
4. ———, *Quasi-exactly solvable quartic potential*, J. Phys. A **31** (1998), no. 14, L273–L277.
5. C. M. Bender, S. Boettcher, and P. N. Meisinger, *PT-symmetric quantum mechanics*, J. Math. Phys. **40** (1999), no. 5, 2201–2229.
6. B. L. Burrows, M. Cohen, and T. Feldmann, *A unified treatment of Schrödinger's equation for anharmonic and double well potentials*, J. Phys. A **22** (1989), no. 9, 1303–1314.
7. V. Buslaev and V. Grecchi, *Equivalence of unstable anharmonic oscillators and double wells*, J. Phys. A **26** (1993), no. 20, 5541–5549.
8. E. Caliceti, S. Graffi, and M. Maioli, *Perturbation theory of odd anharmonic oscillators*, Comm. Math. Phys. **75** (1980), no. 1, 51–66.
9. F. Cannata, M. Ioffe, R. Roychoudhury, and P. Roy, *A new class of PT-symmertric Hamiltonianis with real spectra*, Phys. Lett. A **281** (2001), no. 5-6, 305–310.
10. P. Dorey, C. Dunning, and R. Tateo, *Spectral equivalences, Bethe ansatz equations, and reality properties in PT-symmetric quantum mechanics*, J. Phys. A **34** (2001), no. 28, 5679–5704.
11. N. W. Evans, *Group theory of the Smorodinsky–Winternitz system*, J. Math. Phys. **32** (1991), no. 12, 3369–3375.
12. F. M. Fernández, *Introduction to perturbation theory in quantum mechanics*, CRC Press, Boca Raton, FL, 2001.
13. F. M. Fernández, R. Guardiola, J. Ros, and M. Znojil, *Strong-coupling expansions for the PT-symmetric oscillators* $V(x) = ax + b(ix)^2 + c(ix)^3$, J. Phys. A **31** (1998), no. 50, 10105–10112.
14. J. Friš, V. Mandrosov, Ya. A. Smorodinsky, M. Uhlíř, and P. Winternitz, *On higher symmetries in quantum mechanics*, Phys. Lett. **16** (1965), 354–356.
15. A. P. Hautot, *Schrödinger's equation and special radial electric fields*, Phys. Lett. A **38** (1972), no. 5 ,305–306.
16. G. Lévai, F. Cannata, and A. Ventura, *PT-symmetric potentials and the* so(2, 2) *algebra*, J. Phys. A **35** (2002), no. 24, 5041–5057.
17. A. Mostafazadeh, *Pseudo-supersymmetric quantum mechanics and isospectral pseudo-Hermitian Hamiltonians*, Nuclear Phys. B **640** (2002), no. 3, 419–434.
18. M. A. Rodriguez, *A survey of quasi-exactly solvable models in many-body problems*, in this volume.
19. V. Singh, S. N. Biswas, and K. Datta, *Anharmonic oscillator and the analytic theory of continued fractions*, Phys. Rev. D (3) **18** (1978), no. 6, 1901–1908.

20. P. Tempesta, A. V. Turbiner, and P. Winternitz, *Exact solvability of superintegrable systems*, J. Math. Phys. **42** (2001), no. 9, 4248–4257.
21. A. V. Turbiner, *Quasi-exactly-solvable problems and* sl(2) *algebra*, Comm. Math. Phys. **118** (1988), no. 3, 467–474.
22. A. G. Ushveridze, *Quasi-exactly solvable models in quantum mechanics*, Institute of Physics Publishing, Bristol, 1994.
23. R. R. Whitehead, A. Watt, G. P. Flessas, and M. A. Nagarajan, *Exact solutions of the Schrödinger equation* $(-d^2/dx^2 + x^2 + \lambda x^2/(1+gx^2))\psi(x) = E\psi(x)$, J. Phys. A **15** (1982), no. 4, 1217–1226.
24. P. Winternitz, *On the relation between superintegrability and exact solvability*, in this volume.
25. M. Znojil, *The exact bound-state Ansaetze as zero-order approximants in perturbation theory. I, The formalism and Padé oscillators*, Czechoslovak J. Phys. **41** (1991), 397–408.
26. ———, r^D *oscillators with arbitrary* $D > 0$ *and perturbation expansions with Sturmian*, J. Math. Phys. **38** (1997), no. 10, 5087–5097.
27. ———, *PT symmetric harmonic oscillators*, Phys. Lett. A **259** (1999), no. 3-4, 220–223.
28. ———, *Harmonic oscillator well with a screened Coulombic core is quasi-exactly solvable*, J. Phys. A **32** (1999), no. 24, 4563–4570.
29. ———, *Shape invariant potentials with PT symmetry*, J. Phys. A **33** (2000), no. 7, L61–L62.
30. ———, *Quasi-exactly solvable quartic potentials with centrifugal and Coulombic terms*, J. Phys. A **33** (2000), no. 22, 4203–4211.
31. ———, *Spiked and PT-symmetrized decadic potentials supporting elementary N-plets of bound states*, J. Phys. A **33** (2000), no. 38, 6825–6833.
32. ———, *Solvable PT-symmetric Hamiltonians*, Phys. Atomic Nuclei **65** (2002), no. 6, 1149–1151.
33. ———, *Non-Hermitian SUSY and singular, PT-symmetrized oscillators*, J. Phys. A **35** (2002), no. 9, 2341–2352.
34. ———, *PT symmetry and supersymmetry*, GROUP 24: Physical and Mathematical Aspects of Symmetries (Paris, 2002), Institute of Physics Publishing, Bristol, 2003, pp. ??–??.
35. M. Znojil, F. Cannata, B. Bagchi, and R. Roychoudhury, *Supersymmetry without Hermiticity within PT symmetric quantum mechanics*, Phys. Lett. B **483** (2000), no. 1-3, 284–289.
36. M. Znojil and M. Tater, *Complex Calogero model with real energies*, J. Phys. A **34** (2001), no. 8, 1793–1803.
37. ———, *Exactly solvable three-body Calogero-type model with translucent two-body barriers*, Phys. Lett. A **284** (2001), no. 6, 225–230.

Nuclear Physics Institute, Academy of Sciences of the Czech Republic, 250 68 Řež, Czech Republic

E-mail address: znojil@ujf.cas.cz

Titles in This Series

TITLES IN THIS SERIES

7 **J. Feldman, R. Froese, and L. M. Rosen, Editors,** Mathematical quantum theory I: Field theory and many-body theory, 1994

6 **Guido Mislin, Editor,** The Hilton Symposium 1993: Topics in topology and group theory, 1994

5 **D. A. Dawson, Editor,** Measure-valued processes, stochastic partial differential equations, and interacting systems, 1994

4 **Hershy Kisilevsky and M. Ram Murty, Editors,** Elliptic curves and related topics, 1994

3 **Rémi Vaillancourt and Andrei L. Smirnov, Editors,** Asymptotic methods in mechanics, 1993

2 **Philip D. Loewen,** Optimal control via nonsmooth analysis, 1993

1 **M. Ram Murty, Editor,** Theta functions: From the classical to the modern, 1993